Castings

Castings

John Campbell *OBE FREng*
Professor of Casting Technology,
University of Birmingham, UK

OXFORD   AMSTERDAM   BOSTON   LONDON   NEW YORK   PARIS
SAN DIEGO   SAN FRANCISCO   SINGAPORE   SYDNEY   TOKYO

Butterworth-Heinemann
An imprint of Elsevier Science
Linacre House, Jordan Hill, Oxford OX2 8DP
22 Wheeler Road, Burlington MA 01803

First published 1991
Paperback edition 1993
Reprinted 1993, 1995 (twice), 1997, 1998, 1999, 2000
Second edition 2003

British Library Cataloguing in Publication Data
A catalogue record for this book is available from the British Library

Library of Congress Cataloguing in Publication Data
A catalogue record for this book is available from the Library of Congress

ISBN 0 7506 4790 6

For information on all Butterworth-Heinemann
publications visit our website at www.bh.com

Typeset by Replika Press Pvt. Ltd, India
Printed and bound in Great Britain by MPG Books Ltd, Bodmin, Cornwall

Contents

Preface

Metal castings are fundamental building blocks, the three-dimensional integral shapes indispensable to practically all other manufacturing industries.

Although the manufacturing path from the liquid to the finished shape is the most direct, this directness involves the greatest difficulty. This is because so much needs to be controlled simultaneously, including melting, alloying, moulding, pouring, solidification, finishing, etc. Every one of these aspects has to be correct since failure of only one will probably cause the casting to fail. Other processes such as forging or machining are merely single parts of multi-step processes. It is clearly easier to control each separate process in turn.

It is no wonder therefore that the manufacture of castings is one of the most challenging of technologies. It has defied proper understanding and control for an impressive five thousand years at least. However, there are signs that we might now be starting to make progress.

Naturally, this claim appears to have been made by all writers of textbooks on castings for the last hundred years or so. Doubtless, it will continue to be made in future generations. In a way, it is hoped that it will always be true. This is what makes casting so fascinating. The complexity of the subject invites a continuous stream of new concepts and new solutions.

The author trained as a physicist and physical metallurgist, and is aware of the admirable and powerful developments in science and technology that have facilitated the progress enjoyed by these branches of science. These successes have, quite naturally, persuaded the Higher Educational Institutes throughout the world to adopt *physical metallurgy* as the natural materials discipline required to be taught. *Process metallurgy* has been increasingly regarded as a less rigorous subject, not requiring the attentions of a university curriculum. Perhaps, worse still, we now have

materials science, where breadth of knowledge has to take precedence over depth of understanding. This work makes the case for *process metallurgy* as being a key complementary discipline. It can explain the properties of metals, in some respects outweighing the effects of alloying, working and heat treatment that are the established province of physical metallurgy. In particular, the study of casting technology is a topic of daunting complexity, far more encompassing than the separate studies, for instance, of fluid flow or solidification (as necessary, important and fascinating as such focused studies clearly are). It is hoped therefore that in time, casting technology will be rightly recognized as a complex engineering discipline, worthy of individual attention.

The author has always admired those who have only published what was certain knowledge. However, as this work was well under way, it became clear to me that this was not my purpose. Knowledge is hard to achieve, and often illusive, fragmentary and ultimately uncertain. This book is offered as an exercise in education, more to do with thinking and understanding than learning. It is an exercise in grappling with new concepts and making personal evaluations of their worth, their cogency, and their place amid the scattering of facts, some reliable, others less so. It is about research, and about the excitement of finding out for oneself.

Thus the opportunity has been taken in this revised edition of *Castings* to bring the work up to date particularly in the new and exciting areas of surface turbulence and the recently discovered compaction and unfurling of folded film defects (the bifilms). Additional new concepts of alloy theory relating to the common alloy eutectics Al–Si and Fe–C will be outlined. At the time of writing these new paradigms are not quite out of the realm of speculation, but most areas are now well grounded in about 200 person years of effort in the author's laboratory over the last 12 years. Furthermore, many have been rigorously tested and proved in foundries.

This aspect of quoting confirmation of scientific concepts from industrial experience is a departure that will be viewed with concern by those academics who are accustomed to the apparent rigour of laboratory experiments. However, for those who persevere and grow to understand this work it will become clear that laboratory experiments cannot at this time achieve the control over liquid metal quality that can now be routinely provided in some industrial operations. Thus the evidence from industry is vital at this time. Suitable laboratory experiments can catch up later.

The author has allowed himself the luxury of hypothesis, that a sceptic might brand speculation. Broadly, it has been carried out in the spirit of the words of John Maynard Keynes, 'I would rather be vaguely right than precisely wrong.' This book is the first attempt to codify and present the New Metallurgy. It cannot therefore claim to be authoritative on all aspects at this time. It is an introduction to the new thinking of the metallurgy of cast alloys, and, by virtue of the survival of many of the defects during plastic working, wrought alloys too.

The primary aim remains to challenge the reader to think through the concepts that will lead to a better understanding of this most complex of forming operations, the casting process. It is hoped thereby to improve the professionalism and status of casting technology, and with it the products, so that both the industry and its customers will benefit.

It is intended to follow up this volume *Castings I – Principles* with two further volumes. The next in line is *Castings II – Practice* listing my ten rules for the manufacture of good castings with one chapter per rule. It concentrates on an outline of current knowledge of the theory and practice of designing filling and feeding systems for castings. It is intended as a more practical work. Finally, I wish to write something on *Castings III – Processes* because, having personal experience of many of the casting processes, it has become clear to me that a good comparative text is much needed. I shall then take a rest.

Even so, as I mentioned in the Preface to *Castings*, and bears repeat here, the rapidity of casting developments makes it a privilege to live in such exciting times. For this reason, however, it will not be possible to keep this work up to date. It is hoped that, as before, this new edition will serve its purpose for a time, reaching out to an even wider audience, and assisting foundry people to overcome their everyday problems. Furthermore, I hope it will inspire students and casting engineers alike to continue to keep themselves updated. The regular reading of new developments in the casting journals, and attendance at technical meetings of local societies, will encourage the professionalism to achieve even higher standards of castings in the future.

JC
West Malvern, Worcestershire, UK
1 September 2002

Dedication

I dedicate this book to my wife, Sheila, for her encouragement and support. I recognize that such acknowledgements are commonly made at the beginnings of books, to the extent that they might appear trite, or hackneyed. However, I can honestly say that I had no idea of the awful reality of the antisocial problems reflected by these tributes.

Although it may be true that, following P. G. Wodehouse, without Sheila's sympathy and encouragement this book would have been finished in half the time, it is also true that without such long-suffering efforts beyond the call of duty of any wife, it would never have been finished at all.

Introduction

I hope the reader will find inspiration from the new concepts described in this work.

What is presented is a new approach to the metallurgy of castings. Not everything in the book can claim to be proved at this stage. Ultimately, science proves itself by underpinning good technology. Thus, not only must it be credible but, in addition, it must really work. Perhaps we may never be able to say for certain that it is really true, but in the meantime it is proposed as a piece of knowledge as reliable as can now be assembled (Ziman 2001).

Even so, it is believed that for the first time, a coherent framework for an understanding of cast metals has been achieved.

The bifilm, the folded-in surface film, is the fundamental starting point. It is often invisible, having escaped detection for millennia. Because the presence of bifilms has been unknown, the initiation events for our commonly seen defects such as porosity, cracks and tears have been consistently overlooked.

It is not to be expected that all readers will be comfortable with the familiar, cosy concepts of 'gas' and 'shrinkage' porosity relegated to being mere consequences, simply growth forms derived from the new bifilm defect, and at times relatively unimportant compared to the pre-existing bifilm itself. Many of us will have to relearn our metallurgy of cast metals. Nevertheless, I hope that the reader will overcome any doubts and prejudices, and persevere bravely. The book was not written for the faint-hearted.

As a final blow (the reader needs resilience!), the book nowhere claims that good castings are easily achieved. As was already mentioned in the Preface, the casting process is among the most complex of all engineering production systems. We currently need all the possible assistance to our understanding to solve the problems to achieve adequate products.

For the future, we can be inspired to strive for, and perhaps one day achieve, defect-free cast products. At that moment of history, when the bifilm is banished, we shall have automatically achieved that elusive target – *minimum costs*.

Chapter 1

The melt

Some liquid metals may be really like liquid metals. Such metals may include pure liquid gold, possibly some carbon–manganese steels while in the melting furnace at a late stage of melting. These, however, are rare.

Many liquid metals are actually so full of sundry solid phases floating about, that they begin to more closely resemble slurries than liquids. In the absence of information to the contrary, this condition of a liquid metal should be assumed to apply. Thus many of our models of liquid metals that are formulated to explain the occurrence of defects neglect to address this fact. The evidence for the real internal structure of liquid metals being crammed with defects has been growing over recent years as techniques have improved. Some of this evidence is described below. Most applies to aluminium and its alloys where the greatest effort has been. Evidence for other materials is presented elsewhere in this book.

It is sobering to realize that many of the strength-related properties of liquid metals can only be explained by assuming that the melt is full of defects. Classical physical metallurgy and solidification science, which has considered metals as merely pure metals, is currently unable to explain the important properties of cast materials such as the effect of dendrite arm spacing, and the existence of pores and their area density. These key aspects of cast metals will be seen to arise naturally from the population of defects.

It is not easy to quantify the number of non-metallic inclusions in liquid metals. McClain and co-workers (2001) and Godlewski and Zindel (2001) have drawn attention to the unreliability of results taken from polished sections of castings. A technique for liquid aluminium involves the collection of inclusions by pressurizing up to 2 kg of melt, forcing it through a fine filter, as in the PODFA and PREFIL tests. Pressure is required because the filter is so fine. The method overcomes the sampling problem by concentrating the inclusions by a factor of about 10 000 times (Enright and Hughes 1996 and Simard *et al.* 2001). The layer of inclusions remaining on the filter can be studied on a polished section. The total quantity of inclusions is assessed as the area of the layer as seen under the microscope, divided by the quantity of melt that has passed through the filter. The unit is therefore the curious quantity $mm^2.kg^{-1}$. (It is to be hoped that at some future date this unhelpful unit will, by universal agreement, be converted into some more meaningful quantity such as volume of inclusions per volume of melt. In the meantime, the standard provision of the diameter of the filter in reported results would at least allow a reader the option to do this.)

To gain some idea of the range of inclusion contents an impressively dirty melt might reach $10 \, mm^2.kg^{-1}$, an alloy destined for a commercial extrusion might be in the range 0.1 to 1, foil stock might reach 0.001, and computer discs 0.0001 $mm^2.kg^{-1}$. For a filter of 30 mm diameter these figures approximately encompass the range 10^{-3} (0.1 per cent) down to 10^{-7} (0.1 part per million by volume) volume fraction.

Other techniques for the monitoring of inclusions in Al alloy melts include LIMCA (Smith 1998), in which the melt is drawn through a narrow tube. The voltage drop applied along the length of the tube is measured. The entry of an inclusion of different electrical conductivity into the tube causes the voltage differential to rise by an amount that is assumed to be proportional to the size of the inclusion. The technique is generally thought to be limited to inclusions approximately in the range 10 to 100 μm or so. Although widely used for the casting of wrought alloys, the author regrets that that technique has to be viewed with great reservation. Inclusions in light alloys are often up to 10 mm diameter, as will become clear. Such inclusions do find their way into the LIMCA tube, where they tend to hang, caught up at the mouth of the tube, and rotate into spirals like a flag tied to the mast by only one corner (Asbjornsonn 2001). It is to be regretted that most workers using LIMCA have been unaware of these serious problems.

Ultrasonic reflections have been used from time to time to investigate the quality of melt. The early work by Mountford and Calvert (1959–60) is noteworthy, and has been followed up by considerable development efforts in Al alloys (Mansfield 1984), and Ni alloys and steels (Mountford *et al.* 1992–93). Ultrasound is efficiently reflected from oxide films (almost certainly because the films are double, and the elastic wave cannot cross the intermediate layer of air, and thus is efficiently reflected). However, the reflections may not give an accurate idea of the size of the defects because of the irregular, crumpled form of such defects and their tumbling action in the melt. The tiny mirror-like facets of large defects reflect back to the source only when they happen to rotate to face the beam. The result is a general scintillation effect, apparently from many minute and separate particles. It is not easy to discern whether the images correspond to many small or a few large defects.

Neither Limca nor the various ultrasonic probes can distinguish any information on the types of inclusions that they detect. In contrast, the inclusions collected by pressurized filtration can be studied in some detail. In aluminium alloys many different inclusions can be found. Table 1.1 lists some of the principal types.

Nearly all of these foreign materials will be deleterious to products intended for such products as foil or computer discs. However, for shaped castings, those inclusions such as carbides and borides may not be harmful at all. This is because having been precipitated from the melt, they are usually therefore in excellent atomic contact with the alloy material. These well-bonded non-metallic

phases are thereby unable to act as initiators of other defects such as pores and cracks. Conversely, they may act as grain refiners. Furthermore, their continued good bonding with the solid matrix is expected to confer on them a minor or negligible influence on mechanical properties. (However, we should not forget that it is possible that they may have some influence on other physical or chemical properties such as machinability or corrosion.)

Generally, therefore, this book concentrates on those inclusions that have a major influence on mechanical properties, and that can be the initiators of other serious problems such as pores and cracks. Thus the attention will centre on *entrained surface films*, that exhibit unbonded interfaces with the melt, and lead to a spectrum of problems. Usually, these inclusions will be oxides. However, carbon films are also common, and occasionally nitrides, sulphides and other materials.

The pressurized filtration tests can find all of these entrained solids, and the analysis of the inclusions present on the filter can help to identify the source of many inclusions in a melting and casting operation. However, the only inclusions that remain undetectable but are enormously important are the newly entrained films that occur on a clean melt as a result of surface turbulence. These are the films commonly entrained during the pouring of castings, and so, perhaps, not required for detection in a melting and distribution operation. They are typically only 20 nm thick, and so remain invisible under an optical microscope, especially if draped around a piece of refractory filter that when sectioned will appear many thousands of times thicker. The only detection technique for such inclusions is the lowly *reduced pressure test*. This test opens the films (because they are always double, and contain air, as will be explained in detail in Chapter 2) so that they can be seen. The radiography of the cast test pieces reveals the size, shape and numbers of such important inclusions, as has been shown by Fox and Campbell (2000). The small cylindrical test pieces can be sectioned to yield a parallel form that gives optimum radiographic results. Alternatively, it is more convenient to cast the test pieces with parallel sides. The test will be discussed in more detail later.

1.1 Reactions of the melt with its environment

A liquid metal is a highly reactive chemical. It will react both with the gases above it and the solid material of the crucible that contains it. If there is any kind of slag or flux floating on top of the melt, it will probably react with that too. Many melts also react with their containers such as crucibles and furnace linings.

Table 1.1 Types of inclusions in Al alloys

Inclusion type		Possible origin
Carbides	Al_4C_3	Pot cells from Al smelters
Boro-carbides	Al_4B_4C	Boron treatment
Titanium boride	TiB_2	Grain refinement
Graphite	C	Fluxing tubes, rotor wear, entrained film
Chlorides	$NaCl$, KCl, $MgCl_2$, etc.	Chlorine or fluxing treatment
Alpha alumina	$\alpha\text{-}Al_2O_3$	Entrainment after high-temperature melting
Gamma alumina	$\gamma\text{-}Al_2O_3$	Entrainment during pouring
Magnesium oxide	MgO	Higher Mg containing alloys
Spinel	$MgOAl_2O_3$	Medium Mg containing alloys

The driving force for these processes is the striving of the melt to come into equilibrium with its surroundings. Its success in achieving equilibrium is, of course, limited by the rate at which reactions can happen, and by the length of time available.

Thus reactions in the crucible or furnace during the melting of the metal are clearly seen to be serious, since there is usually plenty of time for extensive changes. The pick-up of hydrogen from damp refractories is common. Similar troubles are often found with metals that are melted in furnaces heated by the burning of hydrocarbon fuels such as gas or oil.

We can denote the chemical composition of hydrocarbons as C_xH_y and thus represent the straight chain compounds such as methane CH_4, ethane C_2H_6 and so on, or aromatic ring compounds such as benzene C_6H_6, etc. (Other more complicated molecules may contain other constituents such as oxygen, nitrogen and sulphur, not counting impurities which may be present in fuel oils such as arsenic and vanadium.)

For our purposes we will write the burning of fuel taking methane as an example

$$CH_4 + 2O_2 = CO_2 + 2H_2O \qquad (1.1)$$

Clearly the products of combustion of hydrocarbons contain water, so the hot waste gases from such furnaces are effectively wet.

Even electrically heated furnaces are not necessarily free from the problem of wet environment: an electric resistance furnace that has been allowed to stand cold over a weekend will have had the chance to absorb considerable quantities of moisture in its lining materials. Most furnace refractories are hygroscopic and will absorb water up to 5 or 10 per cent of their weight. This water is released into the body of the furnace over the next few days of operation. It has to be assumed that the usual clay/graphite crucible materials commonly used for melting non-ferrous alloys are quite permeable to water vapour and/or hydrogen, since they are designed to be approximately 40 per cent porous. Additionally, hydrogen permeates freely through most materials, including steel, at normal metallurgical operating temperatures of around 700°C and above.

This moisture from linings or atmosphere can react in turn with the melt M:

$$M + H_2O = MO + H_2 \qquad (1.2)$$

Thus a little metal is sacrificed to form its oxide, and the hydrogen is released to equilibrate itself between the gas and metal phases. Whether it will, on average, enter the metal or the gas above the metal will depend on the relative partial pressure of hydrogen already present in both of these phases. The molecular hydrogen has to split into atomic

hydrogen before it can be taken into solution, as is described by

$$H_2 = 2[H] \qquad (1.3)$$

The equation predicting the partial pressure of hydrogen in equilibrium with a given concentration of hydrogen in solution in the melt is

$$[H]^2 = kP_{H2} \qquad (1.4)$$

where the constant k has been the subject of many experimental determinations for a variety of gas–metal systems (Brandes 1983; Ransley and Neufeld 1948). It is found to be affected by alloy additions (Sigworth and Engh 1982) and temperature. When the partial pressure of hydrogen P = 1 atmosphere, it is immediately clear from this equation that k is numerically equal to the solubility of hydrogen in the metal at that temperature. Figure 1.1 shows

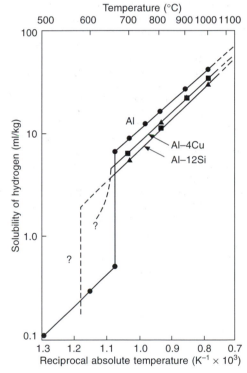

Figure 1.1 *Hydrogen solubility in aluminium and two of its alloys, showing the abrupt fall in solubility on solidification.*

how the solubility of hydrogen in aluminium increases with temperature.

It is vital to understand fully the concept of an equilibrium gas pressure associated with the gas in solution in a liquid. We shall digress to present a few examples to illustrate the concept.

Consider a liquid containing a certain amount of hydrogen atoms in solution. If we place this liquid in an evacuated enclosure then the liquid will find itself out of equilibrium with respect to the environment above the liquid. It is supersaturated with respect to its environment. It will then gradually lose its hydrogen atoms from solution, and these will combine on its surface to form hydrogen molecules, which will escape into the enclosure as hydrogen gas. The gas pressure in the enclosure will therefore gradually build up until the rate of loss of hydrogen from the surface becomes equal to the rate of gain of the liquid from hydrogen that returns, reconverting to individual atoms on the surface and re-entering solution in the liquid. The liquid can then be said to have come into equilibrium with its environment.

Similarly, if a liquid containing little or no gas (and therefore having a low equilibrium gas pressure) were placed in an environment of high gas pressure, then the net transfer would, of course, be from gas phase to liquid phase until the equilibrium partial pressures were equal. Figure 1.2 illustrates the case of three different initial concentrations of hydrogen in a copper alloy melt, showing how initially high concentrations fall, and initially low concentrations rise, all finally reaching the same concentration which is in equilibrium with the environment.

This equilibration with the external surroundings is relatively straightforward to understand. What is perhaps less easy to appreciate is that the equilibrium gas pressure in the liquid is also effectively in operation *inside* the liquid.

This concept can be grasped by considering bubbles of gas which have been introduced into the liquid by stirring or turbulence, or which are adhering to fragments of surface films or other inclusions that are floating about. Atoms of gas in solution migrate across the free surface of the bubbles and into their interior to establish an equilibrium pressure inside.

On a microscopic scale, a similar behaviour will be expected between the individual atoms of the liquid. As they jostle randomly with their thermal motion, small gaps open momentarily between the atoms. These embryonic bubbles will also therefore come into equilibrium with the surrounding liquid.

It is clear, therefore, that the equilibrium gas pressure of a melt applies both to the external and internal environments of the melt.

We have so far not touched on those processes that control the *rate* at which reactions can occur. The kinetics of the process can be vital.

Consider, for instance, the powerful reaction between the oxygen in dry air and liquid aluminium: no disastrous burning takes place; the reaction is held in check by the surface oxide film which forms, slowing the rate at which further oxidation can occur. This is a beneficial interaction with the environment. Other beneficial passivating (i.e. inhibiting) reactions are seen in the melting of magnesium under a dilute SF_6 (sulphur hexafluoride) gas, as described, for instance, by Fruehling and Hanawalt (1969). A further example is the beneficial

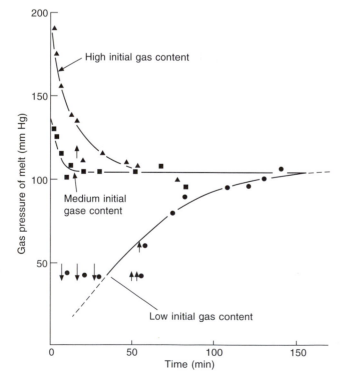

Figure 1.2 *Hydrogen content of liquid aluminium bronze held in a gas-fired furnace, showing how the melt equilibrates with its surroundings. Data from Ostrom et al. (1975).*

the whole world supply of the metal ... ion began. We can therefore safely ... its solubility to zero. Yet everyone ... aluminium and its alloys are full of ... is this possible? The oxides certainly ... been precipitated by reaction with ... lution. Oxygen can only react with ... urthermore, the surface can only access ... the metal if it is entrained, or folded ... nechanical, not a chemical process. ... ce of oxygen in aluminium is thereby ... stood, and will be re-examined ... m many different viewpoints as we ... gh the book.

... ow to the presence of hydrogen in ... his behaves quite differently.

... is calculated from Equation 1.4 ... case for hydrogen solubility in liquid ... demonstrates that on a normal day ... t relative humidity, the melt at 750°C ... ch about 1 ml.kg^{-1} (0.1 ml.100 g^{-1}) ... ydrogen. This is respectably low for ... ial castings (although perhaps just ... high for aerospace standards). Even ... t humidity the hydrogen level will ... tolerable for most applications. This ... for degassing aluminium alloys by ... other than waiting. If originally high ... t will equilibrate by losing gas to its ... as is also illustrated by the copper- ... Figure 1.2).

... sideration of Figure 1.3 indicates that ... d aluminium is in contact with wet ... wet gases, the environment will ... lose to one atmosphere pressure of

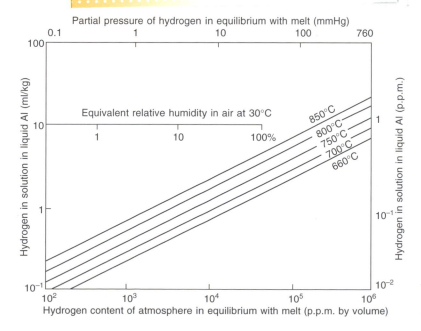

Figure 1.3 *Hydrogen content of liquid aluminium shown as increasing with temperature and the hydrogen content of the environment as hydrogen gas or as water vapour.*

water vapour, causing the concentration of gas in solution to rise to nearer 10 ml.kg^{-1}. This spells disaster for most normal castings. Such metal has been preferred, however, for the production of many non-critical parts, where the precipitation of hydrogen pores can compensate to some extent for the shrinkage on freezing, and thus avoid the problem and expense of the addition of feeders to the casting. Traditional users of high levels of hydrogen in this way are the permanent mould casters of automobile inlet manifolds and rainwater goods such as pipes and gutters. Both cost and the practicalities of the great length to thickness ratio of these parts prevent any effective feeding.

Raising the temperature of the melt will increase the solubility of hydrogen in liquid aluminium. At a temperature of 1000°C the solubility is over 40 ml.kg^{-1}. However, of course, if there is no hydrogen available in its environment the melt will not be able to increase its gas content no matter what its temperature is. This self-evident fact is easy to overlook in practice because there is nearly always some source of moisture or hydrogen, so that, usually, high temperatures are best avoided if gas levels are to be kept under good control. Most aluminium alloy castings can be made successfully at casting temperatures of 700–750°C. Rarely are temperatures in the range 750–850°C actually required, especially if the running system is good.

A low gas content is only attained under conditions of a low partial pressure of hydrogen. This is why some melting and holding furnaces introduce only dry filtered air, or a dry gas such as bottled nitrogen, into the furnace as a protective blanket. Occasionally the ultimate solution of treating the melt in vacuum is employed (Venturelli 1981). This dramatically expensive solution does have the benefit that the other aspects of the environment of the melt, such as the refractories, are also properly dried. From Figure 1.3 it is clear that gas levels in the melt of less than 0.1 ml/kg are attainable. However, the rate of degassing is slow, requiring 30–60 minutes, since hydrogen can only escape from the surface of the melt, and takes time to stir by convection, and finally diffuse out. The time can be reduced to a few minutes if the melt is simultaneously flushed with an inert gas such as nitrogen.

For normal melting in air, the widespread practice of flushing the melt with an inert gas from the immersed end of a lance of internal diameter of 20 mm or more is only poorly effective. The useful flushing action of the inert gas can be negated at the free surface because the fresh surface of the liquid continuously turned over by the breaking bubbles represents ideal conditions for the melt to equilibrate with the atmosphere above it. If the weather is humid the rate of regassing can exceed the rate of degassing.

Systems designed to provide numerous fine bubbles are far more effective. The free surface at the top of the melt is less disturbed by their arrival. Also, there is a greatly increased surface area, exposing the melt to a flushing gas of low partial pressure of hydrogen. Thus the hydrogen in solution in the melt equilibrates with the bubbles with maximum speed. The bubbles are carried to the surface and allowed to escape, taking the hydrogen with them. Such systems have the potential to degas at a rate that greatly exceeds the rate of uptake of hydrogen.

Rotary degassing systems can act in this way. However, their use demands some caution. On the first use after a weekend, the rotary head and its shaft will introduce considerable hydrogen from their absorbed moisture. It is to be expected that the melt will get worse before it gets better. Thus degassing to a constant (short) time is a sure recipe for disaster when the refractories of the rotor are damp. In addition, there is the danger that the vortex at the surface of the melt may carry down air into the melt, thus degrading the melt by manufacturing oxides faster than they can be floated out. This is a common and disappointing mode of operation of a technique that has good potential when used properly. The simple provision of a baffle board to prevent the rotation of the surface, and thus suppress the vortex formation, is highly effective.

When dealing with the rate of attainment of equilibrium in melting furnaces the times are typically 30–60 minutes. This slow rate is a consequence of the large volume to surface area ratio. We shall call this ratio the modulus. Notice that it has dimensions of length. For instance, a 10 tonne holding furnace would have a volume of approximately 4 m^3, and a surface area in contact with the atmosphere of perhaps 10 m^2, giving a modulus of 4/10 m = 0.4 m = 400 mm. A crucible furnace of 200 kg capacity would have a modulus nearer 200 mm.

These values around 300 mm for large bodies of metal contrast with those for the pouring stream and the running system. If these streams are considered to be cylinders of liquid metal approximately 20 mm diameter, then their effective modulus is close to 5 mm. Thus their reaction time would be expected to be as much as 300/5 = 60 times faster, resulting in the approach towards equilibrium within times of the order of one minute. (This same reasoning explains the increase in rate of vacuum degassing by the action of bubbling nitrogen through the melt.) This is the order of time in which many castings are cast and solidified. We have to conclude, therefore, that reactions of the melt with its environment continue to be important at all stages of its progress from furnace to mould.

There is much evidence to demonstrate that the

melt does interact rapidly with the chemical environment within the mould. There are methods available of protecting the liquid by an inert gas during melting and pouring which are claimed to reduce the inclusion and pore content of many alloys that have been tested, including aluminium alloys, and carbon and stainless steels (Anderson *et al.* 1989). Additional evidence is considered in section 4.5.2.

The aluminium–hydrogen system considered so far is a classic model of simplicity. The only gas that is soluble in aluminium in any significant amounts is hydrogen. The magnesium–hydrogen system is similar, but rather less important in the sense that the hydrogen solubility is lower, so that dissolved gas is in general less troublesome. Other systems are in general more complicated as we shall see.

1.1.2 Copper alloys

Copper-based alloys have a variety of dissolved gases and thus a variety of reactions. In addition to hydrogen, oxygen is also soluble. Reaction between these solutes produces water vapour according to (where square brackets indicate an element in solution)

$$2[H] + [O] = H_2O \qquad (1.5)$$

Thus water vapour in the environment of molten copper alloys will increase both hydrogen and oxygen contents of the melt. Conversely, on rejection of stoichiometric amounts of the two gases to form porosity, the principal content of the pores will not be hydrogen and oxygen but their reaction product, water vapour. An excess of hydrogen in solution will naturally result in an admixture of hydrogen in the gas in equilibrium with the melt. An excess of oxygen in solution will result in the precipitation of copper oxide.

Much importance is often given to the so-called steam reaction:

$$2[H] + Cu_2O = 2Cu + H_2O \qquad (1.6)$$

This is, of course, a nearly equivalent statement of Equation 1.5. The generation of steam by this reaction has been considered to be the most significant contribution to the generation of porosity in copper alloys that contain little or no deoxidizing elements. This seems a curious conclusion since the two atoms of hydrogen are seen to produce one molecule of water. If there had been no oxygen present the two hydrogen atoms would have produced one molecule of hydrogen, as indicated by Equation 1.3. Thus the same volume of gases is produced in either case. It is clear therefore that the real problem for the maximum potential of gas porosity in copper is simply hydrogen.

(However, as we shall see in later sections, the presence of oxygen will be important in the nucleation of pores in copper, but only if oxygen is present in solution in the liquid copper, not just present as oxide. The distribution of pores as subsurface porosity in many situations is probably good evidence that this is true. We shall return to consideration of this phenomenon later.)

Proceeding now to yet more possibilities in copper-based materials, if sulphur is present then a further reaction is possible:

$$[S] + 2[O] = SO_2 \qquad (1.7)$$

and for copper alloys containing nickel, an important impurity is carbon, giving rise to an additional possibility:

$$[C] + [O] = CO \qquad (1.8)$$

Systematic work over the last decade at the University of Michigan (see, for instance, Ostrom *et al.* (1981)) on the composition of gases that are evolved from copper alloys on solidification confirms that pure copper with a trace of residual deoxidizer evolves mainly hydrogen. Brasses (Cu–Zn alloys) are similar, but because zinc is only a weak deoxidant the residual activity of oxygen in solution gives rise to some evolution of water vapour. Interestingly, the main constituent of evolved gas in brasses is zinc vapour, since these alloys have a melting point above the boiling point of zinc (Figure 1.4). Pure copper and the tin bronzes evolve mainly water vapour with some hydrogen. Copper–nickel alloys with nickel above 1 per cent have an increasing contribution from carbon monoxide as a result of the promotion of carbon solubility by nickel.

Thus when calculating the total gas pressure in equilibrium with melts of copper-based alloys, for instance inside an embryonic bubble, we need to add all the separate contributions from each of the contributing gases.

The brasses represent an interesting special case. The continuous vaporization of zinc from the free surface of a brass melt carries away other gases from the immediate vicinity of the surface. This continuous outflowing wind of metal vapour creates a constantly renewed clean environment, sweeping away gases which diffuse into it from the melt, and preventing contamination of the local environment of the metal surface with furnace gases or other sources of pollution. For this reason cast brass is usually found to be free from gas porosity.

The zinc vapour burns in the air with a brilliant flame known as zinc flare. Flaring may be suppressed by a covering of flux. However, the beneficial degassing action cannot then occur, raising the danger of porosity, mainly from hydrogen.

The boiling point of pure zinc is 907°C. But the presence of zinc in copper alloys does not cause boiling until higher temperatures because, of course,

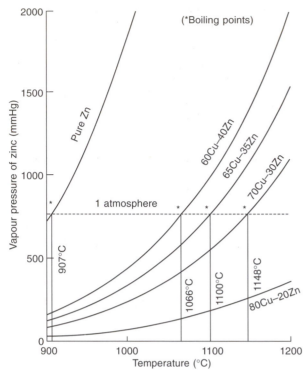

Figure 1.4 *Vapour pressure of zinc and some brasses. Data from Hull (1950).*

the zinc is diluted (strictly, its activity is reduced). Figure 1.4 shows the effects of increasing dilution on raising the temperature at which the vapour pressure reaches one atmosphere, and boiling occurs. The onset of vigorous flaring at that point is sufficiently marked that in the years prior to the wider use of thermocouples foundrymen used it as an indication of casting temperature. The accuracy of this piece of folklore can be appreciated from Figure 1.4. The flaring temperatures increase in step with the increasing copper contents (i.e. at greater dilutions of zinc), and thus with the increasing casting temperatures of the alloys.

Around 1 per cent of zinc is commonly lost by flaring and may need to be replaced to keep within the alloy composition specification. In addition, workers in brass foundries have to be monitored for the ingestion of zinc fumes.

Melting practice for the other copper alloys to keep their gas content under proper control is not straightforward. Below are some of the pitfalls.

One traditional method has been to melt under oxidizing conditions, thereby raising the oxygen in solution in the melt in an attempt to reduce gradually the hydrogen level. Prior to casting, the artificially raised oxygen in solution is removed by the addition of a deoxidizer such as phosphorus, lithium or aluminium. The problem with this technique is that even under good conditions the rate of attainment of equilibrium is slow because of the limited surface areas across which the elements have to diffuse. Thus little hydrogen may

have been removed. Worse still, the original oxidation has often been carried out in the presence of furnace gases, so raising oxygen and (unwittingly) hydrogen levels simultaneously (Equation 1.2) high above the values to be expected if the two dissolved gases were in equilibrium. The addition of deoxidizer therefore still leaves hydrogen at near saturation.

The further problem with this approach is that the deoxidizer precipitates out the oxygen as a suspension of solid oxide particles in the melt, or as surface oxide films. Either way, these by-products are likely to give problems later as non-metallic inclusions in the casting, and, worse still, as nuclei to assist the precipitation of the remaining gases in solution, thus promoting the very porosity that the technique was intended to avoid. In conclusion, it is clear there is little to commend this approach.

A second reported method is melting under reducing conditions to decrease losses by oxidation. Hydrogen removal is then attempted just before casting by adding copper oxide or by blowing dry air through the melt. Normal deoxidation is then carried out. The problem with this technique is that the hydrogen-removal step requires time and the creation of free surfaces, such as bubbles, for the elimination of the reaction product, water vapour. Waiting for the products to emerge from the quiescent surface of a melt sitting in a crucible would probably take 30–60 minutes. Fumes from the fuel-fired furnace would be ever present to help to undo any useful degassing. Clearly therefore, this technique cannot be recommended either!

The second technique described above would almost certainly have used a cover of granulated charcoal over the melt to provide the reducing conditions. This is a genuinely useful way of reducing the formation of drosses (dross is a mixture of oxide and metal, so intimately mixed that it is difficult to separate) as can be demonstrated from the Ellingham diagram (Figure 1.5), the traditional free energy/temperature graph. The oxides of the major alloying elements copper, zinc and tin are all reduced back to their metals by carbon, which preferentially oxidizes to carbon monoxide (CO) at this high temperature. (The temperature at which the metal oxide is reduced, and carbon is oxidized to CO, is that at which the free energies for the formation of CO exceed that of the metal oxide, i.e. CO becomes more stable. This is where the lines cross on the Ellingham diagram.)

However, it is as well to remember that charcoal contains more than just carbon. In fact, the major impurity is moisture, even in well-dried material that appears to be quite dry. An addition of charcoal to the charge at an early stage in melting is therefore relatively harmless because the release of moisture,

and the contamination of the charge with hydrogen and oxygen, will have time to be reversed. In contrast, an addition of charcoal at a late stage of melting will flood the melt with fresh supplies of hydrogen and oxygen that will almost certainly not have time to evaporate out before casting. Any late additions of anything, even alloying additions, introduce the risk of unwanted gases.

Reliable routes to melted metal with low gas content include:

1. Electric melting in furnaces that are never allowed to go cold.
2. Controlled use of flaring for zinc-containing alloys.
3. Controlled dry environment of the melt. Additions of charcoal are recommended if added at an early stage, preferably before melting. (Late additions of charcoal or other sources of moisture are to be avoided.)

In summary, the gases which can be present in the various copper-based alloys are:

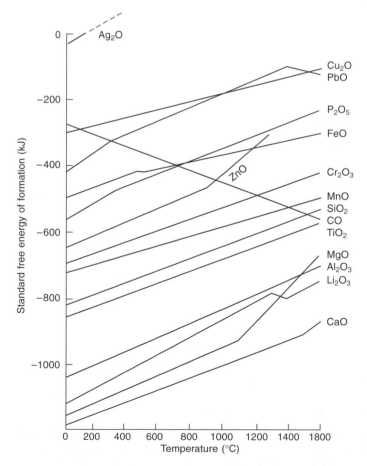

Figure 1.5 *The Ellingham diagram, illustrating the free energy of formation of oxides as a function of temperature.*

Pure copper H_2, H_2O
Brasses, gunmetals H_2, H_2O, Zn, Pb
Cupro-nickels H_2, H_2O, CO, (N_2?)

1.1.3 Iron alloys

Like copper-based alloys, iron-based alloys are also complicated by the number of gases that can react with the melt, and that can cause porosity by subsequent evolution on solidification. Again, it must be remembered that all the gases present can add their separate contributions to the total pressure in equilibrium with the melt.

Oxygen is soluble, and reacts with carbon, which is one of the most important constituents of steels and cast irons. Carbon monoxide is the product, following Equation 1.8.

In steelmaking practice the CO reaction is used to lower the high carbon levels in the pig iron produced by the blast furnace. (The high carbon is the result of the liquid iron percolating down through the coke in the furnace stack. A similar situation exists in the cupola furnace used in the melting of cast iron used by iron foundries.) The oxygen to initiate the CO reaction is added in various forms, traditionally as shovelfuls of granular FeO thrown onto the slag, but in modern steelmaking practice by spectacular jets of supersonic oxygen. The stage of the process in which the CO is evolved is so vigorous that it is aptly called a 'carbon boil'.

After the carbon is brought down into specification, the excess oxygen that remains in the steel is lowered by deoxidizing additions of manganese, silicon or aluminium. In modern practice a complex cocktail of deoxidizing elements is added as an alternative or in addition. These often contain small percentages of rare earths to control the shape of the non-metallic inclusions in the steel. It seems likely that this control of shape is the result of reducing the melting point of the inclusions so that they become at least partially liquid, adopting a more rounded form that is less damaging to the properties of the steel.

Hydrogen is soluble as in Equation 1.3, and exists in equilibrium with the melt as indicated in Equation 1.4. However, a vigorous carbon boil will reduce any hydrogen in solution to negligible levels by flushing it from the melt.

In many steel foundries, however, steel is melted from scrap steel (not made from pig iron, as in steelmaking). Because the carbon is therefore already low, there is no requirement for a carbon boil. Thus hydrogen remains in the melt. In contrast to oxygen in the melt that can quickly be reduced by the use of a deoxidizer, there is no quick chemical fix for hydrogen. Hydrogen can only be encouraged to leave the metal by providing an extremely dry and hydrogen-free environment. If a carbon boil cannot be artificially induced, and if environmental control is insufficiently good, or is too slow, then the comparatively expensive last resort is vacuum degassing. This option is common in the steelmaking industry, but less so in steel melting for the making of shaped castings.

A carbon boil can be induced in molten cast iron, providing the silicon is low, simply by blowing air onto the surface of the melt (Heine 1951). Thus it is clear that oxygen can be taken into solution in cast iron even though the iron already contains high levels of carbon. The reaction releases CO gas at (or actually slightly above) atmospheric pressure. During solidification, in the region ahead of the solidification front, carbon and oxygen are concentrated still further. It is easy to envisage how, therefore, from relatively low initial contents of these elements, they can increase together so as to exceed a critical product [C] · [O] to cause CO bubbles to form in the casting. The equilibrium equation, known as the solubility product, relating to Equation 1.8 is

$$[C] \cdot [O] = kP_{CO} \tag{1.9}$$

We shall return to this important equation later. It is worth noting that the equation could be stated more accurately as the product of the activities of carbon and oxygen. However, for the moment we shall leave it as the product of concentrations, as being accurate enough to convey the concepts that we wish to discuss.

Nitrogen is also soluble in liquid iron. The reaction follows the normal law for a diatomic gas:

$$N_2 = 2[N] \tag{1.10}$$

and the corresponding equation to relate the concentration in the melt [N] with its equilibrium pressure P_{N2} is simply:

$$[N]^2 = kP_{N2} \tag{1.11}$$

As before, the equilibrium constant k is a function of temperature and composition. It is normally determined by careful experiment.

The reactions of iron with its environment to produce surface films of various kinds is dealt with in section 5.5.

1.2 Transport of gases in melts

Gases in solution in liquids travel most quickly when the liquid is moving, since, of course, they are simply carried by the liquid.

However, in many situations of interest the liquid is stationary, or nearly so. This is the case in the boundary layer at the surface of the liquid. The presence of a solid film on the surface will hold the surface stationary, and because of the effect of viscosity, this stationary zone will extend for some distance into the bulk liquid. The thickness of the boundary layer is reduced if the bulk of the liquid is violently stirred. However, within the stagnant liquid of the boundary layer the movement of solutes can occur only by the slow process of diffusion, i.e. the migration of populations of atoms by the process of each atom carrying out one random atomic jump at a time.

Another region where diffusion is important is in the partially solidified zone of a solidifying casting, where the bulk flow of the liquid is normally a slow drift.

In the solid state, of course, diffusion is the only mechanism by which solutes can spread.

There are two broad classes of diffusion processes: one is interstitial diffusion, and the other is substitutional diffusion. Interstitial diffusion is the squeezing of small atoms through the interstices between the larger matrix atoms. This is a relatively easy process and thus interstitial diffusion is relatively rapid. Substitutional diffusion is the exchange, or substitution, of the solute atom for a

similar-sized matrix atom. This process is more difficult (i.e. has a higher activation energy) because the solute atom has to wait for a gap of sufficient size to be created before it can jostle its way among the crowd of similar-sized individuals to reach the newly created space.

Figures 1.6 to 1.8 show the rates of diffusion of various alloying elements in the pure elements, aluminium, copper and iron. Clearly, hydrogen is an element that can diffuse interstitially because of its small size. In iron, the elements C, N and O all behave interstitially, although significantly more slowly than hydrogen.

The common alloying elements in aluminium, Mg, Zn and Cu, clearly all behave as substitutional solutes. Other substitutional elements form well-defined groups in copper and iron.

However, there are a few elements that appear to act in an intermediate fashion. Oxygen in copper occupies an intermediate position. The elements sulphur and phosphorus in iron occupy an interesting

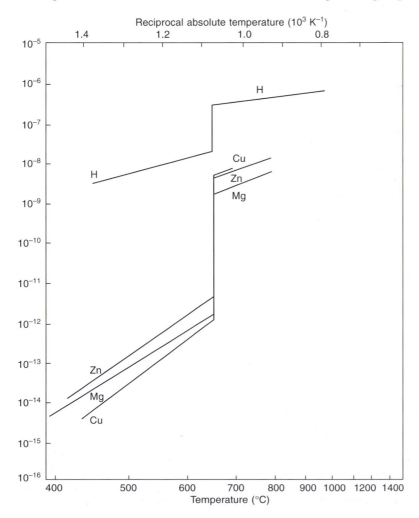

Figure 1.6 *Diffusion coefficients for elements in aluminium.*

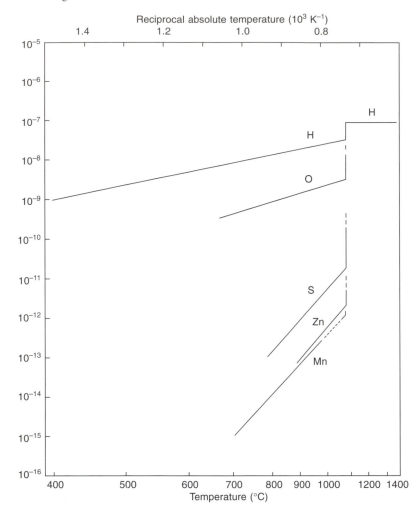

Figure 1.7 *Diffusion coefficients for elements in copper.*

intermediate position; a curious behaviour that does not appear to have been widely noticed.

Figure 1.8 also illustrates the other important feature of diffusion in the various forms of iron: the rate of diffusion in the open body-centred cubic lattice (alpha and delta phases) is faster than in the more closely packed face-centred cubic (gamma phase) lattice. Furthermore, in the liquid phase diffusion is fastest of all, and differences between the rates of diffusion of elements that behave widely differently in the solid become less marked.

These relative rates of diffusion will form a recurrent theme throughout this book. The reader will benefit from memorizing the general layout of Figures 1.6, 1.7 and 1.8.

1.3 Surface film formation

When the hot metal interacts with its environment many of the reactions result in products that dissolve rapidly in the metal, and diffuse away into its interior.

Some of these processes have already been described. In this section we shall focus our attention on the products of reactions that remain on the surface. Such products are usually films.

Oxide films usually start as simple amorphous (i.e. non-crystalline) layers, such as Al_2O_3 on Al, or MgO on Mg and Al–Mg alloys (Cochran *et al.* 1977). Their amorphous structure probably derives necessarily from the amorphous melt on which they nucleate and grow. However, they quickly convert to crystalline products as they thicken, and later often develop into a bewildering complexity of different phases and structures. Many examples can be seen in the studies reviewed by Drouzy and Mascre (1969) and in the various conferences devoted to oxidation (Microscopy of Oxidation 1993). Some films remain thin, some grow thick. Some are strong, some are weak. Some grow slowly, others quickly. Some are heterogeneous and complex in the structure, being lumpy mixtures of different phases.

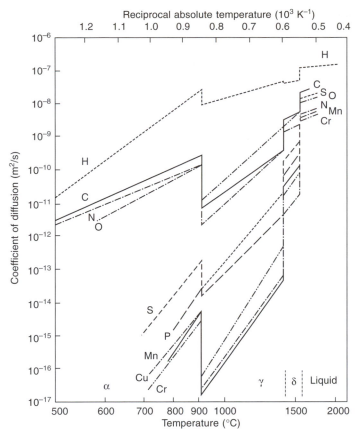

Figure 1.8 *Diffusion coefficients for elements in iron.*

Diffusion data for Figures 1.6 to 1.8.
General: LeClaire A. D. (1984) in Smithells Metals Reference Book *6th edn, Butterworths, London (Brandes E. A., ed.);*
Al(liq): Matrix; Cu, Zn, Mg: Edwards J. B., Hucke E. E., Martin J. J. (1968). Met. Rev. 120, *Parts 1 and 2; H: Physik*
Daten (1976), 5(1); Al(s): Matrix; Cu: Peterson N. L., Rothman S. J. (1970). Phys. Rev., Bi, *3264; H: Outlaw R. A.,*
Peterson D. T., Schmidt F. A. (1982). Scripta Met., 16, *287–292; Cu(s): Matrix; O: Kirscheim R. (1979).* Acta Met., 27,
869; 5. Moya F., Moya-Goutier G. E., Cabane-Brouty F. (1969). Phys. Stat. Solidi, 35, *893; McCarron R. L., Belton G.*
R. (1969). TAIME, 245, *1161–1166; Fe: Matrix;* H:Physik Daten *(1981) 5(13); C*: Physik Daten *(1981) 5(14); N*:
Physik Daten *(1982). 5(15); O*: Physik Daten *(1982). 5, (16); 5, P, Mn, Cu, Cr: LeClaire A. D. (1990). In* Landolt-
Bornstein International Critical Tables. *Berlin: J. Springer, Cr, Mn in liquid: Ono Y., Matsumoto 5. (1975).* Trans Japan
Inst. Met., 16, *415–423.*

The nature of the film on a liquid metal in a continuing equilibrium relationship with its environment needs to be appreciated. In such a situation the melt will always be covered with the film. For instance, if the film is skimmed off it will immediately re-form. A standard foundry complaint about the surface film on certain casting alloys is that 'you can't get rid of it!'

Furthermore, it is worth bearing in mind that the two most common film-forming reactions, the formation of oxide films from the decomposition of moisture, and the formation of graphitic films from the decomposition of hydrocarbons, both result in the increase of hydrogen in the metal. The comparative rates of diffusion of hydrogen and other elements in solution in various metals are shown in Figures 1.6 to 1.8. These reactions will be dealt with in detail later.

In the case of liquid copper in a moist, oxidizing environment, the breakdown of water molecules at the surface releases hydrogen that diffuses away rapidly into the interior. The oxygen released in the same reaction (Equation 1.5), and copper oxide, Cu_2O, that may be formed as a temporary intermediate product, are also soluble, at least up to 0.14 per cent oxygen. The oxygen diffuses and dissipates more slowly in the metal so long as the solubility limit in the melt is not exceeded. It is clear, however, that no permanent film is created under oxidizing conditions. Also, of course, no film forms under reducing conditions. Thus liquid copper is free from film problems in most circumstances. (Unfortunately this may not be true for the case where the solubility of the oxide is exceeded at the surface, or in the presence of certain carbonaceous atmospheres, as we shall see later. It is also untrue for many copper alloys, where the alloying element provides a stable oxide.)

Liquid silver is analogous to copper in that it dissolves oxygen. In terms of the Ellingham diagram (Figure 1.5) it is seen that its oxide, Ag_2O, is just stable at room temperature, causing silver to tarnish (together with some help from the presence of sulphur in the atmosphere to form sulphides), as every jeweller will know! However, the free energy of formation of the oxide is positive at higher temperatures, appearing therefore above zero on the figure. This means that the oxide is unstable at higher temperatures. It would therefore not be expected to exist except in cases of transient non-equilibrium.

Liquid tin is also largely free from films.

The noble metals such as gold and platinum are, for all practical purposes, totally film-free. These are, of course, all metals that are high on the Ellingham diagram, reflecting the relative instability of their oxides, and thus the ease with which they are reduced back to the metal.

Cast iron is an interesting case, occupying an intermediate position in the Ellingham diagram. It therefore has a complicated behaviour, sometimes having a film, whose changing composition converts it from solid to liquid as the temperature falls. Its behaviour is considered in detail in section 5.5 devoted to cast iron.

The light alloys, aluminium and magnesium have casting alloys characterized by the stability of the products of their surface reactions. Although part of the reaction products, such as hydrogen, diffuse away into the interior, the noticeable remaining product is a surface oxide film. The oxides of these light alloys are so stable that once formed, in normal circumstances, they cannot be decomposed back to the metal and oxygen. The oxides become permanent features for good or ill, depending on where they finally come to rest on or in the cast product. This is, of course, one of our central themes.

An interesting detail is that magnesium alloys are known to give off magnesium vapour at normal casting temperatures, the oxide film growing by oxidation of the vapour. This mechanism seems to apply not only for magnesium-based alloys (Sakamoto 1999) but also for Al alloys containing as little as 0.4 weight per cent Mg (Mizuno et al. 1996).

A wide range of other important alloys exist whose main constituents would not cause any problem in themselves, but which form troublesome films in practice because their composition includes just enough of the above highly reactive metals. These include the following.

Liquid lead exhibits a dull grey surface oxide consisting of solid PbO. This interferes with the wetting of soldered joints, giving the electrician the feared 'dry joint', which leads to arcing, overheating and eventual failure. This is the reason for the provision of fluxes to exclude air and possibly provide a reducing environment (resin-based coverings are used; the choride-based fluxes to dissolve the oxide are now less favoured because of their residual corrosive effects). The use of pre-tinning of the parts to be joined is also helpful since tin stays free from oxide at low temperature. The addition of 0.01 per cent Al to lead is used to reduce oxidation losses during melting. However, it would be expected to increase wettability problems. From the Ellingham diagram it is clear that lead can be kept clear of oxide at all temperatures for which it is molten by a covering of charcoal: the CO atmosphere will reduce any PbO formed back to metallic lead. However, we should note that lead solders are being phased out of use for environmental and health reasons.

Zinc alloys: most zinc-based castings are made from pressure die casting alloys that contain approximately 4 per cent Al. This percentage of aluminium is used to form a thin film of aluminium oxide that protects the iron and steel parts of the high pressure die casting machines and the die itself from rapid attack by zinc. From the point of view of the casting quality, the film-formation problem does give some problems, assisting in the occlusion of air and films during the extreme surface turbulence of filling. Nevertheless, these problems generally remain tolerable because the melting and casting temperatures of zinc pressure die casting alloys are low, thus probably restricting the development of films to some extent.

Other zinc-based alloys that contain higher quantities of aluminium, the ZA series containing 8, 12 and 27 per cent Al, become increasingly problematical as film formation becomes increasingly severe, and the alloy becomes increasingly strong, and so more notch sensitive.

Al–Mg alloy family, where the magnesium level can be up to 10 weight per cent, is widely known as being especially difficult to cast. Along with aluminium bronze, those aluminium alloys containing 5–10 per cent Mg share the dubious reputation of being the world's most uncastable casting alloys! This notoriety is, as we shall see, ill-deserved. If well cast, these alloys have enviable ductility and toughness, and take a bright anodized finish much favoured by the food industry, and those markets in which decorative finish is all important.

Aluminium bronze itself contains up to approximately 10 per cent Al, and the casting temperature is of course much higher than that of aluminium alloys. The high aluminium level and high temperature combine to produce a thick and tenacious film that makes aluminium bronze one of the most difficult of all foundry alloys. Some other high strength brasses and bronzes that contain aluminium are similarly difficult.

Ductile irons (otherwise known as spheroidal

graphite or nodular irons) are markedly more difficult to cast free from oxides and other defects when compared to grey (otherwise known as flake graphite) cast iron. This is the result of the minute concentration of magnesium that is added to spherodize the graphite, resulting in a solid magnesium silicate surface film.

Vacuum cast nickel- and cobalt-based high temperature alloys for turbine blades contain aluminium and titanium as the principal hardening elements. Because such castings are produced by investment (lost wax) techniques, the running systems have been traditionally poor. It is usual for such castings to be top poured, introducing severe surface turbulence, and creating high scrap levels. In an effort to reduce the scrap, the alloys have been cast in vacuum. It is quite clear, however, that this is not a complete solution. A good industrial vacuum is around 10^{-4} torr. However, not even the vacuum of 10^{-18} torr that exists in the space of near earth orbit is good enough to prevent the formation of alumina. Theory predicts that a vacuum around 10^{-40} torr is required. The real solution is, of course, not to attempt to prevent the formation of the oxide, but to avoid its entrainment. Thus top pouring needs to be avoided. A well-designed bottom-gated filling system would be an improvement. However, a counter-gravity system of filling would be the ultimate answer.

As an interesting aside, it may be that the film on high temperature Ni-based alloys might actually be AlN. This nitride does not appear to form at the melting temperatures used for Al alloys, despite its apparent great thermal stability, probably for kinetic reasons. However, at the higher temperatures of the Ni-based alloys it may form in preference to alumina. The Ni-based superalloys are well known for their susceptibility to react with nitrogen from the air and so become permanently contaminated. In any case the reaction to the nitride may be favoured even if the rates of formation of the oxide and nitride are equal, simply because air is four-fifths nitrogen.

Steels are another important, interesting and complicated case, often containing small additions of Al as a deoxidizer. Once again, AlN is a leading suspect for film formation in air. Steels are also dealt with in detail later.

Titanium alloys, particularly TiAl, may not be troubled by a surface film at all. Certainly during the hot isostatic pressing (hipping) of these alloys any oxide seems to go into solution. Careful studies have indicated that a cut (and, at room temperature, presumably oxidized) surface can be diffusion bonded to full strength across the joint, and with no detectable discontinuity when observed by transmission electron microscopy (Hu and Loretto 2000). It seems likely, however, that the liquid alloy may exhibit a transient film, like the oxide on copper and silver, and like the graphite film on cast iron in some conditions. Transient films are to be expected where the film-forming element is arriving from the environment faster than it can diffuse away into the bulk. This is expected to be a relatively common phenomenon since the rates of arrival, rates of surface reaction and rates of dissolution

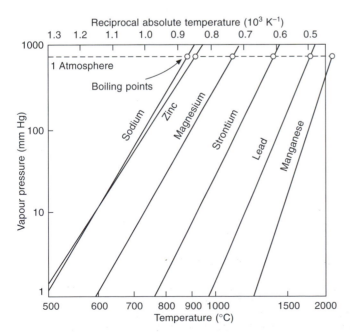

Figure 1.9 *Increase in the pressure of vapour of some more volatile elements as temperature increases. Data from Brandes (1983).*

are hardly likely to be matched in most situations.

In conditions for the formation of a transient film, if the surface happens to be entrained by folding over, although the film is continuously dissolving, it may survive sufficiently long to create a legacy of permanent problems. These could include the initiation of porosity, tearing or cracking, prior to its complete disappearance. In this case the culprit responsible for the problem would have vanished without trace.

In the course of this work we shall see how in a few cases the chemistry of the surface film can be altered to convert the film from a solid to a liquid, thus reducing the dangers that follow from an entrainment event. More usually, however, the film can neither be liquefied nor eliminated. It simply has to be lived with. A surface entrainment event therefore ensures the creation of a defect.

Entrained films form the major defect in cast materials. Our ultimate objective to avoid films in cast products cannot be achieved by eliminating the formation of films. The only practical solution to the elimination of entrainment defects is the elimination of entrainment. The simple implementation of an improved filling system design can completely eliminate the problems caused by entrained films. This apparently obvious solution is so self-evident that it has succeeded in escaping the attention of most of the casting community for the last several thousand years.

A discussion of the techniques to avoid entrainment during the production of cast material is an engineering problem too large to be covered in this book. It has to await the arrival of a second volume planned for this series *Castings II – Practice* listing my ten rules for good castings.

Chapter 2

Entrainment

If perfectly clean water is poured, or is subject to a breaking wave, the newly created liquid surfaces fall back together again, and so impinge and mutually assimilate. The body of the liquid re-forms seamlessly. We do not normally even think to question such an apparently self-evident process.

However, in practice, the same is not true for many common liquids, the surface of which is a solid, but invisible film. Aqueous liquids often exhibit films of proteins or other large molecular compounds.

Liquid metals are a special case. The surface of most liquid metals comprises an oxide film. If the surface happens to fold, by the action of a breaking wave, or by droplets forming and falling back into the melt, the surface oxide becomes entrained in the bulk liquid (Figure 2.1).

The *entrainment process* is a folding action that necessarily folds over the film dry side to dry side.

The submerged surface films are therefore *necessarily always double*.

Also, of course, because of the negligible bonding across the dry opposed interfaces, the defect now *necessarily resembles and acts as a crack*. Turbulent pouring of liquid metals can therefore quickly fill the liquid with cracks. The cracks have a relatively long life, and can survive long enough to be frozen into the casting. We shall see how they have a key role in the creation of other defects during the process of freezing, and how they degrade the properties of the final casting.

Entrainment does not necessarily occur only by the dramatic action of a breaking wave as seen in Figure 2.1. It can occur simply by the contraction of a 'free liquid' surface. In the case of a liquid surface that contracts in area, the area of oxide itself is not able to contract. Thus the excess area is forced to fold. Considerations of buoyancy (in

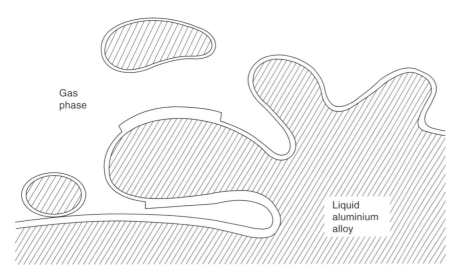

Gas phase

Liquid aluminium alloy

Figure 2.1 *Sketch of a surface entrainment event.*

all but the most rigid and thick films) confirm that the fold will be inwards, and so entrained (Figure 2.2). Such loss of surface is common during rather gentle undulations of the surface, the slopping and surging that can occur during the filling of moulds. Such gentle folding might be available to unfold again during a subsequent expansion, so that the entrained surface might almost immediately detrain once again. This potential for reversible entrainment may not be important, however; it seems likely that much enfolded material will remain, possibly because of entanglement with cores and moulds, or because bulk turbulence may tear it away from the surface and transport it elsewhere.

With regard to all film-forming alloys, accidental entrainment of the surface during pouring is, unfortunately, only to be expected. This normal degradation phenomenon is fundamental to the quality and reliability issues for cast metals, and, because of their inheritance of these defects, they survive, remaining as defects in wrought metals too. It is amazing that such a simple mechanism could have arrived at the twenty-first century having escaped the notice of thousands of workers, researchers and teachers.

Anyway, it is now clear that the entrained film has the potential to become one of the most severely damaging defects in cast products. It is essential, therefore, to understand film formation and the way in which films can become incorporated into a casting so as to damage its properties. These are vitally important issues. They are dealt with below.

It is worth repeating that a surface film is not harmful while it continues to stay on the surface. In fact, in the case of the oxide on liquid aluminium in air, it is doing a valuable service in protecting the melt from catastrophic oxidation. This is clear when comparing with liquid magnesium in air, where the oxide is not protective. Unless special precautions are taken, the liquid magnesium burns with its characteristic brilliant flame until the whole melt is converted to the oxide. In the meantime so much heat is evolved that the liquid melts its way through the bottom of the crucible, through the base of the furnace, and will continue down through a concrete floor, taking oxygen from the concrete

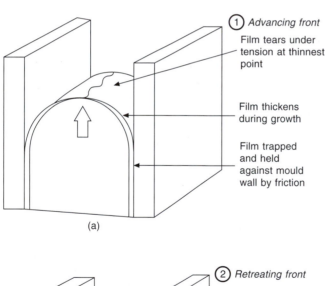

(1) *Advancing front*

Film tears under tension at thinnest point

Film thickens during growth

Film trapped and held against mould wall by friction

(a)

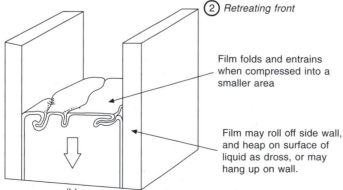

(2) *Retreating front*

Film folds and entrains when compressed into a smaller area

Film may roll off side wall, and heap on surface of liquid as dross, or may hang up on wall.

(b)

Figure 2.2 *Expansion of the surface followed by a contraction, leading to entrainment.*

to sustain the oxidation process until all the metal is consumed. This is the incendiary bomb effect. Oxidation reactions can be impressively energetic!

A solid film grows from the surface of the liquid, atom by atom, as each metal atom combines with newly arriving atoms or molecules of surrounding gas. Thus for an alumina film on the surface of liquid aluminium the underside of the film is in perfect atomic contact with the melt, and can be considered to be well wetted by the liquid. (Care is needed with the concept of *wetting* as used in this instance. Here it refers merely to the perfection of the atomic contact, which is evidently automatic when the film is grown in this way. The concept contrasts with the use of the term *wetting* for the case where a sessile drop is placed on an alumina substrate. The perfect atomic contact may again exist where the liquid covers the substrate, but at its edges the liquid will form a large contact angle with the substrate, indicating, in effect, that it does not wish to be in contact. Technically, the creation of the liquid/solid interface raises the total energy of the system. The *wetting* in this case is said to be poor.)

The problem with the surface film only occurs when it becomes *entrained* and thus submerged in the bulk liquid.

When considering submerged oxide films, it is important to emphasize that the side of the film which was originally in contact with the melt will continue to be well wetted, i.e. it will be in perfect atomic contact with the liquid. As such it will adhere well, and be an unfavourable nucleation site for volume defects such as cracks, gas bubbles or shrinkage cavities. When the metal solidifies the metal–oxide bond will be expected to continue to be strong, as in the perfect example of the oxide on the surface of all solid aluminium products, especially noticeable in the case of anodized aluminium.

The upper surface of the solid oxide as grown on the liquid is of course dry. On a microscale it is known to have some degree of roughness. In fact some upper surfaces of oxide films are extremely rough. Some, like MgO, being microscopically akin to a concertina, others like a rucked carpet or ploughed field, or others, like the spinel Al_2MgO_4, an irregular jumble of crystals.

The other key feature of surface films is the great speed at which they can grow. Thus in the fraction of a second that it takes to cause a splash or to enfold the surface, the expanding surface, newly creating liquid additional area of liquid, will react with its environment to cover itself in new film. The reaction is so fast as to be effectively instantaneous for the formation of oxides.

Other types of surface films on liquid metals are of interest to casters. Liquid oxides such as silicates are sometimes beneficial because they can detrain leaving no harmful residue in the casting. Solid graphitic films seem to be common when liquid metals are cast in hydrocarbon-rich environments. In addition, there is some evidence that other films such as sulphides and oxychlorides are important in some conditions. Fredriksson (1996) describes TiN films on alloys of Fe containing Ti, Cr and C when melted in a nitrogen atmosphere. Nitride films may be common in irons and steels.

In passing, in the usual case of an alloy with a solid oxide film, it is of interest to examine whether the presence of oxide in a melt necessarily implies that the oxide is double. For instance, why cannot a single piece of oxide be simply taken and immersed in a melt to give a single (i.e. non-double) interface with the melt? The reason is that as the piece of oxide is pushed through the surface of the liquid, the surface film on the liquid is automatically pulled down either side of the introduced oxide, coating both sides with a double film, as illustrated schematically in Figure 2.3. Thus the entrainment mechanism necessarily results in a submerged film that is at least double. If the surface film is solid, it therefore always has the nature of a crack.

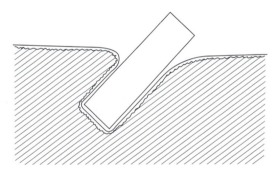

Figure 2.3 *Submerging of a piece of oxide (i.e. the introduction of an exogenous inclusion).*

Finally, it is worth warning about widespread inaccurate and vague concepts that are heard from time to time, and where clear thinking would be a distinct advantage. Two of these are discussed below.

For instance, one often hears about 'the breaking of the surface tension'. What can this mean? Surface tension is a physical force in the surface of the liquid that arises as a result of the atoms of the liquid pulling their neighbours in all directions. On atoms deep in the liquid there is of course no net force. However, for atoms at the surface, there are no neighbours above the surface, these atoms experience a net inward force from atoms below in the bulk. This net inward force is the force we know as surface tension. It is always present. It cannot make any sense to consider it being 'broken'.

Another closely related misconception describes 'the breaking of the surface oxide' implying that this is some kind of problem. However, the surface oxide, if a solid film, is *always* being broken during normal filling, but is being *continuously* reformed as a new surface becomes available. As the melt fills a mould, rising up between its walls, an observer looking down at the metal will see its surface oxide tear, dividing and sliding sideways across the meniscus, eventually becoming the skin of the casting. However, of course, the surface oxide is immediately and continuously re-forming, as though capable of infinite expansion. This is a natural and protective mode of advancement of the liquid metal front. It is to be encouraged by good design of filling systems.

As a fine point of logic, it is to be noted that the tearing and sliding process is driven by the friction of the casting skin, pressed by the liquid against the microscopically rough mould wall. Since this part of the film is trapped and cannot move, and if the melt is forced to rise, the film on the top surface is forced to yield by tearing. This mode of advance

is the secret of success of many beneficial products that enhance the surface finish of castings. For instance, coal dust replacements in moulding sands encourage the graphitic film on the surface of liquid cast irons, as will be detailed later.

As we have explained above, the mechanism of entrainment is the folding over of the surface to create a submerged, doubled-over oxide defect. This is the central problem. The folding action can be macroscopically dramatic, as in the pouring of liquid metals, or the overturning of a wave or the re-entering of a droplet. Alternatively, it may be gentle and hardly noticeable, like the contraction of the surface.

2.1 Entrainment defects

The entrainment mechanism is a folding-in action. Figure 2.4 illustrates how entrainment can result in a variety of submerged defects. If the entrained

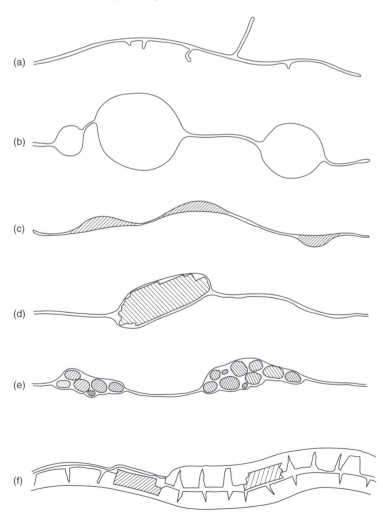

Figure 2.4 *Entrainment defects: (a) a new bifilm; (b) bubbles entrained as an integral part of the bifilm; (c) liquid flux trapped in a bifilm; (d) surface debris entrained with the bifilm; (e) sand inclusions entrained in the bifilm; (f) an entrained old film containing integral debris.*

surface is a solid film the resulting defect is a crack (Figure 2.4a) that may be only a few nanometres thick, and so be invisible to most inspection techniques. The other defects are considered below.

In the case of the folding-in of a solid film on the surface of the liquid the defect will be called a *bifilm*. This convenient short-hand denotes the double film defect. Its name emphasizes its double nature, as in the word *bicycle*. The name is also reminiscent of the type of marine shellfish, the *bivalve*, whose two leaves of its shell are hinged, allowing it to open and close. (The pronunciation is suggested to be similar to bicycle and bivalve, and not with a short 'i', that might suggest the word was 'biffilm'.)

To emphasize the important characteristic crack-like feature of the folded-in defect, the reader will notice that it will be often referred to as a 'bifilm crack', or 'oxide crack'. A typical entrained film is seen in Figure 2.5a, showing its convoluted nature. This irregular form, repeatedly folding back on itself, distinguishes it from a crack resulting from stress in a solid. At high magnification in the scanning electron microscope (Figure 2.5b) the gap between the double film looks like a bottomless canyon. This layer of air (or other mould gas) is always present, trapped by the roughness of the film as it folds over.

Figure 2.6 is an unusual polished section photographed in an optical microscope in the

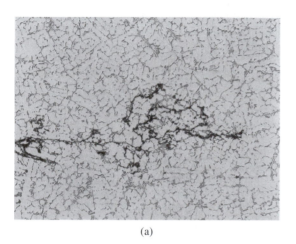

(a)

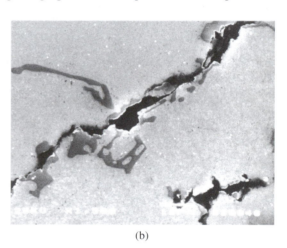

(b)

Figure 2.5 *(a) Convoluted bifilm in Al–7Si–0.4Mg alloy; (b) high magnification of the double film shown above, revealing its canyon-like appearance (Green and Campbell 1994).*

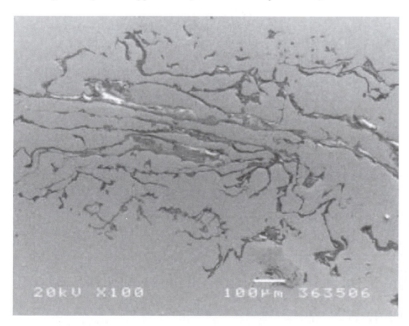

Figure 2.6 *Polished section of Al–7Si–0.4Mg alloy breaking into a bifilm, showing the upper part of the double film removed, revealing the inside of the lower part (Divandari 2000).*

author's laboratory by Divandari (2000). It shows the double nature of the bifilm, since by chance, the section happened to be at precisely the level to take away part of the top film, revealing a second, clearly unbonded, film underneath.

As we have mentioned, the surface can be entrained simply by contracting. However, if more severe disturbance of the surface is experienced, as typically occurs during the pouring of liquid metals, pockets of air can be accidentally trapped by chance creases and folds at random locations in the double film, since the surface turbulence event is usually chaotic. (Waves in a storm rarely resemble sine waves.) The resultant scattering of porosity in castings seems nearly always to originate from the pockets of entrained air. This appears to be the most common source of porosity in castings (so-called 'shrinkage', and so-called 'gas' precipitating from solution are only additive effects that may or may not contribute additional growth). The creation of this source of porosity has now been regularly observed in the study of mould filling using X-ray radiography. It explains how this rather random distribution of porosity typical in many castings has confounded the efforts of computers programmed to simulate only solidification.

Once entrained, the film may sink or float depending on its relative density. For films of dense alloys such as copper-based and ferrous materials, the entrained bifilms float. In very light materials such as magnesium and lithium the films generally sink. For aluminium oxide in liquid aluminium the situation is rather balanced, with the oxide being denser than the liquid, but its entrained air, entrapped between the two halves of the film, often brings its density close to neutral buoyancy. The behaviour of oxides in aluminium is therefore more complicated and worth considering in detail.

Initially, of course, enclosed air will aid buoyancy, assisting the films to float to the top surface of the melt. However, as will be discussed later, the enclosed air will be slowly consumed by the continuing slow oxidation of the surfaces of the crack. Thus the buoyancy of the films will slowly be lost. This behaviour of the bifilm explains a commonly experienced sampling problem, since the consequential distribution of defects in suspension at different depths in aluminium furnaces makes it problematic to obtain good quality metal out of a furnace.

The reason is that although most oxides sink to the bottom of the furnace, a significant density of defects collects just under the top surface. Naturally, this makes sampling of the better quality material in the centre rather difficult.

In fact, the centre of the melt would be expected to have a transient population of oxides that, for a time, were just neutrally buoyant. Thus these films would leave their position at the top, would circulate for a time in the convection currents, finally taking up residence on the bottom as they lost their buoyancy. Furthermore, any disturbance of the top would be expected to augment the central population, producing a shower, perhaps a storm, of defects that had become too heavy, easily dislodged from the support of their neighbours, and which would then tumble towards the bottom of the melt. Thus in many furnaces, although the mid-depth of the melt would probably be the best material, it would not be expected to be completely free from defects.

Small bubbles of air entrapped between films (Figure 2.4b) are often the source of microporosity observed in castings. Round micropores would be expected to decorate a bifilm, the bifilm itself often being not visible on a polished microsection. Samuel and Samuel (1993) report reduced pressure test samples of aluminium alloy in which bubbles in the middle of the reduced pressure test casting are clearly seen to be prevented from floating up by the presence of oxide films.

Large bubbles are another matter, as illustrated in Figure 2.7. The entrainment of larger bubbles is envisaged as possible only if fairly severe surface turbulence occurs. The conditions are dealt with in detail in the next section.

The powerful buoyancy of those larger pockets of entrained air, generally above 5 mm diameter, will give them a life of their own. They may be sufficiently energetic to drive their way through the morass of other films as schematically shown in Figure 2.7. They may even be sufficiently buoyant to force a path through partially solidified regions of the casting, powering their way through the dendrite mesh, bending and breaking dendrites. Large bubbles have sufficient buoyancy to continuously break the oxide skin on their crowns, powering an ascent, overcoming the drag of the bubble trail in its wake. Bubble trails are an especially important result of the entrainment process, and are dealt with later. Large bubbles that are entrained during the pouring of the casting are rarely retained in the casting. This is because they arrive quickly at the top surface of the casting before any freezing has had time to occur. Because their buoyancy is sufficient to split the oxide at its crown, it is similarly sufficient to burst the oxide skin of the casting that constitutes the last barrier between them and the atmosphere, and so escape. This detrainment of the bubble itself leaves the legacy of the bubble trail.

So many bubbles are introduced to the mould cavity by some poor filling system designs that later arrivals are trapped in the tangled mesh of trails left by earlier bubbles. Thus a mess of oxide trails and bubbles is the result. I have called this mixture bubble damage. In the author's experience, bubble damage is the most common defect in

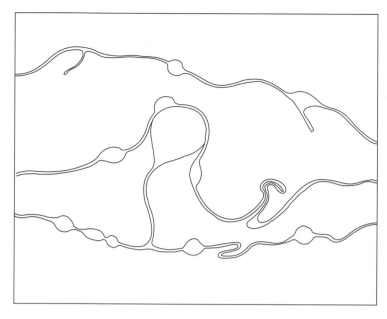

Figure 2.7 *Schematic illustration of bifilms with their trapped microbubbles, and actively buoyant macrobubbles.*

castings, accounting for perhaps 80 per cent of all casting defects. It is no wonder that the current computer simulations cannot predict the problems in many castings. In fact, it seems that relatively few important defects are attributable to the commonly blamed 'gas' or 'shrinkage' origins as expected by traditional thinking.

Pockets of air, as bubbles, are commonly an integral feature of the bifilm, as we have seen. However, because the bifilm was itself an entrainment feature, there is a possibility that the bifilm can form a leak path connecting to the outside world, allowing the bubble to deflate if the pressure in the surrounding melt rises. Such collapsed bubbles are particularly noticeable in some particulate metal matrix composites as shown in the work of Emamy and Campbell (1997), and illustrated in Figures 2.8 and 2.9. The collapsed bubble then becomes an integral part of the original bifilm, but is characterized by a thicker oxide film from its longer exposure to a plentiful supply of air, and a characteristically convoluted shape within the ghost outline of the original bubble.

Larger entrained bubbles are always somewhat crumpled, like a prune. The reason is almost certainly the result of the deformation of the bubble during the period of intense turbulence while the mould is filling. When spherical the bubble would have a minimum surface area. However, when deformed its area necessarily increases, increasing the area of oxide film on its surface. On attempting to regain its original spherical shape the additional area of film is now too large for the bubble, so that the skin becomes wrinkled. Each deformation of the bubble would be expected to add additional area.

(A further factor, perhaps less important, may be the reduction in volume of the bubble as the system cools, and as air is consumed by ongoing oxidation. In this case the analogy with the smaller wrinkled prune, originally a large shiny round plum, may not be too inaccurate.)

The growth of the area of oxide as the surface deforms seems a general feature of entrainment. It is a one-way, irreversible process. The consequent crinkling and folding of the surface is a necessary characteristic of entrained films, and is the common feature that assists to identify films on fracture surfaces. Figure 2.10a is a good example of a thin, probably young, film on an Al–7Si–0.4Mg alloy. Figure 2.10b is a typical film on an Al–5Mg alloy. The extreme thinness of the films can be seen on a fracture surface of an Al–7Si–0.4Mg alloy (Figure 2.11) that reveals a multiply folded film that in its thinnest part measures just 20 nm thick. Older films (not shown) can become thick and granular resembling slabs of rough concrete.

The irregular shape of bubbles has led to them often being confused with shrinkage pores. Furthermore, bubbles have been observed by video X-ray radiography of solidifying castings to form initiation sites for shrinkage porosity; bubbles appear to expand by a 'furry' growth of interdendritic porosity as residual liquid is drawn away from their surface in a poorly fed region of a casting. Such developments further obscure the key role of the bubble as the originating source of the problem.

In addition to porosity, there are a number of other, related defects that can be similarly entrained.

Flux inclusions containing chlorides or fluorides are relatively commonly found on machined surfaces

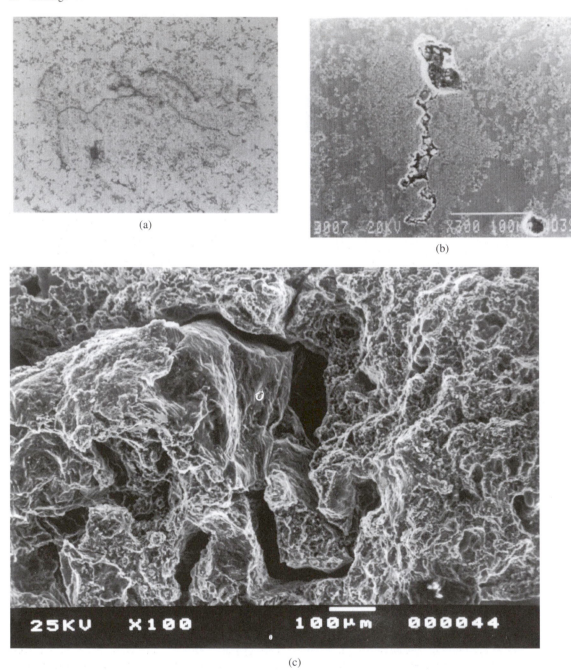

(a)

(b)

(c)

Figure 2.8 *Collapsed bubbles in Al–TiB$_2$ MMC. (a) and (b) show polished microsections of the ghost outlines of bubbles; (c) the resulting bifilm intersecting a fracture surface (Emamy and Campbell 1997).*

of cast components. Such fluxes are deliquescent, so that when opened to the air in this way they absorb moisture, leading to localized pockets of corrosion on machined surfaces. During routine examination of fracture surfaces, the elements chlorine and fluorine are quite often found as chlorides or fluorides on aluminium and magnesium alloys. The most common flux inclusions to be expected are NaCl and KCl.

However, chlorine and fluorine, and their common compounds, the chlorides and fluorides, are insoluble in aluminium, presenting the problem

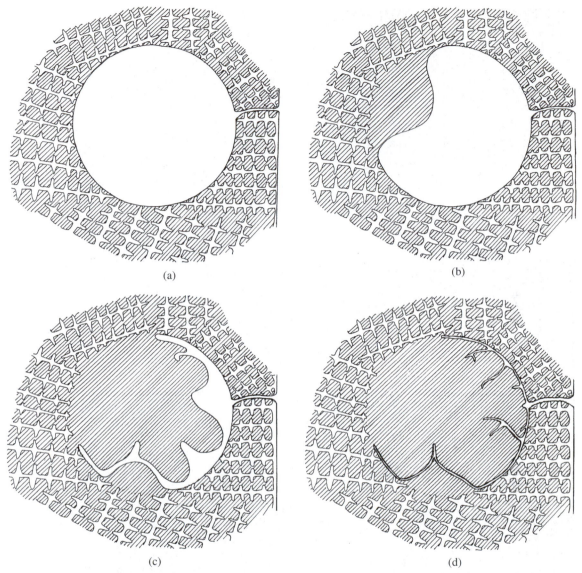

Figure 2.9 *Schematic illustration of the stages in the collapse of a bubble, showing the residual double film and the volume of dendrite-free eutectic liquid.*

of how such elements came to be in such locations. These phenomena can probably only be explained by assuming that such materials were originally on the surface of the melt, but have been mechanically entrained, wrapped in an oxide film (Figure 2.4c). Thus flux inclusions on a fracture surface indicate the presence of a bifilm, probably of considerable area, although the presence of its single remaining half will probably be not easily seen on the fracture.

Also, because flux inclusions are commonly liquid at the temperatures of liquid light alloys, and perfectly immiscible, why do they not spherodize and rapidly float out, becoming re-assimilated at the surface of the melt? It seems that

rapid detrainment does not occur. Once folded in the sandwich of entrained oxide of a bifilm, the package is slow to settle because of its extended area. The consequential long residence times allow transport of these contaminants over long distances in melt transfer and launder systems. In melting systems using electromagnetic pumps, fluxes are known to deposit in the working interiors of pumps, eventually blocking them. It is likely that the inclusions are forced out of suspension by the combined centrifugal and electromagnetic body forces in the pump. It is hardly conceivable that fluxes themselves, as relatively low viscosity liquids, could accomplish this. However, the accretion of a

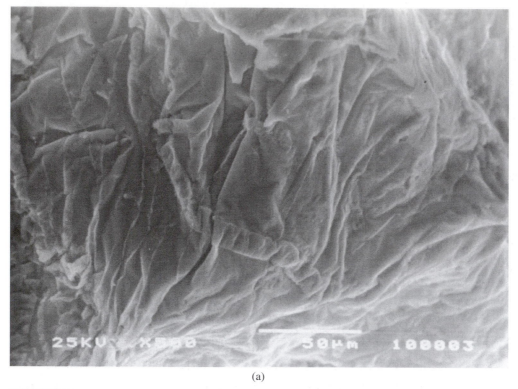

(a)

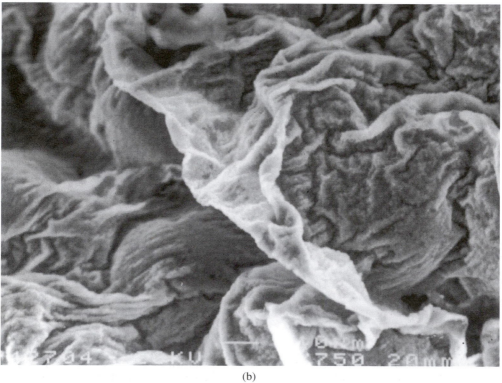

(b)

Figure 2.10 *Fracture surfaces showing (a) a fairly new thin film on an Al–7Si–0.4Mg alloy; (b) a film on an Al–5Mg alloy (courtesy Divandari 2000).*

Figure 2.11 *Multiply folded film on the fracture surface of an Al–7Si–0.4Mg alloy.*

mixture of solid oxide films bonded with a sticky liquid flux would be expected to be highly effective in choking the system.

When bifilms are created by folding into the melt, the presence of a surface liquid flux would be expected to have a powerful effect by effectively causing the two halves of the film to adhere together by viscous adhesion. This may be one of the key mechanisms explaining why fluxes are so effective in reducing the porosity in aluminium alloy castings. The bifilms may be glued shut. At room temperature the bonding by solidified flux may aid strength and ductility to some extent. On the other hand, the observed benefits to strength and ductility in flux-treated alloys may be the result of the reduction in films by agglomeration (because of their sticky nature in the presence of a liquid flux) and flotation. These factors will require much research to disentangle.

Whether fluxes are completely successful to prevent oxidation of the surface of light alloys does not seem to be clear. It may be that a solid oxide film always underlies the simple chloride fluxes and possibly some of the fluoride fluxes.

Even now, some foundries do successfully melt light alloys without fluxes. Whether such practice is really more beneficial deserves to be thoroughly investigated. What is certain is that the environment would benefit from the reduced dumping of flux residues from this so-called cleansing treatment.

There are circumstances when the flux inclusions may not be associated with a solid oxide film simply because the oxide is soluble in the flux. Such fluxes include cryolite ($AlF_3 \cdot 3NaF$) as used to dissolve alumina during the electrolytic production of aluminium, and the family K_2TiF_6, KBF_4, K_3Ta_7 and K_2ZrF_6. Thus the surface layer may be a uniform liquid phase.

For irons and steels the liquid slag layer is expected to be liquid throughout. Where a completely liquid slag surface is entrained Figure 2.12 shows the expected detrainment of the slag and the accidentally entrained gases and entrained liquid metal. Such spherodization of fluid phases is expected to occur in seconds as a result of the high surface energy of liquid metals and their rather low viscosity. A classical and spectacular break-up of a film of liquid is seen in the case of the granulation of liquid ferro-alloys (Figure 2.13). A ladle of alloy is poured onto a ceramic plate sited above the centre of a bath of water. On impact with the plate the jet of metal spreads into a film that thins with distance from the centre. The break-up of the metal film is seen to occur by the nucleation of holes in the film, followed by the thinning of ligaments between holes, and finally by the break-up of the ligaments into droplets.

In contrast, a film of liquid flux on aluminium

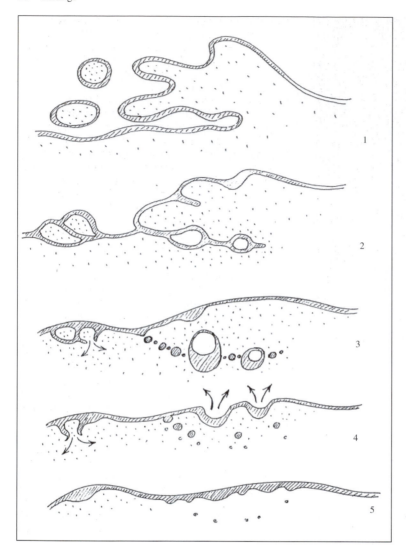

Figure 2.12 *Entrainment of a liquid film, showing the subsequent detrainment of metal, gases and most of the entrained liquid.*

alloys is not, in general, expected to have this simple completely liquid structure. It seems certain that in most instances a solid alumina film will underlie the liquid flux. Thus the film will behave as a sticky solid layer, and will be capable of permanently entraining bubbles and liquid flux. Spherodization of the liquid flux will not be able to occur.

However, successful spherodization of entrained surface materials may not only occur for liquid films. It seems possible that alumina is first entrained into liquid steel as films. However, once entrained the films would be expected to spherodize. This is because the films are extremely thin, and the temperature will be sufficiently high to encourage rapid spherodization driven by the high interfacial energy that is typical of metallic systems. The clusters and stringers of isolated alumina inclusions in rolled steels might have formed by this process,

the rolling action being merely to further separate and align them (Figure 2.14a). This suggestion is supported by observations of Way (2001) that show the blockage of nozzles of steels in continuous casting plants to be the result of arrays of separate granular alumina inclusions (Figure 2.14b). It seems certain that such inclusions were originally alumina films (separate inclusions seem hardly conceivable) formed by the severe surface turbulence inside the nozzle.

Figure 2.4d represents one of many different kinds of exogenous inclusion, such as a piece of refractory from the wall of a furnace or ladle. Exogenous inclusions can only be entrained together with the surface film. Attempts to introduce inoculants of various kinds into aluminium alloys have been made from time to time. All of these attempts are seen to be prejudiced by the problem

Figure 2.13 *Granulation of liquid ferromanganese into water (courtesy Uddeholm Technology, Sweden 2001).*

of penetrating the surface oxide; all particles are necessarily introduced in their own paper bags of oxide, so that the melt never contacts the introduced particle. This behaviour explains the results of Mohanty and Gruzleski (1995) who found that TiB_2 particles added to liquid aluminium did not nucleate grains, but were pushed by dendrites to the grain boundaries. This is the behaviour expected of any particles surrounded by a film of alumina that entraps a layer of air.

Similarly, the vortex technique for the addition of particles to make a particulate metal matrix composite (MMC) is similarly troubled. It is to be expected that the method would drag the oxide film down into the vortex, together with the heaps of particles supported on the film, so that the clumping of particles, now nicely and individually packaged, wrapped in film, is probably unavoidable. The 'clumping' of particles observed in such MMCs survives even long periods of vigorous stirring. Such features are seen in the work by Nai and Gupta (2002) in which the central particles in clumps of stirred-in SiC particles are observed to have fallen out during the preparation of the polished sample, leaving small cave-like structures lined with SiC particles that are somehow still held in place. In addition, whether in clumps or not, during the preparation of the mixture the particles are almost certainly held in suspension by the network of bifilms that will have been introduced, in the same way that networks of films have been observed to hold gas bubbles in suspension (Samuel and Samuel 1993). It seems also likely that the oxide network will reduce the fluidity of the material, and reduce the achievable mechanical properties.

In a further example, the manufacture of aluminium alloy MMCs based on aluminium alloy containing SiC particles introduced under high vacuum were investigated by Emamy and Campbell (1997). These authors found that under the hydrostatic tension produced by poor feeding conditions the SiC/matrix interface frequently decohered. This contrasted with the behaviour of an Al–TiB_2 MMC in which the TiB_2 particles were created by a flux reaction and added to the matrix alloy without ever having been in contact with the air. These particles exhibited much better adhesion at the particle/matrix interface when subjected to tension.

Figure 2.4d illustrates the entrainment of a piece of old oxide, probably from an earlier part of the melting or holding process. Such oxide, if very old, will have grown thick, and no longer justifies a description as a film. It can be crusty or even plate-like. It is simply just another exogenous inclusion. A typical example is seen in Figure 2.10c. If entrained during pouring, it will be wrapped in a new thin film, separated from the old inclusion by an air film, and will therefore act as a new film. If, as more likely, the old oxide was entrained some time previously during the melting or holding process, its entrainment wrapping will have itself become old, having lost the majority of its air layer by continued oxidation and thickening the oxide. In this case it will have become fairly 'inert' in the sense that it cannot easily nucleate pores; all its surfaces will have reacted with the melt and thus have come into good atomic contact, and its air layer will have become largely lost and welded closed.

Sand inclusions in castings are a similar case (Figure 2.4e), but whose mechanism of entrainment is probably rather more complicated. It is not easy to envisage how minute sand grains could penetrate the surface of a liquid metal against the repulsive action of surface tension, and the presence of an oxide film acting as a mechanical barrier. The penetration of the liquid surface would require the grain to be fired at the surface at high velocity, like a bullet. However, of course, such a dramatic mechanism is unlikely to occur in reality. Although the description below appears complicated at first sight, a sand entrainment mechanism can occur easily, involving little energy, as described below.

In a well-designed filling system the liquid metal entirely fills the system, and its hydrostatic pressure acts against the walls of the system to gently support the mould, holding sand grains in place. Thus although the mould surface will become hot, a resin

(a)

(b)

Figure 2.14 *(a) Alumina stringers in deep-etched rolled steel; (b) alumina particles from a blocked tundish nozzle from a steel continuous casting plant (Way 2001).*

binder will often first soften, then harden and strengthen as volatiles are lost. The sand grains in contact with the metal will finally have their binder degraded (pyrolysed) to the point at which only carbon remains. The carbon film remaining on the grains has a high refractoriness. It continues to protect the grains since it in turn is protected from oxidation because, at this late stage, most of the local oxygen has been consumed to form carbon monoxide. The carbon forms a non-wetted interface with the metal, thus enhancing protection from penetration and erosion.

The situation is different if the filling system has been poorly designed, allowing a mix of air to accompany the melt. This problem commonly accompanies the use of filling systems that have oversized cross-section areas, and thus remain unfilled with metal. In such systems the melt can ricochet backwards and forwards in the channel. The mechanical impact, akin to the cavitation effect on a ship's propeller, is a factor that assists erosion. However, other factors are also important. The contact of the melt with the wall of the mould heats the sand surface. As the melt bounces back from the wall air is drawn through the mould surface, and on the return bounce, the air flow is reversed. In this way air is forced backwards and forwards through the heated sand, and so burns away the binder. The burning action is intensified akin to the air pumped by bellows in the blacksmith's forge.

When the carbon is finally burned off the surface of the grains, the sand is no longer bonded to the mould. Furthermore, the oxide on the surface of the melt can now react with the freshly exposed silica surface of the grains, and thus adhere to them. If the melt now leaves the sand surface once again, the liquid surface is now covered with adhering grains of sand. As the surface folds over the grains are thereby enfolded into an envelope of oxide film. Such films can practically always be seen enfolding clusters of sand inclusions. Similarly, sand inclusions are often found on the inner surfaces of bubbles.

Sand inclusions are therefore a sure sign that the filling system design is poor, involving significant surface turbulence. Conversely, sand inclusions can nearly always be eliminated by attention to the filling system, not the strength nor the variety of the sand binder. The best solution lies in the hands of the casting engineer, not the binder chemist.

Finally, Figure 2.4f is intended to give a glimpse of the complexity that is likely to be present in many bifilms. Part of the film is new and thin, forming an asymmetrical thick/thin double film. (Reactions of the melt with the bifilm, as discussed at many points later in this work, may lead to asymmetrical precipitation of reactants if one side of the bifilm is more favourable than the other.) It would be expected (but not illustrated in the figure) that other parts of the defect would be fairly symmetrical thin double films. Elsewhere in the figure both halves are old, thick and heavily cracked. In addition, the old film contains debris that has become incorporated, having fallen onto the surface of the melt at various times in its progress through the melting and distribution system.

Entrained material is probably always rather messy.

2.2 Entrainment processes

2.2.1 Surface turbulence

As liquid metal rises up a vertical plate mould cavity, with the meniscus constrained between the two walls of the mould, an observer looking down from above (suitably equipped with protective shield of course!) will see that the liquid front advances by splitting the surface film along its centre. The film moves to each side, becoming the skin of the casting walls. As this happens, of course, the surface film is continuously renewed so that the process is effectively continuous. (In some circumstances, this vertical advance actually occurs by the action of a horizontal transverse travelling wave. This interesting phenomenon will be discussed in detail later.)

The important feature of this form of advance of the liquid front is that the surface film never becomes entrained in the bulk of the liquid. Thus the casting remains potentially perfect.

A problem arises if the velocity of flow is sufficiently high at some locality for the melt to rise enough above the general level of the liquid surface to subsequently fall back under gravity and so entrap a portion of its own surface. This is the action of a breaking wave. It is represented in Figure 2.15. This process has been modelled by computer simulation by Lin and co-workers (1999).

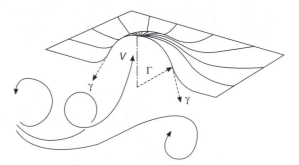

Figure 2.15 *Balance of forces in the surface of a liquid.*

To gain some insight into the physics involved we can carry out the following analysis. The pressure tending to perturb the surface is the inertial pressure $\rho V^2/2$ where ρ is the density of the melt, and V is the local flow velocity. The action of the tension in the surface (the surface tension γ) tends to counter any perturbation by pulling the surface flat. The maximum pressure that surface tension can provide is achieved when the surface is deformed into a hemisphere of radius r, giving a pressure $2\gamma/r$ resisting the inertial pressure attempting to deform the surface. Thus in the limit that the velocity is just sufficiently high to push the surface beyond the hemisphere, and so just exceeding the maximum restraint that the surface tension can provide, we have $\rho V^2/2 = 2\gamma/r$, so that the critical velocity at which the surface will start to suffer surface turbulence (my name for the phenomenon of breaking waves and entrainment of the surface into the bulk) is

$$V_{crit} = 2(\gamma/r\rho)^{1/2} \qquad \textbf{(2.1)}$$

This is a surprisingly accurate result in view of its rough derivation. The approach is unsatisfactory in the sense that the radius r has to be assumed. However, there is little wrong with the physics. The derivation is reproduced here because of its directness and simplicity.

Equation 2.1 is not an especially convenient equation since the critical radius of curvature of the front at which instability occurs is not known independently. However, we can guess that it will not be far from a natural radius of the front as

32   Castings

defined, for instance, by a sessile drop. (The word 'sessile' means 'sitting', and is used in contrast to the 'glissile', or 'gliding' drop.) The sessile drop sitting on a non-wetted surface exhibits a specific height h defined by the balance between the forces of gravity, tending to flatten it, and surface tension, tending to pull the drop together to make it spherical. We can derive an approximate expression for h considering a drop on a non-wetted substrate as illustrated in Figure 2.16. The pressure to expand the drop is the average pressure $\rho gh/2$ acting over the central area hL, giving the net force $\rho gh^2L/2$. For a drop in equilibrium, this force is equal to the net force due to surface tension acting over the length L on the top and bottom surfaces of the drop, $2\gamma L$. Hence

$$\rho gh^2L/2 = 2\gamma L$$

giving $\quad h = 2(\gamma/\rho g)^{1/2}$ $\qquad$ (2.2)

Assuming that $h = 2r$ approximately, and eliminating r from Equation 2.1 we have

$$V_{crit} = 2(\gamma g/\rho)^{1/4} \qquad (2.3)$$

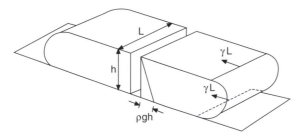

Figure 2.16 *Balance of forces in a sessile drop sitting on a non-wetted substrate.*

This is the more elegant way to derive the critical velocity involving no assumptions and no adjustable parameters. It is analogous to the concept illustrated in Figure 2.17 in which liquid metal rises through a vertical ingate to enter a mould.

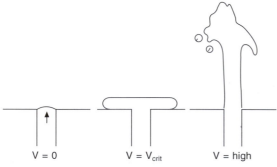

Figure 2.17 *Concept of liquid emerging into a mould at zero, critical and high velocities.*

Clearly, for case (a) when the velocity at the entrance to the gate is zero the liquid is perfectly safe. There is no danger of enfolding the liquid surface. However, of course, the zero velocity filling condition is not useful for the manufacture of castings.

In contrast, case (c) can be envisaged in which the velocity is so high that the liquid forms a jet. The liquid is then in great danger of enfolding its surface during its subsequent fall. Regrettably, this condition is common in castings.

The interesting intermediate case (b) illustrates the situation where the melt travels at sufficient velocity to rise to a level that is just supported by its surface tension. This is the case of a sessile drop. Any higher and the melt would subsequently have to fall back, and would be in danger of enfolding its surface. Thus the critical velocity required is that which is only just sufficient to project the liquid to the height h of the sessile drop. This height is simply that derived from conservation of energy, $mV^2/2 = mgh$, giving $h = V^2/2g$. Substituting this value for the value for the height of the sessile drop above (Equation 2.2) gives the same result for the critical velocity (Equation 2.3).

From Equation 2.2 we can find the height from which the metal can fall, accelerated by gravity, before it reaches the critical velocity given by Equation 2.3. When falling from greater heights the surface of the liquid will be in danger of being entrained in the bulk liquid. The critical fall heights for entrainment of the surface are listed in Table 2.1. (Physical properties of the liquids have been taken at their melting points (Brandes and Brook 1992), since the values change negligibly with temperature, but are somewhat affected by alloying.)

At first sight it seems surprising that nearly all liquids, including water, have critical velocities in the region of 0.3 to 0.5 ms^{-1}. This is a consequence

Table 2.1 The critical heights and velocities of some liquids (densities and surface tensions adopted from Brandes and Brook 1992)

Liquid	Density kgm^{-3}	Surface tension Nm^{-1}	Critical height h mm	Critical velocity ms^{-1}
Ti	4110	1.65	12.8	0.50
Al	2385	0.914	12.5	0.50
Mg	1590	0.559	12.0	0.48
Fe	7015	1.872	10.4	0.45
Ni	7905	1.778	9.6	0.43
Cu	8000	1.285	8.1	0.40
Zn	6575	0.782	7.0	0.37
Pt	19 000	1.8	6.2	0.35
Au	17 360	1.14	5.2	0.32
Pb	10 678	0.468	4.2	0.29
Hg	13 691	0.498	3.9	0.27
Water	1000	0.072	5.4	0.33

of the fact that the velocity is a function of the *ratio* of surface tension and density, and, in general, both these physical constants change in the same direction from one liquid to another. Furthermore, the effect of the one quarter power further suppresses differences. Most of the important engineering metals have critical velocities close to 0.5 ms^{-1} so that this figure is now widely assumed in the calculation of filling systems for castings.

The value of 0.5 ms^{-1} for liquid aluminium has been confirmed experimentally by Runyoro *et al.* (1992). He observed by video the emergence of liquid aluminium at increasing speeds from an ingate in a sand mould. An outline of some of the melt profiles is given in Figure 2.18. At just over the critical velocity, at 0.55 ms^{-1}, the metal falls back only slightly, forming its first fold. At progressively higher velocities the potential for damage during the subsequent fall of the jets is evident.

In further work by Runyoro (1992) shown in Figure 2.19 the dramatic rise in surface cracks identified by dye penetrant testing, and the precipitous fall in bend strength of castings filled above 0.5 ms^{-1} was such a shock at the time of its discovery that these results completely changed the direction of research work in the author's laboratory for the next ten years.

The critical velocity for Cu–10Al alloy at 0.4 ms^{-1} was subsequently confirmed by Helvaee and Campbell (1996), and a similar value for ductile

cast irons shown in Figure 2.20 was reported by the author (Campbell 2000). The critical velocity for aluminium has been explored by computer simulation (Lai *et al.* 2002) in which area of the melt surface was computed as a function of increasing ingate velocity. Assuming the same values for physical constants as used here, a value of precisely 0.5 ms^{-1} was found (Figure 2.21), providing reassuring confirmation of the concept of critical velocity by an independent technique.

2.2.1.1 Weber number

The concept of the critical velocity for the entrainment of the surface by what the author has called *surface turbulence* is enshrined in the Weber number, *We*. This elegant dimensionless quantity is defined as the ratio of the inertial pressure in the melt, assessed as $\rho V^2/2$, with the pressure due to surface tension $\gamma(1/r_1 + 1/r_2)$ where r_1 and r_2 are the radii of curvature of the surface in two perpendicular directions. If only curved in one plane the pressure becomes γ/r, or for a hemisphere where the orthogonal radii are equal, $2\gamma/r$. Thus

$$We = \rho L V^2/\gamma \qquad\qquad (2.4)$$

Only the dimensioned quantities are included in the ratio, the scalar numbers (the factors of 2) are neglected, as is usual for a dimensionless number. The radius term becomes the so-called 'characteristic

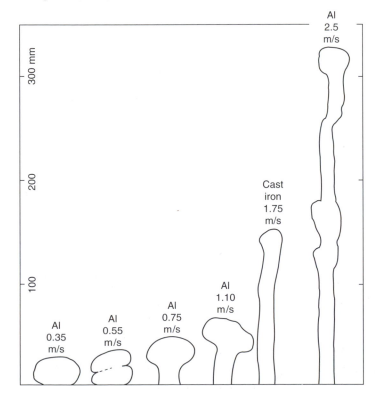

Figure 2.18 *Profile of melts of aluminium and cast iron emerging from an ingate at various speeds (Runyoro et al. 1992).*

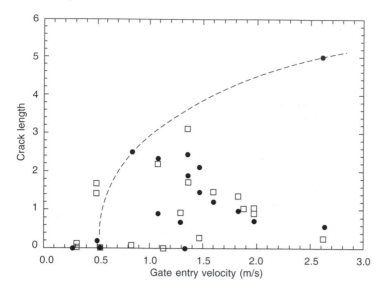

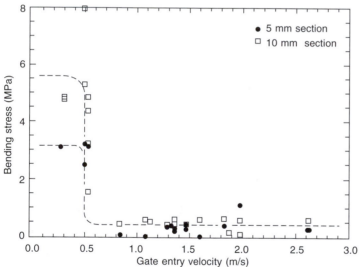

Figure 2.19 *(a) Crack lengths by dye penetrant testing in Al alloys cast at different ingate velocities; and (b) the bend strengths of the same plate castings (Runyoro et al. 1992).*

length' parameter, L. For flows in channels, L is typically chosen to be the hydraulic radius, defined as the ratio of the occupied area of the channel to its wetted perimeter. For a filled cylindrical channel this is half the radius of the channel (diameter/4). For a filled very thin slot, it is approximately half of the slot width. As will be appreciated, the hydraulic radius is a concept closely allied to that of 'geometrical modulus' as commonly used by ourselves as casting engineers.

When $We = 1$ the inertial and surface forces are roughly balanced. From the definition of We in Equation 2.4, and assuming $L = 12.5/2$ mm we find $V = 0.25$ ms^{-1}, a value only a factor of 2 different from the critical velocity found earlier. When it is recalled that the real value of dimensionless numbers

lies in their use to define the nearest order of magnitude (i.e. the nearest factor of ten), this is impressively close agreement. The agreement is only to be expected, of course, since the condition that $We < 1$ defines those conditions where the liquid surface is tranquil, and $We = 1$ defines the conditions in which surface turbulence can start.

The We number has been little used in casting research. This is a major omission on the part of researchers. It is instructive to look at the values that We can adopt in some casting processes.

Work on aluminium bronze casting has shown empirically that these alloys retain their quality when the metal enters the mould cavity through vertical gates at speeds of up to 60 mms^{-1}. Assuming that the radius of the free liquid surface is close to

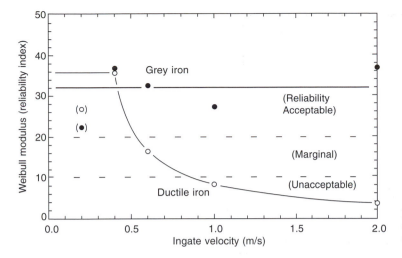

Figure 2.20 *Reliabilities of grey and ductile irons cast at different ingate speeds (Campbell 2000).*

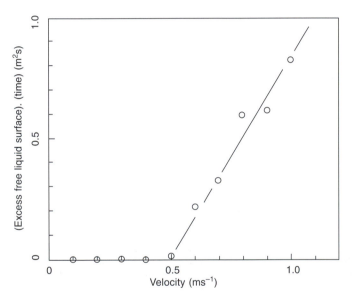

Figure 2.21 *Computer simulation of the filling of a mould with aluminium, showing the excess free area of the melt as a function of ingate velocity (Lai et al. 2002).*

5 mm (corresponding to half of the height of the sessile drop given in Table 2.1), the density is close to 8000 kgm^{-3}, and surface tension is approximately 1.3 Nm^{-1}, then *We* in this case is seen to be approximately 0.1.

In a mould of section thickness of 5 mm fitted with a glass window, Nieswaag observed that a rising front of liquid aluminium just starts to break up into jets above a speed of approximately 0.37 ms^{-1}. This is close to our predicted critical velocity (Table 2.1). Assuming that the radius of the liquid front is 2.5 mm, and taking 2400 kg m^{-3} for the density of liquid aluminium, the corresponding critical value of *We* at which the break-up of the surface first occurs is therefore seen to be approximately 0.9. The users of vertically injected squeeze casting machines now commonly use 0.4

ms^{-1} as a safe filling speed (for instance, Suzuki 1989), corresponding to *We* close to 1.0.

UK work in the 1950s (IBF Technical Subcommittee 1960) showed that cast iron issues from a horn gate as a fountain when the sprue is only about 125 mm high and the gate about 12 mm in diameter (deduced from careful inspection of the photograph of the work). We can calculate, therefore, that *We* is approximately 80 for this well-advanced condition of surface turbulence.

Going on to consider the case of a pressure die casting of an aluminium alloy in a die of section thickness 5 mm and for which the velocity *V* might be 40 ms^{-1}, *We* becomes approximately 100 000. It is known that the filling conditions in a pressure die casting are, in fact, characterized by extreme surface fragmentation, and possibly represent an upper limit which might be expected while making a casting. Owen (1966) found from examination of the surfaces of zinc alloy pressure die castings that the metal advanced into the mould in a series of filaments or jets. These advanced splashes were subsequently overtaken by the bulk of the metal, although they were never fully assimilated. These observations are typical of much work carried out on the filling of the die during the injection of pressure die castings.

In summary, it seems that We numbers in the range 0–1.0 define the range of flow conditions that are free from surface turbulence. Around 1.0 appears to be transitional. By the time the Weber number reaches 100, surface turbulence is certainly a problem, with jumping and splashing to a height of 100 mm or so. However, at values of 100 000 the concept of surface turbulence has probably given way to modes of flow characterized by jets and sprays, and eventually, atomization.

2.2.1.2 Froude number Fr

The Froude number is another useful dimensionless quantity. It assesses the propensity of conditions in the liquid for making gravity waves. Thus it compares the inertial pressures $\rho V^2/2$ that are perturbing the surface with the restraining action of gravity, ρgh, where g is the acceleration due to gravity, and h is the height of the wave. Thus the Froude number is defined as

$$Fr = V^2/gh \qquad (2.5)$$

It is important to emphasize the regimes of application of the We and Fr numbers. The Weber number deals with small-scale surface effects such as ripples, droplets and jets. This is in contrast to the Froude number, which deals with larger-scale gravity waves, including such effects as slopping and surging in more open vessels. We shall see later how Fr is useful to assess conditions for the formation of the hydraulic jump, a troublesome feature of the flow in large runners.

In the meantime, however, the Froude number is useful to assess conditions where the undulation of the surface may cause its surface area to contract, forcing an entrainment condition. This concept has been so far overlooked as a major potential role in entrainment. The continued expansion of the advancing front is a necessary and sufficient condition for the avoidance of entrainment. Conversely, the contraction of the front in the sense of a loss of area of the 'free surface' of the liquid is a necessary and sufficient condition for the entrainment of bifilms.

Interestingly, the Weber number condition, indicating the onset of surface turbulence in the form of a potentially breaking wave, is a sufficient but not necessary condition for entrainment of bifilms. However, it is most probably a necessary condition for the entrainment of bubbles and perhaps some other surface materials. It is also likely that the surface turbulence condition aids the permanent entrainment of material entrained by surface contraction that might otherwise detrain. Thus the surface turbulence condition in the form of the critical velocity condition remains a major influence on casting quality.

2.2.1.3 Reynold's number Re

Many texts deal with turbulence, especially in running and gating systems, citing Reynold's number Re:

$$Re = \rho VL/\mu \qquad (2.6)$$

where V is the velocity of the liquid, ρ is the density of the liquid, L is a characteristic linear dimension of the geometry of the flow path, and μ is the viscosity. The definition of Reynold's number follows from a comparison of inertial pressures $C_d\rho V^2/2$ (where C_d is the drag coefficient) and viscous pressures $\mu V/L$. The reader can confirm that the ratio of these two forces does give the dimensionless number Re (neglecting, of course, the presence of numerical factors $2C_d$). The characteristic length L is, as before, usually defined as the hydraulic radius.

Clearly, the inertial forces are a measure of those effects that would cause the liquid to flow in directions dictated by its momentum. These effects are resisted by the effect of viscous drag from the walls of the channel containing the flow. At values of Re below about 2000, viscous forces prevail, causing the flow to be smooth and laminar, i.e. approximately parallel to the walls. At values of Re higher than this, flow tends to become turbulent; the walls are too far away to provide effective constraint, so that momentum overcomes viscous restraint, causing the flow to degenerate into a chaos of unpredictable swirling patterns.

It is vital to understand, however, that all of this flow behaviour takes place in the bulk of the liquid, underneath the surface. During such turbulence in the bulk liquid, therefore, the surface of the melt may remain relatively tranquil. Turbulence as predicted and measured by Re is therefore strictly *bulk* turbulence. It does not apply to the problem of assessing whether the surface film will be incorporated into the melt. This can only be assessed by the concept of *surface* turbulence associated with We.

Flow in running systems is typically turbulent as defined by Reynold's number. Even for very narrow running systems designed to keep surface turbulence to a minimum, bulk turbulence would still be expected. In this case for a small aluminium alloy casting where $V = 2$ ms^{-1}, in a runner only 4 mm high, hence $L = 2$ mm, Re is still close to 7000. For a fairly small steel casting weighing a few kilograms where the melt is flowing at 2 ms^{-1} in a runner 25 mm high, giving approximately $L = 10$ mm, and assuming $\mu = 5.5 \times 10^{-3}$ Nsm^{-2} and $\rho = 7000$ kgm^{-3}, Re is approximately 20 000. For a large casting a metre or more high and weighing several tonnes it is quickly shown that Re is in the region of 2 000 000. Thus bulk turbulence exists in running systems of all types, and can be severe.

Only in the very narrow pores of ceramic foam filters is flow expected to be laminar. For velocities of 2 ms^{-2} for liquid aluminium in a filter of pore diameter 1 mm $Re = 2000$. For liquid steel, which is somewhat more viscous, $Re = 500$. These values show that the flow is not far from an unstable turbulent condition. Thus only moderate increases in speed of flow would promote turbulence even in these very small channels.

A summary of flow regimes is presented in Figure 2.22. This is a map to guide our thinking and our research about the very different flow behaviour that liquids can adopt when subjected to different conditions of speed and geometry.

2.2.2 Melt charge materials

Problems of the introduction of films from the surfaces of charge materials are most commonly seen in the melting of aluminium and its alloys. Whether or not the films on the surface of the solid charge materials find their way into the liquid metal depends on the type of furnace used for melting.

Furnaces such as tower or shaft designs are designed to provide a classical counterflow heat exchanger process; the charge is preheated during its course down the tower, while furnace waste gases flow upwards. The base of such a tower is usually a gentle slope of refractory on which the charge in the tower is supported. The slope is known as a dry hearth. Such a design of melting unit is not only thermally efficient, but has the advantage that the oxide skins on the surface of the metallic charge materials are mainly left in place on the dry hearth. The liquid metal flows practically at its melting point, out of the skin of the melting materials, down the hearth, travelling through an extended oxide tube, and into the bulk of the melt. Here it needs to be brought up to a useful casting temperature. The oxide tube forms as the liquid flows, and finally collapses, more or less empty, when the flow is finished. The oxide content of melts from this variety of furnace is low. (The melting losses and gas content might be expected to be a little higher in this design of melter, but because of the protective nature of the oxide that forms in environments with very high water content, the reaction between the melt and the waste gases is surprisingly low.)

The additional advantage of a dry hearth furnace

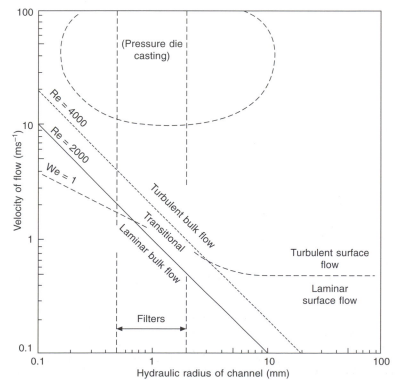

Figure 2.22 *Map of flow regimes in castings.*

for aluminium alloys is that foundry returns that contain iron or steel cast-in inserts (such as the iron liners of cylinder blocks or valve seats in cylinder heads) can be recycled. The inserts remain on the hearth and can, from time to time, be raked clear, together with all the dross of oxide skins from the charge materials. (A dross consists of oxides with entrapped liquid metal. Thus most dross contains between 50 and 80 per cent metal, making the recovery of aluminium from dross economically valuable.)

The benefits of melting in a dry hearth furnace are, of course, eliminated at a stroke by the misguided enthusiasm of the operator, who, thinking he is keeping the furnace clean and tidy, and that the heap of remaining oxide debris sitting on the hearth will all make good castings, shoves the heap into the melt. Unfortunately, it is probably slightly less effort to push the dross downhill, rather than rake it out of the furnace through the dross door. The message is clear, but requires restating frequently. Good technology alone will not produce good castings. Good training and vigilant management remain essential.

Furnaces in which the solid charge materials are added directly into a melting furnace or into a liquid pool produce quite a different quality of metal. The oxide originally on the charge material becomes necessarily submerged, to become part of the melt when the underlying solid melts. In the case of charge materials such as ingots that have been chill cast into metal moulds the surface oxide introduced in this way is relatively thin. However, charges that are made from sand castings that are to be recycled represent a worst case. The oxide film on sand castings has grown thick during the extended cooling period of the casting in the aggressively moist and oxidizing environment of the sand mould. The author has found complete skins of cylinder block castings floating around in the liquid metal. The melt can become so bad as to resemble a slurry of old sacks. Unfortunately this is not unusual.

In a less severe case where normal melting was carried out repeatedly on 99.5 per cent pure aluminium, Panchanathan *et al.* (1965) found that progressively poorer mechanical properties were obtained. By the time the melt had been recycled eight times, the elongation values had fallen from approximately 30 to 20 per cent. This is easily understood if the oxide content of the metal is progressively increased by repeated casting.

2.2.3 Pouring

During the pouring of some alloys, the surface film on the liquid grows so quickly that it forms a tube around the falling stream. The author calls this an oxide flow tube.

A patent dating from 1928 (Beck *et al.* 1928)

describes how liquid magnesium can be transferred from a ladle into a mould by arranging for the pouring lip of the ladle to be as close as possible to the pouring cup of the mould, and to be in a relatively fixed position so that the semi-rigid oxide pipe which forms automatically around the jet is maintained unbroken, and thus protects the metal from contact with the air (Figure 2.23a).

A similar phenomenon is seen in the pouring of aluminium alloys and other metals such as aluminium bronze.

However, if the length of the falling stream is increased, then the shear force of the falling liquid against the inner wall of the tube increases. This drag may become so great that after a second or so the oxide tears, allowing the tube to detach from the lip of the ladle. The tube then accompanies the metal into the mould, only to be immediately

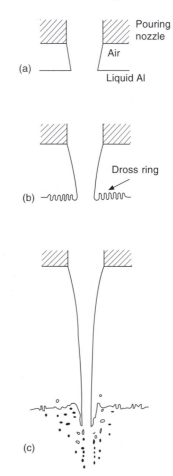

Figure 2.23 *Effect of increasing height on a falling stream of liquid illustrating: (a) the oxide flow tube remaining intact; (b) the oxide flow tubes being successively detached and accumulating to form a dross ring; and (c) the oxide film and air being entrained in the bulk liquid.*

replaced by a second tube, and so on. A typical 10 kg aluminium alloy casting poured in about 10 seconds can be observed to carry an area of between 0.1 and 1.0 m² of oxide into the melt in this way. This is an impressive area of oxide to be dispersed in a casting of average dimensions only $100 \times 200 \times 500$ mm, especially when it is clear that this is only one source of oxides that threatens the casting. The oxide in the original metal, together with the oxides entrained by the surface turbulence of the pour, will be expected to augment the total significantly.

2.2.3.1 The critical fall height

When melts are transferred by pouring from heights less than the critical heights predicted in Table 2.1 (the heights of the sessile drop) there is no danger of the formation of entrainment defects. Surface tension is dominant in such circumstances, and can prevent the folding inwards of the surface, and thus prevent entrainment defects (Figure 2.23a). It is unfortunate that the critical fall height is such a minute distance. Most falls that an engineer might wish to design into a melt handling system, or running system, are nearly always greater, if not vastly greater. However, the critical fall height is one of those extremely inconvenient facts that we casting engineers have to learn to live with.

Why is the critical fall height the same as the height of a sessile drop? It is because the critical velocity V is that required to propel the metal from an ingate to the height at which it is still just supported by surface tension (Figure 2.17). This is the same velocity V that the melt would have acquired by falling from that height; a freely travelling particle of melt starting from the ingate would execute a parabola, with its upward starting and downward finishing velocities identical.

However, even above this theoretical height, in practice the melt may not be damaged by the pouring action. The mechanical support of the liquid by the surface film in the form of its surrounding oxide tube can still provide freedom from entrainment, although the extent of this additional beneficial regime is perhaps not great. For instance, if the surface tension is effectively increased by a factor of 2 or 3 by the presence of the film, the critical height may increase by a factor $3^{1/4} = 1.3$. Thus perhaps 30 per cent or so may be achievable, taking the maximum fall from about 12 to 16 mm for aluminium. This seems negligible for most practical purposes.

At slightly higher speed of the falling stream, the tubes of oxide concertina together to form a dross ring (Figure 2.23b). Although this represents an important loss of metal on transferring liquid aluminium and other dross-forming alloys, it is not clear whether defects are also dragged beneath the

surface and thereby entrained.

At higher speeds still, the dross is definitely carried under the surface of the liquid, together with entrained air, as shown in Figure 2.23c. Turner (1965) has reported that, above a pouring height of 90 mm, air begins to be taken into the melt with the stream, to reappear as bubbles on the surface. This is well above the critical fall heights predicted above, and almost certainly is a consequence of the some stabilization of the surface of the falling jet by the presence of a film. The mechanical rigidity of the tubular film holds the jet in place, and effectively delays the onset of entrainment by the plunging action greatly in excess of the predicted 30 per cent. Clearly, more work is required to clarify the allowable fall heights of different alloys.

In a study of water models, Goklu and Lange (1986) found that the quality of the pouring nozzle affects the surface smoothness of the plunging jet, which in turn influences the amount of air entrainment. They found that the disturbance to the surface of the falling jet is mainly controlled by the turbulence ahead of and inside the nozzle that forms the jet. In a practical instance of a jet plunging at 10 ms⁻¹ into steel held in a 4 m diameter ladle, Guthrie (1989) found that the Weber number was 1.7×10^6 whereas the Froude number was only 2.5. Thus despite very little slopping and surging, the surface forces were being overwhelmed by inertial forces by nearly two million times, causing the creation of a very dirty re-oxidized steel.

In the case of water, of course, the stabilizing action of a film is probably not important, if present at all. It is suggested here that the benefits noted in Turner's results quoted above may derive from the action of the oxide tube rigidizing the surface, damping surface perturbations, creating a smoother falling stream that entrains less air and oxide.

During the pouring of a casting from the lip of a ladle via a weir basin kept properly full of metal, the above benefit will apply: the oxide will probably not enter the casting if the pouring head is sufficiently low, as is achievable during lip pouring. However, in practice it seems that for fall distances of more than perhaps 50 or 100 mm freedom from damage cannot be relied upon.

In fact, the benefits of defect-free pouring are easily lost if the pouring speed into the entry point of the filling system is too high. This is often observed when pouring castings from unnecessary height. In aluminium foundries this is usually by robot. In iron foundries it is commonly via automatic pouring systems from fixed launders sited over the line of moulds. In steel foundries it is common to pour from bottom poured ladles that contain over a metre depth of steel above the exit nozzle (the situation for steel from bottom-teemed ladles is further complicated by the depth of metal in the ladle decreasing progressively). In all types of foundries the surface oxide is automatically entrained and carried into the casting if a simple conical pouring bush is used to funnel the liquid stream into the sprue. In this case, of course, practically all of the oxide formed on the stream will enter the casting. The current widespread use of conical pouring basins has to be changed if casting quality is to be improved.

2.2.4 The oxide lap defect I – surface flooding

The steady, progressive rise of the liquid metal in a mould may be interrupted for a number of reasons. There could be (i) an inadvertent break during pouring, or (ii) an overflow of the melt (called elsewhere in this work a 'waterfall effect') into a deep cavity at some other location in the mould, or (iii) the arrival of the front at a very much enlarged area, thus slowing the rate of rise nearly to a stop. If the melt stops its advance the thickness of the oxide on the melt surface is no longer controlled by the constant splitting and regrowing action. It now simply thickens. If the delay to its advance is prolonged, the surface oxide may become a rigid crust.

When filling restarts (for instance, when pouring resumes, or the overflow cavity is filled) the fresh melt may be unable to break through the thickened surface film. When it eventually builds up enough pressure to force its way through at a weak point, the new melt will flood over the old, thick film, sealing it in place. Because the newly arriving melt will roll over the surface, laying down its own new, thin film, a double film defect will be created. The double film will be highly asymmetrical, consisting of a lower thick film and an upper thin film.

Asymmetric films are interesting, in that precipitates sometimes prefer one film as a substrate for formation and growth, but not the other. An example is briefly described later in the section concerning observations of an oxide flow tube.

The newly arriving melt will only have the pressure of its own sessile drop height as it attempts to run into the tapering gap left between the old meniscus and the mould wall. Thus this gap is imperfectly filled, leaving a horizontal lap defect clearly visible around the perimeter of the casting.

Notice that in this way (assuming oxidizing conditions) we have created an **oxide lap**. If the arrest of the advance of the melt had been further delayed, or if the solidification of the melt had been accelerated (as near a metal chill, or in a metal mould) the meniscus could have lost so much heat that it had become partially or completely solid. In this case the lap would take on the form of a **cold lap** (the name 'cold shut' is recommended to be avoided as being an unhelpful description). The distinction between oxide laps and cold laps is sometimes useful, since whereas both may be eliminated by avoiding any arrest of progress of the liquid front, only the cold lap may be cured by increasing the casting temperature, whereas the oxide lap may become worse.

A further key aspect of the stopping of the front is that the double film defect that is thereby created is a single, huge planar defect, extending completely through the product. Also, its orientation is perfectly horizontal. (Notice it is quite different from the creation of double film defects by surface turbulence. In this chaotic process the defects are random in shape, size, orientation and location in the casting.) Flooding over the surface in this way is relatively common during the filling of castings, especially during the slow filling of all film-forming alloys.

For horizontal surfaces, the unstable advance of the front takes a dendritic form, with narrow streams progressing freely ahead of the rest of the melt. This is because while the molten metal advances quickly in the mould the surface film is being repeatedly burst and moved aside. The faster the metal advances in one location, the thinner and weaker the film, so that the rate of advance of the front becomes less impeded. If another part of the front slows, then the film has additional time to strengthen, further retarding the local rate of advance. Thus in film-forming conditions fast-rising parts of the advancing front rise faster, and slow-moving parts rise slower, causing the advance of the liquid front to become unstable (Campbell 1988). This is the classic type of instability condition that gives rise to a finger-like dendritic form of an advancing front, whether a liquid front, or a solidifying front.

Figure 2.24 shows the filling pattern of a thin-walled box casting such as an automotive sump or

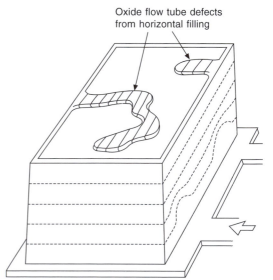

Figure 2.24 *Filling of a thin-walled oil pan casting, showing the gravity-controlled rise in the walls, but unstable flow across horizontal areas.*

oil pan. If the streams continue to flow, so as to fill eventually the whole of the horizontal section, the confluence welds (see section 2.2.5) abutting the oxides on the sides of the streams will constitute cracks through the complete thickness of the casting. When highly strained, such castings are known to crack along the lines of the confluence welds outlining the filling streams.

For the case of vertical filling, when the advance of the front has slowed to near zero, or has actually momentarily stopped, then the strength of the film and its attachment to the mould will prevent further advance at that location. If the filling pressure continues to build up, the metal will burst through at a weak point, flooding over the stationary front. In a particular locality of the casting, therefore, the advance of the metal will be a succession of arrests and floodings, each new flood burying a double oxide film (Figure 2.25).

This very deleterious mode of filling can be avoided by increasing the rate of filling of the mould.

The problem can, in some circumstances, also be tackled by reducing the film-forming conditions. This is perhaps not viable for the very stable oxides such as alumina and titania when casting in air. It

is the reason for vacuum casting those alloys which are troubled by films; although the oxides cannot be prevented from forming by casting in vacuum, their rate of formation is reduced. The thin double film is expected, in principle, to constitute a crack as serious as that of thick double film (because the entrained layer of air is expected to have the same negligible strength). However, there are additional reasons why the thin film may be less damaging. A film that is mechanically less strong is more easily torn and more easily ravelled into a more compact form. Internal turbulence in the melt will tend to favour the settling of the defect into stagnant corners of the mould. Here it will be quickly frozen into the casting before it has chance to unfurl significantly.

Films on cast iron for instance are controllable by casting temperature and by additions to the sand binder to control the environment in the mould (sections 1.1.3 and 5.5.1). Films on some steels are controllable by minor changes to the chemistry of the metal as a result of changes to deoxidation practice (section 5.6).

2.2.5 Oxide lap defect II: the confluence weld

Even in those castings where the metal is melted and handled perfectly, so that no surface film is created and submerged, the geometry of the casting may mean that the metal stream has to separate and subsequently join together again at some distant location. This separation and rejoining necessarily involves the formation of films on the advancing fronts of both streams, with the consequent danger of the streams having difficulty in rejoining successfully. This junction has been called a confluence weld (Campbell 1988). Most complex castings necessarily contain dozens of confluence welds.

The author recalls that in the early days of the Cosworth process, a small aluminium alloy pipe casting was made for very high pressure service conditions. At that time it was assumed that the mould should be filled as slowly as possible, arriving at the top of the pipe just as the melt was freezing to encourage directional feeding. When the pipe was finally cast it looked perfect. It passed radiographic and dye penetrant tests. However, it failed catastrophically under a simulated service test by splitting longitudinally, exactly along its top, where the metal streams were assumed to join. The problem defeated our expert team of casting engineers, but was solved instantly by our foundry manager, George Wright, our very own dyed-in-the-wool foundryman. He simply turned up the filling rate (neglecting the niceties of setting up favourable temperature gradients to assist feeding). The problem never occurred again. Readers will note a moral (or two) in this story.

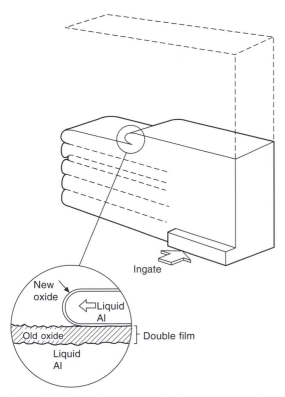

Figure 2.25 *Unstable advance of a film-forming liquid, showing the formation of laps as the interface intermittently stops and restarts by bursting through and flooding over the surface film.*

Figure 2.26 shows various situations where confluence problems occur in castings. Such locations have been shown to be predictable in interesting detail by computer simulation (Barkhudarov and Hirt 1999). The weld ending in a point illustrated in Figure 2.27 is often seen in thin-walled aluminium alloy sand castings; the point often has the appearance of a dark, upstanding pip. The dark colour is usually the result of the presence of sand grains, impregnated with metal. The metal penetration of the mould occurs at this point as a result of the conservation of momentum of the flow, impacted and concentrated at this point. The effect is analogous to the implosion of bubbles on the propeller of a ship: the bubble collapses as a jet, concentrating the momentum of the in-falling liquid. The repeated impacts of the jet fatigue the metal surface, finally causing failure in the form of cavitation damage.

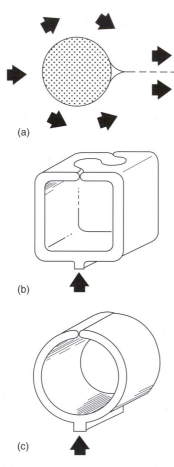

Figure 2.26 *Confluence geometries: (a) at the side of a round core; (b) randomly irregular join on the top of a bottom-gated box; and (c) a straight and reproducible join on the top of a bottom-gated round pipe (Campbell 1988).*

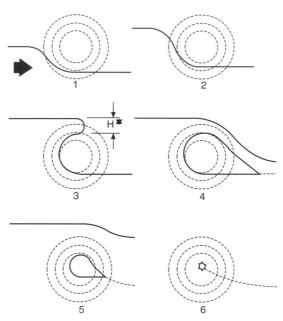

Figure 2.27 *Local thin area denoted by concentric contours in an already thin wall, leading to the creation of a filling instability, and a confluence weld ending in a point discontinuity (Campbell 1988).*

Returning to the issue of the confluence weld, a complete spectrum of conditions can be envisaged:

1. The two streams do not meet at all.
2. The two streams touch, but the joint has no strength.
3. The joint has partial strength.
4. The joint has full strength because the streams have successfully fused, resulting in a joint that is indistinguishable from bulk material.

For conditions (1) and (2) the defects are either obvious, or are easily detected by dye penetrant or other non-destructive tests. If the problem is seen it is usually not difficult to cure as described below. Condition (4) is clearly the target in all cases, but up till now it is not certain how often it has been achieved in practice. This can now also be clarified.

As with many phenomena relating to the mechanical effects of double oxide films, the understanding comes rather straightforwardly from a thought experiment. (Easier and quicker than making castings in the foundry. However, confirmatory experiments will be welcome in due course.) The concept is illustrated in Figure 2.28.

In the case of two liquid fronts that progress towards each other by the splitting and reforming of their surface films, the situation just after the instant of contact is fascinating. At this moment the splitting will occur at the point of contact because the film is necessarily thinnest at this point: no

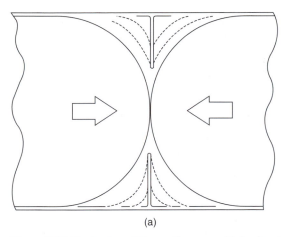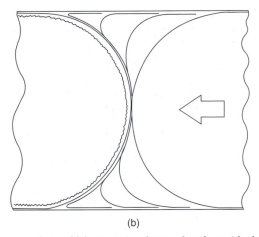

(a)   (b)

Figure 2.28 *Mechanism of the confluence weld, leading to: (a) a perfect weld from moving fronts after the residual thin bifilm has been flattened against the surface of the casting; and (b) a through-thickness crack at a stopped front.*

oxygen can access the microscopic area of contact. As the streams continue to engage, the oxide on the surfaces of the two menisci continue to slide back from the point of contact, but because of the exclusion of oxygen from the contact region, no new film can form here. Remnants of the double film occupy a quarter to a third of the outer part of the casting section, existing as a possible crack extending inward from each surface. This is most unlikely to result in a defect because such films will be thin because of their short growth period. Having little rigidity, being more akin to tissue paper of gossamer lightness, it will be folded against the oxide skin of the casting by the random gales of internal turbulence. There it will attach, adhering as a result of little-understood atomic forces. Any such forces, if they exist, are likely to be only weak. However, the vanishingly thin and weak films will not need strong forces to ensure their capture. Thus, finally, the weld is seen to be perfect. This situation is expected to be common in castings.

The case contrasts with the approach of two liquid fronts, in which one front comes to a stop, but the other continues its advance. In this case the stationary front builds up the thickness of its oxide layer to become strong and rigid. When the 'live' front meets it, the newly arriving film is now pinned in place at the point of contact of the rigid, thick film by friction. Thus the continuously advancing stream expands around the rigidized meniscus, forcing its oxide film to split and expand to allow the advance, causing a layer of new film to be laid down on the old thick substrate. Clearly, a double film defect constituting a crack has been created completely across the wall of the casting. Again, the double film is asymmetrical.

Note once again that for the conditions in which one of the fronts is stationary, the final defect is a lap defect in which the crack is usually in a vertical

plane (although, of course, other geometries can be envisaged). This contrasts with the surface-flooding defect, lap defect type I, where the orientation of the crack is substantially horizontal.

A location in an Al alloy casting where a confluence weld was known to occur was found to result in a crack. When observed under the scanning electron microscope the original thick oxide could be seen trapped against the tops of dendrites that had originally flattened themselves against the double film. The poor feeding in that locality had drained residual liquid away from the defect, sucking large areas of the film deeply into the dendrite mesh. One of the remaining islands of film pinned in its original place by the dendrites is shown in Figure 2.29. The draped appearance suggesting the dragging action of the surrounding film as it was pulled and torn away.

In summary, if the two fronts can be kept 'live' the confluence is expected to be a perfect weld. If one of the fronts stops, the result is a crack.

At first sight there seems little room for partial bonds. However, it is conceivable that even after a double film has formed, given the right conditions, the crack may partially or completely heal.

For instance, in cast irons the double film could be graphitic, and so go into solution in the iron given sufficient time and temperature. Pellerier and Carpentier (1988) are among the few who have reported an investigation into a confluence weld defect in iron. They studied a thin-walled ductile iron casting cast in a mould containing cores bonded with a urethane resin. They found a thin film (but seem not to have noticed whether the film might have been double) of graphite and oxides through the casting at a point where two streams met. The bulk metal matrix structure was ferritic (indicating an initial low carbon content in solution) but close to the film was pearlitic, indicating that some carbon

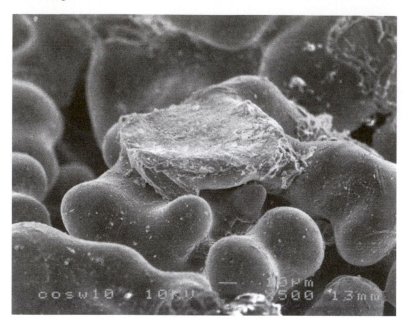

Figure 2.29 *SEM fractograph of a confluence weld at a stopped front. An island of thick oxide remains, the rest having withdrawn into the depths of the dendrite mesh, retreating with residual liquid because of local feeding problems in the casting.*

from the film was going into solution. No mechanical tests were carried out, but the tensile properties across the defect are not expected to be high. At least some of the original double film of graphite seems to have survived in place (and flake graphite is not noted for its strength). The authors did not go on to explore conditions under which the confluence weld could be avoided.

Other conditions in which confluence welds, once formed, might be encouraged to heal are dealt with in the section on the deactivation of defects.

Finally, however, it is clear that the weld problem can be eliminated by keeping the liquid fronts moving. This is simply arranged by casting at a sufficiently high rate. Care is needed of course to avoid casting at too high a rate at which surface turbulence may become an issue. However, providentially, there is usually a comfortably wide operational window in which the fill rate can meet all the requirements to avoid defects.

2.2.6 The oxide flow tube

The oxide flow tube is a major geometrical crack resulting from the entrainment of the oxide around a flowing stream.

The stream might be a falling jet, commonly generated in a waterfall condition in the mould, as in Figure 2.30. It creates the curious defect, the major cylindrical crack. The stream does not need to fall vertically. Streams can be seen that have slid down gradients in such processes as tilt casting when carried out under poor control. Part of the associated flow tube is often visible on the surface of the casting.

Oxide flow tube defect from a fall

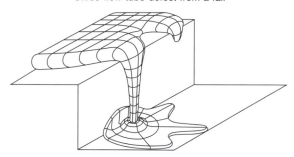

Figure 2.30 *Waterfall effect causing: (i) a stationary top surface; (ii) a falling jet creating a cylindrical oxide flow tube; and (iii) random surface turbulence damage in the lower levels of the casting.*

Alternatively, a wandering horizontal stream can define the flow tube, as is commonly seen in the spread of liquid across a horizontal surface. Figure 2.24 shows how, in a thin horizontal section, the banks of the flowing stream remain stationary while the melt continues to flow. When the flow finally fills the section, coming to rest against the now-rigid banks of the stream, the banks will constitute long meandering bifilms as cracks, following the original line of the flow.

The jets of flow in pressure die castings can be seen to leave permanent legacies as oxide tubes, as seen in section in Figure 2.31.

All these examples illustrate how unconstrained (i.e. free from contact with guiding walls) *gravity filling* or *horizontal filling* both risk the formation

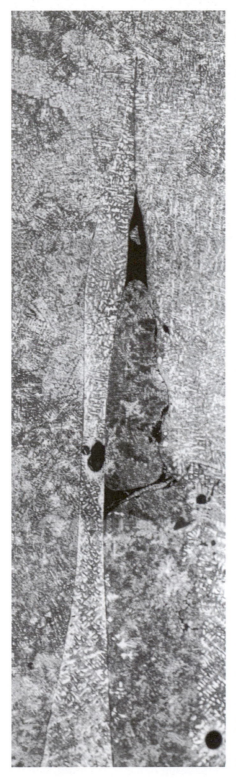

Figure 2.31 *Oxide flow tubes separating regions of the solidifying structure of a high pressure Al alloy die casting (courtesy Ghomashchi and Chadwick 1986).*

of serious defects. (The unconstrained filling of moulds without risk can only be achieved by *counter-gravity*.)

Both of these kinds of streams exhibit surfaces that are effectively stationary, and thus grow a thick oxide. When the rising melt finally entrains such features, the new thin oxide that arrives, rolling up against the old thick oxide, creates a characteristic asymmetrical double film. On such a double film in a vacuum cast Ni-based superalloy, the author has seen sulphide precipitates formed only on one side of the defect, indicating that only one side of the double film was favourable to nucleation and growth (microphotographs were not released for security reasons). Too little work was carried out to know whether the thick or thin side of the bifilm was the active substrate.

In all cases it will be noticed that in such interruptions to flow, where, for any reason, the surface of the liquid locally stops its advance, a large asymmetric double film defect is created. These defects are always large, and always have a recognizable, predictable geometrical form (i.e. they are cylinders, planes, meandering streams, etc.). They are quite different to the double films formed by surface turbulence, which are random in size and shape, and completely unpredictable as a result of their chaotic origin.

2.2.7 Microjetting

In some conditions the advance of the liquid front appears to become chaotic on a microscale. Jets of liquid issue from the front, only to be caught up within a fraction of a second by the general advance of the front, and so become incorporated back into the bulk liquid. The jets, of course, become oxidized, so that the advancing liquid will naturally be expected to become contaminated with a random assortment of tangled double films.

Such behaviour was observed during the casting of Al–7Si–0.4Mg alloy in plaster moulds (Evans *et al.* 1997). In this experiment the wall thickness of the castings was progressively reduced to increase the effect of surface tension to constrain the flow, reducing surface turbulence, and thus increasing reliability. As predicted, this effect was clearly seen as the section was reduced from 6 to 3 mm. However, as the section was reduced further to increase the benefit, instead of the reliability increasing further, it fell dramatically.

At the smaller sections direct video observation of the advancing front revealed that the smooth profile of the meniscus was punctured by cracks, through which tiny jets of metal spurted ahead, only to be quickly engulfed by the following flow. The image could be likened to advancing spaghetti.

It seems likely that the effect is the result of the strength of the oxide film on the advancing front in thin section castings. In thin sections, the limited area of the front limits the number of defects present in the film. The effect seems analogous to the behaviour of metal whiskers, whose remarkable strength derives from the fact that they are too small to contain any significant defects. Following this logic, a small area of film may contain no significant defect, and so may resist failure. Pressure therefore builds up behind the film, until finally it ruptures, the split releasing a jet of liquid. (To explain further, the effect is not observed in thick sections because the greater area of film assures the presence of plenty of defects, so the film splits easily, and the advance of the melt is smooth.)

Similar microjets have been observed to occur during the filling of Al alloy castings via 2 mm thick ingates. Single or multiple narrow jets have been seen to shoot across the mould cavity from such narrow slot ingates (Cunliffe 1994).

The microjetting mechanism of advance of liquid metals has so far only been observed in aluminium alloys, and the precise conditions for its occurrence are not yet known. It does not seem to occur in all narrow channels. The gaseous environment surrounding the flow may be critical to the behaviour of the oxide film and its failure mechanism. Also, the effect may only be observed in conditions where not only the thickness but the width of the channel is also limited, thus discouraging the advance of the front by the steady motion of transverse waves (the unzipping mode of advance to be discussed later).

Where it does occur, however, the mechanical properties of the casting are seriously impaired. The reliability falls by up to a factor of 3 as casting sections reduce from 3 to 1 mm. Unless microjetting can be understood and controlled, the effect might impose an ultimate limit on the reliability of thin section castings. This would be a bitter disappointment, and hard to accept. To avoid the risk of this outcome more research is needed.

2.2.8 The bubble trail

A bubble trail is the name coined in *Castings* (1991) to describe the defect that was predicted to remain in a film-forming alloy after the passage of a bubble through the melt.

Since air, water vapour and other core gases are normally all highly oxidizing to the liquid metal, a bubble of any of these gases will react aggressively, oxidizing the metal as it progresses, and leaving in its wake the collapsed tube of oxide like an old sack. (In the case of graphite film-forming gases, the bubble trail is, of course, expected to be a collapsed graphitic tube.) The inner walls of the trail will come together dry side to dry side, and so be non-adherent, once again constituting a classic form of a double film defect. This particular bifilm has its special characteristic features, as do the other major bifilms, the random defects arising from surface turbulence, and the geometrical defects that result from the various oxide laps.

The mechanism of the expansion of the film forming the crown of the bubble is schematically illustrated in Figure 2.32. The bubble forces its way upwards while splitting the film on its crown that is attempting to hold it back. Only large bubbles have sufficient buoyancy to overcome the resistance to its motion provided by the strength of the film. The film exerts its restraint because it is effectively tethered to the point, often located in the early part of the filling system, where the bubble was first entrained. The expanding region of film on the crown effectively slides around the surface of the bubble, continuing to expand until the equator of the bubble is reached. At this point the area of the film is a maximum. Since the film cannot contract, further

progress of the bubble causes the film to fold and crease. Any spiralling motion of the bubble will additionally tighten the rope-like trail.

Figure 2.32b further illustrates the different sections to be expected along the length of the bubble and its trail, showing the gradual collapsing process that creates the trail.

Divandari (1999) was the first to observe the formation of bubble trails in aluminium castings by X-ray video technique. He introduced air bubbles artificially into a casting, and was subsequently able to pinpoint the location of the trails and fracture the casting to reveal the defect. Figure 2.33a shows the inside of a trail in Al–7Si–0.4Mg alloy. The longitudinally folded film is clear, as is the presence of shrinkage cavities that have expanded away from the defect because the casting was not provided with a feeder. The small amount of shrinkage has sucked back the residual liquid, stretching the film over the dendrites as seen in Figure 2.33b. The

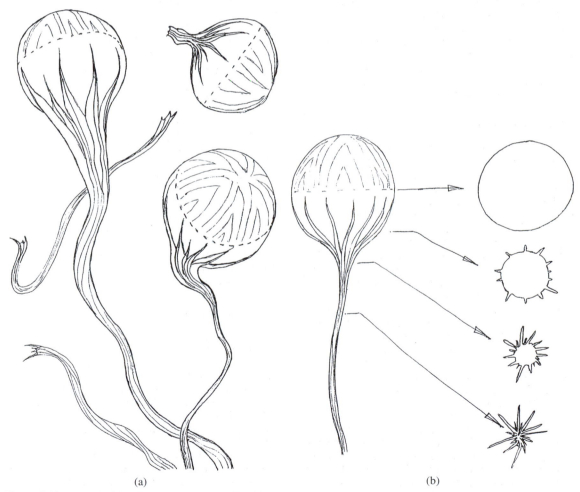

(a)   (b)

Figure 2.32 *(a) Schematic illustrations of rising bubbles and associated trails; (b) cross-sections illustrating the progressive collapse of the bubble trail.*

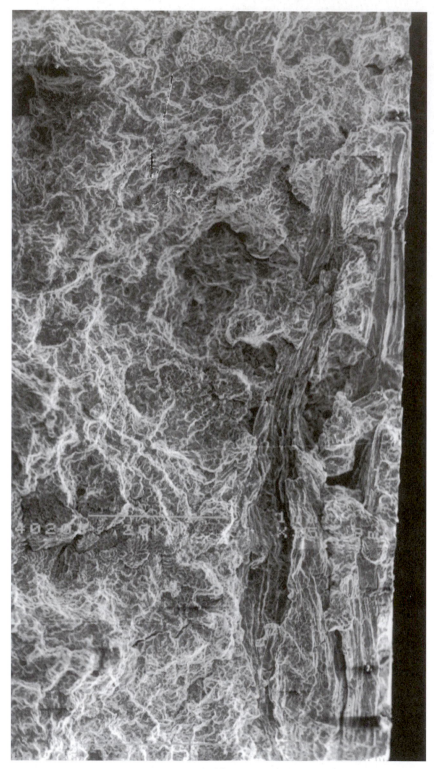

(a)

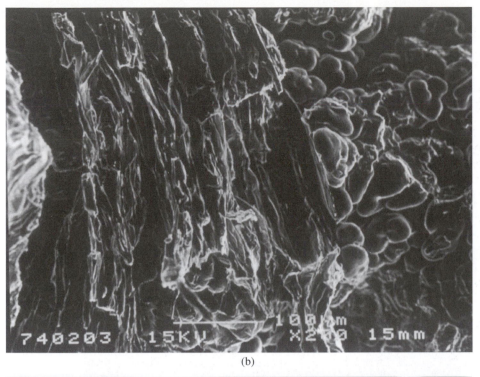

(b)

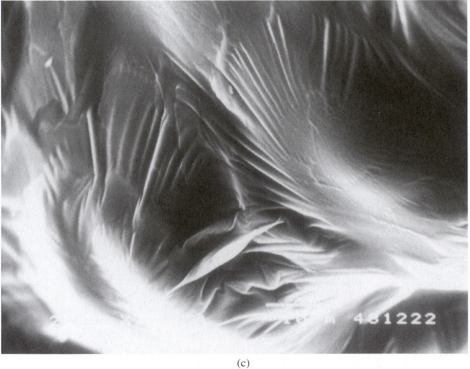

(c)

Figure 2.33 *(a) SEM fractograph of a bubble trail in Al–7Si–0.4Mg alloy; (b) a close-up, including areas of shrinkage probably grown from the trail; (c) the oxide film of the trail draped over dendrites, on the point of being sucked into the mesh because of a shrinkage problem in the casting (Divandari 2000).*

thicker, older part of the film is seen pulled in tight creases, and where torn is seen to be replaced by newer, thinner film.

Later he observed open trails in zinc pressure die castings (Figure 2.34). In this case the trail was almost certainly re-opened by high internal pressure (up to 100 bar) in the bubble and its trail when the die was opened slightly prematurely, releasing the support of the die before solidification was complete. The pressurized bubbles have effectively expanded, opening like powerful springs, while the casting was still in a plastic state.

Figure 2.34 *SEM image of a fracture surface of a zinc pressure die cast component revealing a bubble with an open trail (Divandari 2000).*

A cross-section of a group of bubble trails was observed by Divandari (1999) on a metallographic section of an aluminium alloy (Figure 2.35). The trails were nowhere near the trajectory of the bubbles, illustrating how internal circulations in the casting had transported the damage, ravelling and tumbling the trails to a distant location. (Other sources of large oxides were excluded from this experiment.) The defects strongly resembled interdendritic shrinkage porosity as a result of their concave, cuspoid outlines. However, the dendrite arm size was an order of magnitude smaller than the average repeat distance of the cusps, indicating that dendrites had not defined the outline of this porosity. It is clear that much porosity in castings has been wrongly attributed to shrinkage, and that the bubble trail is actually a common defect, to be expected in most gravity poured castings and many other types of casting.

A scenario such as that shown in Figure 2.36a is common. The bubbles are entrained early in the filling system, arriving as a welter of defects, boiling up with the liquid as it fills the mould cavity. Many of the early bubbles will have sufficient buoyancy

to escape. When the casting is finally solidified the appearance on a radiograph is expected to be something like that shown in Figure 2.36b. Porosity and cracks may be detected in regions above the ingate. The large number of remaining trails is likely to be invisible. This mix of residual bubbles and oxides is usefully termed *bubble damage*. In the gate itself, the trails will have been pushed by dendrites into the centre of the section, and will resemble centreline shrinkage porosity. Close examination, however, will confirm its identity as an assembly of close-packed double films. For many resin-bonded sand castings the internal surfaces of the leaves will have a slightly discoloured appearance, stained by the mould gases.

Figure 2.37a shows a bronze casting that has suffered a highly turbulent filling system. The bubbles entrained by this turbulence can be seen to be grouped along a horizontal line (a site of a poor joint between cope and drag, allowing a cooling fin to create a line of dendrites that had caught many of the bubbles) and towards the top of the casting in the 'skeins of geese' mode that characterizes these defects in bronzes. The skeins are probably the directionality to the visible defects provided by the arrays of invisible oxide bifilms, streaming like tattered banners in the wind. Figure 2.37b is a close-up of the top of the same casting after defects have been revealed by machining.

A radiograph of a nearly pure copper casting that has also suffered bubble entrainment in its filling system is shown in Figure 2.38. This casting was thin-walled and extensive, so that it continued to receive bubbles late into pouring, with the result that they have become trapped in the solidifying casting. The author has seen similar extensive arrays of bubbles in aluminium alloy oil pan castings.

Figure 2.32 shows a bubble that has been torn free from its trail. Such bubbles, with the stump of their trail showing the bubble to be tumbling irregularly as it rises, have been directly observed by X-ray video. The work by Divandari (Figure 2.35) confirms that bubble trails, in filtered melts known to be free from oxide tangles, can detach and float freely, finally appearing at distant locations.

Clearly, the trail, whether still attached to its bubble or not, is a serious threat to the mechanical strength and integrity of the casting.

Bubble trails are known in low-pressure casting if the riser tube is not pressure-tight, allowing bubbles of the pressurizing gas to leak through, rising up the riser tube and so directly entering the casting. The author has observed such a trail in a radiograph of an aluminium alloy casting. In the quiescent conditions after filling, the resulting trail was smooth and straight, rising through the complete height of the casting. When concentrating on the examination of the radiograph for minute traces of porosity or inclusions such extensive geometric

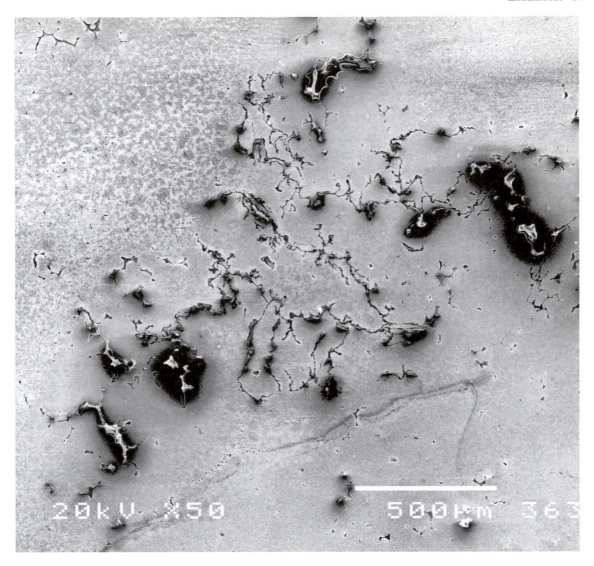

Figure 2.35 *Optical metallographic section of a number of bubble trails (Divandari 2000).*

features are easy to overlook, appearing to be the shadows of integral structural parts of the casting!

A rather serious form of bubble trail is also commonly observed to form from core blows. This will be dealt with in section 6.4.

Finally, it is worth considering what length a bubble trail might reach. If we assume a bubble of 10 mm diameter rising through liquid aluminium, and if aluminium oxide grows to a thickness of 20 nm, it is not difficult to estimate that the 20 per cent oxygen in the air is used up after creating a trail of about 0.5 m. This is quite sufficient to cause a major problem in most castings.

For castings poured in vacuum (actually dilute air of course!) vacuum bubbles are still to be expected to form, although the situation is a little more complicated. The entrained atmosphere in this case will be at a pressure somewhere between 10^{-3} and 10^{-6} atmosphere (1 torr to 10^{-3} torr). (The local vacuum inside the mould at the instant of pouring is likely to be much higher as a result of mould outgassing than the pressure indicated on the vacuum gauge of the furnace.) Thus, considering the case of the vacuum casting of Ni-base superalloys, at a nominal depth of 100 mm in the liquid of density close to 8000 kgm^{-3}, a bubble of 20 mm diameter will collapse down to somewhere in the region of 5 to 0.5 mm diameter before its internal pressure rises to equal that of the surrounding melt. (We are neglecting the somewhat

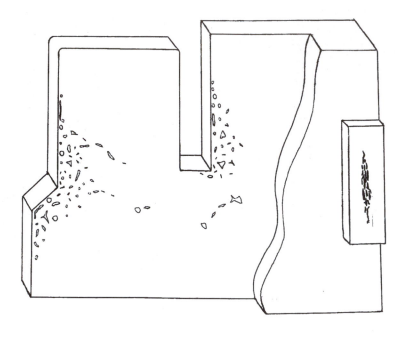

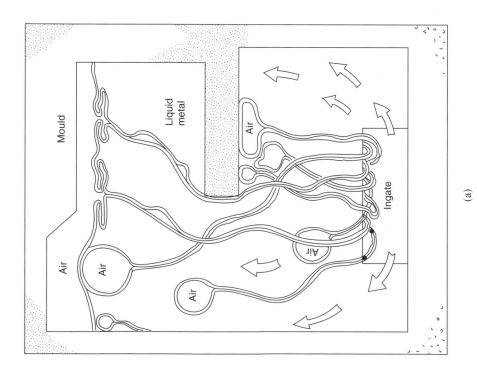

Figure 2.36 (a) Schematic illustration of bubbles entering a casting; and (b) the residual apparent damage consisting of residual bubbles and cracks (the remaining films are here invisible as usual).

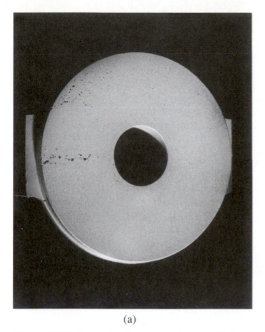
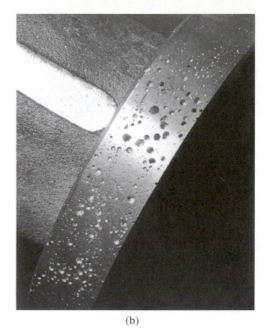

(a)   (b)

Figure 2.37 *Bronze bush casting after machining, revealing entrained air bubbles on (a) the front face and (b) on the top edge of the flange.*

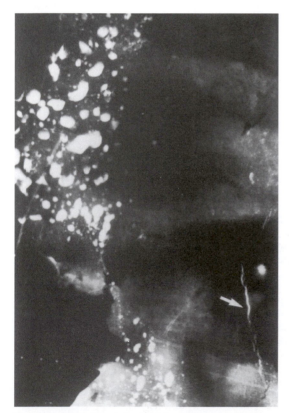

Figure 2.38 *Radiograph of a thin-wall copper statue showing extensive bubble damage.*

balancing effects of (i) a small reduction in size because of surface tension, and (ii) a small increase in size from the expansion of the mould gases, since moulds for investment casting are already at 1000°C or more.) The bubbles will be smaller at greater depths, and larger towards the top of the casting of course. Their considerably reduced concentration of oxygen will reduce the potential lengths of trails, or result in a thickness of oxide much less than 20 nm. It is not easy to predict what form a bubble trail may take in these circumstances, if the bubble is able to rise at all. There seems no shortage of research to do yet.

2.2.9 Marangoni-driven entrainment

This short section is added for completeness. We should bear in mind that not all entrainment arises simply as a result of the outside application of mechanical forces as we have described so far. Where the surface is liquid, and if the surface tension of the liquid is greatly affected by the gas phase in contact with it, there is the potential for Marangoni flow. This is a dynamically unbalanced condition in which a region of high surface tension pulls the surface liquid from an area of low surface tension. Oxygen is highly surface active in liquid iron, and greatly reduces the surface tension. In the conditions where oxygen is blown at the surface in steel making processes, Chung and Cramb (2000) find that the surface flows are so intense that fine droplets of FeO are entrained in the gas phase.

These conditions are unusual, requiring a liquid-free surface, and a highly surface active impinging gas. Most metal interfaces with the environment are, as we have seen, covered with a solid film. Also, the gas phase is usually only weakly, if at all, surface active. Thus, despite its interest, Marangoni entrainment is not expected to be a common feature of metallurgical reactions in cast products.

2.3 Furling and unfurling

Throughout its life, the bifilm undergoes a series of geometrical rearrangements. An understanding of these different forms is essential to the understanding of the properties of castings. The stages in the life cycle of the bifilm are:

1. entrainment by surface turbulence;
2. furling by bulk turbulence; and
3. unfurling in the stillness of the liquid after the filling of the casting is complete.

The verb '*to furl*' is used in the same sense that a sailing boat will *furl* its sails, gathering them in; or it may *unfurl* them, letting them out.

We have seen earlier that if the surface of the liquid suffers turbulence (i.e. the Weber number well in excess of unity) the liquid is likely to enfold our familiar bifilm defects. Their size will probably depend on the strength (i.e. the tear resistance) of the film. For some high duty stainless steels the size might be 100 mm across. For Al–Si alloys containing low levels of Mg the bifilms seem to be more usually from 1 mm up to 10 or 15 mm diameter.

Once submerged, however, the bifilms will be subjected to the conditions of bulk turbulence beneath the surface of the liquid. This is because Reynold's number, *Re*, is nearly always over the critical value of 2000 in liquid metal-filling systems as is clear from Figure 2.22. (Exceptions may be counter-gravity and tilt-pouring systems.) The launching of our delicate, gossamer-thin double film into a maelstrom of vortices in this dense liquid will ensure that it is pummelled and ravelled into a compact, convoluted form almost immediately. In this form it will be effectively reduced in shape and size from a planar crack of diameter 1 to 10 mm to a small ragged ball of diameter in the range 0.1 to 1 mm. These are the sizes of crumpled dross-like defects seen by Fox (2000) and commonly seen on polished metallographic sections of aluminium alloy castings.

It is worth emphasizing that although the bifilm now has a highly contorted form – crumpled, convoluted and ravelled in an untidy random manner – it has not lost its crack-like character. However, its form contrasts sharply with the types of crack that the metallurgist normally associates with stress.

It seems also worth underlining the fact that the double film is unlikely to become separated once again into single films. This is because where the two halves are in contact, interfacial forces will be important, and inter-film friction will ensure that both leaves will move as one. In addition, the continued oxidation, thickening the leaves, will reduce the volume of air in the inter-film gap, reducing the pressure between the films. The exterior atmospheric and metallostatic pressures will thereby pressurize the two halves together, ensuring that the films continue their association as a pseudo-single entity, despite the absence, or near absence, of bonding between the halves.

In this compact form, bifilms are able to pass through filters of normal pore size in the region of 1 mm diameter. Some will be arrested as a result of untidy trails of film that may become caught by the filter, since the compact forms will be winding and unwinding continuously in a random manner in the turbulence. Inside the filter the constraint of the narrow channels promotes laminar flow (Figure 2.22). It is possible that many compact bifilms will be unravelled and flattened against the internal surfaces of the filter if they become hooked up somewhere in a laminar flow stream. This seems likely to be a potentially important filtration mechanism for compacted bifilms.

At practically all other locations, during the whole of the transfer into the mould, the bulk turbulence will ensure that the bifilms are continuously tumbled, entering the mould in a relatively compact form.

Thus the casting initially finds itself with a distribution of compact and convoluted bifilms. Their tensile strength will, on average, be about half that of a pore of the same average diameter as the bifilm ball because of the effect of the mechanical interlinking of the crumpled crack as seen in Figure 2.39. If the convoluted crack is subject to a tensile stress it cannot come apart easily. There is much interlocking of sound material, so that much plastic deformation and shearing will be involved in the failure.

Because of the combination of this small strength with their small size, the compact bifilms are rendered as harmless as possible at this stage.

However, once the filling of the casting is complete, the bifilm now finds itself in a new, quiescent environment. If there is any movement of liquid because of solute or thermally driven convection, such movements are relatively gentle, occurring at low values of *Re*.

Conditions are now right for the bifilm to unfurl, opening, growing by about a factor of 10 on average, to its original size in the range of 1 to 10 mm or so. By this action the defect now re-establishes itself as a planar crack, and so can impair the strength and ductility of the metal to the maximum extent. This astonishing ability of the bifilm to open like a

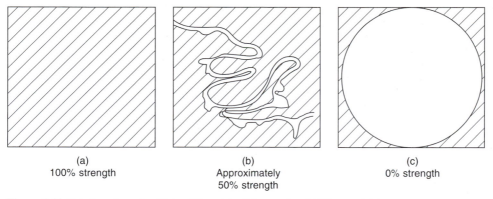

Figure 2.39 *Relative strength of (a) solid matrix, (b) convoluted bifilms and (c) pores.*

flower explains many of the problems of castings, as will become clear throughout this book.

What causes the bifilm to open again?

Interestingly, there are several potential driving forces. Although they will be considered in detail in later sections of the book, they are listed briefly here from recent studies of aluminium alloys where the effects have been most thoroughly studied to date. They include:

1. The precipitation by hydrogen in the gas film between the oxide interfaces, thus inflating the defect. The inflation is likely to take place in two main stages. The first we may call microinflation, with short lengths (defined and separated from adjacent lengths by the sites of folds) inflating individually. When sufficiently inflated, these short lengths will start to exert an unfolding pressure on adjacent lengths. As the inflation proceeds, the bifilm will unfold from fold to fold, length by length. Thus the unfurling action will be jerky and irregular, until the whole bifilm is unfolded. At this stage, of course, if enough hydrogen continues to precipitate, the bifilm will eventually grow into a spherical pore. The stages of growth are illustrated in Figure 2.40a for a simply folded bifilm, and 2.40b illustrates the stages for a more convoluted bifilm. If much gas precipitates, either because there is much gas in solution, or because there is plenty of time for diffusion, thereby aiding the collection of gas from a wide region, the gas pore may eventually outgrow its original bifilm to become a large spherical pore as shown in Figure 2.40c if growing freely in the liquid, or as in Figure 2.40d if growing later during freezing, appearing now as an interdendritic pore. On sectioning the pore it is probable that the originating fragment of bifilm will never be found. Interestingly, at the stage at which the pore starts to outgrow the boundaries of its originating bifilm, it will extend its own oxide film using its residual oxygen gas

(probably derived from the original bifilm). The new area of the defect will then consist of very new thin film. It is possible to imagine that, if all the oxygen were consumed in this way, eventually the pore would grow by adding clean, virgin metal surface. The many round pores in cast metals indicate that the linear dimensions of many bifilms are smaller than the diameters of such pores. It is a reminder of the wide spread of bifilm sizes to be expected.

2. Shrinkage may pull the two films apart. If this happened at an early stage of freezing the defect would grow no differently to a hydrogen pore, inflating from fold to fold, and the final fold eventually growing to be more or less spherical, as seen in Figures 2.40a and c. If the opening took place at a late stage of freezing the defect would attempt to open when surrounded by dendrites as shown in Figure 2.40d. Both of these modes of opening could be driven by gas or shrinkage or both. Thus the rounded and interdendritic forms are not reliable indicators of the driving force for the growth of pores. The round form is a reliable indicator that the pore grew prior to the arrival of dendrites, while the dendritic pore is a reliable indicator that this was a late arrival. No more can be concluded. In the case of the dendritic morphology, withdrawal of the residual liquid (being either pushed by gas from inside the pore, or pulled by reduced pressure outside the pore) would stretch the film over surrounding dendrites, and pull the bifilm halves into the interdendritic regions. The originating films might eventually be sucked out of sight deep into the dendrite mesh.

3. Iron rich phases, particularly the βFe particles, Al_5FeSi, nucleate and grow on the outer, wetted surfaces of the bifilm. Initially, when the βFe crystals are no more than a few nanometres thick, the crystals can follow the curvature and creases of the crumpled double film. However, as the crystals grow in thickness, because the rigidity

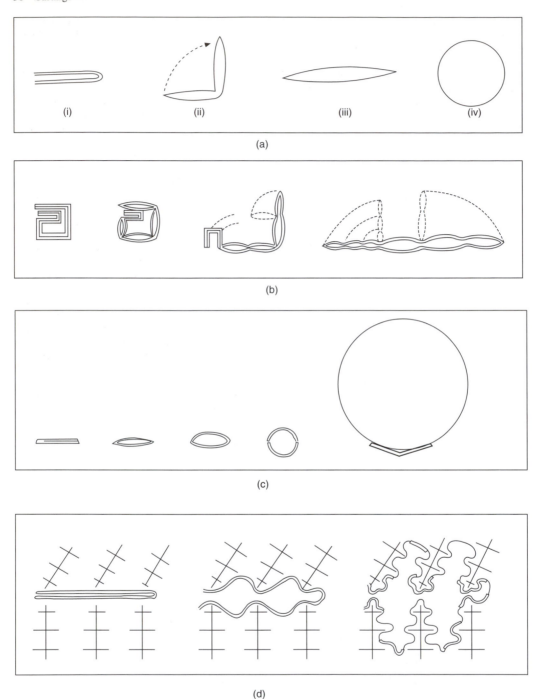

Figure 2.40 *(a to d) Stages of unfurling and inflation of bifilms.*

of a beam in bending mode is proportional to its thickness to the third power, they quickly develop rigidity. Thus as they thicken and rigidize, they *force* the straightening of the bifilm. The result is the familiar βFe plate, straight as a needle on a polished section. Occasional curved βFe plates are probably the result of restraint at the two ends of the plate, so that the plate has been unable to straighten fully, remaining stressed like an archer's bow. In any case, βFe plates often exhibit a crack along their centreline (if βFe has precipitated on both sides of the bifilm) or between the βFe and the matrix, showing apparent decohesion from the matrix (if the βFe has precipitated only on one side of the bifilm). It seems likely that all βFe particles are cracked or decohered in this way because of the presence of the originating bifilm, but the cracks may not be easily seen on a polished section.

4. Oxides are pushed ahead of growing dendrites, with the result that bifilms are automatically unravelled and flattened, effectively organized into planar areas among dendrite arrays, and are pushed into interdendritic and grain boundary regions. This is an important mechanism since it can occur in all cast alloys that have solid surface films. The effect is illustrated schematically in Figure 2.41.

Such a room-temperature fracture surface is seen in Figure 2.42 for an Al–4.5Cu alloy (Mi 2000) in which the fracture surface is covered in a thin alumina film. The heaps of excess film pushed ahead of the dendrites is clearly seen in the central regions of the cast test bar. Similar flattening of bifilms is seen to be driven by a twinned 'feather' dendrite (Figure 2.43a) and a conventional cubic symmetry dendrite (Figure 2.43b) in an Al–Si–Cu alloy. Again,

the piled-up bifilm pushed ahead of the growing crystals is clear. In Figure 2.44 an AlN or Al_2O_3 film is probably the cause of the planar boundaries seen in the fracture surface of a vacuum cast and HIPped Ni-base alloy IN939 (Cox et al. 2000). In the case of both the Al–Cu alloy and the Ni-base vacuum-cast alloy the castings had suffered surface turbulence during filling, but, significantly, the planar boundary features on the fracture surfaces were not observed in control castings made without such turbulence. The planar features were also observed to be associated with poor tensile ductility. In section 5.5.2 further planar defects are discussed for cast irons, and in section 5.6.2 probable examples in steels are proposed.

Clearly, the driving forces for the unfurling of bifilms are all those factors that are already known to the casting metallurgist as precisely those factors that impair ductility; including hydrogen in solution, shrinkage, iron contamination in Al alloys, and large grain size.

However, we need to bear in mind that it will take time for the unfurling processes to occur. In a casting that is frozen quickly, the bifilms are frozen into the casting in their compact form, with the result that most casting alloys that are chilled rapidly are strong and ductile, as all foundry engineers know.

Conversely, castings that have suffered lengthy solidification times exhibit shrinkage problems,

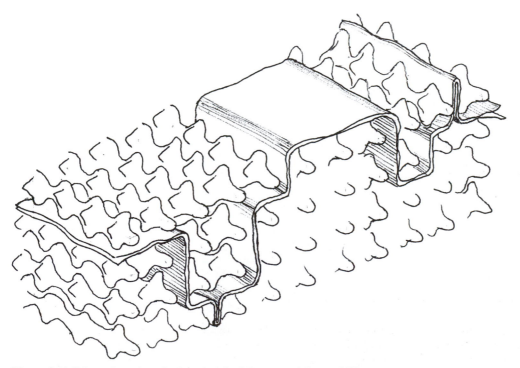

Figure 2.41 *Schematic action of advancing dendrites to straighten a bifilm.*

higher levels of gas porosity, or large grain size. In addition, all exhibit reduced properties, particularly reduced ductility. The beneficial action of rapid freezing to enhance ductility is, once again, well known to the casting metallurgist.

In summary, it is clear that all the factors that refine DAS (dendrite arm spacing) improve the mechanical properties of cast aluminium alloys. This is not primarily the result of action of the DAS alone, as has been commonly supposed. The

DAS is merely a measure of the local freezing time, which is the time available for the inflation and unfurling of defects. If the growth of area of the defect was not bad enough, its inflation, causing its surfaces to separate, transforming it into a pore, makes the defect even more damaging. Thus its strength falls further as seen in Figure 2.39. It is both (i) the growth of area and (ii) the growth of volume of the defects that combine to inflict so much damage to the mechanical properties as

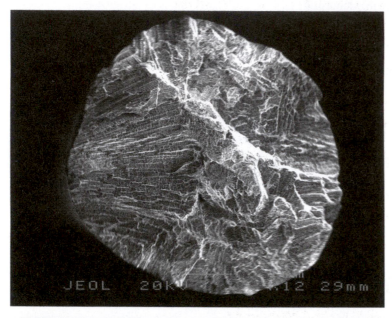

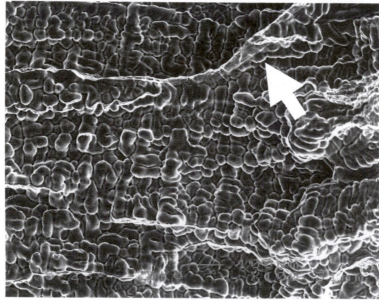

(a)

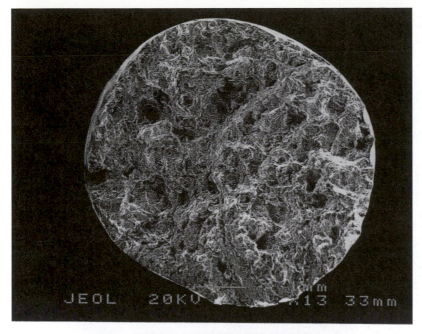

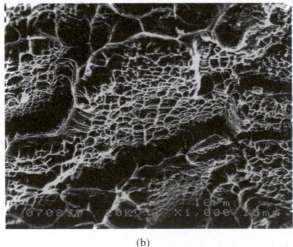

(b)

Figure 2.42 *(a) Tensile fracture surface of an Al–4.5Cu alloy that had suffered an entrainment effect observed by X-ray video radiography, showing large new bifilms straightened by large grains, and heaps of excess bifilms pushed into central areas, and a close-up showing a thin doubly folded area (arrowed). (b) Elsewhere in the same casting showing a fracture surface of quietly filled material and an area of ductile dimpled fracture (courtesy Mi 2000).*

solidification time increases. Increases in DAS are not the cause of the loss of ductility, the DAS is merely an indicator of the time involved.

Returning to the pivotal role of the entrainment defects, it seems clear from practical work in foundries that although the design of the filling system is extremely important, efforts to control the surface turbulence in the mould cavity itself

are the most important factor to yield reliable properties.

This seems likely to be the result of the less severe surface bulk turbulence in the mould cavity. Reynold's numbers in a nicely filled mould are in the region of 10^3, so that little energy is available to make the films more compact. Thus any bifilm entrained in the mould will probably remain

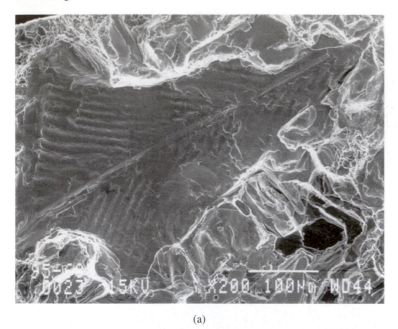

(a)

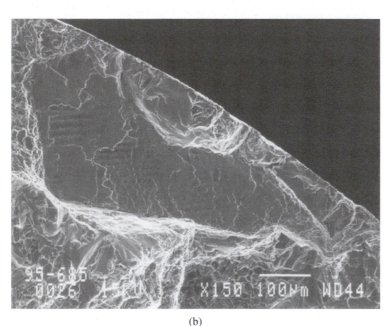

(b)

Figure 2.43 *Fracture surface of fatigued A319 alloy showing straightened oxides by the growth of (a) a twinned 'feather' crystal and (b) a conventional cubic symmetry dendrite (courtesy J. Boileau, Ford Labs 2000).*

substantially open, maximizing its area. This in itself will cause maximum damaging effect, but will be additionally enhanced by the hydrogen diffusion to the enlarged area, inflating the defect, and so further reducing any load carrying capacity it may once have enjoyed. The highest velocities and maximum turbulence is in the sprue, runner and gates, where Reynold's numbers of 10^4 to 10^5 are easily attained, indicating severe bulk turbulence. Bifilms entrained in the running system will

therefore be highly compacted, and, at least initially, have less damaging consequences for mechanical properties.

It is almost certain that in some circumstance the bifilms will be unable to unfurl despite the action of some or all of the above driving forces. This is because they will sometimes be glued shut. Such glues will include liquid fluxes. If present on the surface of the melt, they will find themselves folded into the bifilm, and so form a sticky centre

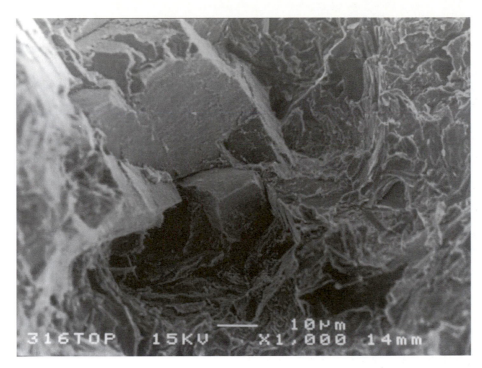

Figure 2.44 *Tensile fracture surface of a vacuum-cast and HIPped Ni-base alloy, showing apparently brittle grain boundaries after having been cast turbulently.*

to the sandwich, holding the bifilm closed, but the excess flux weeping or squeezed to its outer edges will aid the pinning of folds. The extreme thinness of bifilms will mean that even weak glues such as molten fluxes will be effective welds. Such action may explain part of the beneficial effect of fluxes. The melt may not be significantly cleaned by the flux, but its inclusions merely rendered less damaging by being unable to unfurl.

A similar action may explain the effect of low melting point additions such as bismuth to aluminium alloys. Although Papworth and Fox (1998) attribute the benefits of the Bi addition to the disruption of the integrity of alumina films, it seems more likely that a low melting point Bi-rich liquid will be inert towards the oxide, and that its action is passive, being merely an adhesive.

While bifilms remain a feature of our processing techniques for aluminium alloys, fluxes or low melting point metals might be a useful ploy, even if it is effectively 'papering over the cracks' of our current metallurgical inadequacies.

For the future, the very best cast material will be made free from bifilms in foundries specifically designed to deliver perfect quality metal. Such concepts are already on the drawing board and may be realized soon. When this utopia is achieved we shall be able to make castings with any DAS (as will be explained later), solidified fast or slow, all with perfect properties. It is a day to look forward to.

2.4 Deactivation of entrained films

Once submerged, the inside of the bifilm will act as a thin reservoir of air. Air will exist in pockets, bubbles trapped between the films, and in folds of the films, especially for films that have grown somewhat thicker, because their greater rigidity will bridge larger cavities among the folds. In addition, most oxides are microscopically rough, so that the opposed surfaces themselves will constitute part of the reservoir, and assist to convey the gas to all parts of the inter-layer. A particularly rough oxide that takes a corrugated form is MgO as seen on Al–5Mg alloy (Figure 2.10b).

Nevertheless, of course, the bifilm is not likely to know that it is submerged. The oxides on its interior surfaces will therefore be expected to continue their growth regardless. Thus, gradually its layer of air will be consumed as the oxygen is used up. Subsequently, the nitrogen will be likely to be converted to nitrides.

In aluminium alloys, although the nitride of aluminium is not so stable as the oxide, and will therefore not form in competition with the oxide, it is very stable at these temperatures. It is therefore likely to grow, if only slowly. At the temperatures of molten ferrous alloys, of course, aluminium nitride forms rapidly. Thus the layer of air is expected to disappear in time, but depending on the alloy and its impurities, at different rates. The 1 per cent remnant of argon is, however, likely to continue to remain as a gas because of its nearly perfect insolubility in molten metals.

During the final stages of the disappearance of the gas layer the opposed oxides would be expected to grow together to some extent. Any resultant bonding may not be particularly strong, but is likely to confer some improvement in strength compared to the layer of air.

If it is true that the gas layer will eventually be consumed, and if some diffusion bonding has occurred, then the most deleterious effects of the bifilm will have been removed. In effect, the potential of the defect for causing leaks, or nucleating bubbles, cavities, or cracks, will have been greatly reduced. This has been confirmed by Nyahumwa *et al.* (1998 and 2000) who found that porosity in the fatigue fracture surfaces of Al–7Si–0.4Mg alloy appeared to be linked often with new oxides, but not old oxides.

Further evidence for the growing together of bifilms with time is provided by the work of Huang and co-workers (2000) who examined fragments of bifilms detached by ultrasonic vibration from polished metallographic sections. These revealed that the bifilms of alumina, in the process of transforming to spinel, appeared to be extensively fused together. Figure 2.45 is a reinterpretation of their observations.

The automatic deactivation of inclusions with time may be relatively fast for new, thin films. This would explain the action of good running systems, where, after the fall of the liquid in the down-runner, the metal is allowed to reorganize for perhaps a second or so before entering the gates into the mould cavity. In general, such castings do appear to be good, although no formal study has been conducted to explore this effect.

However, the deactivation of inclusions seems

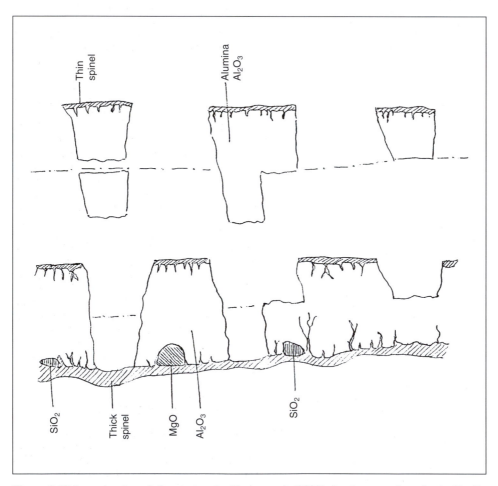

Figure 2.45 *Interpretation of observations by Huang et al. (2000) showing a section of a double film, ultrasonically fragmented from the surface of an Al–7Si–Mg alloy.*

certain to be encouraged in cases where the metal is subjected to pressure, since mechanical properties of the resulting castings are so much improved, becoming significantly more uniform and repeatable, and pores and cracks of all types appear to be eliminated. Even if the air film is not completely eliminated, it seems inevitable that the oxygen and nitrogen will react at a higher rate and so be more completely consumed, and any remaining gases will be compressed. In addition, if the pressure is sufficient, liquid metal may be forced through the permeable oxide so as to fuse with, and thus weld to, metal on the far side of the film. The crack-like discontinuity would then have effectively been 'stitched' together, or possibly 'tack welded'. Treatments that would promote such assisted deactivation include:

1. The solidification of the casting under an applied pressure. This is well seen in the case of squeeze casting, where pressures of 50–150 MPa (500 – 1500 atmospheres) are used. However, significant benefits are still reported for sand and investment castings when solidified under pressures of only 0.1 to 0.7 MPa (1 to 7 atmospheres), easily obtainable from a normal compressed-air supply. Berry and Taylor (1999) review these attempts. It is interesting to reflect that the reduced pressure test, commonly used as a porosity test in aluminium alloy casting, further confirms the importance of this effect. It uses exactly the same principle but in the opposite direction: the pressure on the solidifying casting is lowered to maximize porosity so that it can be seen more easily (Dasgupta et al. 1998 and Fox and Campbell 2000).
2. Hipping is a solid-state deactivation process, which, by analogy with the liquid state, appears to offer clues concerning the mechanism of deactivation processes in the liquid. The hot isostatic pressing (hipping) of (solid!) castings is carried out at temperatures close to their melting point to soften the solid as far as possible. Pressures of up to 200 MPa (2000 atmospheres) are applied in an attempt to compress flat all the internal volume defects and weld together the walls of the defects by diffusion bonding. It is clear that in the case of hipping aluminium alloys, the aluminium oxide that encases the gas film will not weld to itself, since the melting point of the oxide is over 2000°C and a temperature close to this will be required to cause any significant diffusion bonding. The fact that hipping is successful at much lower temperatures, for instance approximately 530°C in the case of Al–7Si–0.4Mg alloy, indicates that some additional processes are at work. For instance, the diffusion bonding of oxides may only occur in the presence of reactions in the oxide. In the Al–7Si–0.4Mg

alloy, the first oxide to form will probably be a variety of alumina, and so very stable and unlikely to bond. However, after time at temperature, the alumina absorbs Mg from the matrix alloy, converting the oxide to a spinel structure. During the atomic rearrangements required for the reconstruction of the lattice, diffusion bonding will be favoured. This mechanism seems likely to explain the resistance of Al–5Mg and Al–10Mg alloys to improvement by hipping. The oxide involved in this case is MgO. This, like alumina, is stable and unlikely to take part in any bonding action itself, and does not have the benefit of a further lattice transformation reaction to encourage bonding. (This extreme reluctance to bond may explain part of why bifilms in Al–Mg alloys are so effective in creating consequential defects such as porosity and hot tears, etc.)
3. Finally, it should be noted that if the entrained solid film is partially liquid, some kind of bonding is likely to occur much more readily, as has been noted in section 2.3. This may occur in light alloy systems where fluxes have been used in cleaning processes, or contamination may have occurred from traces of chloride or fluoride fluxes from the charge materials or from the crucible. Such fluxes will be expected to cause the surfaces of the oxide to adhere by the mechanism of viscous adhesion when liquid, and as a solid binder when cold. The beneficial action of fluxes in the treatment of liquid Al alloys may therefore not be the elimination of bifilms but their partial deactivation by assisting bonding. Similarly, as we have noted previously, the observations by Papworth (1998) might indicate that Bi acts as a kind of solder in the bifilms of Al alloys. The metal exists as a liquid to 271°C, so that when subjected to the pressure of squeeze casting, it would be forced to percolate and in-fill bifilms. Their resistance to deformation and separation would be expected to be improved if only by their generally increased mechanical rigidity (i.e. not by any chemical bonding effect). If proved to be true, this effect could be valuable. In the case of higher temperature liquid alloys, particularly steels, some oxides act similarly, forming low melting point eutectic mixtures. These systems will be discussed in detail later.

2.5 Soluble, transient films

Although many films such as alumina on aluminium are extremely stable, and completely insoluble in the liquid metal, there are some alloy/film combinations in which the film is soluble.

Transient films are to be expected in many cases in which the arrival of film-forming elements (such as oxygen or carbon) exceeds the rate at which the elements can diffuse away into the bulk metal. If then entrained, the film may be folded in to create a crack, a hot tear, or initiate shrinkage porosity. However, after the initiation of this secondary defect, the originating bifilm then quietly goes into solution, never to be seen again.

Such a case is commonly seen in grey cast irons. If a lustrous carbon film forms on the iron and is entrained in the melt, after some time it will dissolve. The rate of dissolution will be rather slow because the iron is already nearly saturated with carbon. Thus the entrained film may last just long enough to initiate some other longer lived and more serious defect, prior to its disappearance. The longer the time available the greater is the chance that the film will go into solution. Thus entrained lustrous carbon films are usually never seen in heavy section grey iron castings.

Oxide films on titanium alloys, including titanium aluminide alloys, appear to be soluble. However, Mi *et al.* (2002) have produced evidence that films do occur, and can be seen in some circumstances by SEM (scanning electron microscope) in castings. Previously Hu and Loretto (2000) had shown conclusively that even thick oxide films on TiAl alloys go into solution during hot isostatic pressing (hipping), leaving no metallographic trace. Thus in finished Ti alloy casting, all of which are usually hipped as part of the standard production process, entrained films are never seen. Only their consequentially created defects remain (if, for instance, the porosity is connected to the surface, and so is not closed by hipping).

2.6 Detrainment

In a liquid metal subjected to surface turbulence, there are a number of defects that can be eliminated from the melt without difficulty. These detrainment events take a variety of forms.

If the oxide surface film is particularly strong, it is possible that even if entrained by a folding action of the surface, the folded film may not be free to be carried off by the flow. It is likely to be attached to part of the surface that remains firmly attached to some piece of hardware such as the sides of a launder, or the wall of a sprue. Thus the entrained film might be detrained, being pulled clear once again. This detrainment process is so fast that the film hardly has time to consider itself entrained.

Even if not completely detrained, a strong film, strongly attached to the wall of the mould, may simply remain hanging in place, flapping in the flow in the melt delivery system, but fortunately remain harmless to the casting.

Beryllium has been added at levels of only 0.005 per cent to reduce oxidation losses on Al–Mg alloys.

However, on attempting to eliminate Be for environmental and health reasons, difficulties have been found in the successful production of wrought alloys by continuous (direct chill) casting. The beneficial action of Be that was originally unsuspected, but now highlighted, is thought to be the result of the strengthening of the film by the addition of the low levels of Be, thus encouraging the hanging up of entrained films in the delivery system to the mould rather than their release into the flowing stream.

Large bubbles have sufficient buoyancy to break their own oxide film, and the casting surface oxide skin (once again, the two films constituting a double oxide barrier of course) between them and the outside world at the top of the casting. They can thus detrain. If successful, this detrainment is not without trace, however, because of the presence of the bubble trail that remains to impair the casting.

Small bubbles of up to about 5 mm diameter have more difficulty to detrain. They are commonly trapped immediately under the top skin of the casting, having insufficient buoyancy to break both the film on the top of the melt, together with their own film. Thus they are unable to allow their contents to escape to atmosphere. They are the bane of the machinist, since they lodge just under the oxide skin at the top of a casting, and become visible only after the first machining cut. In iron castings requiring heat treatment, the surface oxidizes away to reveal the underlying bubbles. Similarly, shot blasting will also often reveal such defects.

The most complete and satisfactory detrainment is achieved by liquid surface layers such as fluxes and slags. This is because, once entrained, these phases spherodize, and therefore float out with maximum speed. On arrival at the liquid surface the liquid droplets are simply reassimilated in the surface liquid layer, and disappear.

2.7 Evidence for bifilms

The evidence for bifilms actually constitutes the entire theme of this book. Nevertheless, it seems worthwhile to devote some time to highlight some of the more direct evidence.

It is important to bear in mind that the double oxide film defects are everywhere in metals. We are not describing occasional single 'dross' or 'slag' defects or other occasional accidental exogenous types of inclusions. The bifilm defect occurs naturally, and in copious amounts, every time a metal is poured. Many metals are crammed with bifilm defects. The fact that they are usually so thin has allowed them to evade detection for so long. Until recently, the ubiquitous presence of these very thin double films has not been widely accepted because no single metal quality test has been able to resolve such thin but extensive defects.

Even so, over the years there have been many significant observations.

A clear example is seen in the use of the reduced pressure test for aluminium alloys. The technique is also known, with slight variations in operating procedure, as the Straube Pfeiffer test (Germany), Foseco Porotec test (UK) and IDECO test (Germany). At low gas contents many operators have been puzzled by the appearance of hairline cracks, often extending over the whole section of the test casting. They have problems in understanding the cracks since the test is commonly viewed as a check of hydrogen porosity. However, as gas content rises, the defects expand to become lens-shaped, and finally, if expansion continues, become completely spherical, fulfilling at last the expectation of their appearance as hydrogen pores. The effect is almost exactly that shown in Figure 2.40. Such an effect has been widely observed by many foundry people many times. An example is presented by Rooy (1992).

In a variant of the test to determine the quantity known as the *density index*, two small samples of a melt are solidified in thin-walled steel crucibles in air and under a partial vacuum respectively. A comparison of the densities of the samples solidified in air and vacuum gives the so-called density index. However, in this simple form the quantity is not particularly reproducible. The comparison is complicated as a result of the development of shrinkage porosity in the sample solidified in air. A better comparison is found by taking the lower half of the air-solidified sample and discarding the top half containing the shrinkage porosity. The sound base is then compared to the sample frozen under vacuum. This gives an unambiguous assessment of the porosity due to the combined effect of gas and bifilms. Without bifilms the hydrogen cannot precipitate, leading to a sound test casting, and giving the curious (and of course misleading) impression that the hydrogen can be 'filtered' out of liquid aluminium.

A recent novel development of the reduced pressure test has been made that allows direct observation of bifilms (Fox and Campbell 2000). The rationale behind the use of this test is as follows. The bifilms are normally impossible to see by X-ray radiography when solidified under 1 atmosphere pressure. If, however, the melt is subjected to a reduced pressure of only 0.1 atmosphere, the entrained layer of air should expand by ten times. Under 0.01 atmosphere the layer should expand 100 times, etc. In this way it should be possible to see the entrained bifilms by radiography. A result is shown in Figure 2.46.

In this work a novel reduced pressure test machine was constructed so that tests could also be carried

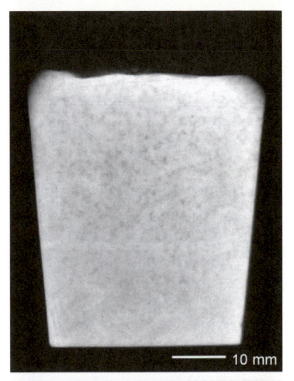

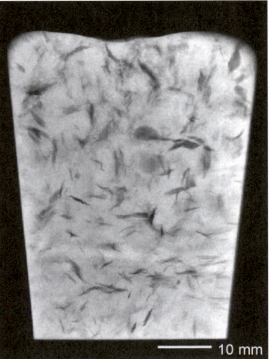

Figure 2.46 *Radiograph of reduced pressure test samples of as-melted Al–7Si–0.4Mg alloy solidified under pressures from (a) 1 atmosphere and (b) 0.01 atmosphere (Fox and Campbell 2000).*

out using chemically bonded sand moulds to make test castings as small slabs with overall dimensions approximately 50 mm high, 40 mm wide and 15 mm thick. The parallel faces of the slabs allowed X-ray examination without further preparation.

Figure 2.46 shows radiographs of plate castings from a series of tests that were carried out on metal from a large gas-fired melting furnace in a commercial foundry. Figure 2.46a shows a sample that was solidified in air indicating evidence of fine-scale porosity appearing as dark, faint compact images of the order of 1 millimetre in diameter. At progressively lower test pressures the compact 'pores' unfold and grow into progressively longer and thicker streaks, finally reaching 10 to 15 mm in length at 0.01 atmosphere (Figure 2.46b).

The 'streak-like' appearance of the porosity is due to an edge-on view of an essentially planar defect (although residual creases of the original folds are still clear in some images). The fact that these defects are shown in such high contrast at the lowest test pressure suggests that they almost completely penetrate the full 15 mm thickness of the casting, and may only be limited in size by the 15 mm thickness of the test mould. The more extensive areas of lower density porosity are a result of defects lying at different angles to the major plane of the casting.

At 0.01 atmosphere the thickness of bifilms as measured on the radiographs for those defects lying in the line of sight of the radiation was in the range 0.1 to 0.5 mm. This indicates that the original thickness of bifilms at 1 atmosphere was approximately 1 to 5 μm.

These samples containing large bifilms are shown here for clarity. They contrast with more usual samples in which the bifilms appear to be often less than 1 mm in size, and are barely visible on radiographs at 1 atmosphere pressure. The work by Fox and Campbell (2000) on increasing the hydrogen content of such melts in the RPT at a constant reduced pressure typically reveals the inflation of clouds of bifilms, first becoming unfurled and slightly expanded by the internal pressure of hydrogen gas, and finally resulting in the complete inflation of the defects into expanded spheres at high hydrogen levels.

A much earlier result was so many years ahead of its time that it remained unappreciated until recently. In 1959 at Rolls-Royce, Mountford and Calvert observed the echoes of ultrasonic waves that they directed into liquid aluminium alloys held in a crucible. What appeared to be an entrapped layer of air was observed as a mirror-like reflection of ultrasound from floating debris. (Reflections from other fully wetted solid phases would not have been so clear; only a discontinuity like a crack, a layer of air, could have yielded such strong echoes.) Some larger particles could be seen to rotate, reflecting

like a beacon when turning face-on after each revolution. Immediately after stirring, the melts became opaque with a fog of particles. However, after a period of 10 to 20 minutes the melt was seen to clear, with the debris forming a layer on the base of the crucible. If the melt was stirred again the phenomenon could be repeated.

Stirred melts were found to give castings containing oxide debris together with associated porosity. It is clear that the macroscopic pores observed on their polished sections appear to have grown from traces of micropores observable along the length of the immersed films. Melts that were allowed to settle and then carefully decanted from their sediment gave castings clear of porosity.

Other interesting features that were observed included the precipitation of higher-melting-point heavy phases, such as those containing iron and titanium, on to the floating oxides as the temperature was lowered. This caused the oxides to drop rapidly to the bottom of the crucible. Such precipitates were not easy to get back into suspension again. However, they could be poured during the making of a casting if a determined effort was made to disturb the accumulated sludge from the bottom of the container. The resulting defects had a characteristic appearance of large, coarse crystals of the heavy intermetallic phase, together with entrained oxide films and associated porosity. These observations have been confirmed more recently by Cao and Campbell (2000) on other Al alloys.

It is clear, therefore, from all that has been presented so far that a melt cannot be considered to be merely a liquid metal. In fact, the casting engineer must think of it as a slurry of various kinds of debris, mostly bifilms of various kinds, all with entrained layers of air or other gases.

In a definitive piece of research into the fatigue of filtered and unfiltered Al–7Si–0.4Mg alloy by Nyahumwa et al. (1998 and 2000), test bars were cast by a bottom-filling technique and were sectioned and examined by optical metallography. The filtered bars were relatively sound. However, for unfiltered castings, extensively tangled networks of oxide films were observed to be randomly distributed in almost all polished sections. Figure 2.5 shows an example of such a network of oxide films in which micropores (assumed to be residual air from the chaotic entrainment process) were frequently observed to be present. In these oxide film networks, it was observed that oxide film defects constitute cracks showing no bond developed across the oxide/oxide interface. In the higher magnification view of Figure 2.5 the width between the two dry surfaces of folded oxide film is seen to vary between 1 and 10 μm, in confirmation of the low pressure test results described above. However, widths of cracks associated with pores were usually found to be substantially greater than 10 μm, in places approaching 1 mm. Here the crack had opened sufficiently to be considered as a pore.

A polished section of a cast aluminium alloy breaking into a tangled bifilm is presented in Figure 2.6. The top part of the folded film comes close to the sectioned surface in some places, and has peeled away, revealing the inside surface of the underlying remaining half of the bifilm.

The detachment of the top halves of bifilms to reveal the underlying half is a technique used to find bifilms by Huang *et al.* (2000). They subjected polished surfaces of aluminium alloy castings to ultrasonic vibration in a water bath. Parts of bifilms that were attached only weakly were fatigued off, revealing strips or clouds of glinting marks and patches when observed by reflected light. They found that increasing the Si content of the alloys reduced the lengths of the strips and the size of the clouds, but increased the number of marks. The addition of 0.5 and 1.0 Mg reduced both the number and size of marks. Their fascinating polished sections of the portions of the bifilms that had detached revealed fragmentary remains of the double films of alumina apparently bonded together in extensive patches, appearing to be in the state of partially transforming to spinel (Figure 2.45).

The scanning electron microscope (SEM) has been a powerful tool that has revealed much detail of bifilms in recent years. One such example by Green (1995) is seen in Figure 2.11, revealing a film folded many times on the fracture surface of an Al–7Si–0.4Mg alloy casting. Its composition was confirmed by microanalysis to be alumina. The thickness of the thinnest part appeared to be close to 20 nm. It was so thin that despite its multiple folds the microstructure of the alloy was clearly visible through the film.

Finally, there are varieties of bifilms in some castings that are clear for all to see. These occur in lost foam castings, and are appropriately known as fold defects. Some of these are clearly pushed by dendrites into interdendritic spaces of the as-cast structure (Tschapp *et al.* 2000). The advance of the liquid into the foam is usually sufficiently slow that the films grow thick and the defects huge, and are easily visible to the unaided eye. Other clear examples, but on a finer scale, are seen in high pressure die castings. Ghomashchi (1995) has recorded that the solidified structure is quite different on either side of such features. For instance, the jets of metal that have formed the casting are each surrounded by oxides (their 'oxide flow tubes' as discussed in section 2.2.6) seen in Figure 2.31. Between the various flowing jets, each bounded by its film, the boundaries naturally and necessarily come together as double films, or bifilms. They form effective barriers between different regions of the casting.

As an 'opposite' or 'inverted' defect to a flow tube, a bubble trail constitutes a long bifilm of rather special form. The passage of air bubbles though aluminium alloy melts has been observed by video radiography (Divandari 1998). The bubble trail has been initially invisible on the video radiographic images. However, the prior solidification of the outer edges of the casting imposed a tensile stress on the interior of the casting that increased with time. At a critical stress the bifilm appeared. It flashed into view in a fraction of a second, expanding as a long crack, following the path taken by the bubbles, through what had appeared previously to be featureless solidifying metal.

The evidence for bifilms has been with us all for many years.

2.8 The significance of bifilms

Although the whole of this work is given over to the concept of bifilms, so that, naturally, much experimental evidence is presented as a matter of course, this short section lists the compelling logic of the concept.

Since the folded oxides and other films constitute cracks in the liquid, and are known to be of all sizes and shapes, they can become by far the largest defects in the final casting. They can easily be envisaged as reaching from wall to wall of a casting, causing a leakage defect in a casting required to be leak-tight, or causing a major structural weakness in a casting requiring strength or fatigue resistance.

In addition to constituting defects in their own right, if they are given the right conditions during the cooling of the casting, the loosely encapsulated gas film can act as an excellent initiation site for the subsequent growth of gas bubbles, shrinkage cavities, hot tears, cold cracks, etc. The nucleation and growth of such consequential damage will be considered in later sections.

Entrainment creates bifilms that:

1. may never come together properly and so constitute air bubbles immediately;
2. alternatively, they may be opened (to become thin cracks, or opened so far as to become bubbles) by a number of mechanisms:
   (a) precipitation of gas from solution creating gas porosity;
   (b) hydrostatic strain, creating shrinkage porosity;
   (c) uniaxial strain, creating hot tears or cold cracks;
   (d) in-service stress, causing failure in service.

Thus bifilms can be seen to simplify and rationalize the main features of the problems of castings. For those who wish to see the logic laid out formally

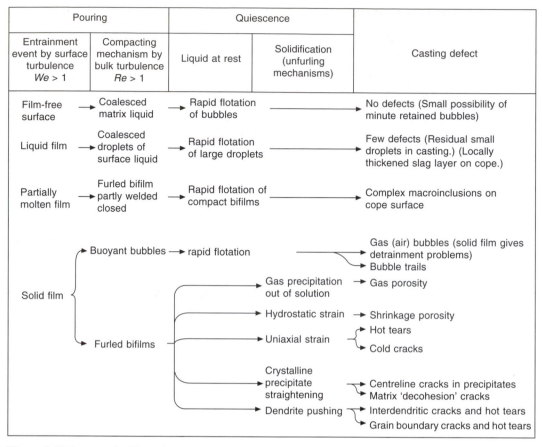

Pouring		Quiescence		
Entrainment event by surface turbulence $We > 1$	Compacting mechanism by bulk turbulence $Re > 1$	Liquid at rest	Solidification (unfurling mechanisms)	Casting defect
Film-free surface	→ Coalesced matrix liquid	→ Rapid flotation of bubbles		No defects (Small possibility of minute retained bubbles)
Liquid film	→ Coalesced droplets of surface liquid	→ Rapid flotation of large droplets		Few defects (Residual small droplets in casting.) (Locally thickened slag layer on cope.)
Partially molten film	→ Furled bifilm partly welded closed	→ Rapid flotation of compact bifilms		Complex macroinclusions on cope surface
Solid film	Buoyant bubbles → rapid flotation			Gas (air) bubbles (solid film gives detrainment problems) Bubble trails
	Furled bifilms		Gas precipitation out of solution	→ Gas porosity
			Hydrostatic strain	→ Shrinkage porosity
			Uniaxial strain	Hot tears Cold cracks
			Crystalline precipitate straightening	Centreline cracks in precipitates Matrix 'decohesion' cracks
			Dendrite pushing	Interdendritic cracks and hot tears Grain boundary cracks and hot tears

Figure 2.47 *Framework of logic linking surface conditions, flow and solidification conditions to final defects.*

this is done in Figure 2.47 for metals (i) without films, such as gold, (ii) with films that are liquid, (iii) with films that are partially liquid, and (iv) with films that are fully solid.

Note that the defects on the right of Figure 2.47 cannot, in general, be generated without starting from the bifilm defect on the left. The necessity for the bifilm initiator follows from the near impossibility of generating volume defects by other mechanisms in liquid metals, as will be discussed in sections 6.1 and 7.2. The classical approach using nucleation theory predicts that nucleation of any type (homogeneous or heterogeneous) is almost certainly impossible. Only surface-initiated porosity appears to be possible without the action of bifilms. In contrast to the difficulty of homogeneous or heterogeneous nucleation of defects, the initiation of defects by the simple mechanical action of the opening of bifilms requires nearly zero driving force; it is so easy that in all practical situations it is the only initiating mechanism to be expected.

We are therefore forced to the fascinating and enormously significant conclusion that in the absence of bifilms castings cannot generate defects

that reduce strength or ductility.

(This hugely important fact has to be tempered only very slightly, since porosity can also be generated easily by surface initiation if a moderate pressurization of the interior of the casting is not provided by adequate feeders. However, of course, adequate feeding of the casting is normally accepted as a necessary condition for soundness. This is the one technique that is widely applied, and we can therefore assume its application here.)

The author has the pleasant memories of the early days (circa 1980) of the development of the Cosworth process, when the melt in the holding furnace had the benefit of days to settle since production at that time did not occupy more than a few shifts per week. The melt was therefore unusually free from oxides, and the castings were found to be completely free from porosity. As the production rate increased during the early years the settling time was progressively reduced to only a few hours, causing a disappointing reappearance of microporosity. This link between melt cleanness and freedom from porosity is well known. One of the first demonstrations of this fact was the simple

Table 2.2 Possible bifilm defects in different alloy systems

Alloy type	Possible defect type
Al-Si alloys	Centreline and matrix decohesion cracks in plate-like intermetallics (Si particles, Fe-rich precipitates, etc.) Planar hot tears with dendrite raft morphology of fracture surface.
Flake cast irons	Nitrogen fissures
Ductile irons	Plate fracture (spiking) defect
Steels	Rock-candy fractures on AlN at grain boundaries
	Type II sulphide phenomena
Ni-base superalloys (vacuum cast)	Intergranular facets on fracture surface Initiation of stray grains and high-angle grain boundaries in single crystals

N.B. The causes of defects in the cases of the higher temperature alloys, irons, steels and Ni-based alloys are based only on circumstantial (although strong) evidence at the time of writing.

and classic experiment by Brondyke and Hess (1964) that showed that filtered metal exhibited reduced porosity.

An important point to note is that the subsequently generated defect, which may be large in extent, may be simply initiated by and grow from a small bifilm. On the other hand, the bifilm itself may be large, so that any consequential defect such as a pore or a hot tear actually *is* the bifilm, but simply opened up. In the latter case no growth of area of the subsequent defect is involved, only separation of the two halves of the bifilm. Both situations seem possible in castings.

Standing back for a moment to view the larger scene of the commercial supply of castings, it is particularly sobering that there is a proliferation of standards and procedures throughout the world to control the observable defects such as gas porosity and shrinkage porosity in castings. Although once widely known as 'Quality Control' (QC) the practice is now more accurately named 'Quality Assurance' (QA). However, as we have seen, the observable porosity and shrinkage defects are often negligible compared to the likely presence of bifilms, which are difficult, if not impossible, to detect with any degree of reliability. They are likely to be more numerous, more extensive in size, and have more serious consequences.

The significance of bifilms is clear and worth repeating. They are often not visible to normal detection techniques, but can be more important than observable defects. They are often so numerous and/or so large that they can control the properties of castings, sometimes outweighing the effects of alloying and heat treatment.

The conclusion is inescapable: it is more important to specify and control the casting process to avoid the formation of bifilms than to employ apparently rigorous QA procedures, searching retrospectively (and possibly without success) for any defects they may or may not have caused.

Chapter 3

Flow

Getting the liquid metal out of the crucible or melting furnace and into the mould is a critical step when making a casting: it is likely that most casting scrap arises during the few seconds of pouring of the casting.

The series of funnels, pipes and channels to guide the metal from the ladle into the mould constitutes our liquid metal plumbing, and is known as the running system. Its design is crucial; so crucial, that this important topic requires treatment at length. This is promised in *Castings II – Practice*. This second volume will describe the practical aspects of making castings. It will be required reading for all casting engineers.

Although volume II is not yet written, we shall nevertheless assume that the reader has read and learned *Castings Practice* from cover to cover. As a result, the reader will have successfully introduced the melt into the running system, so that the system is now nicely primed, having excluded all the air, allowing the melt to arrive at the gate, ready to burst into the mould cavity.

The question now is, 'Will the metal fill the mould?'

Immediately after the pouring of a new casting, colleagues, sceptics and hopefuls assemble around the mould to see the mould opened for the first time. There is often a hush of expectation amid the foundry din. The casting engineer who designed the filling system, and the pourer, are both present. They are about to have their expertise subjected to the ultimate acid test. There is a question asked every time, reflecting the general feeling of concern, and asked for the benefit of defusing any high expectations and preparing for the worst. 'Is it all there?'

This is the aspect of flow dealt with in this section. The nature of the flow is influenced once again by surface films, both those on the surface and those entrained, and by the rate of heat flow and the metallurgy of solidification. In different ways these factors all limit the distance to which the metal can progress without freezing. We shall examine them in turn.

Careful application of casting science should allow us now to know that not only will the casting be all there, but it will be all right.

3.1 Effect of surface films on filling

3.1.1 Effective surface tension

When the surface of the liquid is covered with a film, especially a strong solid film, what has happened to the concept of a surface tension of the liquid?

It is true that when the surface is at rest the whole surface is covered by the film, and any tension applied to the surface will be borne by the surface film (not the surface of the liquid. Actually, there will be a small contribution towards the bearing of the tension in the surface by the effect of the interfacial tension between the liquid and the film, but this can probably be neglected for most practical purposes.) This is a common situation for the melt when it is arrested by capillary repulsion at the entrance to a narrow section. Once stopped, the surface film will thicken, growing into a mechanical barrier holding back the liquid.

This situation is commonly observed when multiple ingates are provided from a runner into a variety of sections, as in some designs of fluidity test. The melt fills the runner, and is arrested at the entrances to the narrower sections, the main liquid supply diverting to fill the thicker sections that do not present any significant capillary repulsion. During this period, the melt grows a stronger film on the thinner sections, with the result that when the heavier sections are filled, and the runner pressurized, the thin sections require an additional tension in their surfaces to overcome the tensile strength of the film before the metal can burst through. For this reason fluidity tests with multiple sections from a single runner are always found to give an effective surface tension typical of a stationary surface, being two or three times greater than the surface tension of the liquid. Results of such tests are described in section 3.3.4.

Turning now to the dynamic situation where the front of the melt is moving, new surface is continuously being created as the old surface is pinned against the mould wall by friction, becoming the outer skin of the casting (as in an unzipping type of propagation as described below). The film on the advancing surface continuously splits, and is continuously replaced. Thus any tension in the surface of the melt will now be supported by a strong chain (the surface film) but with weak links (the fresh liquid metal) in series. The expansion of the surface is therefore controlled by the weak link, the surface tension of the liquid, in this instance. The strong solid film merely rides as pieces of loose floating debris on the surface. Thus normal surface tension applies in the case of a dynamically expanding surface, as applies, for instance, to the front of an advancing liquid.

During the turbulent filling of a casting the dynamic surface tension is the one that is applicable, since a new casting surface is being created with great rapidity. It is clear that the critical velocities for liquid metals calculated using the dynamic surface tension actually agree accurately with experimental determinations, lending confidence to the use of surface tension of the liquid for expanding liquid surfaces.

3.1.2 The rolling wave

Lap type defects are rather commonly observed on castings that have been filled slowly (Figure 2.25). It was expected therefore that a lap type defect would be caused by the melt rolling over the horizontal, oxidized liquid surface, creating an extensive horizontal double film defect (Figure 3.1b). Interestingly, an experiment set up to investigate the effect (Evans *et al.* 1997) proved the expectation wrong.

As a background to the thinking behind this search, notice the difference between the target of the work and various similar defects. The authors were not looking for (i) a cold lap, otherwise known as a cold shut, since no freezing had necessarily occurred. They were not searching for (ii) a randomly incorporated film as generated by surface turbulence, nor (iii) a rolling backwards wave seen in runners, where the tumbling of the melt over a fast underjet causes much turbulent entrainment of air and oxides.

This was a careful study of several aluminium alloys, over a wide range of filling speeds. It seems conclusive that a rolling surface wave to cause an oxide lap does not exist in most situations of interest to the casting engineer. Although Loper and Newby (1994) do appear to claim that they observe a rolling wave in their experiments on steel the description of their work is not clear on this point. It does seem that they observed unzipping waves (see below). A repeat of this work would be useful.

The absence of the rolling wave at the melt surface of aluminium alloys is strong evidence that the kind of laps shown in Figure 2.25 must be cold laps (the old name 'cold shut' is an unhelpful piece of jargon, and is not recommended). Rolling waves that form cold laps in aluminium alloy castings can probably only form when the metal surface has developed sufficient strength by solidification to support the weight of the wave. Whether this is a general rule for all cast metals is not yet clear. It does seem to be true for steels, and possibly aluminium alloys, continuously cast into direct chill moulds as described in the following section.

3.1.3 The unzipping wave

Continuing our review of the experiment by Evans *et al.* (1997) to investigate surface waves, as the meniscus slowed on approach towards the top of the mould, an unexpected discontinuous filling behaviour was recorded. The front was observed to be generally horizontal and stationary, and its upward advance occurred by the propagation of a transverse wave that started at the up-runner, and propagated across the width of the plate (Figure 3.1) until reaching the most distant point. The speed of propagation of the waves was of the order of $100 \, \text{mms}^{-1}$. Reflecting waves were observed to bounce back from the end wall. Waves coming and going appeared to cross without difficulty, simply adding their height as they passed.

What was unexpected was the character of the waves. Instead of breaking through and rolling over the top of the surface, the wave broke through from underneath, and propagated by splitting the surface oxide as though opening a zip (Figure 3.1c).

The propagation of these meniscus-unzipping waves was observed to be the origin of faint lines on the surface of the casting that indicated the level of the meniscus from time to time during the filling process. They probably occurred by the transverse wave causing the thickened oxide on the meniscus to be split, and subsequently displaced to lie flat against the surface of the casting. The overlapping and tangling of these striations appeared to be the result of the interference between waves and out-of-phase reflections of earlier waves.

The surface markings are, in general, quite clear to the unaided eye, but are too faint to be captured

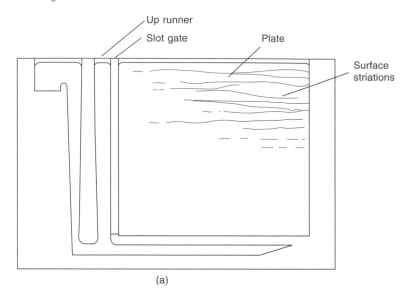

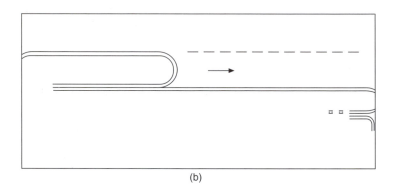

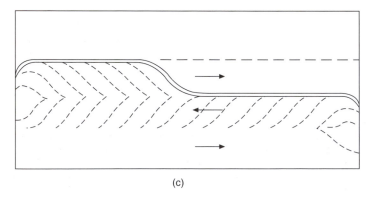

Figure 3.1 *(a) Side gated plate casting used to explore transverse wave effects (Evans et al. 1997); (b) rolling wave that may only occur on a partially frozen surface; and (c) an unzipping wave that may leave no internal defect, only a feint superficial witness on the casting surface.*

on a photograph. A general impression is given by the sketch in Figure 3.1. The first appearance of the striations seems to occur when the velocity of rise of the advancing meniscus in liquid 99.8 per cent purity Al falls to 60 ± 20 mm/s or below. At

that speed the advance of the liquid changes from being smooth and steady to an unstable discontinuous mode. Most of the surface of the melt is pinned in place by its surface oxide, and its vertical progress occurs only by the passage of successive horizontal transverse waves. At the wave front the surface oxide on the meniscus is split, being opened out and laid against the surface of the casting, where it is faintly visible as a transverse striation.

Since this early work, the author has seen the unzipping wave travelling in a constant direction around the circumference of a cylindrical feeder, spiralling its way to the top. Even more recently, the surface of continuously cast cylindrical ingots of 300 mm diameter have been observed to be covered with spirals. Some of these are grouped, showing that there were several waves travelling helically around the circumference, leaving the trace of a 'multi-start thread'. Clearly, different alloys produced different numbers of waves, indicating a different strength of the oxide film. The cylindrical geometry represents an ideal way of studying the character of the wave in different alloys. Such work has yet to be carried out.

In the meantime, it is to be noted that there are a great many experimental and theoretical studies of the meniscus marks on steels. Particularly fascinating are the historical observations by Thornton (1956). He records the high luminosity surface oxide promoting a jerky motion to the meniscus, and the radiant heat of the melt causing the boiling of volatiles from the mould dressing, creating a wind that seemed to blow the oxide away from the mould interface. The oxide and the interfacial boiling are also noted by Loper and Newby (1994).

Much more work has been carried out recently on the surface ripples on continuously cast steels. Here the surface striations are not merely superficial. They often take the form of cracks, and have to be removed by scalping the ingot before any working operations can be carried out. It seems that in at least some cases, as a result of the presence of the direct chill mould, the meniscus does freeze, promoting a rolling wave, and a rolled-in double oxide crack. Thus the defect is a special kind of cold lap.

An example is presented from some microalloyed steels that are continuously cast. Cracking during the straightening of the cast strand has been observed by Mintz and co-workers (1991). The straightening process occurs in the temperature range 1000 down to 700°C, which coincides with a ductility minimum in laboratory hot tensile tests. The crack appears to initiate from the oscillation marks on the surface of the casting, and extends along the grain boundaries of the prior austenite to a distance of at least 5 to 8 mm beneath the surface. The entrainment event in this case is the rolling over of the (lightly oxidized) meniscus on to the heavily oxidized and probably partly solidified meniscus in contact with the mould. Evidence that entrainment occurs at this point is provided by the inclusion of traces of mould powder in the crack. The problem is most noticeable with microalloyed grades containing niobium. The fracture surfaces of laboratory samples of this material are found to be faceted by grain boundaries, and often contain mixtures of AlN, NbN and sulphides.

Other typical early researches are those by Tomono (1983) and Saucedo (1983). The problems of the solidifying meniscus are considered by Takeuchi and Brimacombe (1984). Later work is typified by that of Thomas (1995) who has considered the complexities introduced by the addition of flux powder (which, when molten, acts as a lubricant) to the mould, and the effects of the thermal and mechanical distortion of the solidified shell.

All this makes for considerable complication. However, the role of the non-metallic surface film on the metal being entrained to form a crack seems in general to have been overlooked as a potential key defect-forming mechanism. In addition, the liquid steel melt in the mould cannot be seen under its cover of molten flux, so that any wave travelling around the inside of the mould, if present, is perfectly concealed. It may be that we are still some way from fully understanding the surface features of continuously cast steels.

3.2 Effect of entrained films on filling

It was found by Groteke (1985) that a filtered aluminium alloy was 20 per cent more fluid than a 'dirty' melt. This finding was confirmed by Adufeye and Green (1997) who added a ceramic foam filter to the filling system of their fluidity mould. They expected to find a reduction in fluidity because of the added resistance of the filter. To their surprise, they observed a 20 per cent increase of fluidity.

It seems, therefore, in line with common sense, that the presence of solids in the liquid causes filling problems for the casting. In particular, the entry of metal from thick sections into thin sections would be expected to require good-quality liquid metal, because of the possible accumulation and blocking action of the solids, particularly films, at the entry into the thin section. Thin sections attached to thick, and filled from the thick section, are in danger of breaking off rather easily as a result of layers of films across the junction. Metal flash on castings will often break off cleanly if there is a high concentration of bifilms in a melt. Flash adhering in a ductile way might therefore be a useful quick test of the cleanness and resulting properties of the alloy. As an aside, an abrupt change of section, of the order of 10 mm down to 1 or 2 mm, is such an efficient film concentrating location that it is a good way to locate and research the structure of films by simply breaking off the thin section.

It is instructive to compare the fluidity observed for semi-solid (strictly 'partly solid') and metal matrix composite (MMC) cast materials. Such materials have a viscosity that may be up to 10 to 50 times that of the parent liquid. However, their fluidity as measured by the spiral or other fluidity test is hardly impaired. In fact, quite extensive and relatively thin-walled castings can be poured by gravity without too much difficulty. This result should really not be so surprising or counter-intuitive. This is because semi-solid alloys and MMCs (i) contain a dispersion of uniformly small solid phases, and (ii) viscosity *per se* does not even appear as a factor in the simple equations for fluidity. Thus we would not expect the fluidity of MMCs to be significantly reduced because of their high viscosity.

In contrast, dirty alloys suffer from a wide dispersion of sizes of inclusions. It is almost certainly the large films that are effective in bridging, and therefore blocking, the entrances to narrow sections.

In support of this conclusion, it is known from experience, if not from controlled experiment, that melts of the common aluminium alloy Al–7Si–0.4 Mg that have been subjected to treatments that are expected to increase the oxide content have suffered reduced fluidity as a result. The treatments include:

1. Repeated remelting.
2. High content of foundry returns (especially sand cast materials) in the charge.
3. The pouring of excess metal from casting ladles back into the bulk metal in the casting furnace, particularly if the fall height is allowed to be great.
4. Recycling in pumped systems where the returning melt has been allowed to fall back into the bulk melt (recirculating pumped systems that operate entirely submerged probably do not suffer to nearly the same extent, providing the air is not entrained from the surface).
5. Degassing with nitrogen from a submerged lance for an extended period of time (for instance, a 1000 kg furnace subjected to degassing for several days.

These treatments, if carried out to excess, can cause the melt to become so full of dispersed films that the liquid assumes the consistency of a concrete slurry. As a consequence the fluidity has been low. More serious still, the mechanical properties, particularly the ductility, of the resulting castings has been nothing short of catastrophic.

Treatments of the melts to reduce their oxide content by flushing with argon, or treatments with powdered fluxes introduced in a carrier gas via an immersed lance, are reported to return the situation to normality. Much work is required in this important area to achieve a proper understanding of these problems.

It is as well to keep in mind that a wide spectrum of sizes and thicknesses of oxide film probably exist. With the benefit of the hindsight provided by much intervening research, a speculative guess dating from the first edition of *Castings* (1991) is seen to be a reasonable description of the real situation (Table 3.1). The only modification to the general picture is the realization that the very new films can be even thinner than was first thought. Thicknesses of only 20 nm seem common for films in many aluminium alloys. The names 'new' and 'old' are a rough and ready attempt to distinguish the types of film, and have proved to be a useful way to categorize the two main types of film.

3.3 Fluidity (maximum fluidity length) L_f

The ability of the molten metal to continue to flow while it continues to lose temperature and even while it is starting to solidify is a valuable feature of the casting process. There has been much research into this property, with the result that a quantitative concept has evolved, which has been called '*fluidity*'. In terms of casting alloys, the fluidity, denoted here as L_f, is defined as the maximum distance to which the metal will flow in a standard mould. Thus fluidity is simply a length, measured, for instance, in millimetres. (The use of the foundry term *fluidity* should not be confused with its use in physics, where fluidity is defined as the reciprocal of viscosity.)

A second valuable quantitative concept that has often been overlooked was introduced by Feliu (1962). It is the parameter L_c, which Feliu called the critical length. Here, however, we shall use the name continuous-fluidity length, to emphasize its relation to L_f, to which we could similarly give the

Table 3.1 Forms of oxide in liquid aluminium alloys

Growth time	Thickness	Type	Description	Possible source
0.01–1 s	1 nm–1 μm	New	Confetti-like fragments	Pour and mould fill
10 s to 1 min	10 μm	Old 1	Flexible, extensive films	Transfer ladles
10 min to 1 hr	100 μm	Old 2	Thicker films, less flexible	Melting furnace
10 hr to 10 days	1000 μm	Old 3	Rigid lumps and plates	Holding furnace

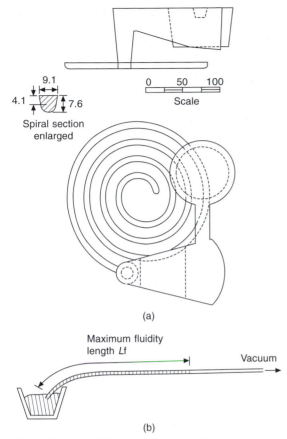

9.1

4.1   7.6

Spiral section
enlarged

0   50   100

Scale

(a)

Maximum fluidity
length *Lf*

Vacuum

(b)

Figure 3.2 *Typical fluidity tests for (a) foundry and (b) laboratory use.*

full descriptive name of maximum-fluidity length (although, in common with general usage, we shall keep the name 'fluidity' for short!). In this section we shall confine our attention to L_f, dealing with the equally important L_c in a following section.

Although fluidity (actually maximum fluidity, remember!) has therefore been measured as the (maximum) length to which the metal will flow in a long horizontal channel, this type of mould is inconveniently long for regular use in the foundry. If the channel is wound into a spiral, then the mould becomes compact and convenient, and less sensitive to levelling errors. Small pips on the pattern at regular spacings of approximately 50 mm along the centreline of the channel assist in the measurement of length. Figure 3.2 shows a typical fluidity spiral. Also shown is a horizontal channel that has been used for laboratory investigations into the fluidity of metals in narrow glass tubes. In this case the transparent mould allows the progress of the flow to be followed from start to finish.

There has been much work carried out over decades on the fluidity of a variety of casting alloys in various types of fluidity test moulds, mostly of

the spiral type. Fundamental insights have been gained mainly from the great bulk of work carried out at MIT under the direction of Merton C. Flemings. He has given useful reviews of his work several times, notably in 1964, 1974 and 1987.

The work of the many different experimenters in this field is difficult to review in any comprehensive way because almost each new worker has in turn introduced some new variant of the spiral test. Other workers have gone further, introducing completely new tests with the emphasis on the casting of thin sections.

Here we attempt to review the data once again, with the aim of emphasizing the unity of the subject because of the basic, common underlying concepts. The various fluidity tests of the spiral, tube, or strip types are shown not to be in opposition. On the contrary, if proper allowance is made for the surface tension and modulus effects, these very different tests are shown to give exactly equivalent information.

It will become clear that the fluidity of metals is mainly controlled by the effects of fluid mechanics, solidification and surface tension. This is a reassuring demonstration of good science.

3.3.1 Mode of solidification

Flemings (1974) demonstrated that the fluidity of pure metals and eutectics that freeze at a single temperature is different to that of alloys that freeze over a range of temperatures. These two different solidification types we shall call skin freezing and pasty freezing for short.

For a skin freezing material the mode of solidification of the stream in a fluidity test appears to be, as one might expect, by planar front solidification from the walls of the mould towards the centre (Figure 3.3). The freezing occurs at some point along the length of the channel, after the

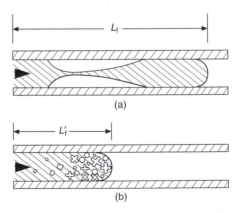

Figure 3.3 *Flow arrest: (a) in pure metals by complete solidification; and (b) in long-freezing-range alloys by partial solidification.*

metal has lost its initial superheat (the excess temperature above its melting point). The solidified region actually migrates downstream somewhat as the leading edge of the material is remelted by the incoming hot metal, and refreezing occurs further downstream. We shall return to this phenomenon in a later section. However, neglecting these details just for the moment, it is clear that the stream can continue to flow until the moment at which the freezing fronts meet, closing the flow channel. Note that this choking of flow happens far back from the flow front. In addition, solidification needs to be 100 per cent complete at this location for flow to stop. Assuming the liquid has an approximately constant velocity V, then if its freezing time in that section is t_f, we have the flow distance:

$$L_f = V \cdot t_f \qquad (3.1)$$

This is an approximate equation first proposed by Flemings and co-workers (1963). It overestimates L_f because the velocity does reduce somewhat as the channel becomes constricted, giving a smaller measured fluidity. Nevertheless Equation 3.1 is good enough for many purposes, and we shall assume its applicability here.

The pattern of solidification of this short freezing range material can be confirmed from the evidence of polished cross-sections of the castings. Columnar crystals are seen to grow from the sides of the mould, angled inwards against the flow, and meeting in the middle of the section. Confirmation of the presence of liquid cut off by the meeting of the solidification fronts is clear from the shrinkage pipe that forms at the liquid front, and grows back towards the point at which the flow is blocked.

The pattern of solidification of longer freezing range materials can be inferred from metallography in a similar way. It turns out to be quite different to that of the skin freezing material described above. The dendrites growing from the mould wall at an early stage of freezing are fragmented by the flow. The stream therefore develops as a slurry of tumbling dendritic crystals. They are carried in the central flow too close to the liquid front, where, when the amount of solid in suspension exceeds a critical percentage, the dendrites start to interlock, making the mixture unflowable. There is some evidence that this critical concentration depends somewhat on the applied head of metal (Flemings *et al.* 1961). However, for most practical situations this appears to lie between approximately 20 and 50 per cent. The arrest of flow occurs therefore when there is only approximately 20 to 50 per cent solidification. Thus for long freezing range alloys Equation 3.1 becomes

$$L_f = 0.2V \cdot t_f \quad \text{to} \quad 0.5V \cdot t_f \qquad (3.2)$$

This factor of roughly 2 to 5 between the fluidities

of short and long freezing range alloys seems to be a common feature of all alloy systems, and yet does not seem to have been widely recognized. For instance, Figure 3.4 illustrates the profound effect of a small amount of Sn on pure Al. A few mass per cent of Sn reduces the fluidity of Al by a factor of 3 or 4.

A similar effect is seen in Figure 3.5 illustrating the lead/tin system at 50°C above its liquidus. The curve joining the data points is based on the fluidity of the long freezing range alloys being approximately one-fifth of the short freezing range alloys when lead-rich, from the available data points. There are no equivalent data points on the tin-rich side of the diagram, so here we have guessed that the longer freezing range alloys might have only approximately half fluidity, because the freezing range is narrower, and the one data point is consistent with this. More experimental points would have removed this uncertainty.

The fluidity of the eutectic in Figure 3.5 appears to be over 50 per cent higher than the straightforward method of mixtures of its components (i.e. the straight line joining the fluidities of the two pure metals). This may be understandable in terms of the sum of two effects. (i) The pure metals may be exhibiting some dendritic growth, probably due to the presence of some impurity. This would suppress the fluidity of the pure constituents, but not necessarily the eutectic (especially if the 'impurities' consisted of its pure constituents). (ii) The determination of fluidity of alloy at a constant

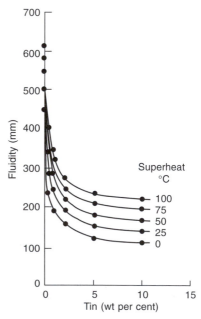

Figure 3.4 *Variation of fluidity with composition of Al–Sn alloys. Data from Feliu et al. (1960).*

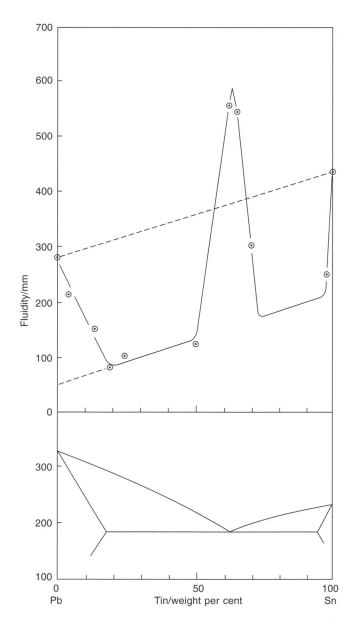

Figure 3.5 *Fluidity of lead–tin alloys at 50°C superheat determined in the glass tube fluidity test. Original data from Ragone et al. (1956).*

temperature automatically increases the fluidity of the eutectic because of its low melting point. The effect is clearly seen in the Al–Zn system (Figure 3.6). This issue will be considered in more detail later.

Portevin and Bastien (1934) suggest from their results that the shape of precipitating solid crystals affects the flow of the remaining liquid; smooth crystals of solidifying intermetallic compounds creating less friction than dendritic crystals. Their famous result for the Sb–Cd system poured at a constant superheat of 100°C as shown in Figure 3.7 illustrates this convincingly. The greater fluidity of intermetallic compounds and eutectics with

respect to their pure constituent elements may be a universal fact. (Once again, as for the Al–Zn system, we shall see more evidence that effective superheat enhances this advantage of eutectics further, although reduces the advantage to intermetallic compounds.)

Incidentally, it is worth dwelling on other details of this classic work. For instance, the peritectic reaction shown in Figure 3.7 at approximately 63 per cent Cd was found not to exist in later work on this binary phase diagram (Brandes and Brook 1992). Thus the small step in fluidity, carefully depicted at this point, seems likely to be attributable to experimental error. Furthermore, at the Sb-rich

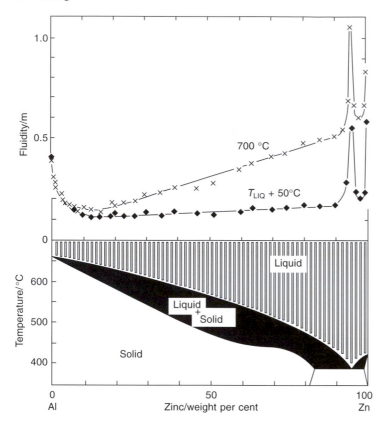

Figure 3.6 *Al–Zn alloys poured into a straight cast iron channel showing fluidity at constant superheat, and the enhanced fluidity of the eutectic when cast at constant temperature (Lang 1972).*

end of the figure, the fall in fluidity will probably be steeper than that shown, so that the plateau minimum will be reached sooner than the limit of solid solution as assumed by the authors. This will be as a result of non-equilibrium freezing. However, these are minor quibbles of an otherwise monumental and enduring piece of work, years ahead of its time.

We can usefully generalize these results in the following way. Figure 3.8 shows a schematic illustration of a simple binary eutectic system. The fluidity at zero superheat is linear with composition for the skin freezing metals and alloys (the pure metals A and B and the eutectic), whereas it is assumed to be about half of this for the long freezing alloys. The resulting relationship of fluidity with composition is therefore seen to be the cusped line denoted (i). This would be the expected form of the fluidity/composition relation for a constant superheat.

If, however, the alloys are cast at a constant temperature, T_c, there is an additional contribution to fluidity from the superheat that now varies with composition as indicated in Figure 3.8 (ii). To find the total fluidity as a function of composition curves (i) and (ii) are added (the addition in these illustrative examples neglects any differences of scale between the two contributions). The effect is to greatly

enhance the fluidity at the eutectic composition.

The same exercise has been carried out for a high melting point intermetallic compound, AB, in Figure 3.9. The disappointingly low fluidity predicted for the intermetallic explains some of the problems that are found in practice with attempts to make shaped castings in such alloys. It explains why intermetallic compounds are rarely used as natural casting alloys.

The expected poor fluidity of intermetallics underlines the suggestion by Portevin and Bastien that the high fluidity of intermetallics that they observe, although at constant superheat as seen in Figure 3.7, is nevertheless in fact strongly influenced by some other factor, such as the shape of the solidifying crystals.

Another interesting and important lesson is to be gained from the predictions illustrated in Figures 3.8 and 3.9. The peaks in fluidity are impressively narrow. Thus for certain eutectic alloys the fluidity is awesomely sensitive to small changes to composition, particularly when it is realized that Figures 3.8 and 3.9 are simple relationships. Many alloy systems are far more complicated, and much steeper cusps of fluidity with compositional changes are to be expected. Dramatic changes of fluidity performance are to be expected with only minute changes in composition. Perhaps this is the reason

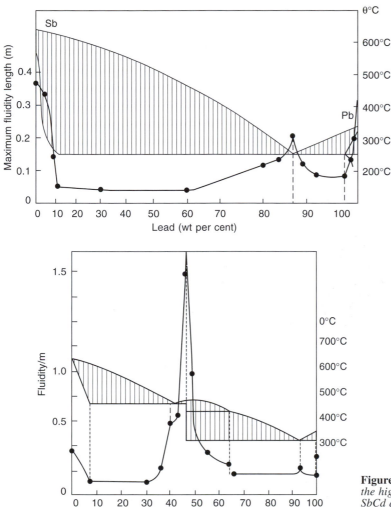

Figure 3.7 *Fluidity of Sb–Cd alloys showing the high fluidity of the intermetallic phase SbCd determined by casting at a constant superheat into a cast iron spiral mould (Portevin and Bastien 1934).*

why some castings can sometimes be filled, and at other times not; some batches of alloy cast well, but some nominally identical alloys cast poorly. Clearly, to maximize reproducibility we must improve the targeting of peak performance. The unattractive alternative is to forsake the peaks and devise more easily reproducible alloys in the troughs of mediocre fluidity. A good compromise would be to target less narrow peaks, where the penalties of missing the peak are less severe.

The approach illustrated in Figures 3.8 and 3.9 can be used on more complex ternary alloy systems. In more complex multi-component alloy systems the understanding that pure metals and eutectics exhibit a fluidity at least twice that of their long freezing range intermediate alloys was the key that allowed even sparse and apparently scattered data on alloy systems to be rationalized. This was the

background of Figure 3.10 (a, b and c) that showed results for the Al–Cu–Si systems that could not be drawn by its investigators, but was only unravelled in a subsequent independent effort (Campbell 1991). However, the author learned only later, while the 1991 paper was in the process of publication, that Portevin and Bastien had already accomplished a similar evaluation of the Sn–Pb–Bi ternary system as early as 1934. In this prophetic work they had constructed a three-dimensional wax model of their fluidity response as a function of composition, showing ridges, peaks and valleys.

In practice, it is important to realize that the improved fluidity of short versus long freezing range materials forms the basis of much foundry technology.

The enormous use of cast iron compared to cast steel is in part a reflection of the fact that cast iron

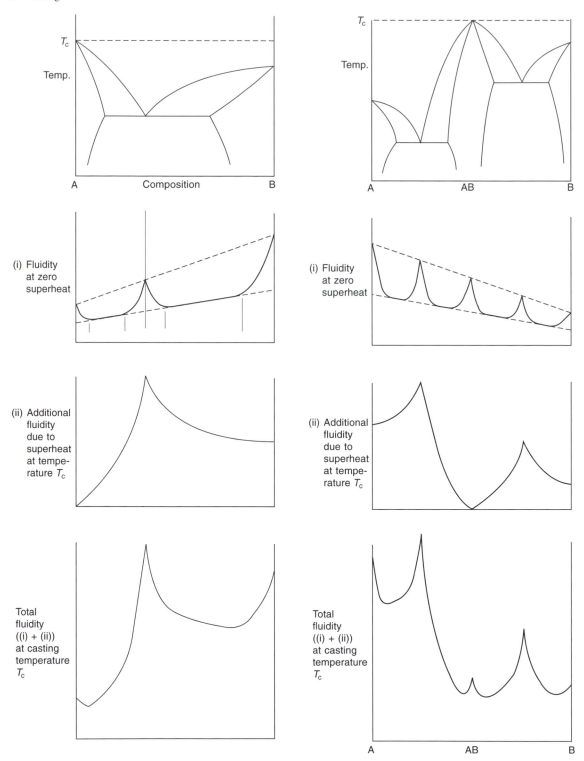

Figure 3.8 *Schematic illustration of the different behaviours of eutectics when tested at constant superheat or constant temperature.*

Figure 3.9 *Schematic illustration of the different behaviours of intermetallics when tested at constant superheat or constant temperature.*

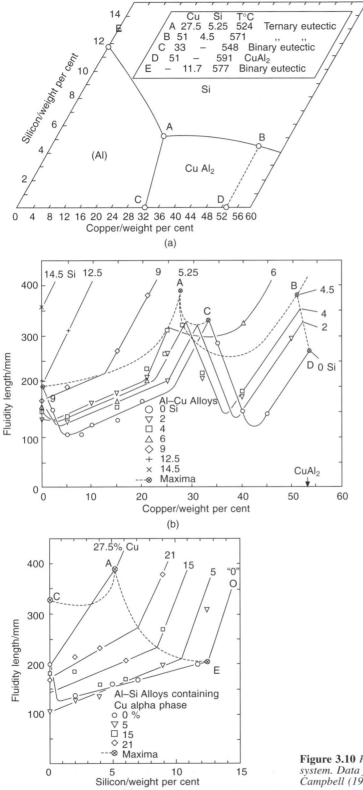

Figure 3.10 *Phase diagram and fluidity of the Al–Si–Cu system. Data from Garbellini et al. (1990); interpretation Campbell (1991).*

is a eutectic or near-eutectic alloy and so has excellent fluidity. Steel, in general, has a longer freezing range and relatively poor fluidity. Additionally, of course, the higher casting temperatures of steel greatly increase the practical problems of melting and casting, and cause the liquid metal to lose its heat at a faster rate than iron, further reducing its fluidity by lowering its effective fluid life t_f.

Fluidity can be affected by changes in composition of the alloy in other ways. For instance, the effect of phosphorus additions to grey iron are well known: the wonderful artistic castings of statues, fountains, railing and gates produced in the nineteenth and early twentieth centuries were made in high-phosphorus iron because of its excellent fluidity. The effect is quantified in Figure 3.11a. The powerful effect of phosphorus on cast iron is solely the result of its action to reduce its freezing point. This is proved by Evans (1951), who found that when plotted as a function of superheat (the casting temperature minus the liquidus), the phosphorus addition hardly affects the fluidity (Figure 3.11b).

Flemings has taken up this point, suggesting that the good fluidity of cast irons compared to steels is only a function of the higher superheat which can be used for cast iron. However, this is only part of the truth. Figure 3.11 shows data replotted from early work by Evans (1951) on grey iron and Andrew et al. (1936) on steels. The reasonably linear plots of fluidity against casting temperature are important and interesting in themselves. However, they can be redrawn as in Figure 3.12. Here the horizontal flat portions of the curves show that the long freezing range alloys have constant fluidity at a given superheat, in agreement with Flemings and confirming the similar effect of phosphorus that we have already witnessed; any increase in fluidity is only the result of increases in superheat as the composition changes. However, as either the pure metal, or the eutectic at 4.3 per cent carbon, is approached there is clear evidence of enhanced flowability, showing that there is an additional effect at work here, almost certainly relating to the mode of solidification as we have discussed above. This effect was found as long ago as 1932 by Berger. The Al–Zn alloy system investigated by Lang in 1972 shows a similar special enhancement of fluidity when cast at constant superheat as seen in Figure 3.6.

The effect of freezing range on fluidity is not confined to metals. Bastien et al. (1962) have shown that the effect is also clearly present in molten nitrate mixtures and in mixtures of organic compounds. It is to be expected that the effect will be significant in other solidifying systems such as water-based solutions and molten ceramics, etc.

3.3.2 Effect of velocity

The velocity is explicit in the equations (3.1 and 3.2) for fluidity. It is all the more surprising therefore that it has not been the subject of greater attention. However, most workers in the field have used fluidity tests with relatively small head heights, usually not exceeding 100 mm, so that velocities during test are usually fixed, and rather modest, under 2 ms^{-1}.

It might seem reasonable to expect that fluidity would increase linearly with increase in velocity. However, this is a mistake. It is this erroneous belief that causes many casting engineers to assume that to fill thin-section casting the melt has to be thrown into the mould at maximum speed. These attempts to improve filling usually fail to produce good castings because of the complicating effects of bulk and surface turbulence.

For instance, as head height h of a casting is increased in a filling system free from friction effects, we might expect that because $V = (2gh)^{1/2}$, fluidity would increase proportionally to $h^{0.5}$. However, the resistance to flow from turbulence in

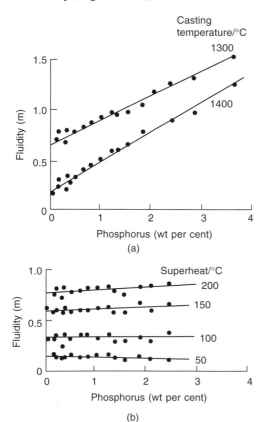

Figure 3.11 *Effect of phosphorus on the fluidity of grey iron plotted (a) as a function of casting temperature, and (b) as superheat above the liquidus (Evans 1951).*

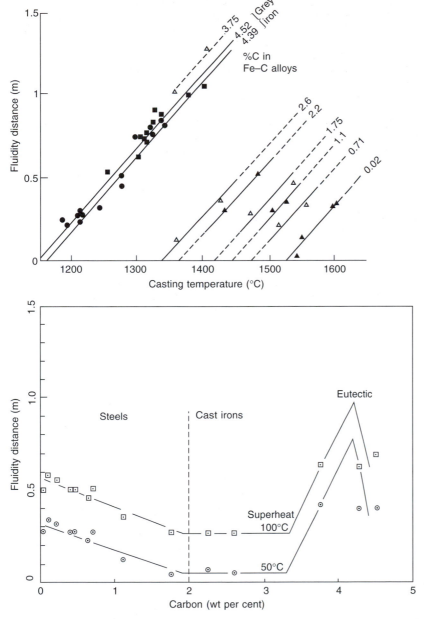

Figure 3.12 *(a) Effect of temperature on the fluidity of Fe–C alloys; (b) the same data plotted as a function of composition. Data for grey iron from Evans (1951), and for Fe–C alloys from Berger (1932) and Andrew et al. (1936).*

the bulk of the liquid rises according to $(1/2)\rho V^2$. Thus losses rise rapidly with increased velocity V, causing it to be increasingly difficult to create high velocities, particularly in narrow channels. In fact Tiryakioglu and co-authors (1993) find experimentally that fluidity of 355 aluminium alloy rises at a much lower rate, that seems proportional to only $h^{0.1}$. This is almost certainly the result of the rapidly increasing drag effects on the motion of the liquid as velocity increases. Thus attempts to increase fluidity by increasing pressurization and/

or velocity are subject to the law of rapidly diminishing returns.

Furthermore, at high speeds, it becomes increasingly difficult to keep surface turbulence under control. Thus the jetting and splashing of the flow entrains air, and creates oxide laps. The oxide laps occur because of the disintegration of the flow front. Splashes or surges of melt ahead of the general liquid front have to await the arrival of the main flow in order to be reassimilated into the casting. During this time their surface oxidizes, so that reassimilation is difficult, or impossible, resulting in an oxide lap. If the arrival of the main flow takes longer, the awaiting material might freeze. Thus the splash and its droplets might become a cold lap.

In contrast, at more modest speeds, if the melt is introduced so that the liquid front remains smooth and intact, the air is not mixed in, but is pushed ahead, and the front has no isolated fragments. The very integrity of the front acts to keep itself warm. Both oxide laps and cold laps are avoided. The casting is as sound as the filling system can make it.

In confirmation of these problems, Zadeh and Campbell (2001) have demonstrated how the fall of melts down sprues 500 mm high have resulted in good fluidity lengths, with apparently good castings judging by their external appearance. On examination by X-ray radiography, however, the castings were found to be full of entrained bubbles and films. It was necessary to introduce bubble traps and ceramic filters to regain control over the integrity of the flow to give good quality castings. Having done this, the fluidity distance was then no better than if the melt had been poured from 100 mm.

Thus, in general, up to small head heights of the order of 100 mm fluidity will be expected to rise linearly with increase in speed of flow. However, further increases in speed with current filling system designs appear to be counterproductive if castings relatively free from defects are required.

In summary of the effects of velocity therefore, it is essential to fill the mould without surface turbulence for the majority of alloy systems. The importance of this fact cannot be overstated. There is no substitute for a good filling system that will give a controlled advance of the liquid front.

3.3.3 Effect of solidification time t_f

From Equation 3.1 it is clear that the greater the solidification time, t_f, the further the metal will run before freezing. This is a far more satisfactory way to improve the filling of castings than increasing the velocity, since the surface turbulence problems can therefore be kept under control.

For conditions in which the mould material controls the rate of heat loss from the casting, such as in sand casting, quoting Flemings (1963), the freezing time is approximately:

$$t_f = \frac{\pi}{16 K_m \rho_m C_m} \left\{ \frac{a \rho_s H}{T_f - T_0} \right\}^2 \qquad \textbf{(3.3)}$$

where symbols are defined at the end of the section.

For conditions of freezing in chill moulds where the interface resistance to heat flow is dominant, we may quote the alternative analytical equation

for the freezing time due to Flemings:

$$t_f = \rho_s HV/\{h(T_m - T_0)A\} \qquad \textbf{(3.4)}$$

The main factors that increase t_f are clear in these equations. The factors are dealt with individually in the following subsections.

3.3.3.1 Modulus

Greaves in 1936 made the far-sighted suggestion that fluidity length was a function of (volume/area) = modulus m. Much later, Feliu (1964) found the modulus squared relation to be approximately true, both for maximum and continuous fluidity results. (The agreement of Feliu's results with the square relation is improved if the effect of surface tension is taken into account as shown later in section 3.4)

The effect of the thickness or shape of the channel on the time of freezing of the metal in the channel is nicely accounted for by Chvorinov's rule (implicit in Equations 3.3 and 3.4). We can define the modulus m as the ratio 'volume/cooling area' of the casting, which is in turn of course equivalent to the ratio 'area/cooling perimeter' of the cross-section of a long channel. Remember that modulus has dimensions of length, and is most conveniently quoted, therefore, in millimetres. Restating Equation 3.3 gives

$$t_f = k_m m^2 \qquad \textbf{(3.5)}$$

The relation applies where the heat flow from the casting is regulated by the thermal conductivity of the mould (as is normal for a chunky casting in a reasonably insulating sand mould). The value of the constant k_m can be determined by experiment in some convenient units, such as s.mm^{-2}. To gain some idea of the range of k_m in normal castings: for steel in green sand moulds it is approximately 0.40; for Al–7Si alloys in dry sands it is 5.8, and for Al–3Cu–5Si it is 11.0 s.mm^{-2}. It is worth noting that Equation 3.5 applies nicely to chunky castings in sand moulds, and other reasonably insulating moulds such as investment moulds.

For the case where interfacial heat transfer dominates heat flow, as in chill moulds, or thin walled sand castings, Equation 3.4 gives

$$t_f = k_i m/h \qquad \textbf{(3.6)}$$

where k_i is a constant, and h is the rate of flow of heat across the air gap for a given temperature difference across the gap, and for a given area of interface. The parameter h is known as the heat-transfer coefficient. We shall see later how h starts to dominate in thin section castings, so that Equation 3.6 is often found to apply in thin section sand castings.

Restating Equation 3.1 in terms of the relations 3.5 and 3.6, we have for sand moulds

$$L_f/m = kVm \qquad (3.7)$$

and for dies

$$L_f/m = k'V/h \qquad (3.8)$$

We now have two simple but powerful formulae to assist in the prediction of whether a mould will fill. Interestingly, the Equation 3.7 for dies (and thin section sand castings) shows that the ratio of fluidity to modulus, L_f/m, is a constant for a given mould material. For instance, in Al alloys cast into sand moulds the *fluidity to section thickness* ratio is around 100 (i.e. a section thickness of 4 mm should allow the metal to flow 400 mm prior to arrest by freezing). This is a really useful relation. For instance, if the ratio of thickness to distance to be run is only 50 the casting is no problem to make. When the ratio is 100 the casting is not to be underestimated; it is near the limits. When the ratio is 150 the casting cannot be made by flowing metal through the mould cavity from a single gate entry point. A completely different filling system is required. This is therefore a useful guide as a quick check which can be made at the drawing stage as to whether a design of casting might be castable or not. We can restate the critical *fluidity to section thickness ratio = 100* in terms of the more universal *fluidity to modulus ratio = 200*, allowing us to transfer this useful ratio to any shape of interest to a fair approximation.

In his experiments on sand moulds, Feliu (1964) appeared to find a relation between fluidity and (modulus)$^{3/2}$ for thin section sand castings in the range 0.7 to 2 mm modulus (section thickness 1.4 to 4 mm). However, he did not allow for the effect of surface tension in reducing the effective head of metal. When this is allowed for, as will be illustrated in the exercise in the following section, it indicates that a linear relation between fluidity and section thickness as suggested by Equation 3.8 actually applies reasonably well. Barlow (1970) confirms that over a range of Pyrex tubes from 1 to 4 mm diameter the fluidity of both Al–10Si and Cu–10Al is closely linear with channel section. It seems therefore that Equation 3.8 has rather general validity. Clearly, interfacially controlled heat flow is perhaps rather more common than is generally thought, especially in thin sections, where the freezing times are very short.

The value of the concept of modulus, m, is that any shape of channel can be understood and compared with any other shape. For instance, straight tubes can be compared with trapezoidal spirals, or with thin flat strips. All that matters for the control of solidification time is the modulus (we are neglecting for the moment the refinements that allow for the slight inaccuracies of the simple modulus approach).

3.3.3.2 Heat transfer coefficient

A reduction in the rate of heat transfer will benefit fluidity (as will become clearer below, and when we look into the relation between the rate of heat transfer and the rate of solidification in Chapter 5). For this reason insulating ceramic coatings are applied to all gravity and low-pressure dies. The work of Rivas and Biloni (1980) illustrates the benefit to fluidity from the application of a white oxide-based coating to a steel die. This is confirmed by Hiratsuka *et al.* (1994) from work in which metal is drawn into tubes of quartz, copper and stainless steel.

For sand moulds, acetylene black is applied from a sooty flame (Flemings *et al.* 1959) giving very substantial increases in fluidity, by a factor of 2 or 3. This dramatic improvement in fluidity is used in some precision sand foundries, allowing thin-walled castings to be filled that could not otherwise be cast. Although attempts have been made to develop other fluidity-enhancing coatings, none has even approached a similar effectiveness. The unfortunate greasy black pollution caused by the soot in the foundry is unwelcome, causing the whole foundry environment to become blackened by soot. However, this severe disadvantage continues to be tolerated while alternatives are unavailable. The action of the carbon deposit was for many years thought to be an insulating effect; the carbon acting to reduce the heat transfer coefficient. This is no longer thought to be true as explained later.

Cupini and Prates (1977) investigated the effect of hexachlorethane mould coatings. The release of chlorine from the breakdown of this chemical was found to refine the grain size of 99.44Al from the range of between 2 and 20 mm down to 0.2 mm. The effect seemed to be the result of the microscopic disturbance of the surface of the casting, where the dendritic grains were starting to grow.

This effect was further investigated by Southin and Romeyn (1980) who found a considerably more complicated behaviour. They found that for aluminium alloys up to 1 per cent Cu the use of carbon black or hexachlorethane mould coatings *reduced* the fluidity because the planar (or cellular) growth front appeared to be stable, so that strong, compact freezing fronts easily stopped the flow at an earlier stage as a result of an *enhanced* thermal transfer of the carbon. At higher copper contents the fluidity was increased because the dendritic front was fragmented, and the dendritic fragments were carried along in the flowing stream, giving a fine grain size and good fluidity. The authors suggested that these results were only explained if (i) the rate of heat transfer was *increased*; and (ii) the coating mechanically destabilized the growing dendrites. This action was thought to occur by the unstable support of the growing dendrites, with the mechanical collapse of the coating as gas was evolved. This would destabilize, recrystallize and release dendrites. The release mechanism is probably that proposed by Vogel *et al.* (1977) in which, after the plastic deformation at the dendrite root, recrystallization occurred, with the result that the newly formed high angle grain boundary was favourably wetted and therefore penetrated by the melt, allowing the dendrite to detach. Thus instead of the dendrites growing into the stream to obstruct its flow, they were released to tumble along with the stream. The increased rate of heat loss, although usually reducing fluidity, was in this case more than countered by the effect of the coating on the grain behaviour.

3.3.3.3 Superheat

Increases in casting temperature benefit fluidity in a direct way, as shown, for example, in Figure 3.12 and many other examples in this book. In fact, all investigations of fluidity have confirmed that the fluidity increases linearly with the superheat (defined here as the excess of casting temperature over liquidus temperature).

The effect of superheat is therefore valuable in itself. In addition, however, it can help us to understand the special fluid properties of eutectics, and the poor flow capabilities of most high melting point intermetallics as explained earlier in this section.

As a reminder, in Figure 3.6 obtained by Lang (1972) for the Al–Zn system, at constant superheat the fluidity of the eutectic is almost exactly that expected from the rule of mixtures, interpolating between the fluidities of the pure elements Al and Zn. (A rule of mixtures would have predicted the fluidity of the pure elements and the eutectic to lie on a straight line.) When determined at a constant temperature of 700°C, however, the eutectic now has the advantage of a large effective superheat, and the fluidity of the eutectic is correspondingly enhanced, becoming significantly higher than that of either Al or Zn as explained in the derivation of Figure 3.8

3.3.3.4 Latent heat H

The latent heat given up on solidification will take time to diffuse away, thereby delaying solidification, and extending fluidity. Much has been made of this point (see, for instance, Arnold *et al.* 1963) in explaining the claims for the good fluidity of the hypereutectic Al–Si alloys since pure silicon has a latent heat of solidification 4.65 times greater than that of aluminium (data taken from Brandes and Brook 1992).

Equation 3.3 is not likely to be particularly accurate. Nevertheless, when used in a comparative

way, it is likely to give somewhat better results. Thus to find the improvement, for instance, in changing from pure Al to pure Si, we can see that the comparative freezing times are simply given by the ratio:

$$\frac{t_{Si}}{t_{Al}} = \left\{ \frac{T_{Al} - T_0}{T_{Si} - T_0} \right\}^2 \left\{ \frac{\rho_{Si}}{\rho_{Al}} \right\}^2 \left\{ \frac{H_{Si}}{H_{Al}} \right\}^2 \quad \textbf{(3.9a)}$$

$$= \{0.460\}^2 \times \{0.867\}^2 \times \{4.65\}^2$$

$$= 0.21 \times 0.75 \times 22 \quad \textbf{(3.9b)}$$

$$= 3.4 \quad \textbf{(3.9c)}$$

Thus from Equation 3.9b, pure Si would have been expected to have 22 times the fluidity of pure aluminium as a result of its higher latent heat. However, its greater rate of heat loss, seen to be nearly 5 times faster as a result of its higher freezing temperature, reduces this significantly. The low density of silicon also reduces the effect somewhat further. The final result, Equation 3.9c, is that pure silicon would be expected to have 3.4 times the fluidity of pure aluminium. (Rather paradoxically, this conclusion will only be accurate at low concentrations of Si in Al because the other terms in Equation 3.3 then are not significantly changed, and thus properly cancel to give Equation 3.9 accurately.)

We can now construct a relation to examine the effect of the latent heat of Si on the fluidity of Al–Si alloys in sand moulds as in Figure 3.13. It is clear that although the upward trend assignable to the latent heat effect is noticeable, the real contribution to the increased fluidity of the so-called hypereutectic alloys is mainly the result of the move of the equilibrium eutectic to higher Si content corresponding to a non-equilibrium eutectic. This conclusion appears to be supported by additional data by Lang (1972) in which he studies the effect of the addition of Na to the alloys. This stabilizes the eutectic and creates clear peaks corresponding to an effective eutectic at higher approximately 14.5 per cent Si. The existence of a maximum in fluidity at about 14.5 per cent Si has been confirmed by a number of researchers, for instance Pan and Hu (1996) and Adefuye (1997). In addition, the presence of a peak can be interpolated in the work of Parland (1987) who missed the peak, but whose data points on either side corroborate its existence. Furthermore, work on the Al–Cu–Si system (Figure 3.10) confirms the non-equilibrium eutectic at 14.5 per cent Si but indicates that the maximum in the fluidity is not a cuspoid peak, but is characterized by only a change in slope of the fluidity response surface, following the theoretical predicted line in Figure 3.13 (Campbell 1991).

The undercooling behaviour of this eutectic, encouraging the simultaneous precipitation of both

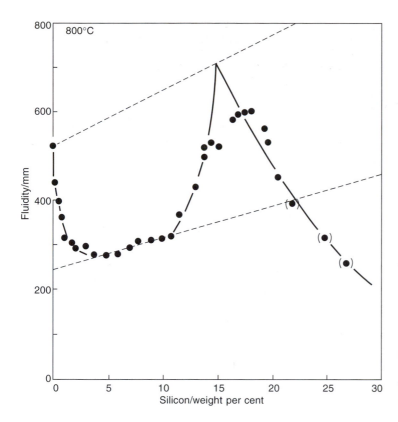

Figure 3.13 *Fluidity of Al–Si alloys showing the theoretical increase in fluidity as Si is increased, as a result of the high latent heat of freezing of Si. Experimental data from Lang (1972).*

aluminium dendrites and primary silicon, are additional complications. Further work is clearly required to clarify the behaviour of this important eutectic system.

The use of the Al–17Si alloy, with phosphorus additions to aid the nucleation of primary silicon particles, is well known for its use on automotive engine blocks, even though less use has been made of the alloy for blocks than it appears to deserve. It has, however, found many other uses for which it was never designed, and for which it is not perhaps really optimum, for instance for gearbox housings and other castings as a probable result of its availability. The more rational use of alloys would be welcome.

It is a further curious illogicality that most of the common Al–Si casting alloys are in the range of 4 to 10 per cent Si, and are therefore concentrated in the region of mediocre fluidity properties. Contrary to the claims of practically every writer on this subject, it is quite clear that the silicon *reduces* the fluidity at these compositions. (Almost certainly a confusion has arisen with the well-known, but less easily defined concept of *castability*. The improved castability of Al–Si alloys is probably a result of the benefits of their greater quantity of eutectic, and possibly a more benign oxide film, compared to many other Al alloy systems.)

The real benefit of silicon in relation to fluidity

is only seen nearer to 14.5 per cent Si, corresponding, probably, to the composition of the non-equilibrium (i.e. the 'real') eutectic. It is possible to modify these so-called hypereutectic alloys with sodium or strontium to stabilize the eutectic form of the silicon (avoiding the usual addition of phosphorus, and so avoiding the usual chunky primary silicon particles that give difficulties for machining). However, these compositions have not yet been promoted by suppliers nor have been taken up by the users (Smith 1981). Casting compositions with promise of excellent fluidity, mechanical properties and machinability remain to be developed. They promise exciting potential.

3.3.3.5 Mould temperature

For most casting processes mould temperature is fixed at or close to room temperature. For such processes, there is little or nothing that can be done to mould temperature to affect fluidity.

The intermediate temperatures of gravity and high-pressure dies, usually at around 300 to 400°C, do contribute modestly to the increase of fluidity in these casting processes. Die temperatures are sometimes raised a little to gain a little extra filling capability. However, the natural working temperature of a die is only changed by a significant, and therefore unpopular, technical effort.

However, for investment casting the ceramic shell allows a complete range of temperatures to be chosen without difficulty. From Equation 3.3 it is seen that the freezing time is proportional to the difference between the freezing point of the melt and the temperature of the mould. The few tests of this prediction are reasonably well confirmed (for instance, Campbell and Olliff 1971).

One important prediction is that when the mould temperature is raised to the melting point of the alloy, the fluidity becomes infinite; i.e. the melt will run for ever! Actually, of course, this self-evident conclusion needs to be tempered by the realization that the melt will run until stopped by some other force, such as gravity, surface tension or the mould wall! All this corresponds to common sense. Even so, this elimination of fluidity limitations is an important feature widely used in the casting of thin-walled aluminium alloy investment castings, where it is easy to cast into moulds held at temperatures in excess of the freezing point of the alloys at approximately 600°C. Single crystal turbine blades in nickel-based alloys are also cast into moulds heated to 1450°C or more, again well above the freezing point of the alloy.

Any problems of fluidity are thereby avoided. Having this one concern removed, the founder is then left with only the dozens of additional important factors that are specified for the casting. Solving one problem completely is a help, but still leaves plenty of challenges for the casting engineer!

3.3.4 Effect of surface tension

If metals wetted the moulds into which they were cast, then the metal would be drawn into the mould by the familiar action of capillary attraction, as water wets and thus climbs up a narrow bore glass tube.

In general, however, metals do not wet moulds. In fact mould coatings and release agents are designed to resist wetting. Thus the curvature of the meniscus at the liquid metal front leads to capillary repulsion; the metal experiences a back pressure resisting entry into the mould. The back pressure due to surface tension, P_{ST}, can be quantified by the simple relation, where r and R are the two orthogonal radii which characterize the local shape of the surface, and γ is the surface tension:

$$P_{ST} = 2\gamma\{(1/r) + (1/R)\} \qquad (3.11)$$

When the two radii are equal, $R = r$, as when the metal is in a cylindrical tube, then the liquid meniscus takes on the shape of a sphere, and Equation 3.5 takes on the familiar form:

$$P_{ST} = 2\gamma/r \qquad (3.12)$$

Alternatively, if the melt is filling a thin, wide strip,

so that R is large compared with r, then $1/R$ becomes negligible and back pressure becomes dominated by only one radius of curvature, since the liquid meniscus now approximates the shape of a cylinder:

$$P_{ST} = \gamma/r \qquad (3.13)$$

At the point at which the back pressure due to capillary repulsion equals or exceeds the hydrostatic pressure, $\rho g h$, to fill the section, the liquid will not enter the section. This condition in the thin, wide strip is

$$\rho g h = \gamma/r \qquad (3.14)$$

This simple pressure balance across a cylindrical meniscus is useful to correct the head height, to find the net available head pressure for filling a thin-walled casting. In the case of the filling of a circular section tube (with a spherical meniscus) do not forget the factor of 2 for both the contributions to the total curvature as in Equation 3.12. In the case of an irregular section, an estimate may need to be made of both radii, as in Equation 3.11.

The effect of capillary repulsion, repelling metal from entering thin sections, is clearly seen by the positive intercept in Figure 3.14 for a medium alloy steel and a stainless steel, in Figure 3.15 for an aluminium alloy, in Figure 3.16 for cast iron and in Figure 3.21 for a zinc alloy. Thus the effect appears to be quite general, as would be expected. The effective surface tension can be worked out in all these cases from an equation such as 3.14. In each case it is found to be around twice the value to be expected for the pure metal in a vacuum. Again, this high effective value is to be expected as explained in section 3.1.1.

In larger round or square sections, where the radii R and r both become large, in the range of 10 to 20 mm, the effects of surface tension become sufficiently small to be neglected for most purposes. Large sections are therefore filled easily.

3.3.4.1 Some practical aspects

In the filling of many castings the sections to be filled are not uniform; the standard complaint in the foundry is 'the sections are thick and thin'. This does sometimes give its problems. This is especially true where the sections become so thin in places that they become difficult to fill because of the resistance presented by surface tension. Aerofoils on propellers and turbine blades are typical examples.

To investigate the filling of aerofoil sections that are typical of many investment casting problem shapes, an aerofoil test mould was devised as shown in Figure 3.18. (This test mould also included some tensile test pieces whose combined volume interfered to some extent with the filling of the aerofoil itself; in later work the tensile test pieces

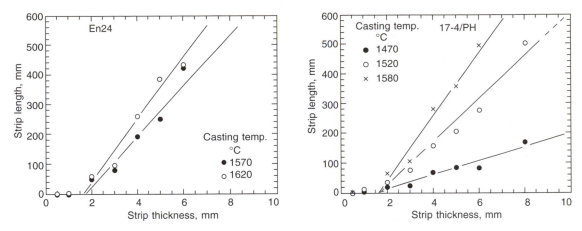

Figure 3.14 *(a) Fluidity data for a low alloy steel, and (b) for a stainless steel poured in a straight channel, furan-bonded sand mould (Boutorabi et al. 1990).*

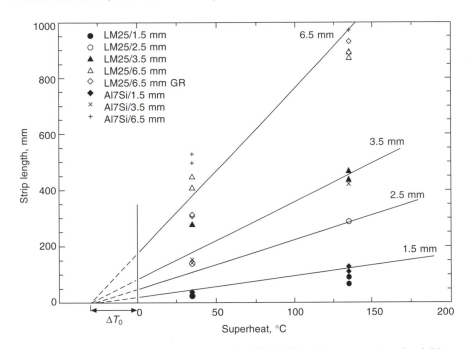

Figure 3.15 *Fluidity of a variety of Al–7Si and Al–7Si–0.4Mg alloys, one grain refined GR, showing linear behaviour with section thickness and casting temperature (Boutorabi et al. 1990).*

were removed, giving considerably improved reproducibility of the fluidity test.)

Typical results for a vacuum-cast nickel-based superalloy are given in Figure 3.19 (Campbell and Olliff 1971). Clearly, the 1.2 mm section fills more fully than the 0.6 mm section. However, it is also clear that at low casting temperature the filling of both sections is limited by the ability of the metal to flow prior to freezing. At these low casting temperatures the fluidity improves as temperature increases, as expected.

However, above a metal casting temperature of approximately 1500°C further increases of temperature do not further improve the filling. As the metal attempts to enter the diminishing sections of the mould, the geometry of the liquid front is closely defined as a simple cylindrical surface. Thus it is not difficult to calculate the thickness of the mould at any point. Half of this thickness is taken as the radius of curvature of the liquid metal meniscus (Figure 3.20). It is possible to predict, therefore, that the degree of filling is dictated by

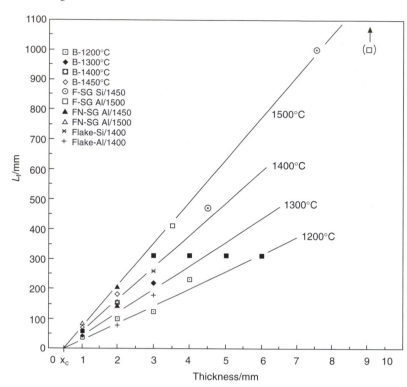

Figure 3.16 *Fluidity of a variety of grey and ductile cast irons showing linear behaviour with section thickness and casting temperature (Boutorabi et al. 1990).*

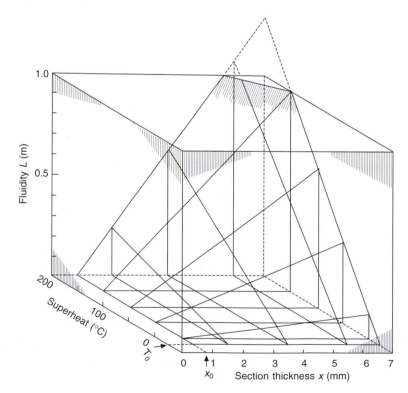

Figure 3.17 *Fluidity results represented from Figure 3.15 for Al–7Si alloys.*

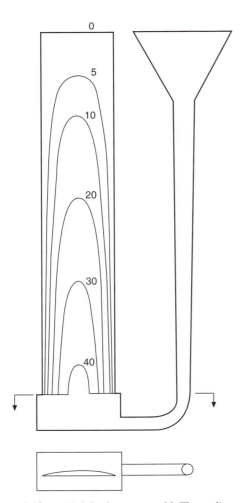

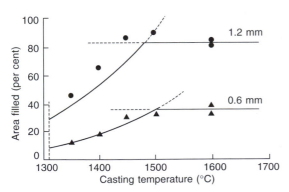

Figure 3.18 *Aerofoil fluidity test mould. The outlines of the cast shape are computed for increasing values of γ/ρgh, units in millimetres (Campbell and Olliff 1971).*

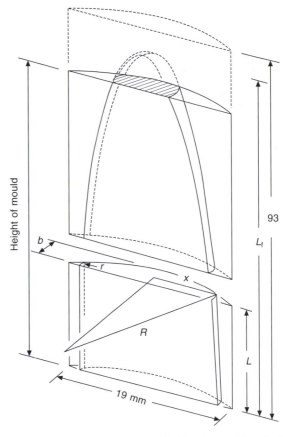

Figure 3.20 *Geometry of the aerofoil fluidity test (Campbell and Olliff 1971).*

Figure 3.19 *Results from the aerofoil fluidity test (Campbell and Olliff 1971) (lines denote theoretical predictions; points are experimental data).*

the local balance at every point around the perimeter of the meniscus between the filling pressure due to

the metal head and the effective back pressure due to the local curvature of the metal surface. In fact, if momentarily overfilled because of the momentum of the metal as it flowed into the mould, the repulsion effect of surface tension would cause the metal to 'bounce back', oscillating either side of its equilibrium filling position, finally settling at its balanced, equilibrium state of fullness.

The authors of this work emphasize the twin aspects of filling such thin sections; flowability limited by heat transfer, and fillability limited by surface tension.

At low mould and/or metal temperatures, the first type of filling, flowability, turns out to be simply classical fluidity as we have discussed above. Metallographic examination of the structures of aerofoils cast at lower temperatures showed columnar grains grown at an angle into the direction of flow, typical of solidification occurring while the metal was flowing. The flow length was controlled by solidification, and thus observed to be a function of superheat and other thermal factors, as we have seen.

The second type of filling, fillability, occurs at higher mould and/or metal temperatures where the heat content of the system is sufficiently high that solidification is delayed until after filling has come to a stop. Studies of the microstructure of the castings confirm that the grains are large and randomly oriented, as would be expected if the metal were stationary during freezing. Filling is then controlled by a mechanical balance of forces. The mode of solidification and further increases of temperature of the metal and the mould play no part in this phase of filling.

In a fluidity test of simpler geometry consisting of straight strips of various thickness, the linear plots of fluidity L_f versus thickness x and superheat ΔT_s are illustrated in Figures 3.15 and 3.16 for Al–7Si alloy and cast iron in sand moulds. It is easy to combine these plots giving the resultant three-dimensional pyramid plot shown in Figure 3.17. The plot is based on the data for the Al alloy in Figure 3.15. In terms of the pressure head h, and the intercepts ΔT_0 and x_0 defined on fluidity plots 3.15 and 3.16, the equation describing the slightly skewed surface of the pyramid is

$$L_f = C(\Delta T_s + \Delta T_0)(x - (2\gamma/\rho gh)) \quad (3.15)$$

Where C is a constant with dimensions of reciprocal temperature. For the Al alloy, C is found from Figure 3.15 to have a value of about 1.3 ± 0.1 K^{-1}, $\Delta T_0 = 30 \pm 5$, $\gamma = 2$ Nm^{-1} allowing for contribution of oxide film to the surface tension, $\rho = 2500$ kgm^{-3}, $g = 10$ ms^{-2} and $h = 0.10$ m. We can then write an explicit equation for fluidity (mm) in terms of superheat (degrees Celsius) and section thickness (mm):

$$L_f = 1.3(\Delta T_s + 30)(x - 1.6)$$

For a superheat $\Delta T_s = 100°C$ and section thickness $x = 2$ mm we can achieve a flow distance $L_f = 68$ mm for Al–7Si in a sand mould. If the head h were increased, fluidity would be higher, as indicated by Equation 3.15 (but noting the limitations discussed in section 3.3.2).

As we have seen, in these thin section moulds both heat transfer and surface tension contribute to limit the filling of the mould, their relative effects differ in different circumstances. This action of both effects causes the tests to be complicated, but, as we have seen, not impossible to interpret. Further practical examples of the simultaneous action of heat transfer and surface tension will be considered in the next section.

3.3.5 Comparison of fluidity tests

Kondic (1959) proposed the various thin section cast strip tests (called here the Voya Kondic (VK) strip test) as an alternative because it seemed to him that the spiral test was subject to unacceptable scatter (Betts and Kondic 1961).

For a proper interpretation of all types of strip test results they need to be corrected for the back pressure due to surface tension at the liquid front. As we have seen, this effectively reduces the available head pressure applied from the height of the sprue. The resulting cast length will correspond to that flow distance controlled by heat transfer, appropriate to that effective head and that section thickness. These results are worked through as an example below.

Figure 3.21 shows the results by Sahoo and Whiting (1984) on a Zn–27Al alloy cast into strips, 17 mm wide, and of thickness 0.96, 1.27, 1.58 and 1.88 mm.

The results for the ZA27 alloy indicate that the minimum strip thickness that can be entered by the liquid metal using the pressure head available in this test is 0.64 ± 0.04 mm. Using Equation 3.14, assuming that the metal head is close to 0.1 m, $R = 17/2$ mm and $r = 0.64/2$ mm, and liquid density close to 5720 kgm^{-3}, we obtain the surface tension $\gamma = 1.90$ Nm^{-1}. (If the $R = 17/2$ curvature is neglected, the surface tension then works out to be 1.98 Nm^{-1} and therefore is negligibly different for our purpose.) This is an interesting value, over double that found for the surface tension of pure Zn or pure Al. It almost certainly reflects the presence of a strong oxide film.

It suggests that the liquid front was, briefly, held up by surface tension at the entry to the thin sections, so that an oxide film was grown that assisted to hold back the liquid even more. The delay is typical of castings where the melt is given a choice of routes, but all initially resisting entry, so that the sprue and runner have to fill completely before pressure is raised sufficiently to break through the surface oxide. If the melt had arrived without choices, and without any delay to pressurization, the melt would probably have entered with a resistance due only to surface tension. In such a condition, γ would be expected to have been close to 1.0 Nm^{-1}.

It suggests that, to be safe, values of at least double the surface tension be adopted when allowing for the possible loss of metal head in filling thin section castings. This factor is discussed at greater length in section 3.1.1.

The ability to extrapolate back to a thickness that will not fill is a valuable feature of the VK fluidity strip test. It allows the estimation of an effective surface tension. This cannot be derived from tests, such as the spiral test, that only use one flow channel. The knowledge of the effective surface tension is essential to allow the comparison of the various fluidity tests that is suggested below.

The data from Figure 3.21 is cross-plotted in Figure 3.22 at notional strip thickness of 1.0, 1.5

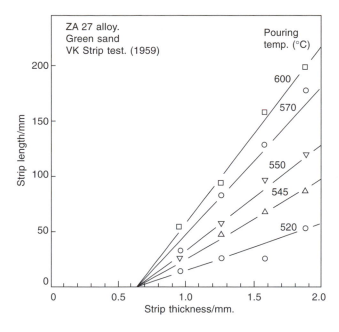

Figure 3.21 *Fluidity of ZA27 alloy cast in greensand using the VK fluidity strip test (2) using data from Sahoo and Whiting (1984).*

and 2.0 mm. (These rounded values are chosen simply for convenience.) The individual lengths in each section have been plotted separately, not added together to give a total as originally suggested by Kondic. (Totalling the individual lengths seems to be a valid procedure, but does not seem to be helpful, and simply adds to the problem of disentangling the results.) Interestingly all the results extrapolate back to a common value for zero fluidity at the melting point for the alloy, 490°C. This is a surprising finding for this alloy. Most alloys extrapolate to a finite fluidity at zero superheat because the metal still takes time to give up its latent heat, allowing the metal time to flow. The apparent zero fluidity at the melting point in this alloy requires further investigation.

Also shown in Figure 3.22 are fluidity spiral results. An interesting point is that, despite his earlier concerns, I am sure VK would have been reassured that the percentage scatter in the data was not significantly different to the percentage scatter in the strip test results.

The further obvious result from Figure 3.22 shows how the fluidity length measurements of the spiral are considerably higher than those of the strip tests. In a qualitative way this is only to be expected because of the great difference in the cross-sections of the fluidity channels. We can go further, though, and demonstrate the quantitative equivalence of these results.

In Figure 3.23, the spiral and strip results are all reduced to the value that would have been obtained if the spiral and the strip tests all had sections of 2 mm × 17 mm.

This is achieved by reducing the spiral results by a factor 4.44 to allow for the effect of surface tension and modulus, making the results equivalent to those in the 2 mm thick cast strip. The 2 mm section results remain unchanged of course. The 1.5 and 1.0 mm results are increased by factors 1.75 and 4.12 respectively. These adjustment factors are derived below.

Taking Equation 3.1 (Equation 3.2 can be used in its place, since we are to take ratios), together with Equations 3.5 and 3.6, and remembering that the velocity is given approximately by $(2gH)^{1/2}$ then we have for sand moulds:

$$L_f = km^n(2gH)^{1/2}$$
$$= km^n(2g(H - (\gamma/r\rho g)))^{1/2} \quad \textbf{(3.16)}$$

where n is 1 for interface controlled heat flow, such as in metal dies and thin sand moulds, and n is 2 for mould control of heat flow, such as in thick sand moulds.

Returning now to the comparison of fluidity tests, then by taking a ratio of Equation 3.16 for two tests numbered 1 and 2, we obtain:

$$\frac{L_1}{L_2} = \left(\frac{m_1}{m_2}\right)^n \left\{\frac{H_1 - (\gamma/r_1\rho g)}{H_2 - (\gamma/r_2\rho g)}\right\}^{1/2} \quad \textbf{(3.17)}$$

For the work carried out by Sahoo and Whiting on both the spiral and strip tests, the ratio given in Equation 3.17 applies as accurately as possible, since the liquid metal and the moulds were the same in each case. Assuming the moduli were 1.74 and 0.985 mm respectively, and the radii were 4

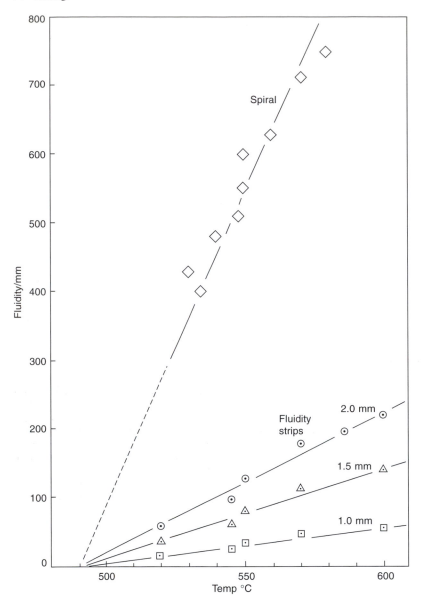

Figure 3.22 *Results of Figure 3.21 replotted to show the effect of superheat explicitly, as though from strips of section thickness 1.0, 1.5 and 2.0 mm, together with results of the spiral fluidity test.*

and 1 mm respectively, $\gamma = 1.9$ Nm^{-1}, and $\rho = 5714$ kgm^{-3}, and the height of the sprue in each case approximately 0.1 m, it follows

$$\frac{L_{f1}}{L_{f2}} = \left\{\frac{1.74}{0.895}\right\}^2 \left\{\frac{0.1 - 0.00847}{0.1 - 0.0339}\right\}^{1/2}$$

$$= 3.77 \times 1.18$$

$$= 4.44$$

The calculation is interesting because it makes clear that the largest contribution towards increased fluidity in these thin section castings derives from

their modulus (i.e. their increased solidification time). The effect of the surface tension is less important in the case of the comparison of the spiral with the 2 mm section. If the spiral of modulus 1.74 mm had been compared with a thin section fluidity test piece of only 1 mm thick, then:

$$L_{f1}/L_{f2} = 13.6 \times 1.68 = 9.25$$

Thus although the surface tension factor has risen in importance from 1.18 to 1.68, the effect of freezing time is still completely dominant, rising from 3.77 to 13.6.

The dominant effect of modulus over surface

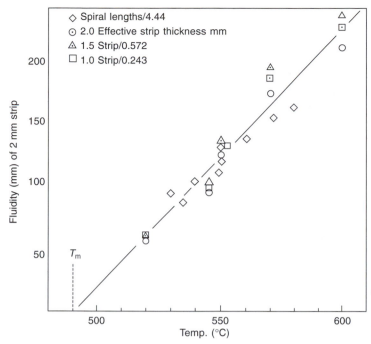

Figure 3.23 *Data from the spiral and strip tests shown in Figure 9, reduced by the factors shown to simulate results as though all the tests had been carried out in a similar size mould, of section 2 mm × 17 mm. All results are seen to agree, confirming the validity of the comparison.*

tension appears to be a general phenomenon in sand moulds as a result of the (usually) small effect of surface tension compared to the head height.

The accuracy with which the spiral data is seen to fit the fluidity strip test results for the Zn–27Al alloy when all are adjusted to the common section thickness of 2 mm × 17 mm (Figure 3.23) indicates that, despite the arguments that have raged over the years, both tests are in fact measuring the same physical phenomenon, which we happen to call fluidity, and both are in agreement.

3.4 Continuous fluidity

In a series of papers published in the early 1960s Feliu introduced a concept of the volume of flow through a section before flow was arrested. He carried out this investigation on, among other methods, a spiral test pattern, moulded in green sand. He made a number of moulds, cutting a hole through the drag by hand to shorten the spiral length, and repeated this for several moulds at various lengths. The metal that poured through the escape

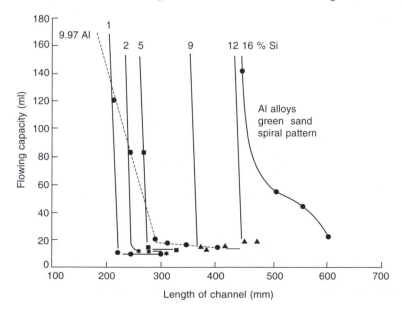

Figure 3.24 *Flow capacity of a channel as a function of length of the channel (Feliu 1962)*

holes was collected in a crucible placed underneath and weighed together with the length of the cast spiral. As the flow distance was progressively reduced, he discovered that at a critical flow distance the metal would continue flowing indefinitely (Figure 3.24). Clearly, any metal that had originally solidified in the flow channel was subsequently remelted by the continued passage of hot metal.

The conditions for remelting in the channel so as to allow continuous flow are illustrated in Figure 3.25. The concept is essential to the understanding of running systems, whose narrow sections would otherwise prematurely block with solidified metal. It is also clearly important in those cases where a casting is filled by running through a thin section into more distant heavy sections.

Because of its importance, I have coined the name 'continuous fluidity length' for this measurement of a flow distance for which flow can continue to take place indefinitely. It contrasts with the normal fluidity concept, which, to be strict, should perhaps be more accurately named as 'maximum fluidity length'.

The results by Feliu shown in Figure 3.24 seem typical. The maximum fluidity length has a finite value at zero superheat. This is because the liquid metal has latent heat, at least part of which has to be lost into the mould before the metal ceases to

flow. Continuous fluidity, on the other hand, has zero value until the superheat rises to some critical level. (Note that in Figures 3.26 to 3.28, the liquidus temperature T_m has been reduced from that of the pure metal by 5 to 10°C to allow for the presence of impurities).

Figures 3.26 to 3.28 display three zones: (i) a zone in which the flow distance is sufficiently short, and/or the temperature sufficiently high, that flow continues indefinitely; (ii) a region between the maximum and the continuous fluidity thresholds where flow will occur for increasingly long periods as distance decreases, or temperature rises; and (iii) a zone in which the flow distance cannot be achieved, bounded on its lower edge by the maximum fluidity threshold.

Examining the implications of these three zones in turn: Zone (iii) is the regime in which most running systems operate; Zone (ii) is the regime in many castings, particularly if they have thin walls; Zone (i) is the regime of bitter experience of costly redesigns, sometimes after all the budget has been expended on the patternwork, and it is finally acknowledged that the casting cannot be made. Fluidity really can therefore be important to the casting designer and the founder.

The author is aware of little other experimental work relating to continuous fluidity. An example worth quoting because of its rarity is that of Loper and LeMahieu on white irons in greensand dating from 1971. (Even so, the interested reader should take care to note that freezing time is not measured directly in this work.)

There is a nice computer simulation study carried out at Aachen University (Sahm 1998) that confirms the principles outlined here. More work is required in this important but neglected field.

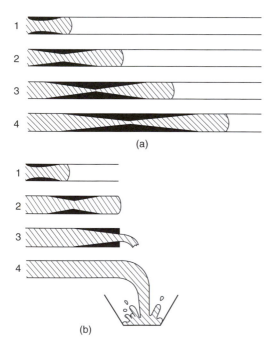

Figure 3.25 *Concepts of (a) maximum fluidity length showing the stages of freezing leading to the arrest of the flow in a long mould; and (b) the continuous flow that can occur if the length of the mould does not exceed a critical length, defined as the continuous fluidity length.*

3.5 Glossary of symbols

a thickness of plate section casting
c_m specific heat of mould
g acceleration due to gravity
h height, or heat transfer coefficient
H latent heat of solidification
L_f (maximum) fluidity length
m modulus (volume/cooling surface area)
P pressure
r,R orthogonal radii of the liquid meniscus
t_f freezing time
T_f freezing temperature
T_0 initial mould temperature
v velocity
γ surface tension
κ_m thermal conducivity of mould
ρ_m density of mould
ρ_s density of solid metal casting

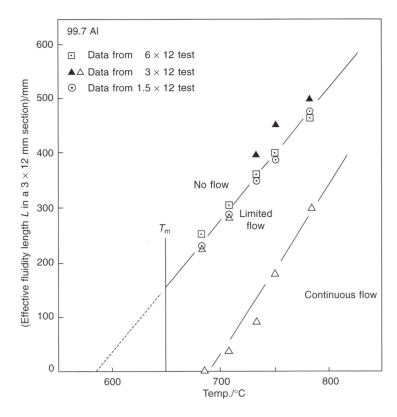

Figure 3.26 *Maximum and continuous fluidity data by Feliu (13) for 99.7Al cast into greensand moulds of sections 6 × 12, 3 × 12 and 1.5 × 12 mm, all reduced as though cast only in a section 3 × 12 mm.*

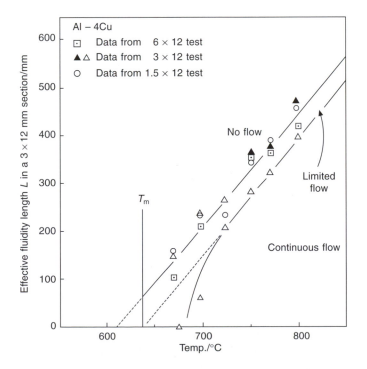

Figure 3.27 *Data for Al–4Cu alloy by Feliu (13) recalculated as though only from section 3 × 12 mm.*

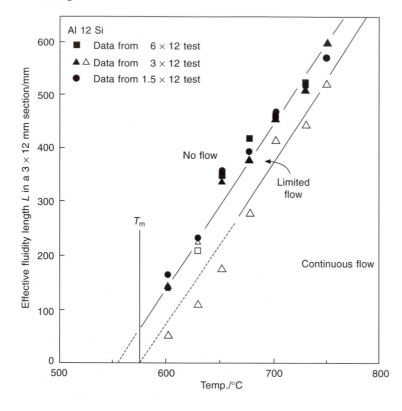

Figure 3.28 *Data for Al–12Si alloy by Feliu (13) recalculated as though only from section 3 × 12 mm.*

Chapter 4

The mould

When the molten metal enters the mould, the mould reacts violently. Frenzied activity crowds into this brief moment of the birth of the casting: buckling, outgassing, pressurization, cracking, explosions, disintegration and chemical attack. The survival of a saleable casting is only guaranteed by the strenuous efforts of the casting engineer to ensure that the moulding and casting processes are appropriate, and are under control.

Only those aspects of the interaction with the mould are considered that introduce defects or otherwise influence the material properties of the casting. Those actions that result, for instance, in the deformation of the casting are not treated here. They will be considered in later volumes.

4.1 Inert moulds

Very few moulds are really inert towards the material being cast into them. However, some moulds are very nearly so. This is especially true at lower temperatures.

For instance, with the cast iron or steel (permanent) mould used in the gravity die casting or low-pressure die casting of aluminium, the mould is coated with an oxide wash. The metal and mould are practically inert towards each other. Apart from the normal oxidation of the surface of the casting by the air, there are no significant chemical reactions. This is a significant benefit of metal moulds that is often overlooked. The die does suffer from thermal fatigue, usually after thousands of casts. This limit to die life can be an important threat to surface finish as the die ages, or occasionally results in catastrophic failure, with disastrous effects on production, because dies take time to replace. Such failure is commonly associated with heavy sections of the casting, such as a heavy boss on a plate. The material of the die in this region suffers from

repeated transformation to austenite and back again. The large volume change accompanying this reaction corresponds to a massive plastic strain of several per cent, so that the steel (or cast iron) suffers thermal fatigue.

The usefulness of a relatively inert mould is emphasized by the work of Stolarczyk (1960), who measured approximately 0.5 per cent porosity in gunmetal casting into steel-lined moulds, compared with 3.5 per cent porosity for identical test bars cast in greensand moulds.

Dies in pressure die casting are hardly inert, partly because of the gradual dissolution of the die, but mainly because of the overwhelming effect of the evaporation of the die-dressing material. This may be an oil- or water-based suspension of graphite sprayed on to the surface of the die, and designed to cool and lubricate the die between shots. The gases found in pores in pressure die castings have been found to be mainly products of decomposition of the die lubrication, and the volume of gases found trapped in the casting has been found to correspond very nearly to the volume of the die cavity.

A little-known problem is the boiling of residual coolant trapped inside joints of the die. Thus as liquid metal is introduced into the die the coolant, especially if water-based, will boil. If there is no route for the vapour to escape via the back of the die, vapour may be forced into the liquid metal as bubbles. If this happens it is likely that at least some of these bubbles will be permanently trapped as blowholes in the casting. This problem is expected to be common to pressure die and squeeze casting processes.

The recent approach to the separation of the cooling of the pressure die casting die from its lubrication is seen as a positive step toward solving this problem. The approach is to use more effective cooling by built-in cooling channels, whereas lubrication is achieved by the application of minute additions of waxes or other materials to the shot sleeve.

For light alloys and lower-temperature casting materials, investment moulds are largely inert. Interestingly, dry sand moulds (i.e. greensand moulds that have been dried in an oven) have been found to be similar, as shown by Locke and Ashbrook (1950).

Carbon-based and graphite dies have been found useful for zinc alloys. However, their lives are short for the casting of aluminium alloys because of the degradation of the carbon by oxidation. Carbon-based moulds are used for the casting of titanium alloys in vacuum. Oxidation of the mould is thereby reduced, but the contamination of the surface of the titanium casting with carbon is severe, promoting the formation of an outer layer of the alloy where the alpha-phase is stabilized. This surface layer is known in titanium castings as the alpha-case. It usually has to be removed by machining or chemical dissolution.

4.2 Aggregate moulds

Sand moulds were almost always made with silica sand, apart from a number of places in the world where silica was unobtainable such as in some parts of Scandinavia where olivine is used. Other foundries have used chromite and zircon for their useful physical properties. However, recently, the problems with the traditional silica mould have driven moves towards many different kinds of particulate materials, some natural minerals and some synthetic. Acknowledging this move from silica sand it is appropriate to call particulate moulding materials 'aggregates' rather than sands. Even so, the terms will be seen to be somewhat interchangeable. Both are used in this section.

In greensand or chemically bonded moulds the chemical interactions increase in number and severity with increasing temperature, as we can appreciate if we work our way up the temperature spectrum of casting alloys:

1. Low-melting-point lead and zinc alloy liquids generally cast at temperatures up to 500°C are too cool to cause significant reactions.
2. Magnesium and aluminium have casting temperatures commonly up to 750°C. They react with water vapour and various other organics to produce the solid oxide skin and free hydrogen that can diffuse into the melt. These reactions continue for some time after solidification and during cooling. This source of hydrogen is likely to be important in growing pores that are located just under the casting surface. We shall return to the subject of the growth of subsurface porosity

later. Despite this reactivity at the surface of the liquid metal, it is worth noting that the temperature at which light alloys are cast does not lead to extensive breakdown of the chemical constituents of the mould, as is clear from Figures 4.1 and 4.2.
3. Copper-based melts up to 1300°C take part in several important reactions.
4. Irons up to 1400°C and steels up to 1600°C are especially reactive in many ways.
5. Titanium and zirconium in the range 1600 to 1700°C are so reactive they are problematic to cast into moulds of any type. Reactions with most moulding materials cause the troublesome alpha-case on titanium alloys, in which the alpha-titanium phase is stabilized by interstitials (oxygen and/or carbon) absorbed from the breakdown of the mould.

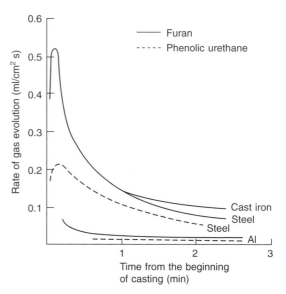

Figure 4.1 *Measured gas evolution rates from castings of aluminium, iron, and steel, in chemically bonded sand moulds (Bates and Monroe 1981).*

The reactions occur with both the mould surface itself and with the atmosphere formed in the mould during filling with the hot metal. These are all dealt with below.

4.2.1 Transformation zones

The evaporation of water in greensand moulds has been the subject of much research. Clearly, as the hot metal heats the surface of the mould, the water (and other volatiles) will be boiled off, migrating away from the mould face, only to condense again in the deeper, cooler parts of the mould. As the heat continues to diffuse in, the water migrates

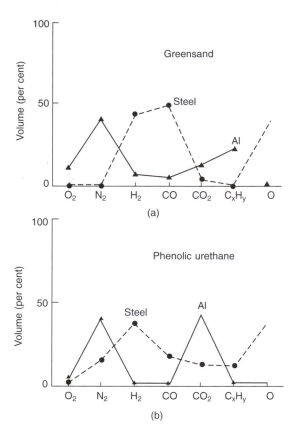

Figure 4.2 *Composition of mould gases (a) from greensand (Chechulin 1965) and (b) from phenolic urethane (Bates and Monroe 1981).*

further. Dry and wet zones travel through the mould like weather systems in the atmosphere.

Looking at these in detail, four zones can be distinguished, as shown in Figure 4.3:

1. The dry zone, where the temperature is high and all moisture has been evaporated from the binder. It is noteworthy that this very high temperature region will continue to retain a relatively stagnant atmosphere composed of nearly 100 per cent water vapour. However, of course, some of this will be reacting at the casting surface to produce oxide and free hydrogen.
2. The vapour transport zone, essentially at a uniform temperature of 100°C, and at a roughly constant content of water, in which steam is migrating away from the casting.
3. The condensation zone, where the steam recondenses. This zone was for many years the subject of some controversy as to whether it was a narrow zone or whether it was better defined as a front. The definitive theoretical model by Kubo and Pehlke (1986) has provided an answer where direct measurement has proved

difficult; it is in fact a zone, confirming the early measurements by Berry *et al.* (1959). This zone gets particularly wet. The raised water content usually greatly reduces the strength of greensand moulds, so that mechanical failure is most common in this zone.
4. The external zone where the temperature and water content remain as yet unchanged.

It is worth taking some space to describe the structure of the dry sand zone.

When casting light alloys and other low-temperature materials, the dry sand layer has little discernible structure.

However, when casting steel it becomes differentiated into various layers that have been detailed from time to time (e.g. Polodurov 1965; Owusu and Draper 1978). Counting the mould coating as number zero, these are:

0. Dressing layer of usually no more than 0.5 mm thickness, and having a dark metallic lustre as a result of its high content of metal oxides.
1. Sinter cake zone, characterized by a dark brown or black colour. It is mechanically strong, being bonded with up to 20 per cent fayalite, the reaction product of iron oxide and silica sand. The remaining silica exists as shattered quartz grains partially transformed to tridymite and cristobalite, which is visible as glittering crystals (explaining the origin of the name cristobalite). This layer is largely absent when casting grey iron at ordinary casting temperatures.
2. Light-grey zone, with few cracked quartz grains and little cristobalite. What iron oxides are present are not alloyed with the silica grains. This zone is only weakly bonded and disintegrates on touch.
3. Charred zone, of dark-grey colour, of intermediate strength, containing unchanged quartz grains but significant levels of iron oxide. Polodurov speculates that this must have been blown into position by mould gases.

The changes in form of the silica sand during heating are complicated. An attempt to illustrate these relations graphically is included in Figure 4.3. This complexity, and particularly the expansion accompanying the phase change from alpha to beta quartz, has prompted a number of foundries to abandon silica sand in favour of more predictable moulding aggregates. This move is expected to become more widespread in future.

4.2.2 Evaporation and condensation zones

As the heat diffuses from the solidifying casting into the mould (Figure 4.4), the transformation zones migrate into the mould. We can follow the progress

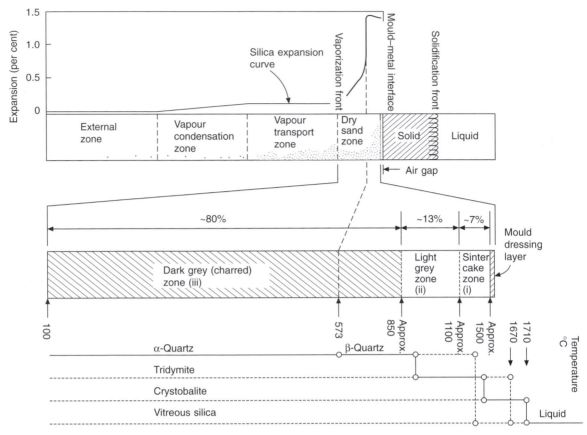

Figure 4.3 *Structure of the heated surface of a greensand mould against a steel casting; and the forms of silica (after Sosman 1927), with solid lines denoting stable states, and broken lines unstable states.*

of the advance of the zones by considering the distance d that a particular isotherm reaches as a function of time t. The solution to this simple one-dimensional heat-flow problem is:

$$d = (Dt)^{1/2} \qquad (4.1)$$

where D is the coefficient of diffusion. In the case of the evaporation front, the isotherm of interest is that at 100°C. We can see from Figures 4.4 and 4.5 that the value of k is close to 1 mm^2 s^{-1}. This means that the evaporation front at 1 s has travelled 1 mm, at 100 s has traveled 10 mm, and requires 10 000 s (nearly three hours!) to travel 100 mm. It is clear that the same is true for aluminium as well as steel. (This is because we are considering a phenomenon that relies only on the rate of heat flow in the mould – the metal and its temperature is not involved.)

For the condensation zone the corresponding value of D is approximately 3 mm^2 s^{-1}, so that the position of the front at 1, 100 and 10 000 s is 1.7, 17 and 170 mm respectively.

These figures are substantiated to within 10 or 20 per cent by the theoretical model by Tsai *et al.*

(1988). This work adds interesting details such as that the rate of advance of the evaporation front depends on the amount of water present in the mould, higher water contents making slower progress. This is to be expected, since more heat will be required to move the front, and this extra heat will require extra time to arrive. The extra ability of the mould to absorb heat is also reflected in the faster cooling rates of castings made in moulds with high water content. Measurements of the thermal conductivity of various moulding sands by Yan *et al.* (1989) have confirmed that the apparent thermal conductivity of the moisture-condensation zone is about three or four times as great as that of the dry sand zone.

An earlier computer model by Cappy *et al.* (1974) also indicates interesting data that would be difficult to measure experimentally. They found that the velocity of the vapour was in the range of 10–100 mms^{-1} over the conditions they investigated.

Their result for the composition and movement of the zones is given in Figure 4.6. Kubo and Pehlke calculate flow rates of 20 mm^{-1}. These authors go on to show that moisture vaporizes not only at the

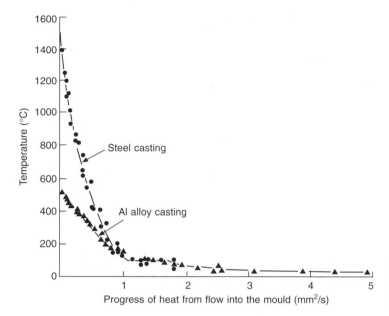

Figure 4.4 *Temperature distribution in a greensand mould on casting an aluminium alloy (Ruddle and Mincher 1949–50) and a steel (Chvorinov 1940).*

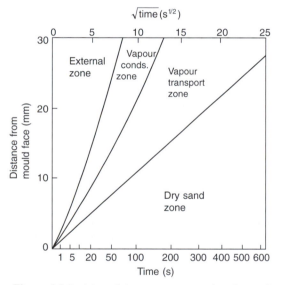

Figure 4.5 *Position of the vapour zones after the casting of aluminium in a greensand mould. Data from Kubo and Pehlke (1986).*

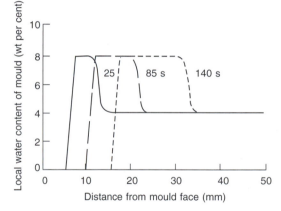

Figure 4.6 *Water content of the vapour transport zone with time and position. Smoothed computed results of Cappy et al. (1974).*

evaporation front, but also in the transportation and condensation zones. Even in the condensation zone a proportion of the water vaporizes again at temperatures below 100°C (Figure 4.5).

The pressure of water vapour at the evaporation front will only be slightly above atmospheric pressure in a normal greensand mould. However, because the pressure must be the same everywhere in the region between the mould/metal interface and the evaporation front, it follows that the dry sand zone must contain practically 100 per cent

water vapour. This is at first sight surprising. However, a moment's reflection will show that there is no paradox here. The water vapour is very dry and hot, reaching close to the temperature of the mould/metal interface. At such high temperatures water vapour is highly oxidizing. There is no need to invoke theories of additional mechanisms to get oxygen to this point to oxidize the metal – there is already an abundance of highly oxidizing water vapour present (the breakdown of the water vapour also providing a high-hydrogen environment, of course, to enter the metal, and to increase the rate of heat transfer in the dry sand zone).

Kubo and Pehlke (1986) confirm that gas in the dry sand and transportation zones consists of nearly 100 per cent water vapour. In the condensation zone the percentage of air increases, until it reaches 100 per cent air in the external zone.

It is found that similar evaporation and condensation zones are present for other volatiles in the greensand mould mixture. Marek and Keskar (1968) have measured the movement of the vapour transport zone for benzene and xylene. The evaporation and condensation fronts of these more volatile materials travel somewhat faster than those of water. When such additional volatiles are present they will, of course, contribute to the 1 atm of gas pressure in the dry zone, helping to dilute the oxidizing effect of water vapour, and helping to explain part of the beneficial effect of such additives. In the following section we will see how many organics decompose at these high temperatures, providing a deposit of carbon, which further assists, in the case of such metals as cast iron, in preventing oxidation and providing a non-wetting mould surface of sand grains that have been coated with carbon.

It is to be expected that vapour transport zones will also be present to various degrees in chemically bonded sands. The zones will be expected to have traces of water mixed with other volatiles such as organic solvents. Little work appears to have been carried out for such binder systems, so it is not easy to conclude how important the effects are, if any. In general, however, the volatiles in such dry sand systems usually total less than 10 per cent of the total volatiles in greensand, so that the associated condensation zones will be expected to be less than one-tenth of those occurring in greensand. It may be, therefore, that they will be unimportant. However, at the time of writing we cannot be sure. It would be nice to know.

All of the above considerations on the rate of advance of the moisture assume no other flows of gases through the mould. This is probably fairly accurate in the case of the drag mould, where the flow of the liquid metal over the surface of the mould effectively seals the surface against any further ingress of gases. A certain amount of convection is expected in the mould, but this will probably not affect the conditions in the drag significantly.

In the vertical walls of the mould, however, convection may be significant. Close to the hot metal, hot gases are likely to diffuse upwards and out of the top of the mould, their place being taken by cold air being drawn in from the surroundings at the base of the mould, or the outer regions of the cope.

General conditions in the cope, however, are likely to be more complicated. It was Hofmann in 1962 who first emphasized the different conditions experienced during the heating up and outgassing of the cope. He pointed out that the radiated heat from the rising melt would cause the cope surface of the mould to start to dry out before the moment

of contact with the melt. During this pre-contact period two different situations can arise:

1. If the mould is open, as the cope surface heats up the water vapour can easily escape through the mould cavity and out via the opening (Figure 4.7). The rush of water vapour through an open feeder can easily be demonstrated by holding a piece of cold metal above the opening. It quickly becomes covered with condensate. The water vapour starts its life at a temperature of only 100°C. It is therefore a relatively cool gas, and is thus most effective in cooling the surface of the mould as it travels out through the surface of the cope on its escape route.

2. If the mould is closed, the situation is quite different. The air being displaced and expanded by the melt will force its way through the mould, carrying away the vapour from the interface (Figure 4.7). The rate of flow of air is typically in the range $10–100 \, l^{-1} \, m^{-2}$ (the reader is encouraged to confirm this for typical castings and casting rates). This is in the same range of flow rate as the transport of vapour given in computer models. Thus if the casting rate is relatively low, then the vapour transport zone is likely to be relatively unaffected, although perhaps a little accelerated in its progress. When the casting rate is relatively high, then the vapour transport zone will be effectively blown away, diluted with the gale of air so that no condensation can occur. Because the water vapour is driven

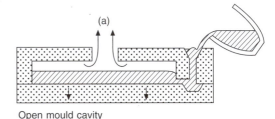

Open mould cavity

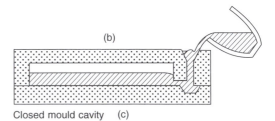

Closed mould cavity   (c)

Figure 4.7 *Three conditions of vapour transport in moulds: (a) free evaporation from the cope; (b) evaporation from the cope confined by the enclosed mould cavity; and (c) evaporation from the drag confined by the cover of liquid metal (Hofmann 1962).*

away from the surface and into the interior of the mould, its beneficial cooling effect at the surface is not felt, with the result that the surface reaches much higher temperatures, as is seen in Figure 4.8. The prospect of the failure of the cope surface by expansion and spalling of the sand is therefore much enhanced.

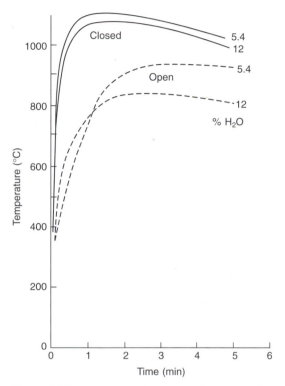

Figure 4.8 *Temperature in the cope surface seen to be significantly lowered by open moulds and high moisture levels. Data from Hofmann (1962).*

However, the rate of heating of the surface by radiation from the melt, particularly for iron and steel castings, can be reduced by a white mould coat, such as a zircon- or alumina-based mould wash. This is a useful technique that is now widely applied for large castings of iron and steel.

One final aspect of vapour transport in the mould is worth a note. There has been much discussion over the years about the contribution of the thermal transpiration effect to the flow of gases in moulds.

Although it appears to have been widely disputed, the effect is certainly real. It follows from the kinetic theory of gases, and essentially is the effect of heated gases diffusing away from the source of heat, allowing cooler gases to diffuse up the temperature gradient. In this way it has been argued that oxygen from the air can arrive continuously at the casting to oxidize the surface to a greater degree than would normally have been expected.

Williams (1970) described an experiment that demonstrated this effect. He took a sample of clay approximately 50 mm long in a standard 25 mm diameter sand sampling tube. When one end was heated to 1000°C and the other was at room temperature, he measured a pressure difference of 10 mmHg if one end was closed, or a flow rate of 20 ml per minute if both ends were open. If these results are typical of those that we might expect in a sand mould, then we can make a comparison as follows. The rate of thermal transpiration is easily shown to convert to $0.53 \, l^{-1} \, m^{-2}$ for the conditions of temperature gradient and thickness of sample used in the experiment. From the model of Cappy *et al.* (1974), we obtain an estimate of the rate of transport of vapour of $100 \, l^{-1} \, m^{-2}$ at approximately this same temperature gradient through a similar thickness of mould. Thus thermal transpiration is seen to be less than 1 per cent of the rate of vapour transport. Additional flows like the rate of volume displacement during casting, and the rate of thermal convection in the mould, will further help to swamp thermal transpiration.

Thermal transpiration does seem to be a small contributor to gas flow in moulds. It is possible that it may be more important in other circumstances. More work is required to reinstate it to its proper place, or lay it to rest as an interesting but unimportant detail.

4.3 Mould atmosphere

On the arrival of the hot metal in the mould, a rich soup of gases boils from the surface of the mould and the cores. The air originally present in the cavity dilutes the first gases given off. This is quickly expelled through vents or feeders, or may diffuse out through the cope. Subsequently, the composition of the mould gas is relatively constant.

In the case of steel being cast into greensand moulds, the mould gas mixture has been found to contain up to 50 per cent hydrogen (Figure 4.2). The content of hydrogen depends almost exactly on the percentage water in the sand binder. (Dry sand moulds have practically no hydrogen.) Other changes brought about by increased moisture in the sand were a decrease in oxygen, an increase in the CO/CO_2 ratio, and the appearance of a few per cent of paraffins. The presence of cereals in the binder was found to provide some oxygen, even though the concentration of oxygen in the atmosphere fell because of dilution with other gases (Locke and Ashbrook 1950). Chechulin (1965) describes the results for greensand when aluminium alloys, cast irons and steels are cast into them. His results are given in Figure 4.2a. Irons and steels produce rather similar mould atmospheres, so only his results for steel are presented.

The high oxygen and nitrogen content of the atmosphere in the case of moulds filled with aluminium simply reflects the high component of residual air (originally, of course, at approximately 20 per cent oxygen and 80 per cent nitrogen). The low temperature of the incoming metal is insufficient to generate enough gas and expand it to drive out the original atmosphere. This effective replacement of the atmosphere is only achieved in the case of iron and steel castings.

The atmosphere generated when ferrous alloys are cast into chemically bonded sand moulds is, perhaps rather surprisingly, not so different from that generated in the case of greensand (Figure 4.2b). The mixture consists mainly of hydrogen and carbon monoxide.

The kinetics of gas evolution were studied by Scott and Bates (1975), who found that hydrogen evolution peaked within 4 to 5 minutes for most chemical binders. However, for the sodium silicate binder a rapid burst of hydrogen was observed, which peaked in less than 1 minute.

Lost-foam casting, where the mould cavity is filled with polystyrene foam (the 'full mould' process), is a special case. Here it is the foam that is the source of gases as it is vaporized by the molten metal. At aluminium casting temperatures the polymerized styrene merely breaks down into styrene, but little else happens, as is seen in Figure 4.9. It seems that the liquid styrene soaks into the ceramic surface coating on the foam, so that the permeability of the coating will temporarily fall to zero. This unhelpful behaviour probably accounts for many of the problems suffered by aluminium alloy castings made by the lost-foam process.

At the casting temperatures appropriate for cast iron, more complete breakdown occurs, with the generation of hydrogen and methane gases, and considerable quantities of free carbon. The carbon deposits on the advancing metal front, possibly as

carbon black, but quickly transforms into graphite. It is possible that the carbon may deposit directly in the form of graphite. However it may occur, the deposit is widely known as 'lustrous carbon'. Once formed, the layer is rather stable at iron-casting temperatures, and can therefore lead to serious defects if entrained in the metal. The problem has impeded the successful introduction of this technology on a wider scale. For steel casting the temperature is sufficiently high to cause the carbon to be taken into solution. Steel castings of low or intermediate carbon content are therefore contaminated by pockets of high-carbon alloy. This problem has prevented lost-foam technology in the form of the full mould process being used for low- and medium-carbon steel castings.

Lost-foam iron castings are not the only type of ferrous castings to suffer from lustrous-carbon defects. The defect is also experienced in cast iron made in phenolic urethane-bonded moulds, and at times can be a serious headache. The absence of carbon is therefore a regrettable omission from the work reported in Figure 4.2. At the time Chechulin carried out this study, the problem would not have been known.

It seems reasonable to expect that carbon may also be produced from the pyrolysis (meaning the decomposition by heat in the absence of oxygen, in contrast to burning, which is the decomposition by high temperature oxidation) of other binder systems. More work is required to check this important point.

4.3.1 Outgassing pressure

The sudden heat from the liquid metal causes the volatile materials in the mould to evaporate fiercely. In greensand moulds and many other binder systems the main component of this volatilization is water. Even in so-called dry-binder systems there is usually

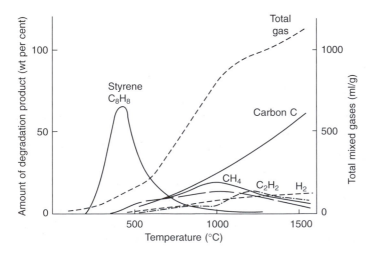

Figure 4.9 *Products of decomposition of expanded polystyrene (Goria et al. 1986).*

enough water to constitute a major contribution to the total volume of liberated gas. On contact with the hot metal, much of the water is decomposed to hydrogen as is seen in the high hydrogen contents of analysed mould gases (Scott and Bates 1975).

In the case of the mould, the generation of copious volumes of gas is usually not a problem. The gas has plenty of opportunity to diffuse away through the bulk of the mould. The pressure build-up in a greensand cavity during mould filling is normally only of the order of 100 mm water gauge (0.01 atm) according to measurements by Locke and Ashbrook (1972). This corresponds to merely 10 mm or so head pressure of liquid iron or steel. However, even this rather modest pressure might be unusually high because their experimental arrangement corresponded to a closely fitting steel moulding box, and escape for mould gases only via the cope. Even so, in greensand systems where the percentage of fines and clay and other constituents is high, the permeability of the mould falls to levels at which the ability of the mould volatiles to escape becomes a source of concern. The venting of the mould by needling with wires is a time-honoured method of reintroducing some permeability.

Chemically bonded moulds are usually of no concern from the point of view of generating a back pressure during the filling of the mould. This is because the sand is usually bought in as ready washed, cleaned and graded into closely similar sizes (a 'three pan sand'). In addition, only a few volume per cent of binder is used, leaving an open, highly permeable bonded mould. A single measurement by the author using a water manometer showed a pressure rise during the filling of a cylinder head mould of less than 1 mm water gauge. Even this negligible rise seemed to decay to nothing within a second or so.

In the case of cores, however, once the core is covered by liquid metal, the escape of the core gases is limited to the area of the core prints, if the metal is not to be damaged by the passage of bubbles through it. Furthermore, the rate of heating of the core is often greater than that of the mould because it is usually surrounded on several sides by hot metal, and the volume of the core is, of course, much less. All these factors contribute to the internal pressure within the core rising rapidly to high values.

Many authors have attempted to provide solutions to the pressure generated within cores. However, there has until recently been no agreed method for monitoring the rate or quantity of evolved gases that corresponds with any accuracy to the conditions of casting. A result of one method by Naro and Pelfrey (1983) is shown in Figure 4.10. (This method is an improvement on earlier methods in which the water and other volatiles would condense in the pipework of the measuring apparatus, reducing the

apparent volume of gases, and thereby invalidating the experimental results.) The really important quantity given by these curves is the *rate* of evolution of gas. The rates, of course, are equal to the slope of the curves in Figure 4.10, and are presented in Figure 4.11. Only a few results are presented for clarity. It is sufficient to note that the rates of outgassing are very different for different chemical systems.

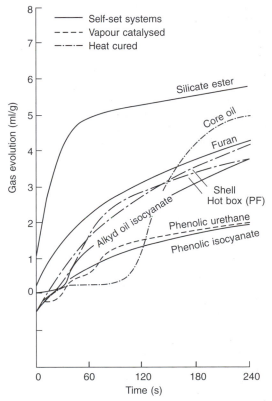

Figure 4.10 *Gas evolution from various binder systems using an improved test procedure that includes the contribution from water and other volatiles (Naro and Pelfrey 1983).*

Taking this recent method of estimating Q, the rate of volume of gas generated from a given weight of core in $mlg^{-1}s^{-1}$ (or preferably the identical-sized unit $l.kg^{-1}s^{-1}$) as being of tolerable relevance to the real situation in castings, we can construct a core outgassing model. We shall roughly follow the method originally pioneered by Worman and Nieman (1973).

We first need to define the concept of permeability. This is a measure of the ease with which a fluid (the mould gas in our case) can flow through a porous material. Permeability P_e is defined as the rate of gas flow Q (as a volume per unit time) through a permeable material of area A and

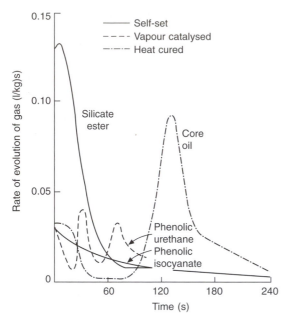

Figure 4.11 *Rates of gas evolution from various sand binders based on the slopes of the curves shown in Figure 2.21.*

length L and driven by a pressure difference ΔP:

$$P_e = QL/A \ \Delta P$$

The SI units of P_e are quickly seen to be:

$$[P_e \text{ units}] = [\text{litre/s}] \ [\text{m}]/[\text{m}^2][\text{Pa}]$$
$$= 1 \ \text{s}^{-1} \ \text{m}^{-1} \ \text{Pa}^{-1}$$

Consider now our simple model of a core shown in Figure 4.12. The measured volume of gas evolved

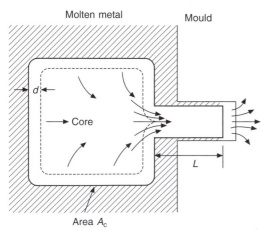

Figure 4.12 *Core model, showing heated layer thickness d outgassing via its print. (In this particular case, metal flash along the sides of the print forces gas to exit only from the end area A.)*

per second from a kilogram of core material is Q. If we allow for the fact that this will have been measured at temperature T_1, usually above 100°C (373 K) to avoid condensation of moisture, and the temperature in the core at the point of generation is T_2, then the volume of gas produced in the core is actually QT_2/T_1 where T is measured in K. For the casting of light alloys the temperature ratio T_2 /T_1 (in K remember) is about 3, whereas for steels it is nearly 6.

If we multiply this by the weight of sand heated by the liquid metal, then we obtain the total volume of gas evolved per second from the core. Thus if the heated layer is depth d, the core area A_c and density ρ, then the volume of gas evolved per second is $QdA_c\rho T_2/T_1$. If the core is surrounded by hot metal, this volume of gas has to diffuse to the print and force its way through the length L of the print of area Ap. We shall assume that the pressure drop experienced by the gas in diffusing through the bulk of the core is negligible in comparison with the difficulty of diffusing through the print. Considering then the permeability definition only for the pressure drop along the print, we obtain the pressure in the core (above the ambient pressure at the outside tip of the print):

$$P = QdA_c \ \rho \cdot LT_2 \ /ApT_1P_e \qquad \textbf{(4.2)}$$

This simple model emphasizes the direct role of permeability P_e and of Q, the rate of gas evolution. It is to be noted that the high casting temperature for steels is seen in Figure 4.1 to result in values for Q approximately twice those for aluminium alloys. Thus cores in steel castings will be twice as likely to create blows than cores in aluminium alloy castings. For this reason, an enclosed core that would give no problems in an aluminium alloy casting may cause blows when the same pattern is used to make an equivalent bronze or iron casting.

Our model also highlights the various geometrical factors of importance. In particular, the area ratio of the core and the print, Ac/Ap, is a powerful multiplier effect, and might multiply the pressure by anything between 10 and 100 times for different core shapes. Also emphasized is the length L of the core print. If the print is a poor fit then L may be unnecessarily lengthened by the flashing of the metal into the print so as to enclose the flow path in an even longer tunnel. If the liquid metal completely surrounds the end of the print too, then, of course, all venting of gases is prevented. Gases are then forced to escape through the molten metal, with consequential bubble damage to the casting.

The important practical conclusions for good core design to be drawn from the model are:

1. High permeability.
2. Core sand binder with low volatile content and/ or low rate of gas evolution.

3. Large area prints.
4. Good fit of prints.

The provision of a vent such as a drilled hole along the length of the print will effectively reduce L to zero; the model predicts that the internal pressure in the core will then be eliminated (the only remaining pressure will, of course, be that to overcome the resistance to flow through the core itself). The value of vents in reducing blowing from cores has been emphasized by many workers. Caine and Toepke (1966), in particular, estimate that a vent will reduce the pressure inside a core by a large factor, perhaps 5 or 10. This is an important effect, easily outweighing all other methods of reducing outgassing pressure in cores.

Vents can be moulded into the core, formed from waxed string. The core is heated to melt out the wax, and the string can then be withdrawn prior to casting. This traditional practice was often questioned as possibly being counterproductive, because of the extra volatiles from the wax that, on melting, soaks into the core. Such fears are seen to be happily unfounded. The technique is completely satisfactory because the presence of the vent completely overrides the effect of the extra volatile content of the core.

A final prediction from the model is the effect of temperature. In theory a lowering of the casting temperature will lower the internal core pressure. However, this is quickly seen to be a negligible effect within the normal practical limits of casting temperatures. For instance, a large change of 100 K in the casting temperature of an aluminium alloy will change the pressure by a factor of approximately 100/900. This is only 11 per cent. For irons and steels the effect is smaller still. It can therefore be abandoned as a useful control measure.

We shall now move on to some further general points.

Cores are almost never made from greensand because the volatile content (particularly water, of course) is too high and the permeability is too low. In addition, the cores would be weak and unable to support themselves on small prints; they would simply sag. If greensand is used at all then it is usually dried in an oven, producing 'dry sand' cores (their name should be more accurately 'dried sand' cores). These are relatively free of volatiles, and are mechanically strong, but retain the poor permeability of the original greensand. They therefore usually require additional venting. This is usually time consuming and labour intensive.

Sand cores are therefore nowadays generally made from clean, washed and dried silica sand that is closely graded in size to maintain as high a level of permeability as is possible. The limit to the size of sand grains and the permeability is set by the requirements of the casting to avoid penetration by

the metal, and the production of internal surfaces of the casting that are unacceptably rough.

These cores are bonded with a chemical binder that is cured by heat or chemical reaction to produce a rigid, easily handled shape. The numerous different systems in use all have different responses to the heat of the casting process, and produce gases of different kinds, in different amounts, at different times, and at different rates (Figures 4.10 and 4.11). For instance, the silicate ester produces most of its gas early, whereas the core oil shows a rapid but limited early evolution, and then a considerable delay before a second, more severe outgassing. These results are not to be taken as absolute in any sense. The manufacturers' products are changing all the time for a variety of reasons: health and safety; economics; commercial; changes in world markets and supplies of raw materials, etc. Thus binder formulations change and new systems are being developed all the time. At present the phenolic isocyanate–urethane systems are among the lowest overall producers of volatiles, which explains their current wide use as intricate cores, for instance in the case of water jackets for automobile cylinder heads and blocks.

Part of the reason for the historical success enjoyed by the phenolic urethane binders is their high strength, which means that the addition levels needed to achieve an easily handled core are low. This is one of the important factors in explaining their position near the bottom of Figure 4.10; the volume of gas evolved is, of course, proportional to the amount of binder present. This self-evident fact is clearly substantiated in the work of Scott et al. (1978), shown in Figure 4.13. (If allowance is made for the fact that these workers used a core sample size of 150 ml, corresponding to a weight of approximately 225 g, then the rate of evolution

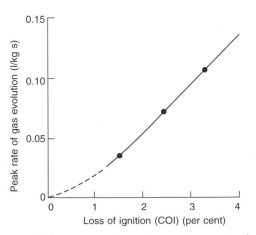

Figure 4.13 *Increase in the peak rate of outgassing as loss on ignition (LOI) increases. Data recalculated from Scott et al. (1978).*

measurements converted to l. kg^{-1} s^{-1} agree closely with those presented in Figure 4.11. This is despite the significant differences in the techniques. The data in Figure 4.11 may therefore be of more universal application than is apparent at first sight.)

4.3.2 Mould gas explosions

The various reactions of the molten metal with the volatile constituents of the mould, particularly the water in many moulding materials, would lead to explosive reactions if it were not for the fact that the reactions are dampened by the presence of masses of sand. Thus although the reactions in the mould are fierce, and not to be underestimated, in general they are not of explosive violence because the 90 per cent or more of the materials involved are inert (simply sand and possibly clay) and have considerable thermal inertia. Outgassing reactions are therefore rather steady and sustained.

These considerations do not apply to the mould cavity itself.

In the mould cavity the gases from the outgassing of the mould may contain a number of potentially flammable or explosive gases. These include a number of vapours such as hydrocarbons such as methane, other organics such as alcohols, and a number of reaction products such as hydrogen and carbon monoxide.

Because of the presence of these gases, explosions sometimes occur and sometimes not. The reasons have never been properly investigated. This is an unsatisfactory situation because the explosion of a mould during casting can be a nasty event. The author has witnessed this in furan-bonded boxless moulds when casting an aluminium alloy casting weighing over 50 kg: there was a muffled explosion, and large parts of the sand mould together with liquid metal flew apart in all directions. After several repeat performances the casters developed ways of pouring this component at the end of long-handled ladles, so as to keep as far away as possible. The cause always remained a mystery. Everyone was relieved when the job came to an end.

Explosions in and around moulds containing iron or steel castings are relatively common. One of the most common is from under the mould, between the mould and its base plate, after the casting has solidified, so that there is less danger either to personnel or casting.

With subsequent experience, and in the absence of any other suggestions, the following is suggested as a possible cause of the problem in the case of the light alloy casting.

Explosions can, of course, only happen when the flammable components of the gas mix with an oxidizing component such as oxygen from the air. The mixing has to be efficient, which suggests that turbulence is important. Also, the mix often has to

be within close compositional limits, otherwise either no reaction occurs, or only slow burning takes place. The limits for the carbon monoxide, oxygen and inert (carbon dioxide and nitrogen) gas mixtures are shown in Figure 4.14.

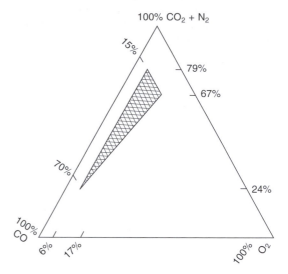

Figure 4.14 *Shaded region defines the explosive regime for carbon monoxide, oxygen and a mixture of carbon dioxide and nitrogen (Ellison and Wechselblatt 1966).*

In the author's experience, the mixing with air, which is essential for explosions, only occurs in moulds in certain conditions. These are moulds that are (i) open to air because of open feeder heads, or (ii) poured with oversize sprues that allow the ingress of air, or (iii) the use of double pouring, using two sprues, where the start of pour is not easy to synchronize, with the result that air is taken down one sprue at the time that metal enters down the other. Thus eliminating open feeders by the use either of blind feeders or chills promises to be a useful step. The provision of a properly calculated single sprue should also help.

'What happens to the air already in the mould?' is the next question. In a single-sprue system filling quiescently from the bottom upwards, the outgassing of the mould and cores will provide a spreading blanket of gas over the liquid. There will be almost no air in this cover, so that no burning or explosion can occur. The air will be displaced ahead, and will diffuse out of the upper parts of the mould. Where the flammable gas blanket meets the air it is expected to be cool, well away from the liquid metal. Thus any slight mixture that will occur at the interface between these layers of gases is not likely to ignite to cause an explosion.

In the case of the casting poured from two sprues, the second stream of metal might arrive to spark the spreading front of gases from the first stream.

In the case of an open feeder, the cold downdraught of fresh air into the mould is likely to penetrate and mix with the flammable blanket, and be sufficiently close to the molten metal to be ignited. A poorly designed turbulent filling system will undo all the good described above. The splashing of hot droplets and jets of metal through the vapour blanket, mixing it with the air, will give ideal conditions to spark an explosion.

That the event occurs from time to time in a random manner is to be expected. It is partly as a result of the randomness introduced by the turbulent mixing, and partly the sensitivity of the composition of the mixture, since Figure 4.14 confirms that only a limited compositional range is explosive.

4.4 Mould surface reactions

4.4.1 Pyrolysis

When the metal has filled the mould the mould becomes hot. A common misconception is to assume that the sand binder then burns. However, this is not true. It simply becomes hot. There is no oxygen to allow burning. What little oxygen is available is quickly consumed in a minor transient burning reaction. However, this quickly comes to a stop. What happens then to the binder is not burning, but pyrolysis.

Pyrolysis is the decomposition of compounds, usually organic compounds, simply by the action of heat. Oxygen is absent, so that no burning (i.e. high temperature oxidation) takes place. Pyrolysis of various kinds of organic binder components to produce carbon is one of the more important reactions that take place in the mould surface. This is because carbon is poorly wetted by many liquid metals. Thus the formation of carbon on the grains of sand, as a pyrolysed residue of the sand binder, produces a non-wetted mould surface, so assisting to create an improved surface finish to the casting (although, as will become clear, the effect of the surface film on the melt is probably more important).

This non-wetting feature of residual carbon on sand grains is at first sight curious, since carbon is soluble in many metals, and so should react and should therefore wet. Cast iron would be a prime candidate for this behaviour. Why does this not happen?

In the case of ductile iron, sand cores do not need a core coat, since the solid magnesium oxide-rich film on the surface is a mechanical barrier that prevents penetration of the metal into the sand.

In the case of grey cast iron in a greensand mould the atmosphere may be oxidizing, causing the melt surface to grow a film of liquid silicate. This is highly wetting to sand grains, so that the application of a core coating such as a refractory wash may be necessary to prevent penetration of the metal into the mould.

However, in the case of grey iron cast in a mould rich in hydrocarbons (i.e. greensand with heavy additions of coal dust, or certain resin-bonded sands) metal penetration is prevented when the hydrocarbons in the atmosphere of the mould decompose on the surface of the hot liquid metal to deposit a film of solid carbon on the liquid. Thus the reason for the robust non-wetting behaviour is that a solid carbon film on the liquid contacts a solid carbon layer at the mould surface. This twin aspect of non-wettability is considered further below.

For the casting of iron, powdered coal additions, or coal substitutes, are usually added to greensands, to improve surface finish in this way, providing a carbon layer to both the sand grains and the liquid surface. The reactions in the pyrolysis of coal were originally described by Kolorz and Lohborg (1963):

1. The volatiles are driven out of the coal to form a reducing atmosphere.
2. Gaseous hydrocarbons break down on the surface of the liquid metal. (Kolorz and Lohborg originally thought the hydrocarbons broke down on the sand grains to form a thin skin of graphite, but this is now almost certain to be not true.)
3. The coal swells and on account of its large expansion is driven into the pores of the sand. This plastic phase of the coal addition appears to plasticize the binder temporarily and thereby eases the problems associated with the expansion of the sand, allowing its expansion to be accommodated without the fracture of the surface. As the temperature increases further and the final volatiles are lost, the mass becomes rigid, converting to a semi-coke. The liquid metal is prevented from contacting and penetrating the sand by this in-filling of carbon, which acts as a non-wetted mechanical barrier.

Kolorz and Lohborg recommend synthetically formulated coal dusts with a high tendency to form anthracitic carbon, of good coking capacity and with good softening properties. They recommend that the volatile content be near 30 per cent, and sulphur less than 0.8 per cent if no sulphur contamination of the surface is allowable. If some slight sulphurization is permissible, then 1.0–1.2 per cent sulphur could be allowed.

In the case of phenolic urethane and similar organic chemical binders based on resin systems, the thermal breakdown of the binder assists the formation of a good surface finish to cast irons and other metals largely in the manner described above. The binder usually goes through its plastic stage prior to rigidizing into a coke-like layer. The much smaller volume fraction of binder, however, does not provide for the swelling of the organic phase to seal the pores between grains. So, in principle, the sand remains vulnerable to penetration by the liquid metal. The second aspect of non-wettability is discussed below.

4.4.2 Lustrous carbon film

The carbonaceous gases evolved from the binder complete their breakdown at the surface of the advancing front of liquid metal, giving up carbon and hydrogen to the advancing liquid front. For steel, the carbon dissolves quickly and usually causes relatively little problem. For cast iron, the carbon dissolves hardly at all, because the temperature is lower and the metal is already nearly saturated with carbon. Thus the carbon accumulates on the surface as a film, taking time to dissolve in the iron. The time for dissolution seems to be about the same as the time for mould filling and solidification. Thus the film has a life sufficiently long to affect the flow of the liquid.

As the film rolls out on the surface of the mould, it confers a non-wetting behaviour on the liquid itself because the liquid is effectively sealed in a non-wetting skin. The skin forms a mechanical barrier between the liquid and the mould. The barrier is laid down as the liquid progresses because of friction between the liquid and the mould, the friction effectively stretching the film and tearing it at the meniscus where it immediately re-forms to continue the process. The laying down of the film as the liquid progresses is analogous to the laying down of the tracks of a track-laying vehicle. The strength and rigidity of the carbon film helps the liquid surface to bridge unsupported regions between sand grains or other imperfections in the mould surface. By this mechanism the surface of the casting can be smoothed.

The mechanism for the improvement of surface finish can only operate effectively if the progress of the meniscus is steady and controlled, i.e. in the absence of surface turbulence.

However, if the carbon film becomes entrained in the liquid because of surface turbulence, becoming a carbon bifilm, it can constitute a serious crack-like defect. In heavy section castings the entrained defect has time to dissolve, and so is less of a problem. Binder systems that produce lustrous carbon should only be used for light section castings with running systems that can guarantee a low level of surface turbulence.

4.4.3 Sand reactions

Other reactions in the mould surface occur with the sand grains themselves. The most common of sand reactions is the reaction between silica (SiO_2) and iron oxide (wustite, FeO) to produce fayalite (Fe_2SiO_4). This happens frequently at the high temperatures required for the casting of irons and steels. It causes the grains to fuse and collapse as they melt into each other, because the melting point of fayalite is only 1205°C. The reacted grains adhere to the surface of the casting because of the presence of the low-melting-point liquid 'glue'. This is known as burn-on.

The common method of dealing with this problem is to prevent the iron oxidizing to form FeO in the first place. This is usually achieved by adding reducing agents to the mould material, such as powdered coal to greensand, or aluminium-powder additions to mould washes and the like. The problem is also reduced in other sands that contain less silica, such as chromite sand. However, the small amounts of silica which are present can still give trouble in steel castings, where the extreme temperature causes the residual silica to fuse with the clay. At these temperatures even the chromite itself may break down, releasing FeO or even Fe. Metal penetration usually follows as the grains melt into each other, and the mould surface generally collapses. The molten, fused mass is sometimes known as 'chromite glaze'. It is a kind of burn-on, and is difficult to remove from steel castings (Petro and Flinn 1978). Again, carbon compounds added to the moulding material are useful in countering this problem (Dietert et al. 1970).

4.4.4 Mould contamination

There are a few metallic impurities that find their way into moulding sands as a result of interaction with the cast metal. We are not thinking for the moment of the odd spanner or tonnes of iron filings from the steady wearing away of the sand plant. (Such ferrous contamination is retrieved in most sand plants by the provision of a powerful magnet located at some convenient point in the recirculating sand system.) Nor are we thinking of the pieces of tramp metal such as flash and other foundry returns. Our concern is with the microscopic traces of metallic impurities that lead to a number of problems, particularly because of the need to protect the environment from contamination.

Foundries that cast copper-based alloys containing lead find that their moulding sand becomes contaminated with lead (Mondloch et al. 1987). The lead is almost certainly lost from the casting by evaporation from the surface after casting. The vapour deposits among the sand grains in the mould as either particles of metallic alloy, or reacts with the clay present, particularly if this is bentonite, to produce Pb–Al silicates. If there is no clay present, as in chemical binder systems such as a furan resin, then no reaction is observed so that metallic lead remains (Ostrom et al. 1982). Thus ways of reducing this problem are: (i) the complete move, where possible in simple castings, to metal moulds; (ii) the complete move, where possible, from lead-containing alloys; or (iii) the use of chemical binders, together with the total recycling of sand in-house. This policy will contain the problem, and the separation of metallic lead from the dry sand in the recycling plant will provide a modest economic resource.

There has been a suggestion that iron can evaporate from the surface of a ferrous casting in the form of iron carbonyl $Fe(CO)_5$. This suggestion appears to have been eliminated on thermodynamic grounds; Svoboda and Geiger (1969) show that the compound is not stable at normal pressures at the temperature of liquid iron. Similar arguments eliminated the carbonyls of nickel, chromium and molybdenum. These authors survey the existing knowledge of the vapour pressures of the metal hydroxides and various sub-oxides but find conclusions difficult because the data is sketchy and contradictory. Nevertheless they do produce evidence that indicates vapour transport of iron and manganese occurs by the formation of the sub-oxides $(FeO)_2$ and $(MnO)_2$. The gradual transfer of the metal by a vapour phase, and its possible reduction back to the metal on arrival on the sand grains coated in carbon, might explain some of the features of metal penetration of the mould, which is often observed to be delayed, and then occur suddenly. More work is required to establish such a mechanism.

The evaporation of manganese from the surface of castings of manganese steel is an important factor in the production of these castings. The surface depletion of manganese seriously reduces the surface properties of the steel. In a study of this problem, Holtzer (1990) found that the surface concentration of manganese in the casting was depleted to a depth of 8 mm and the concentration of manganese silicates in the surface of the moulding sand was increased.

Figure 1.9 confirms that the vapour pressure of manganese is significant at the casting temperature of steel. However, the depth of the depleted surface layer is nearly an order of magnitude larger than can be explained by diffusion alone. It seems necessary to assume, therefore, that the transfer occurs mainly while the steel is liquid, and that some mixing of the steel is occurring in the vicinity of the cooling surface.

It is interesting that a layer of zircon wash on the surface of the mould reduces the manganese loss by about half. This seems likely to be the result of the thin zircon layer heating up rapidly, thereby reducing the condensation of the vapour. In addition, it will form a barrier to the progress of the manganese vapour, keeping the concentration of vapour near the equilibrium value close to the casting surface. Both mechanisms will help to reduce the rate of loss.

Gravity die casters that use sand cores (semi-permanent moulds) will be all too aware of the serious contamination of their moulds from the condensation of volatiles from the breakdown of resins in the cores. The build-up of these products can be so severe as to cause the breakage of cores, and the blocking of vents. Both lead to the scrapping of castings. The blocking of vents in permanent moulds is the factor that controls the length of a production run prior to the mould being taken out of service for cleaning. It is an advantage of sand moulding that is usually overlooked.

4.4.5 Mould penetration

Levelink and Berg have investigated and described conditions (Figure 4.15) in which they claimed that iron castings in greensand moulds were subject to a problem that they suggested was a water explosion. This led to a severe but highly localized form of mould penetration by the metal.

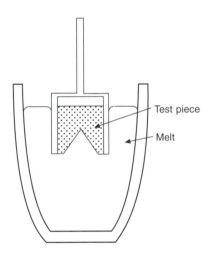

Figure 4.15 *Water hammer (momentum effect) test piece (Levelink and Berg 1971).*

However, careful evaluation of their work indicates that it seems most likely that they were observing a simple conservation of momentum effect. As the liquid metal fills the last volume of the mould it accelerates into the decreasing space, with the result that high shock pressure is generated, and sand penetration by the metal occurs. The effect is similar to a cavitation damage event associated with the collapse of bubbles against the ship's propeller. The oxides and bubbles that were present in many of their tests seem to be the result of entrainment in their rather poor filling system, and not associated with any kind of explosion.

The impregnation of the mould with metal in last regions to fill is commonly observed in all metals in sand moulds. A pressure pulse generated by the filling of a boss in the cope will often also cause some penetration in the drag surface too. The point discontinuity shown in Figure 2.27 will be a likely site for metal penetration into the mould. If the casting is thin-walled, the penetration on the front face will also be mirrored on its back face. Such surface defects in thin-walled aluminium alloy castings in sand moulds are unpopular, because the silvery surface of an aluminium alloy casting is spoiled by these dark spots of adhering sand, and thus will require the extra expense of blasting with shot or grit.

Levelink and Berg (1968) report that the problem is increased in greensand by the use of high-pressure moulding. This may be the result of the general rigidity of the mould accentuating the concentration of momentum (weak moulds will yield more generally, and thus dissipate the pressure over a wider area). They list a number of ways in which this problem can be reduced:

1. Reduce mould moisture.
2. Reduce coal and organics.
3. Improve permeability or local venting; gentle filling of mould to reduce final filling shock.
4. Retard moisture evaporation at critical locations by local surface drying or the application of local oil spraying.

The reduction in the mechanical forces involved by reduced pouring rates or by local venting are understandable as reducing the final impact forces. Similarly, the use of a local application of oil will reduce permeability, causing the air to be compressed, acting as a cushion to decelerate the flow more gradually.

The other techniques in their list seem less clear in their effects, and raise the concern that they may possibly be counterproductive! It seems there is plenty of scope for additional studies to clarify these problems.

Work over a number of years at the University of Alabama, Tuscaloosa (Lane *et al.* 1996), has clarified many of the issues relating to the penetration of sand moulds by cast iron. Essentially, this work concludes that hot spots in the casting, corresponding to regions of isolated residual liquid, are localized regions in which high pressures can be generated by the expansion of graphite. The pressure can be relieved by careful provision of 'feed paths' to allow the excess volume to be returned to the feeder. The so-called 'feed paths' are, of course, allowing residual liquid to escape, working in reverse of normal feeding. If feed paths are not provided, and if the hot spot region intersects the metal/mould interface, then the pressure is relieved by the residual melt forcing its way out to penetrate the mould.

Naturally, any excess pressure inside the casting will assist in the process of mould penetration. Thus large steel castings are especially susceptible to mould penetration because of the high metallostatic pressure. This factor is in addition to the other potential high-temperature reactions listed above. This is the reason for the widespread adoption in steel foundries of the complete coating of moulds with a ceramic wash.

4.5 Metal surface reactions

Easily the most widely occurring and most important metal/mould reaction is the reaction of the metal with water vapour to produce a surface oxide and hydrogen, as discussed in Chapter 1.

However, the importance of the release of hydrogen and other gases at the surface of the metal, leading to the possibility of porosity in the casting, is to be dealt with in Chapter 6. Here we shall devote ourselves to the many remaining reactions. Some are reviewed by Bates and Scott (1977). These and others are listed briefly below.

4.5.1 Oxidation

Oxidation of the casting skin is common for low carbon equivalent cast irons and for most low carbon steels. It is likely that the majority of the oxidation is the result of reaction with water vapour from the mould, and not from air, which is expelled at an early stage of mould filling as shown earlier. Carbon additions to the mould help to reduce the problem.

The catastrophic oxidation of magnesium during casting, leading to the casting (and mould) being consumed by fire, is prevented by the addition of so-called inhibitors to the mould. These include sulphur, boric acid and other compounds such as ammonium borofluoride. More recently, much use has been made of the oxidation-inhibiting gas, sulphur hexafluoride (SF_6), which is used diluted to about 2 per cent in air or other gas to prevent the burning of magnesium during melting and casting. However, since its identification as a powerful ozone-depleting agent, SF_6 is being discontinued for good environmental reasons. A return is being made to dilute mixtures of SO_2 in CO_2 and other more environmentally friendly atmospheres are now under development.

Titanium and its alloys are also highly reactive. Despite being cast under vacuum into moulds of highly stable ceramics such as zircon, alumina or yttria, the metal reacts to reduce the oxides, contaminating the surface of the casting with oxygen, stabilizing the alpha-phase of the alloy. The 'alpha-case' usually has to be removed by chemical machining.

4.5.2 Carburization

Mention has already been made of the problem of casting titanium alloy castings in carbon-based moulds. The carburization of the surface again results in the stabilization of the alpha-phase, and requires to be subsequently removed.

The difficulty is found with stainless steel of carbon content less than 0.3 per cent cast in resin-bonded (Croning) shell moulds (McGrath and Fischer 1973). The carburization, of course, becomes more severe the lower the carbon content of the steel. Also, the problem is worse on drag than on cope faces.

Carbon pick-up is the principal reason why low carbon steel castings are not produced by the lost-foam process. The atmosphere of styrene vapour, which is created in the mould as the polystyrene decomposes, causes the steel to absorb carbon (and presumably hydrogen). The carbon-rich regions of the casting are easily seen on an etched cross-section as swathes of pearlite in an otherwise ferritic matrix.

In controlled tests of the rate of carburization of low carbon steel in hydrocarbon/nitrogen mixtures at 925°C (Kaspersma and Shay 1982) methane was the slowest and acetylene the fastest of the carburizing agents tested, and hydrogen was found to enhance the rate, possibly by reducing adsorbed oxygen on the surface of the steel.

4.5.3 Decarburization

At high ratios of H_2/CH_4, hydrogen decarburizes steel at 925°C (Kaspersma and Shay 1982). This may be the important reaction in the casting of steel in greensand and resin-bonded sand moulds.

In the investment casting of steel, the decarburization of the surface layer is particularly affected because atmospheric oxygen persists in the mould as a consequence of the inert character of the mould, and its permeability to the surrounding environment. Doremus and Loper (1970) have measured the thickness of the decarburized layer on a low carbon steel investment casting and find that it increases mainly with mould temperature and casting modulus. The placing of the mould immediately after casting into a bin filled with charcoal helps to recarburize the surface. However, Doremus and Loper point out that there is a danger that if the timing and extent of recarburization is not correct, the decarburized layer will still exist below!

In iron castings the decarburization of the surface gives a layer free from graphite. This adversely affects machinability, giving pronounced tool wear, especially in large castings such as the bases of machine tools. The decarburization seems to be mainly the result of oxidation of the carbon by water vapour since dry moulds reduce the problem.

An addition of 5 or 6 per cent coal dust to the mould further reduces it. The reaction seems to start at about the freezing point of the eutectic, about 1150°C, and proceeds little further after the casting has cooled to 1050°C (Rickards 1975) (Figure 4.16).

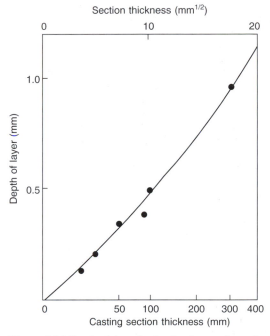

Figure 4.16 *Depth of decarburization in grey iron plates cast in greensand. Data from Rickards (1975).*

4.5.4 Sulphurization

The use of moulds bonded with furane resin catalysed with sulphuric and/or sulphonic acid causes problems for ferrous castings because of the pick-up of sulphur in the surface of the casting. This is especially serious for ductile iron castings, because the graphite reverts from spheroidal back to flake form in this high sulphur region. This has a serious impact on the fatigue resistance of the casting.

4.5.5 Phosphorization

The use of moulds bonded with furane resin catalysed with phosphoric acid leads to the contamination of the surfaces of ferrous castings with phosphorus. In grey iron the presence of the hard phosphide phase in the surface causes machining difficulties associated with rapid tool wear.

4.5.6 Surface alloying

There has been some Russian (Fomin *et al.* 1965) and Japanese (Uto and Yamasaki 1967) work on the alloying of the surface of steel castings by the provision of materials such as ferrochromium or ferromanganese in the facing of the mould. Because the alloyed layers that have been produced have been up to 3 or 4 mm deep, it is clear once again that not only is diffusion involved but also some additional transport of added elements must be taking place by mixing in the liquid state. Omel'chenko further describes a technique to use higher-melting-point alloying additions such as titanium, molybdenum and tungsten, by the use of exothermic mixes. Predictably enough, however, there appear to be difficulties with the poor surface finish and the presence of slag inclusions. Until this difficult problem is solved, the technique does not have much chance of attracting any widespread interest.

4.5.7 Grain refinement

The use of cobalt aluminate ($CoAl_2O_4$) in the primary mould coat for the grain refinement of nickel and cobalt alloy investment castings is now widespread. The mechanism of refinement is not yet understood. It seems unlikely that the aluminate as an oxide phase can wet and nucleate metallic grains. The fact that the surface finish of grain-refined castings is somewhat rougher than that of similar castings without the grain refiner indicates that some wetting action has occurred. This suggests that the particles of $CoAl_2O_4$ decompose to some metallic form, possibly $CoAl$. This phase has a melting point of 1628°C. It would therefore retain its solid state at the casting temperatures of Ni-based alloys. In addition it has an identical face-centred-cubic crystal structure. On being wetted by the liquid alloy it would constitute an excellent substrate for the initiation of grains. The effect is limited to a depth of about 1.25 mm in a Co–Cr alloy casting (Watmough 1980) and is limited to low casting temperatures (as is to be expected; there can be no refinement if all the CoAl particles are either melted or dissolved).

The addition of cobalt to a mould coat is also reported to grain-refine malleable cast iron (Bryant and Moore 1971), presumably for a similar reason.

The use of zinc in a mould coat to achieve a similar aim in iron castings must involve a quite different mechanism, because the temperature of liquid iron greatly exceeds not only the melting point, but even the boiling point of zinc! It may be that the action of the zinc boiling at the surface of the solidifying casting may disrupt the formation of the dendrites, detaching them from the surface so that they become freely floating nuclei within the melt. Thus the grain refining mechanism in this case is grain multiplication rather than nucleation. The effect seems analogous to that described in section 3.3.3.2 for acetylene black and hexachlorethane coatings on moulds.

4.5.8 Miscellaneous

Boron has been picked up in the surfaces of stainless steel castings from furane-bonded moulds that contain boric acid as an accelerator (McGrath and Fischer 1973).

Tellurium is sometimes deliberately added as a mould wash to selected areas of a grey iron casting. Tellurium is a strong carbide former, and will locally convert the structure of the casting from grey to a fully carbidic white iron. This action is said to be taken to reduce local internal shrinkage problems, although its role in this respect seems difficult to understand. It has been suggested that a solid skin is formed rapidly, equivalent to a thermal chill (Vandenbos 1985). The effect needs to be used with caution: tellurium and its fumes are toxic, and the chilled region causes machining difficulties.

The effect of tellurium converting grey to white irons is used to good purpose in the small cups used for the thermal analysis of cast irons. Tellurium is added as a wash on the inside of the cup. During the pouring of the iron it seems to be well distributed into the bulk of the sample, not just the surface, so that the whole test piece is converted from grey to white iron. This simplifies the interpretation of the cooling curve, allowing the composition of the iron to be deduced.

Chapter 5

Solidification structure

In this chapter we consider how the metal changes state from the liquid to the solid, and how the solid develops its structure, together with its pore structure due to the precipitation of gas.

In a later chapter we consider the problems of the usual volume deficit on solidification, and the so-called shrinkage problems that lead to a different set of void phenomena, sometimes appearing as porosity.

This highlights the problem for the author. The problem is how to organize the descriptions of the complex but inter-related phenomena that occur during the solidification of a casting. This book could be organized in many different ways. For instance, naturally, the gas and shrinkage contributions to the overall pore structure are complementary and additive.

The reader is requested to be vigilant to see this integration. I am conscious that while spelling out the detail in a didactic dissection of phenomena, emphasizing the separate physical mechanisms, the holistic vision for the reader is easily lost.

5.1 Heat transfer

5.1.1 Resistances to heat transfer

The hot liquid metal takes time to lose its heat and solidify. The rate at which it can lose heat is controlled by a number of resistances described by Flemings (1974). We shall follow his clear treatment in this section.

The resistances to heat flow from the interior of the casting are:

1. The liquid.
2. The solidified metal.
3. The metal/mould interface.
4. The mould.

5. The surroundings of the mould.

All these resistances add, as though in series, as shown schematically in Figure 5.1.

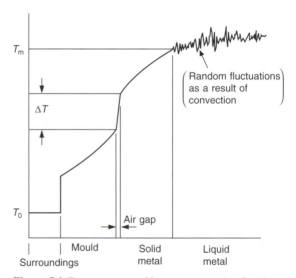

Figure 5.1 *Temperature profile across a casting freezing in a mould, showing the effect of the addition of thermal resistances that control the rate of loss of heat.*

As it happens, in nearly all cases of interest, resistance (1) is negligible, as a result of bulk flow by forced convection during filling and thermal convection during cooling. The turbulent flow and mixing quickly transport heat and so smooth out temperature gradients. This happens quickly since bulk flow of the liquid is fast, and the heat is transported out of the centre of large ingots and castings in a time that is short compared to that required by the remaining resistances, whose rate is controlled by diffusion.

In many instances, resistance (5) is also negligible in practice. For instance, for normal sand moulds the environment of the mould does not affect solidification, since the mould becomes hardly warm on its outer surface by the time the casting has solidified inside. However, there are, of course, a number of exceptions to this general rule, all of which relate to various kinds of thin-walled moulds, which, because of the thinness of the mould shell, are somewhat sensitive to their environment. Iron castings made in Croning shell moulds (the Croning shell process is one in which the sand grains are coated with a thermosetting resin, which is cured against a hot pattern to produce a thin, biscuit-like mould) solidify faster when the shell is thicker, or when the shell is thin and backed up with steel shot. Conversely, the freezing of investment shell castings in steel is delayed by a backing to the shell of granular refractory material preheated to high temperature, but is accelerated by being allowed to radiate heat away freely to cool the surroundings. Iron and steel dies for the casting of aluminium alloys cool faster when the backs of the dies are cooled by water.

Nevertheless, despite such useful ploys for coaxing greater productivity, it remains essential to understand that in general the major fundamental resistances to heat flow from castings are items (2), (3) and (4). For convenience we shall call these resistances 1, 2 and 3.

The effects of all three simultaneously can nowadays be simulated with varying degrees of success by computer. However, the problem is both physically and mathematically complex, especially for castings of complex geometry.

There is therefore still much understanding and useful guidance to be obtained by a less ambitious approach, whereby we look at the effect of each resistance in isolation, considering only one dimension (i.e. unidirectional heat flow). In this way we can define some valuable analytical solutions that are surprisingly good approximations to casting problems. We shall continue to follow the approach by Flemings.

5.1.1.1 Resistance 1: The casting

It has to be admitted that this type of freezing regime is not common for metal castings of high thermal conductivity such as the light alloys or Cu-based alloys.

However, it would nicely describe the casting of Pb–Sb alloy into steel dies for the production of battery grids and terminals; the casting of steel into a copper mould; or the casting of hot wax into metal dies as in the injection of wax patterns for investment casting. It would be of wide application in the plastics industry.

For the unidirectional flow of heat from a metal poured exactly at its melting point T_m against a mould wall initially at temperature T_0, the transient heat flow problem is described by the partial differential equation, where α_s is the thermal diffusivity of the solid:

$$\frac{\partial T}{\partial t} = \alpha_s \frac{\partial^2 T}{\partial x^2} \quad (5.1)$$

The boundary conditions are $x = 0$, $T = T_0$; at $x = S$, $T = T_m$, and at the solidification front the rate of heat evolution must balance the rate of conduction down the temperature gradient, i.e.:

$$H\rho_s \left(\frac{\partial S}{\partial t} \right) = K_s \left(\frac{\partial T}{\partial x} \right)_{x=s} \quad (5.2)$$

where K_s is the thermal conductivity of the solid, H is the latent heat of solidification, and for which the solution is:

$$S = 2\gamma \sqrt{\alpha_s t} \quad (5.3)$$

The reader is referred to Flemings for the rather cumbersome relation for γ. The important result to note is the parabolic time law for the thickening of the solidified shell. This agrees well with experimental observations. For instance, the thickness S of steel solidifying against a cast iron ingot mould is found to be:

$$S = at^{1/2} - b \quad (5.4)$$

where the constants a and b are of the order of 3 and 25 respectively when the units are millimetres and seconds. The result is seen in Figure 5.2.

The apparent delay in the beginning of solidification shown by the appearance of the constant b is a consequence of the following: (i) the turbulence of the liquid during and after pouring, resulting in the loss of superheat from the melt, and so slowing the start of freezing, and (ii) the finite interface resistance further slows the initial rate of heat loss. Initially the solidification rate will be linear, as described in the next section (and hence giving the initial curve in Figure 5.2 because of this plot using the square root of time). Later, the resistance of the solidifying metal becomes dominant, giving the parabolic relation (shown, of course, as a straight line in Figure 5.2 because of the plot using the square root plot of time).

5.1.1.2 Resistance 2: The metal/mould interface

In many important casting processes heat flow is controlled to a significant extent by the resistance at the metal/mould interface. This occurs when both the metal and the mould have reasonably good rates of heat conductance, leaving the boundary between the two the dominant resistance. The interface

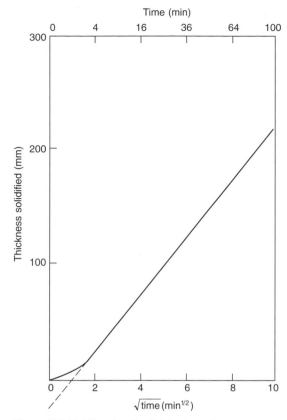

Figure 5.2 *Unidirectional solidification of pure iron against a cast iron mould coated with a protective wash (from Flemings 1974).*

becomes overriding in this way when an insulating mould coat is applied, or when the casting cools and shrinks away from the mould (and the mould heats up, expanding away from the metal), leaving an air gap separating the two. These circumstances are common in the die casting of light alloys.

For unidirectional heat flow the rate of heat released during solidification of a solid of density ρ_s and latent heat of solidification H is simply:

$$q = -\rho_s H A \frac{\partial S}{\partial t} \qquad (5.5)$$

This released heat has to be transferred to the mould. The heat transfer coefficient h across the metal/mould interface is simply defined as the rate of transfer of energy q (usually measured in watts) across unit area (usually a square metre) of the interface, per unit temperature difference across the interface. This definition can be written:

$$q = -hA(T_m - T_0) \qquad (5.6)$$

assuming the mould is sufficiently large and conductive not to allow its temperature to increase

significantly above T_0, effectively giving a constant temperature difference $(T_m - T_0)$ across the interface. Hence equating 5.5 and 5.6 and integrating from $S = 0$ at $t = 0$ gives:

$$S = \frac{h(T_m - T_0)}{\rho_s H} \bullet t \qquad (5.7)$$

It is immediately apparent that since shape is assumed not to alter the heat transfer across the interface, Equation 5.7 may be generalized for simple-shaped castings to calculate the solidification time t_f in terms of the volume V to cooling surface area A ratio (the geometrical modulus) of the casting:

$$t_f = \frac{\rho_s H}{h(T_m - T_0)} \bullet \frac{V}{A} \qquad (5.8)$$

All of the above calculations assume that h is a constant. As we shall see later, this is perhaps a tolerable approximation in the case of gravity die (permanent mould) casting of aluminium alloys where an insulating die coat has been applied. In most other situations h is highly variable, and is particularly dependent on the geometry of the casting.

The air gap

As the casting cools and the mould heats up, the two remain in good thermal contact while the casting interface is still liquid. When the casting starts to solidify, it rapidly gains strength, and can contract away from the mould. In turn, as the mould surface increases in temperature it will expand. Assuming for a moment that this expansion is homogeneous, we can estimate the size of the gap d as a function of the diameter D of the casting:

$$d/D = \alpha_c \{T_f - T\} + \alpha_m \{T_{mi} - T_0\}$$

where α is the coefficient of thermal expansion, and subscripts c and m refer to the casting and mould respectively. The temperatures T are T_f the freezing point, T_{mi} the mould interface, and T_0 the original mould temperature.

The benefit of the gap equation is that it shows how straightforward the process of gap formation is. It is simply a thermal contraction–expansion problem, directly related to interfacial temperature. It indicates that for a casting a metre across which is allowed to cool to room temperature the gap would be expected to be of the order of 10 mm at each of the opposite sides. This is a substantial gap by any standards!

Despite the usefulness of the elementary formula in giving some order-of-magnitude guidance on the dimensions of the gap, there are a number of interesting reasons why this simple approach requires further sophistication.

In a thin-walled aluminium alloy casting of section only 2 mm the room temperature gap would be only 10 μm. This is only one-twentieth of the size of an average sand grain of 200 μm diameter. Thus the imagination has some problem in visualizing such a small gap threading its way amid the jumble of boulders masquerading as sand grains. It really is not clear whether it makes sense to talk about a gap in this situation.

Woodbury and co-workers (2000) lend support to this view for thin wall castings. In horizontally sand cast aluminium alloy plates of 300 mm square and up to 25 mm thickness, they measured the rate of transfer of heat across the metal/mould interface. They confirmed that there appeared to be no evidence for an air gap. Our equation would have predicted a gap of 25 μm. This small distance could easily be closed by the slight inflation of the casting because of two factors: (i) the internal metallostatic pressure provided by the filling system (no feeders were used), and (ii) the precipitation of a small amount of gas; for instance, it can be quickly shown that 1 per cent porosity would increase the thickness of the plate by at least 70 μm. Thus the plate would swell by creep under the combined internal pressure due to head height and the growth of gas pores with minimal difficulty. The 25 μm movement from thermal contraction would be so comfortably overwhelmed that a gap would probably never have chance to form.

Our simple air gap formula assumes that the mould expands homogeneously. This may be a reasonable assumption for the surface of a greensand mould, which will expand into its surrounding cool bulk material with little resistance. A rigid, chemically bonded sand will be subject to more restraint, thus preventing the surface from expanding so freely. The surface of a metal die will, of course, be most constrained of all by the surrounding metal at lower temperature, but the higher conductivity of the mould will raise the temperature of the whole die more uniformly, giving a better approximation once again to homogeneous expansion.

Also, the sign of the mould movement for the second half of the equation is only positive if the mould wall is allowed to move outwards because of small mould restraint (i.e. a weak moulding material) or because the interface is concave. A rigid mould and/or a convex interface will tend to cause inward expansion, reducing the gap, as shown in Figure 5.3. It might be expected that a flat interface will often be unstable, buckling either way. However, Ling and co-workers (2000) found that both theory and experiment agreed that the walls of their cube-like mould poured with white cast iron distorted outwards in the case of greensand moulds, but inwards in the case of the more rigid chemically bonded moulds.

There are further powerful geometrical effects

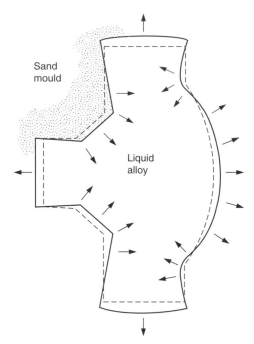

Figure 5.3 *Movement of mould walls, illustrating the principle of inward expansion in convex regions and outward expansion in concave regions.*

to upset our simple linear temperature relation. Figure 5.4 shows the effect of linear contraction during the cooling of a shaped casting. Clearly, anything in the way of the contraction of the straight lengths of the casting will cause the obstruction to be forced hard against the mould. This happens in the corners at the ends of the straight sections. Gaps cannot form here. Similarly, gaps will not occur around cores that are surrounded with metal, and on to which the metal contracts during cooling. Conversely, large gaps open up elsewhere. The situation in shaped castings is complicated and is only just being tackled with some degree of success by computer models.

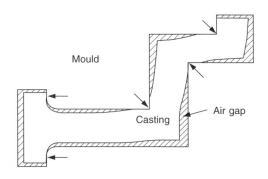

Figure 5.4 *Variable air gap in a shaped casting: arrows denote the probable sites of zero gap.*

Richmond and Tien (1971) and Tien and Richmond (1982) demonstrate via a theoretical model how the formation of the gap is influenced by the internal hydrostatic pressure in the casting, and by the internal stresses that occur within the solidifying solid shell. In Richmond *et al.* (1990) Richmond goes on to develop his model further, showing that the development of the air gap is not uniform, but is patchy. He found that air gaps were found to nucleate adjacent to regions of the solidified shell that were thin, because, as a result of stresses within the solidifying shell, the casting/mould interface pressure first dropped to zero at these points. Conversely, the casting/mould interface pressure was found to be raised under thicker regions of the solid shell, thereby enhancing the initial non-uniformity in the thickness of the solidifying shell. Growth becomes unstable, automatically moving away from uniform thickening. This rather counter-intuitive result may help to explain the large growth perturbations that are seen from time to time in the growth fronts of solidifying metals. Richmond reviews a considerable amount of experimental evidence to support this model. All the experimental data seem to relate to solidification in metal moulds. It is possible that the effect is less severe in sand moulds.

Attempts to measure the gap formation directly (Isaac *et al.* 1985; Majumdar and Raychaudhuri 1981) are extremely difficult to carry out accurately. Results averaged for aluminium cast into cast iron dies of various thickness reveal the early formation of the gap at the corners of the die where cooling is fastest, and the subsequent spread of the gap to the centre of the die face. A surprising result is the reduction of the gap if thick mould coats are applied. (The results in Figure 5.5 are plotted as straight lines. The apparent kinks in the early opening of the gap reported by these authors may be artefacts of their experimental method.)

It is not easy to see how the gap can be affected by the thickness of the coating. The effect may be the result of the creep of the solid shell under the internal hydrostatic pressure of the feeder. This is more likely to be favoured by thicker mould coats as a result of the increased time available and the increased temperature of the solidified skin of the casting. If this is true then the effect is important because the hydrostatic head in these experiments was modest, only about 200 mm. Thus for aluminium alloys that solidify with higher heads and times as long or longer than a minute or so, this mechanism for gap reduction will predominate. It seems possible, therefore, that in gravity die casting of aluminium the die coating will have the major influence on heat transfer, giving a large and stable resistance across the interface. The air gap will be a small and variable contributor. For computational purposes, therefore, it is attractive

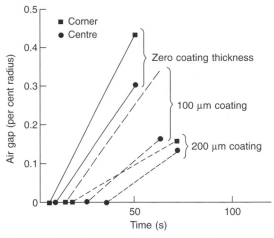

Figure 5.5 *Results averaged from various dies (Isaac et al. 1985), illustrating the start of the air gap at the corners, and its spread to the centre of the mould face. Increased thickness of mould coating is seen to delay solidification and to reduce the growth of the gap.*

to consider the great simplification of neglecting the air gap in the special case of gravity die casting of aluminium.

In conclusion, it is worth mentioning that the name 'air gap' is perhaps a misnomer. The gap will contain almost everything except air. As we have seen previously, mould gases are often high in hydrogen, containing typically 50 per cent. At room temperature the thermal conductivity of hydrogen is approximately 6.9 times higher than that of air, and at 500°C the ratio rises to 7.7. Thus, the conductivity of a gap at the casting/mould interface containing a 50:50 mixture of air and hydrogen at 500°C can be estimated to be approximately a factor of 4 higher than that of air. In the past, therefore, most investigators in this field have probably chosen the wrong value for the conductivity of the gap, and by a substantial margin!

The heat-transfer coefficient

The authors Ho and Pehlke (1984) from the University of Michigan have reviewed and researched this area thoroughly. We shall rely mainly on their work in this section.

When the metal first enters the mould the macroscopic contact is good because of the conformance of the molten metal. Gaps exist on a microscale between high spots as shown in Figure 5.6. At the high spots themselves, the high initial heat flux causes nucleation of the metal by local severe undercooling (Prates and Biloni 1972). The solid then spreads to cover most of the surface of the casting. Conformance and overall contact between the surfaces is expected to remain good during all of this early period, even though the

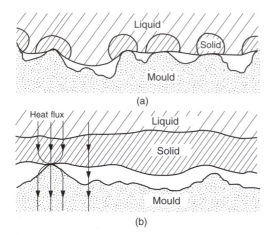

Figure 5.6 *Metal/mould interface at an early stage when solid is nucleating at points of good thermal contact. Overall macroscopic contact is good at this stage (a). Later (b) the casting gains strength, and casting and mould both deform, reducing contact to isolated points at greater separations on non-conforming rigid surfaces.*

mould will now be starting to move rapidly because of distortion.

After the creation of a solidified layer with sufficient strength, further movements of both the casting and the mould are likely to cause the good fit to be broken, so that contact is maintained across only a few widely spaced random high spots (Figure 5.6b).

The total transfer of heat across the interface may be written as the sum of three components:

$$h_t = h_s + h_c + h_r$$

where h_s is the conduction through the solid contacts, h_c is the conduction through the gas phase, and h_r is that transferred by radiation. Ho and Pehlke produce analytical equations for each of these contributors to the total heat flux. We can summarize their findings as follows:

1. While the casting surface can conform, the contribution of solid–solid conduction is the most important. In fact, if the area of contact is enhanced by the application of pressure, then values of h_t up to 60 000 $Wm^{-2}K^{-1}$ are found for aluminium in squeeze casting. Such high values are quickly lost as the solid thickens and conformance is reduced, the values falling to more normal levels of 100–1000 $Wm^{-2}K^{-1}$ (Figure 5.7).
2. When the interface gap starts to open, the conduction through solid contacts becomes negligible. The point at which this happens is clear in Figure 5.7b. (The actual surface temperature of the casting and the chill in this figure are reproduced from the results calculated by Ho and Pehlke.) The rapid fall of the casting surface temperature is suddenly halted, and reheating of the surface starts to occur. An interesting mirror image behaviour can be noted in the surface temperature of the chill, which, now out of contact with the casting, starts to cool. The estimates of heat transfer are seen to simultaneously reduce from over 1000 to around 100 $Wm^{-2}K^{-1}$ (Figure 5.7c).
3. After solid conduction diminishes, the important mechanism for heat transfer becomes the conduction of heat through the gas phase. This is calculated from:

$$h_c = k/d$$

where k is the thermal conductivity of the gas and d is the thickness of the gap. An additional correction is noted by Ho and Pehlke for the case where the

Table 5.1 Mould and metal constants

Material	Melting point (°C)	Liquid–solid contraction (%)	Specific heat (J.Kg K)			Density (kg/m³)			Thermal conductivity (J/m K s)		
			Solid		Liquid	Solid		Liquid	Solid		Liquid
			20°C	m.p.	m.p	20°C	m.p.	m.p.	20°C	m.p.	m.p.
Pb	327	3.22	130	(138)	152	11680	11020	10678	39.4	(29.4)	15.4
Zn	420	4.08	394	(443)	481	7140	(6843)	6575	119	95	9.5
Mg	650	4.2	1038	(1300)	1360	1740	(1657)	1590	155	(90)?	78
Al	660	7.14	917	(1200)	1080	2700	(2550)	2385	238	–	94
Cu	1084	5.30	386	(480)	495	8960	8382	8000	397	(235)	166
Fe	1536	3.16	456	(1130)	795	7870	7265	7015	73	14)?	–
Graphite	–	–	1515	–	–	2200	–	–	147		
Silica sand	–	–	1130	–	–	1500	–	–	0.0061	–	
Investment (Mullite)			750	–	–	1600	–	–	0.0038	–	–
Plaster	–	–	840	–	–	1100	–	–	0.0035	–	–

References: Wray (1976); Brandes (1991); Flemings (1974)

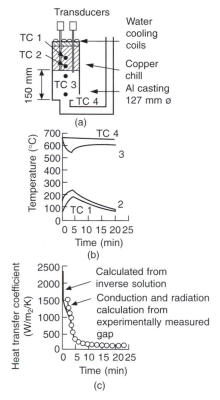

Figure 5.7 *Results from Ho and Pehlke (1984) illustrating the temperature history across a casting/chill interface, and the inferred heat transfer coefficient.*

gap is smaller than the mean free path of the gas molecules, which effectively reduces the conductivity. Thus heat transfer now becomes a strong function of gap thickness. As we have noted above, it will also be a strong function of the composition of the gas. Even a small component of hydrogen will greatly increase the conductivity.

For the case of light alloys, Ho and Pehlke find that the contribution to heat transfer from radiation is of the order of 1 per cent of that due to conduction by gas. Thus radiation can be safely neglected at these temperatures.

Heat transfer coefficients have been calculated by Hallam *et al.* (2000) for the case of Al alloy gravity die (permanent mould) castings. They demonstrate excellent predictions based on the assumption that the resistance of the die coating is mainly due to the gas voids between the casting and the coating surface. Thus the character of the coating surface was a highly influential factor in determining the heat transfer across the casting/mould interface.

For higher-temperature metals, results by Jacobi (1976) from experiments on the casting of steels in different gases and in vacuum indicate that radiation

becomes of increasing importance to heat transfer at these higher temperatures.

5.1.1.3 Resistance 3: The mould

The rate of freezing of castings made in silica sand moulds is generally controlled by the rate at which heat can be absorbed by the mould. In fact, compared to many other casting processes, the sand mould acts as an excellent insulator, keeping the casting warm. However, of course, ceramic investment and plaster moulds are even more insulating, avoiding premature cooling of the metal, and aiding fluidity to give the excellent ability to fill thin sections for which these casting processes are renowned. It is regrettable that the extremely slow cooling can contribute to rather poorer mechanical properties.

Considering the simplest case of unidirectional conditions once again, and metal poured at its melting point T_m against an infinite mould originally at temperature T_0, but whose surface is suddenly heated to temperature T_m at $t = 0$, and that has thermal diffusivity α_m, we now have:

$$\frac{\partial T}{\partial t} = \alpha_m \frac{\partial^2 T}{\partial x^2} \qquad (5.9)$$

Following Flemings, the final solution is:

$$S = \frac{2}{\sqrt{\pi}} \underbrace{\left(\frac{T_m - T_0}{\rho_s H} \right)}_{\text{metal}} \underbrace{\sqrt{K_m \rho_m C_m}}_{\text{mould}} \sqrt{t} \qquad (5.10)$$

This relation is most accurate for the highly conducting non-ferrous metals aluminium, magnesium and copper. It is less good for iron and steel, particularly those ferrous alloys that solidify to the austenitic (face-centred cubic) structure that has especially poor conductivity.

Note that at a high temperature heat is lost more quickly, so that a casting in steel should solidify faster than a similar casting in grey iron. This perhaps surprising conclusion is confirmed experimentally, as seen in Figure 5.8.

Low heat of fusion of the metal, H, similarly favours rapid freezing because less heat has to be removed. Therefore despite their similar freezing points, magnesium castings freeze faster than similar castings in aluminium.

The product $K_m \rho_m C_m$ is a useful parameter to assess the rate at which various moulding materials can absorb heat. The reader needs to be aware that some authorities have called this the heat diffusivity, and this definition was followed in *Castings* (Campbell 1991). However, originally the definition of heat diffusivity b was $(K_m \rho_m C_m)^{1/2}$ as described for instance by Ruddle (1950). In subsequent years the square root seems to have

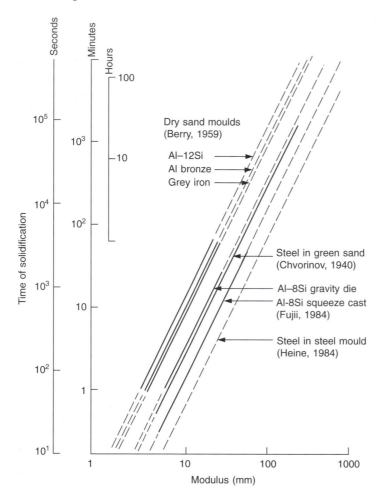

Figure 5.8 *Freezing times of plate-shaped castings in different alloys and moulds.*

been overlooked in error. Ruddle's definition is therefore accepted and followed here. However, of course, both b and b^2 are useful quantitative measures. What we call them is merely a matter of definition. (I am grateful to John Berry of Mississippi State University for pointing out this fact. As a

further aside from Professor Berry, the units of b are even more curious than the units of toughness; see Table 5.2.)

For simple shapes, if we assume that we may replace S with V_s/A where V_s is the volume solidified at a time t, and A is the area of the metal/mould

Table 5.2 Thermal properties of mould and chill materials at approximately 20°C

Material	Heat Diffusivity $(K\rho C)^{1/2}$ $(Jm^{-2} K^{-1}s^{-1/2})$	Thermal Diffusivity $K/\rho C$ (m^2s^{-1})	Heat Capacity per unit volume ρC $(JK^{-1}m^{-3})$
Silica sand	3.21×10^3	3.60×10^{-9}	1.70×10^6
Investment	2.12×10^3	3.17×10^{-9}	1.20×10^6
Plaster	1.8×10^3	3.79×10^{-9}	0.92×10^6
Iron (pure Fe)	16.2×10^3	20.3×10^{-6}	3.94×10^6
Graphite	22.1×10^3	44.1×10^{-6}	3.33×10^6
Aluminium	24.3×10^3	96.1×10^{-6}	2.48×10^6
Copper	37.0×10^3	114.8×10^{-6}	3.60×10^6

interface (i.e. the cooling area of the casting), then when $t = t_f$ where t_f is the total freezing time of a casting of volume V we have:

$$\frac{V}{A} = \frac{2}{\sqrt{\pi}}\left(\frac{T_m - T_0}{\rho_s H}\right)\sqrt{K_m \rho_m C_m}\ \sqrt{t_f} \quad \textbf{(5.11)}$$

and so:

$$t_f = B(V/A)^2 \quad \textbf{(5.12)}$$

where B is a constant for given metal and mould conditions.

Equation 5.12 is the famous Chvorinov rule. Convincing demonstrations of its accuracy have been made many times. Chvorinov himself showed in his paper published in 1940 that it applied to steel castings from 12 to 6 000 kg weight made in greensand moulds. This superb result is presented in Figure 5.9. Experimental results for other alloys are illustrated in Figure 5.8.

Chvorinov's rule is one of the most useful guides to the student. It provides a powerful general method of tackling the feeding of castings to ensure their soundness.

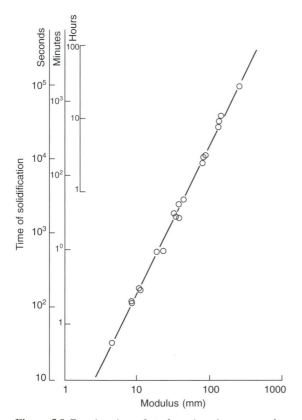

Figure 5.9 *Freezing time of steel castings in greensand moulds as a function of modulus (Chvorinov 1940). Some results for other metal/mould systems have been summarized in Figure 5.8.*

However, the above derivation of Chvorinov's rule is open to criticism in that it uses one-dimensional theory but goes on to apply it to three-dimensional castings. In fact, it is quickly appreciated that the flow of heat into a concave mould wall will be divergent, and so will be capable of carrying away heat more rapidly than in a one-dimensional case. We can describe this exactly (without the assumption of one-dimensional heat flow), following Flemings once again:

$$\frac{\partial T}{\partial t} = \alpha_m\left(\frac{\partial^2 T}{\partial r^2} + \frac{n\partial T}{r\partial r}\right) \quad \textbf{(5.13)}$$

where $n = 0$ for a plane, 1 for a cylinder, and 2 for a sphere. The casting radius is r. The solution to this equation is:

$$\frac{V}{A} = \left(\frac{T_m - T_0}{\rho_s H}\right)$$
$$\times\left(\frac{2}{\sqrt{\pi}}\sqrt{K_m \rho_m C_m}\ \sqrt{t_f} + \frac{nK_m t_f}{2r}\right)$$
$$\textbf{(5.14)}$$

The effect of the divergency of heat flow predicts that for a given value of the ratio V/A (i.e. a given modulus) a sphere will freeze quickest, the cylinder next and the plate last. Katerina Trbizan (2001) provides a useful study, confirming these relative freezing rates for these three shapes. For aluminium in sand moulds, Equation 5.14 indicates these differences to be close to 20 per cent. This is the reason for the safety factor 1.2 recommended when applying Chvorinov's feeding rule since the feeding rules tacitly assume that all shapes with the same modulus freeze at the same time.

It should be noted that the simple Chvorinov link between modulus and freezing time is capable of great sophistication. One of the great exponents of this approach has been Wlodawer (1966), who produced a famous volume devoted to the study of the problem for steel castings. This has been a source book for the steel castings industry ever since.

A final aspect relating to the divergency of heat flow is important. For a planar freezing front, the rate of increase of the solidified metal is parabolic, gradually slowing with thickness, as described by equations such as 5.3 and 5.4 relating to one-dimensional heat flow. However, for more compact shapes such as cylinders, spheres, cubes, etc. the heat flow from the casting is three-dimensional. Thus initially for such shapes, when the solidified layer is relatively thin, the solid thickens parabolically. However, when little liquid remains in the centre of the casting, the extraction of heat in all three directions greatly accelerates the rate of freezing. Santos and Garcia (1998) show that the effect is general. Whereas in a slab casting the velocity of the front slows progressively with distance according to the well-known parabolic law, for cylinders and spheres the growth rate is similar until the front has progressed to about 40 per cent of the radius. From then onwards the front accelerates rapidly (Figure 5.10).

This increase of the rate of freezing in the interior of many castings explains the otherwise baffling observation of 'inverse chill' as seen in cast irons. Normal intuition would lead the caster to expect fast cooling near the surface of the casting, and this is true to a modest degree in all castings. From this point onwards the front slows progressively in uniform plate-like sections, but speeds up dramatically in bars and cylinders, causing grey iron to change to carbidic white iron in the centre of the casting. The accelerated rate has been demonstrated experimentally by Santo and Garcia on a Zn–4Al alloy by measurement of the increasing fineness of dendrite arm spacing towards the centre of a cylindrical casting.

5.1.2 Increased heat transfer

In practice, the casting engineer can manipulate the rate of heat extraction from a casting using a number of tricks. These include the placement of chill blocks in the mould, adjacent to the casting, or fins attached to the casting to increase the surface area through which heat can be dissipated. These techniques will be described in detail in Volume II of this series.

5.1.3 Convection

Convection is the bulk movement of the liquid under the driving force of density differences in the liquid. In section 5.3.4 we shall consider the problems raised by convection driven by solutes; heavy solutes cause the liquid to sink, and the lighter solutes cause flotation. In this section we shall confine our discussion simply to the effects of temperature: hot liquid will expand, becoming less dense, and will rise; cool liquid will contract, becoming denser, and so will sink.

The existence of convection has been cited as important because it affects the columnar to equiaxed transition (Smith et al. 1990). There may be some truth in this. However, in most castings, grain structure is much less important than soundness, and it seems to be little known that convection can give severe soundness problems.

The problems of convective flow create serious problems in counter-gravity filling systems. Figures 5.11 and 5.12 illustrate how, after the mould cavity is completely filled, the temperature gradient in the mould is as wrong as it could be: the hot metal is at the bottom and the cold metal at the top. As the casting starts to solidify, the cold liquid metal drifts downward, draining into the riser tube. Here it is replaced by hot metal flowing up the heated riser tube and into the casting. This freshly reheated metal can remelt a channel through the pasty zone. If the heat input to the furnace at the base of the riser tube is sufficient then a circulation is set up which can become infinitely perpetuating; the rate

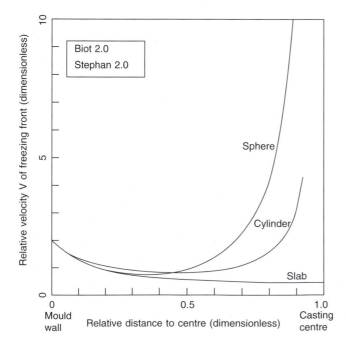

Figure 5.10 Acceleration of the freezing front in compact castings as a result of 3-D extraction of heat (sphere and cylinder curves calculated from Santos and Garcia 1998).

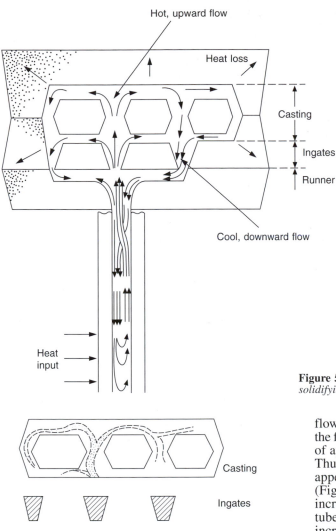

Hot, upward flow

Heat loss

Casting

Ingates

Runner

Cool, downward flow

Heat input

Figure 5.11 *Convection-driven flow within a solidifying low-pressure casting.*

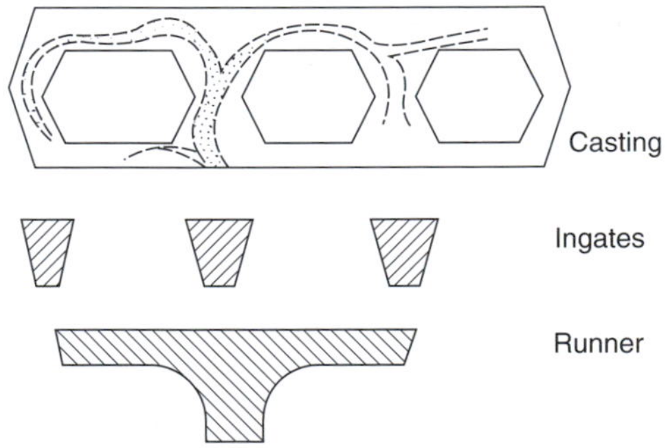

Casting

Ingates

Runner

Figure 5.12 *Remnants of the convective plumes in a casting, defining regions of coarse structure and porosity.*

of heat input from below equals the rate of heat loss from the casting. Flow therefore continues indefinitely, long after the casting should have completely frozen. The result is that when the casting is removed from the casting machine, two things can happen:

1. In the worst case the liquid can drain completely from the flow path, leaving a hollow tunnel through the casting.
2. In the best case, perhaps because of a cranked delivery system or other impediment to the free

flow of liquid out of the casting (such as a filter), the flow path then freezes, but without the benefit of applied pressure or any extra feeding liquid. Thus it becomes a porous region of the casting, appearing to be a region of shrinkage porosity (Figure 5.12). The casting engineer will then increase the size of the feed path from the riser tube to the poorly fed region in an attempt to increase the feeding. With the enhanced ease of convection, and enhanced ease of subsequent emptying of the flow path, the problem merely gets worse!

This was the nightmare problem that blighted the Cosworth development in its early years, almost causing the company to fail. At the time the problem was baffling since many castings could be cast perfectly, but certain not-so-different designs could not be made without severe porosity. (The problem was completely solved some years later by the development of the rollover technique following casting. This is dealt with in Volume II.)

Thermal convection is not only a problem in low-pressure, uphill-filling systems. It is probably common in any casting that takes a long time to freeze. This is because the circulation pattern takes time to build up and time to carve out a significant flow channel.

Thus it is common in investment castings of steels and nickel-based alloys, especially when these are cast into hot moulds at temperatures near 1000°C, and even more when these moulds are backed by insulating material, all at this high temperature.

Figure 5.13a shows a typical problem casting where the side feeder constitutes a heat source and a circulation path. The result is that the casting becomes too hot at the top, gaining for itself an effectively higher modulus and extended freezing time. A shrinkage-type defect in the top of the heavy section of the casting follows, even though the feeder appears to be correctly sized to feed the casting.

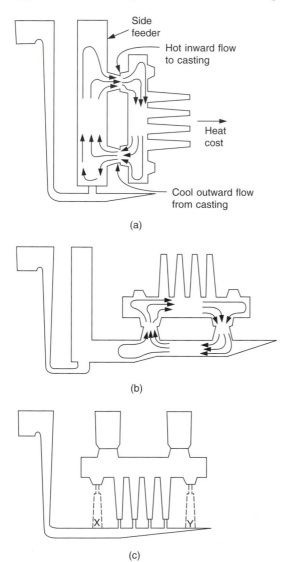

(a)

(b)

(c)

Figure 5.13 *Encouragement of thermal convection by (a) side feeding; (b) bottom feeding; (c) its elimination by top feeding.*

Turning the casting on its back and feeding from underneath (Figure 5.13b), pressurizing via an auxiliary feeder, is similarly problematical since the sprue will freeze early and thus not continue to pressurize. This system is similar to the low-pressure case shown in Figure 5.12. The choice of ingate/ feeder through which the metal decides to rise or fall is probably random, being sometimes at one gate and sometimes at the other in the absence of other influences. The situation of cold dense metal overlying hot lighter liquid is simply unstable and can 'flip' over in either direction. The direction of flip is, of course, highly sensitive to initial perturbations such as the residual effect of the flow induced during filling, or the presence of the heat centre in the heavier runner nearer the sprue, or the fact that the runner may not be perfectly balanced so that more flow has occurred via the far ingate, heating that ingate preferentially. In a metastable density regime a bubble blowing off a core can be a powerful trigger, precipitating a rapid slide into instability.

As we have seen, the counter-gravity geometries can sometimes continue to convect indefinitely. In comparison, the convective flows inside gravity-filled castings are usually not so serious, since without the external heat source, they only continue until the feeder finally solidifies. However, even this may greatly prolong the local solidification time of the casting with the result that, at best, properties are locally impaired, and localized gas porosity will have had increased time to develop. At worst, shrinkage porosity may occur because of the transfer of the remaining solidifying liquid out of the casting and into the feeding system.

The only reliable solution to avoid convection is to place the heavy sections at the top and feed downwards using gravity. This is a stable feeding orientation. Thus the casting shown in Figure 5.13c will have enjoyed optimum conditions of filling uphill and feeding downhill. This is a universally applicable condition for reliable castings.

The optional provision of additional gates x and y to provide some hot feed metal directly below the feeders is attractive, but raises the potential for convective problems, if x and y allow convective paths to form. If x and y are narrowed, so as to freeze off early, convection may be avoided, and this mode of filling may become quite efficient.

In general, however, filling the feeders by flow through the casting has the double disadvantage of (i) heating the casting and (ii) cooling the liquid that finally reaches the feeders. The feeding system is therefore necessarily inefficient. This is a problem from which there is often no escape for static casting processes. (An upsprue/feeder system is possible for some products. This solution is described in Volume II.)

The solution to this problem is the inversion of

the casting immediately after pouring. The filling system is preheated by the flow of metal, and, after inversion, becomes the feeding system. This is an ultimate and powerful solution, universally recommended if completely reliable castings are required.

Finally, the casting engineer needs to be constantly vigilant against problems caused by convection. Convection problems require a trained eye on the lookout for circulation paths that contain hot (or heated) and cool (or freezing) regions. Uphill filling systems are sometimes impaired, whereas uphill feeding systems are usually greatly troubled, often to the point of being insoluble.

5.1.4 Remelting

When considering the solidification of castings it is easy to think simply of the freezing front as advancing. However, there are many times when the front goes into reverse! Melting is common in castings and needs to be considered at many stages.

On a microscale, melting is known to occur at different points on the dendrite arms. In a temperature gradient along the main growth direction of the dendrite the secondary arms can migrate down the temperature gradient by the remelting of the hot side of the arms and the freezing of the cold side. Allen and Hunt (1979) show how the arms can move several arm spacings. Similar microscopic remelting occurs as the small arms shrink and the larger arms grow during dendrite arm coarsening, as will be discussed later.

Slightly more serious thermal perturbations can cause the secondary dendrite arms to become detached when their roots are remelted (Jackson *et al.* 1966). The separated secondaries are then free to float away into the melt to become nuclei for the growth of equiaxed grains. If, however, there is too much heat available, then the growth front stays in reverse, with the result that the nuclei vanish, having completely remelted!

On a larger scale in the casting, the remelting of large sections of the solidification front can occur. This can happen as heat flows are changed as a result of changes in heat transfer at the interface, as the casting flexes and moves in the mould, changing its contact points and pressures at different locations and at different times. It is likely that this can happen as parts of the mould, such as an undersized chill, become saturated with heat, while cooling continues elsewhere. Thus the early rapid solidification in that locality is temporarily reversed.

Local remelting of the solid is seen to occur as a result of the influx of fresh quantities of heat from convective flows because of filling. The so-called flow lines seen on the radiographs of magnesium alloy castings are clearly a result of the local washing away of the solidification front, as a curving river can erode its outer bank.

The existence of continuous fluidity is a widely seen effect resulting directly from the remelting of the solid material that has formed in the filling system, keeping the metal flowing despite an unfavourable modulus. Without the benefits of this phenomenon it would be difficult to make castings at all!

Other convective flows produced by solute density gradients in the freezing zone take time to get established. Thus channels are formed by the remelting action of low-melting-point liquid flowing at a late stage of the freezing process. The A and V segregated channels in steel ingots, and freckle defects in nickel- and cobalt-based alloys, are good examples of this kind of defect.

5.2 Development of matrix structure

5.2.1 General

The liquid phase can be regarded as a randomly close-packed heap of atoms, in ceaseless random thermal motion, with atoms vibrating, shuffling and jostling a meandering route, shoulder to shoulder, among and between their neighbours.

In contrast, the solid phase is an orderly array, or lattice, of atoms arranged in more or less close-packed rows and layers. Atoms arranged in lattices constitute solid bodies we call crystals. The body-centred-cubic (bcc) lattice of alpha-iron (Figure 5.14a) known as ferrite is rather less close packed than the face-centred-cubic (fcc) lattice of the gamma-phase, known as austenite. Figure 5.15 shows only a single 'unit cell' of the lattice. The concept of the lattice is that it repeats such units, replicating the symmetry into space millions of times in all directions. Macroscopic lattices are often seen in castings as crystals that have sizes from 1 μm to 100 mm, representing arrays 10^3 to 10^8 atoms across.

The transition from liquid to solid, the process of solidification, is not always easy, however. For instance, in the case of glass the liquid continues to cool, gradually losing the thermal motion of its atoms, to the point at which it becomes incapable of undergoing sufficient atomic rearrangements for it to convert to a lattice. It has therefore become a supercooled liquid, capable of remaining in this state for ever.

Metals, too, are sometimes seen to experience this reluctance to convert to a solid, despite on occasions being cooled hundreds of degrees Celsius below their equilibrium freezing temperature. This is easily demonstrated for clean metals in a clean container.

If and when the conversion from liquid to solid

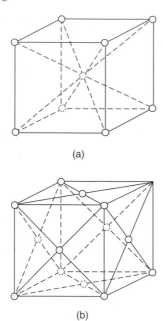

(a)

(b)

Figure 5.14 *Body-centred cubic form of α-iron (a) exists up to 910°C. Above this temperature, iron changes to (b) the face-centred-cubic form. At 1390°C the structure changes to δ-iron, which is bcc once again (a).*

occurs, it is by a process first of nucleation, and then of growth.

Nucleation is the process of the aggregation of clusters of atoms that represent the first appearance of the new phase. The difficulties of achieving a stable cluster are considered briefly in section 5.2.2. *Growth* is, self-evidently, getting bigger. However, this process is subject to factors that encourage or discourage it. Again these will be dealt with later.

In fact, the complexities of the real world dictate not only that the main solid phase appears during solidification, but also that alloys and impurities concentrate in ways to trigger the nucleation and growth of other phases. These include solid and liquid phases that we call second phases or inclusions, and gas or vapour phases which we call gas or shrinkage pores. It is convenient to treat the solid and liquid phases together as condensed (i.e. practically incompressible) matter that we shall consider in this chapter. The gas and vapour phases, constituting the non-condensed matter such as gas and shrinkage porosity respectively, will be treated separately in Chapters 6 and 7.

For those readers who are enthusiastic about nucleation theory, there are many good formal accounts. A readable introduction relating to the solidification of metals is presented in Flemings (1974). We shall consider only a few basic aspects here; enough to enable us to understand how the structures of castings originate.

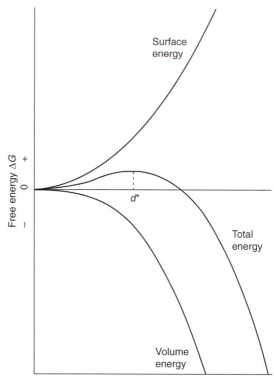

Figure 5.15 *Surface and volume energies of an embryo of solid growing in a liquid give the total energy as shown. Below the critical size d* any embryos will tend to shrink and disappear. Above d* increasing size reduces the total energy, so growth will be increasingly favoured, becoming a runaway process.*

5.2.2 Nucleation of the solid

As the temperature of a liquid is reduced below its freezing point, nothing happens at first. This is because in clean liquids the conversion to the solid phase involves a nucleation problem. We can gain an insight into the nature of the problem as follows.

As the temperature falls, the thermal agitation of the atoms of the liquid reduces, allowing small random aggregations of atoms into crystalline regions. For a small cubic cluster of size d the net energy to form this new phase is reduced in proportion to its volume d^3 and the free energy per unit volume ΔG_v. At the same time, however, the creation of new surface area $6d^2$ involves extra energy because of the interfacial energy γ per unit area of surface. The net energy to form our little cube of solid is therefore:

$$AG = 6d^2\gamma - d^3\Delta G_v \qquad \textbf{(5.15)}$$

Figure 5.15 shows that the net energy to grow the embryo increases at first, reaching a maximum. Embryos that do not reach the maximum require more energy to grow, so normally they will shrink and redissolve in the liquid.

Only when the temperature is sufficiently low to allow a chance chain of random additions to grow an embryo above the critical size will further growth be encouraged by a reduction in energy; thus growth will enter a 'runaway' condition. The temperature at which this event can occur is called the homogeneous nucleation temperature. For metals like iron and nickel it is hundreds of degrees Celsius below the equilibrium freezing point.

Such low temperatures may in fact be attained when making castings, because the liquid only contacts the mould wall at high spots, bridging elsewhere because of the effect of surface tension (Figure 5.6). At such microscopic points of contact, for moulds with high conductivity, the cooling may be so intense that homogeneous nucleation may occur. Nucleation is not likely to be a heterogeneous event in this situation because the liquid will be contained in its oxide skin, and so will not be in actual atomic contact with the surface of the mould. (We shall see later that oxides, such as some crucible materials, are not good nuclei to initiate nucleation heterogeneously.)

It is more common for the liquid to contain other solid particles in suspension on which new embryo crystal can form. In this case the interfacial energy component of Equation 5.15 can be reduced or even eliminated. Thus the presence of foreign nuclei in a melt can give a range of heterogeneous nucleation temperatures; the more effective nuclei requiring less undercooling. In the limit, a liquid may start to freeze at a temperature practically at its equilibrium freezing point in the presence of very favourable foreign nuclei.

It is important to realize that not all foreign particles in liquids are favourable nuclei for the formation of the solid phase. In fact it is likely that the liquid is indifferent to the presence of much of this debris. Only rarely will particles be present that reduce the interfacial energy term in Equation 5.15. Thus, as far as most metals are concerned, oxides are not good nuclei. This is probably a result of their covalent lattice structure. It is worth noting that it makes no difference whether the crystal structures of the oxide and metal are closely matched. The oxides are not wetted. This indicates that their electronic contribution to the interfacial energy with the metal is not favourable for nucleation.

Materials with more metallic properties are better nuclei. For the nucleation of steels, these include some borides, nitrides and carbides, as described in section 5.6.6. For Al alloys, an intermetallic compound, $TiAl_3$, is the key inoculant, together with TiB_2. This will become clear in the section devoted to the individual alloy systems.

5.2.3 Growth of the solid

It seems, however, that the action of an effective grain refining addition to a melt is not only that of nucleating new grains. An important secondary role is that of inhibiting the rate of growth of grains, thereby allowing more to nucleate.

The addition of TiB_2 to Fe–3Si alloys (Campbell and Bannister 1975) had a profound grain refinement action, almost certainly enhanced by the thick layer of borides that surrounded each grain. The mechanical properties were expected to be seriously impaired by this brittle grain boundary phase, illustrating that not all grain refinement of cast alloys is beneficial.

For more complex systems, where many solutes are present, the rate of growth of grains is assumed by Greer and colleagues (2001) to be controlled by the rate at which solute can diffuse through the segregated region ahead to the advancing front. They carry out a detailed exercise for aluminium alloys, using a *growth restriction parameter*. This exercise is reported in section 5.4.3.

Turning now to a consideration of the growth of columnar grains into the casting, the primary solid will spread relatively quickly through the undercooled liquid in contact with the face of the mould. Having formed a solid skin, how does it then continue its progress into the melt?

Progress will only occur at all if heat is extracted through the solid, cooling the freezing front below the equilibrium freezing point. The actual amount of undercooling is usually several degrees Celsius. If the rate of heat extraction is increased, the temperature of the solidification front will fall further, and the velocity of advance, V_s, of the solid will increase correspondingly.

For pure metals, as the driving force for solidification increases so the front is seen to go through a series of transitions. Initially it is planar; at higher rates of advance it develops deep intrusions, spaced rather regularly over the front. These are parts of the front that have been left far behind. This type of growth is called cellular growth. At higher velocities still, the cells grow into rapidly advancing projections, sometimes of complex geometry (Figure 5.16). These tree-like forms have given them the name dendrites (after the Greek word for tree, *dendros*).

For the more important case of alloys, however, the three growth forms are similarly present (Figure 5.16). However, the driving force for instability is a kind of effective undercooling that arises because of the segregation of alloying elements ahead of the front. The presence of this extra concentration of alloying elements reduces the melting point of the liquid. If this reduction is sufficient to reduce the melting point to below the actual local temperature, then the liquid is said to be locally constitutionally undercooled (that is, effectively undercooled because of a change in the constitution of the liquid).

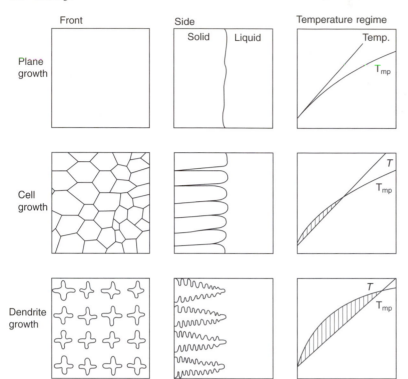

Figure 5.16 *Transition of growth morphology from planar, to cellular, to dendritic, as compositionally induced undercooling increases (equivalent to G/V being reduced).*

Figure 5.17 shows how detailed consideration of the phase diagram can explain the relatively complicated effects of segregation during freezing. It is worth examining the logic carefully.

The original melt of composition C_0 starts to freeze at the liquidus temperature T_L. The first solid to appear has composition kC_0 where k is known as the partition coefficient. This usually has a value less than 1 (although the reader needs to be aware of the existence of the less common but important cases where k is greater than 1). For instance, for $k = 0.1$ the first solid has only 10 per cent of the concentration of alloy compared to the original melt; the first metal to appear is therefore usually rather pure.

In general, k defines how the solute alloy partitions between the solid and liquid phases. Thus:

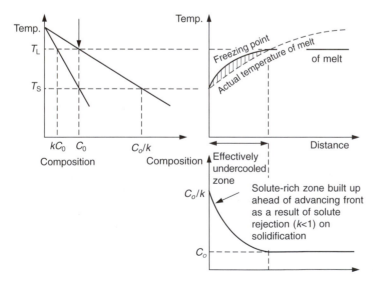

Figure 5.17 *Link between the constitutional phase diagram for a binary alloy, and constitutional undercooling on freezing.*

$$k = C_S/C_L \qquad (5.16)$$

For solidus and liquidus lines that are straight, k is accurately constant for all compositions. However, even where they are curved, the relative matching of the curvatures often means that k is still reasonably constant over wide ranges of composition. When k is close to 1, the close spacing of the liquidus and solidus lines indicates little tendency towards segregation. When k is small, then the wide horizontal separation of the liquidus and solidus lines warns of a strongly partitioning alloying element.

On forming the solid that contains only kC_0 amount of alloy, the solute remaining in the liquid has to be rejected ahead of the advancing front. Thus although the liquid was initially of uniform composition C_0, after an advance of about a millimetre or so the composition of the liquid ahead of the front builds up to a peak value of C_0/k. The effect is like that of a snow plough. This is the steady-state condition shown in Figure 5.17.

In common with all other diffusion-controlled spreading problems, we can estimate the spread of the solute layer ahead of the front by the order-of-magnitude relation for the thickness d of the layer. If the front moves forward by d in time t, this is equivalent to a rate V_s. We then have:

$$d = (Dt)^{1/2} \quad \text{where } V_s = d/t$$

so:

$$d = D/V_s \qquad (5.17)$$

where D is the coefficient of diffusion of the solute in the liquid. It follows that constitutional undercooling will occur when the temperature gradient, G, in the liquid at the front is:

$$G \le -\frac{T_L - T_S}{D/V_S} \qquad (5.18)$$

or

$$G/V_S \le -(T_L - T_S)/D \qquad (5.19)$$

from Figure 5.17, assuming linear gradients. Again, from elementary geometry which the reader can quickly confirm, assuming straight lines on the equilibrium diagram, we may eliminate T_L and T_S and substitute C_0, k and m, where m is the slope of the liquidus line, to obtain the equivalent statement:

$$\frac{G}{V_S} \le \frac{mC_0(1-k)}{kD} \qquad (5.20)$$

which is the solution derived from more rigorous diffusion theory by Chalmers in 1953, nicely summarized by Flemings (1974). This famous result marked the breakthrough in the history of the understanding of solidification by the application of physics. It marked the revolution from *qualitative* description to *quantitative prediction*. Computers have encouraged an acceleration of this new thinking.

Figure 5.16 illustrates how the progressive increase in constitutional undercooling causes progressive instability in the advancing front, so that the initial planar form changes first to form cells, and with further instability ahead of the front will be finally provoked to advance as dendrites.

Notice that the growth of dendrites is in response to an *instability condition* in the environment ahead of the growing solid, not the result of some influence of the underlying crystal lattice (although the crystal structure will subsequently influence the details of the shape of the dendrite). In the same way stalactites will grow as dendrites from the roof of a cave as a result of the destabilizing effect of gravity on the distribution of moisture on the roof. Icicles are a similar example; their forms being, of course, independent of the crystallographic structure of ice. Droplets running down windowpanes are a similar unstable-advance phenomenon that can owe nothing to crystallography. There are numerous other natural examples of dendritic advance of fronts that are not associated with any long-range crystalline internal structure. It is interesting to look out for such examples. Remember also the converse situation that the planar growth condition also effectively suppresses any influence that the crystal lattice might have. It is clear therefore that the constitutional undercooling, assessed by the ratio G/R, is the factor that measures the degree of stability of the growth conditions, and so controls the type of growth front, *not*, primarily, the crystal structure.

Figure 5.18 shows a transition from planar, through cellular, to dendritic solidification in a low-alloy steel that had been directionally solidified in a vertical direction. The speeding up of the solidification front has caused increasing instability. Figures 5.19 and 5.20 show different types of dendritic growth. Both types are widely seen in metallic alloy systems. In fact, dendritic solidification is the usual form of solidification in castings.

A columnar dendrite nucleated on the mould wall of a casting will grow both forwards and sideways, its secondary arms generating more primaries, until an extensive 'raft' has formed (Figure 5.21). All these arms will be parallel in terms of the internal alignment of their atomic planes. Thus on solidification the arms will 'knit' together with almost atomic perfection, forming a single-crystal lattice known as a grain. A grain may consist of thousands of dendrites in a raft. Alternatively a grain may consist merely of a single primary arm, or, in the extreme, merely an isolated secondary arm.

The boundaries formed between rafts of different orientation, originating from different nucleation

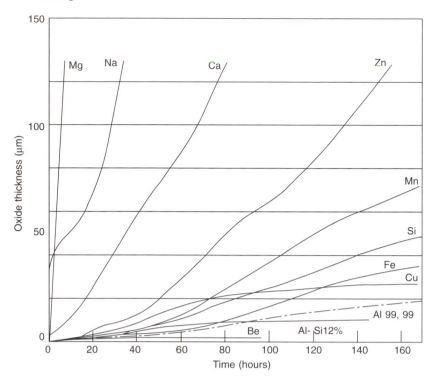

Figure 5.33 *Growth of oxide on Al and its alloys containing 1 atomic % alloying element at 800°C. Data from (Theile 1962).*

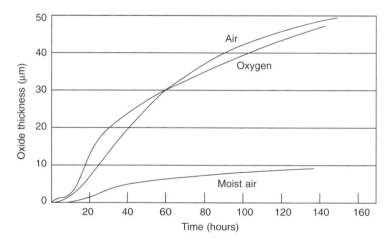

Figure 5.34 *Growth of oxide on 99.9Al at 800°C in a flow of oxygen, and dry and moist air. Data from Theile (1962).*

beans, is an essential concept for the understanding of its behaviour.

We have already seen that progressive magnesium additions change the oxide from alumina to spinel, and finally to magnesia. A cursory study of the periodic table to gain clues of similar behaviour that might be expected from other additives quickly indicates a number of likely candidates. These include the other group IIA elements, the alkaline earth metals including beryllium, calcium, strontium and barium, and the neighbouring group IA elements, the alkali metals lithium, sodium and potassium. The Ellingham diagram (Figure 1.5) also confirms that these elements have similarly stable oxides, so stable in fact that alumina can be reduced back to aluminium and the new oxide take its place on the surface of the aluminium alloy. The disruption or wholesale replacement of the protective alumina or spinel film may have important consequences for the melt.

In the case of additions of beryllium at levels of only 0.005 per cent, the protective qualities of the film on Al–Mg alloy melts is improved, with the result that oxidation losses are reduced as Figure 5.33 indicates. However, low-level additions of Be have been found to be important for the successful production of wrought alloys by continuous (direct chill) casting possibly because of a side effect of the strengthening of the film as discussed in section 2.6.

Strontium is added to Al–Si alloys to refine the structure of the eutectic in an attempt to confer additional ductility to the alloy. However, strontium, like magnesium, seems also to form a spinel, its oxide combining with that of aluminium. In addition, the resistance to tearing of the film is probably also increased, affecting the entrainment process. Because of this additional powerful effect on the oxide film, the action of Sr as an addition to Al alloys is complicated. It is therefore dealt with separately in section 5.4.4.

Sodium is also added to modify the microstructure of the eutectic silicon in Al–Si alloys. In this case the effect on the existing oxide film is not clear, and requires further research. Sodium will have much less of an effect in sensitizing the melt to the effect of moisture because it is less reactive than strontium. In addition, sodium is lost from the melt by evaporation because the melt temperatures used with aluminium alloys, typically in the range 650 to 750°C, approach its boiling point of 883°C, the wind of sodium vapour acting to sweep hydrogen away from the environment of the melt. Both the reduced reactivity and the vaporization would be expected to reduce any hydrogen problems associated with Na treatment compared to Sr treatment, corresponding to general foundry experience.

However, Wightman and Fray (1983) find that all alloys that vaporize disrupt the film and increase the rate of oxidation. The additions they tested included sodium, selenium and (above 900°C) zinc (Figure 1.9). The disruption of the film acts in opposition to the benefit of the wind of vapour purging the environment in the vicinity of the melt. Thus the total effect of these opposing influences is not clear. It may be that at these low concentrations of solute any beneficial wind of vapour is too weak to be useful, allowing the disruption of the film to be the major effect. However, the overall rate is in any case likely to be dominated by the reduced reactivity of Na compared with Sr.

Experience of handling liquid aluminium alloys in industrial furnaces indicates that the character of the oxide film is changed when sodium, strontium or magnesium is added. For instance, as magnesium metal is added to an Al–Si alloy, the surface oxide on the melt is seen to take on a glowing red hue that spreads out from the point that the addition is made. This appears to be an effect of emissivity, not of temperature. Also, the oxide appears to become thicker and stronger. The beneficial effects found for the improved ductility of Al–Si alloys treated with sodium or strontium may be due not only therefore to the refined silicon particle size. This well-known metallurgical phenomenon, much favoured by textbooks on casting, may have only a minor role. The changed strength and distribution of entrained oxide films could be the major effect. Careful research would clarify the issue. In the meantime, the possibilities are discussed in section 5.4.4.

5.4.2 Entrained inclusions

When the surface of the melt becomes folded in, the doubled-over films take on a new life, setting out on their journey as bifilms. The scenario has been discussed in some detail in Chapter 1 in the description of liquid metal as a slurry of defects, and in terms of the details of the entrainment processes in Chapter 2.

As an overview of these complicated effects, Figure 5.35 gives an example of the kinds of populations of defects that may be present. This figure is based on a few measurements by Simensen (1993) and on some shop floor experiences of the author. Thus it is not intended to be any kind of accurate record, it is merely one example. Some melts could be orders of magnitude better or worse than the figures shown here. However, what is overwhelmingly impressive are the vast differences that can be experienced. Melts can be very clean (1 inclusion per litre) or dirty (1000 inclusions per cubic millimetre). This difference is a factor of a thousand million. It is little wonder that the problem of securing clean melts has presented the industry with a practically insoluble problem for so many years. These problems are only now being resolved in some semi-continuous casting plants, and even in this case, many of these plants are not operating particularly well. It is hardly surprising therefore that most foundries for shaped castings have much to achieve. There is much to be gained in terms of increased casting performance and reliability.

5.4.3 Nucleation and growth of the solid (grain refinement)

The addition of titanium in various forms into aluminium alloys has been found to have a strong effect in nucleating the primary aluminium phase. It is instructive to consider the way in which this happens.

Titanium in solution in the liquid metal at a

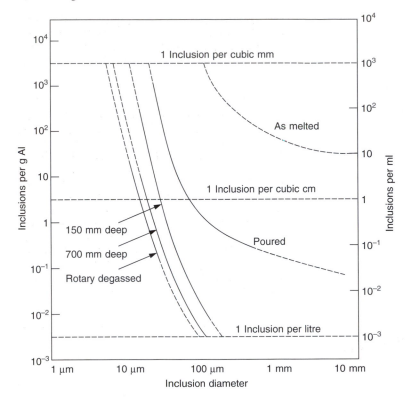

Figure 5.35 *An example of inclusion content in an Al alloy.*

concentration above about 0.15 weight per cent would be expected to be precipitated as $TiAl_3$ in the peritectic reaction (Figure 5.36). There is no doubt that $TiAl_3$ is an active nucleus for aluminium because $TiAl_3$ is found at the centres of aluminium grains, and there is a well-established orientation relationship between the lattices of the two phases (Davies *et al.* 1970).

However, there are two major problems: we have to ask the questions (i) 'How does the $TiAl_3$ phase itself nucleate?' and (ii) 'Why does the $TiAl_3$ phase exist below the 0.15 per cent Ti, where it should be unstable, and should go into solution as indicated on the equilibrium phase diagram' (Figure 5.36). Results of several researchers shown in Figure 5.37 illustrate that the effect of titanium in the grain refinement of aluminium starts at much lower concentrations, below 0.01 titanium.

How titanium can be effective at concentrations lower than 0.15 per cent has remained a mystery in Al–Ti liquid solutions until the epoch-making research by Schumacher and Greer (1993 and 1994). These researchers carried out their studies on an amorphous aluminium alloy, as an analogue of the liquid state. Nucleation is more easily observed since the kinetics of reaction are 10^{16} times slower. Using TEM (transmission electron microscopy) they observed that $TiAl_3$ was present as adsorbed layers on TiB_2 crystals, and so its existence was stabilized

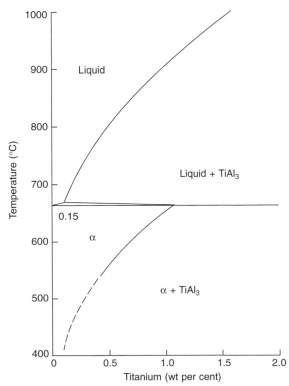

Figure 5.36 *Binary Al–Ti phase diagram.*

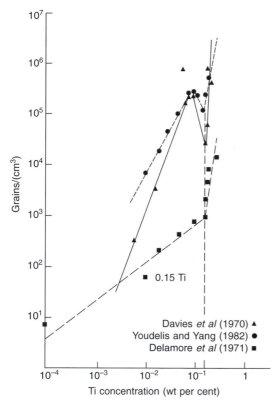

Figure 5.37 *Increase in grain refinement with increasing titanium addition, especially at the peritectic 0.15 Ti.*

at lower levels of Ti than would be expected from the phase diagram, and it was thereby effective in nucleating aluminium.

It is worth noting, however, that in real castings (as contrasted with laboratory experiments), the action of titanium in grain refinement may be rather different. In this case the titanium is added as a master alloy that already contains $TiAl_3$ particles in suspension. As the master alloy disperses in the bulk melt, the $TiAl_3$ particles will tend to dissolve, and sink to the bottom of the furnace. However, this takes time of the order of an hour or so, allowing plenty of time in practice for the treatment and casting of the melt. The addition of boron to the master alloy greatly increases the effectiveness of grain refinement. Although it was thought that the TiB_2 might act in several ways (review: McCartney 1989), it seems most likely that the stabilization of $TiAl_3$ on its surface might be the most important factor, lengthening the life of the active $TiAl_3$ particles.

Effective grain refinement, however, seems to require more than simply the nucleation of new grains. A second important factor is the suppression of their growth. For complex systems, where many solutes are present, the rate of growth of grains is

assumed by Greer and colleagues (2001) to be controlled by the rate at which solute can diffuse through the segregated region ahead to the advancing front. They use a *growth restriction parameter Q* defined as

$$Q = m(k - 1)C_0 \qquad (5.22)$$

Although they note that this relation should be modified by the rate of diffusion D of the solute, these factors are not well known for many solutes. They therefore assume that the rates of diffusion are fairly constant for all solutes of interest in aluminium (this is not far from the truth for the substitutional solutes shown in Figure 1.6). Thus the potentially more accurate summation of the effects of different solutes in solution, weighted inversely by their diffusivities as proposed by Hodaj and Durand (1997), is neglected in favour of a simple summation of Q values. The effect is shown in Figure 5.38. Initially, grain size clearly decreases with increasing total values of Q.

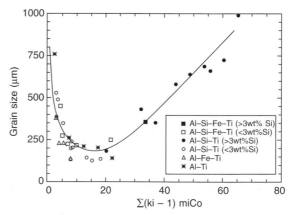

Figure 5.38 *Effect of the growth restriction factor on the grain size of various Al alloys (Greer et al. 2001).*

The subsequent apparent growth of grains with increasing Q above about 20 is thought to be (Greer 2002) the result of the special effect of Si in 'poisoning' the grain refinement action of Ti at Si contents over 3 per cent. The higher Q data is defined only by alloys with Si contents above 3 per cent. For other solutes at high Q, particularly Cu, it is thought that the grain size remains small. It seems that the attainment of fine grain size in Al–Si casting alloys has fundamental limitations, the attainable sizes being 5 to 10 times larger than those in some other casting alloys and in most wrought alloys.

Other nucleation and growth effects are happening during the solidification of many Al alloys as a result of the many solutes that are present, both intended and unintentional.

As an example of one of these, Cao and Campbell

(2000) discovered that βFe plates (Al₅FeSi intermetallic) in Al–Si alloys precipitated on the wetted outside surfaces of bifilms. Initially, the βFe precipitate is sufficiently thin that it can follow the folds of the bifilm. On a fracture surface the iron-rich phase can be clearly seen through the thin oxide film that represents one half of the bifilm (Figure 5.39). At this early stage it is faithfully following the undulations of the oxide film. However, as the βFe particle thickens, the particle becomes increasingly rigid, taking on its preferred crystalline form, and so forces the film to straighten. Finally, the bifilm is often seen as a crack aligned along the centre of the βFe particle, or along the matrix/particle interface if the βFe happened to nucleate only on one side of the bifilm (Figure 5.40).

In Figure 5.41a the bifilm on one side of the particle has been pulled away by gas precipitation or by shrinkage forces, opening a pore on one side of the βFe plate. Because it has been common to observe an association between pores and βFe particles, it has in the past been assumed that the βFe particles blocked the movement of feed liquid along interdendritic channels, and so caused shrinkage porosity. However, this seems most unlikely, in view of the three-dimensional access routes for feed liquid, and in view of the strong probability that pores probably cannot be formed without bifilms.

5.4.4 Modification of the Al–Si eutectic by Na and Sr

When Al–Si alloys are solidified the eutectic silicon is seen on polished sections to consist of coarse plates with sharp edges. These have usually been thought to be detrimental to mechanical properties,

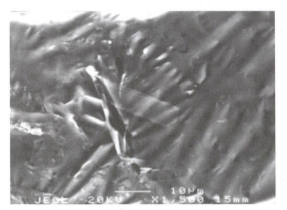

Figure 5.39 *SEM image of an iron-rich particle nucleated on the underside of a thin alumina film imaged by (a) secondary electrons and (b) back-scattered electrons (Cao 2000).*

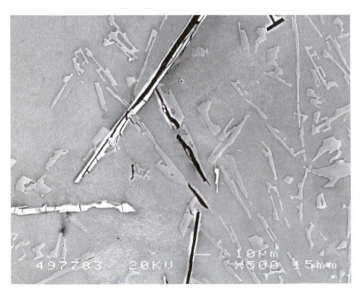

Figure 5.40 *Platelets of βFe and eutectic silicon particles in an Al–Si alloy showing associated bifilm cracks. The transverse cracks probably result from rucks in one of the films (Cao 2000).*

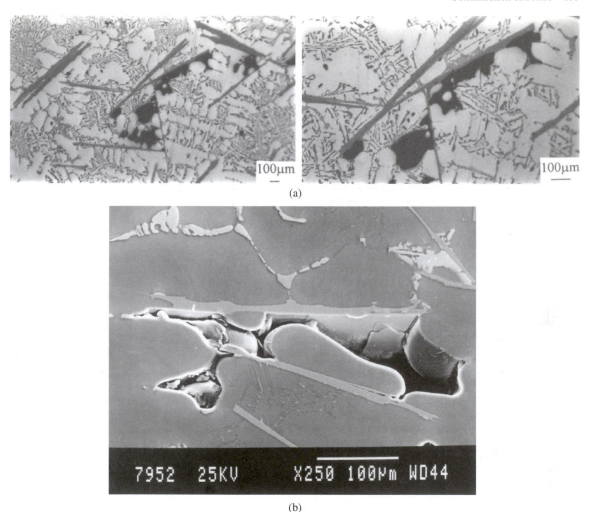

(a)

(b)

Figure 5.41 *Platelets of βFe in an Al-Si alloy showing pores opened by shrinkage or gas, initiated by the bifilm (a) courtesy Cao (2000); (b) courtesy Samuel et al. (2001), Internat. J. Cast Metals Research.*

being assumed to act as crack initiators. For this reason, for alloys containing above about 5 to 7 per cent Si, the addition of sodium or strontium to the melt has been favoured to refine the eutectic silicon phase.

(As an aside, it seems highly probable that the apparent initiation of cracks by large silicon particles was the result of the unsuspected presence of bifilms on which the particles had formed. Thus the whole *raison d'être* of modification may have been misguided. If so, it represents a massive and continuing loss of effort and resources in the aluminium casting industry. Research is required to confirm or negate this important point. However, we do not have any real facts to justify further speculation on this sobering possibility at this stage.)

The action of modification itself was first explained by Flood and Hunt in 1981. They

interrupted the solidification of unmodified and Na-modified Al–Si eutectic alloy to study the form of the growth front in a slice of a cylindrical casting. Their results are summarized and included as part of Figure 5.42. Added into Figure 5.42 is an illustration of the effect of Sr modification. In the case of the unmodified alloy the growth form appeared dendritic, with some nucleation apparently ahead of the freezing front. In the case of the addition of sodium, planar growth of the eutectic front was stabilized, with an effective length of solidification front a factor of 17 times shorter than the dendritic front. Thus for the same given quantity of heat extracted by the mould over a given area, the interface in the case of the planar front would advance at 17 times the rate, and therefore have a much finer structure than the unmodified alloy.

Interestingly, the form of the freezing front for

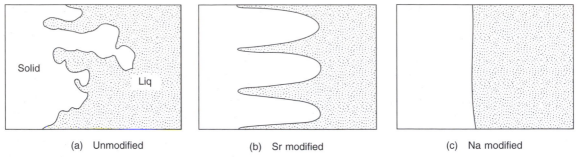

(a) Unmodified   (b) Sr modified   (c) Na modified

Figure 5.42 *Portions of the growth front of the Al–Si eutectic in (a) the unmodified condition, and modified with (b) strontium and (c) sodium (based partly on Flood and Hunt 1981).*

the Sr-modified eutectic alloy is cellular in the freezing conditions commonly found in sand castings. This intermediate condition explains the intermediate performance of Sr. The cellular pattern, resembling a honeycomb, is seen occasionally on the surface of Sr-modified castings that have suffered some degree of poor feeding, encouraging the loss of some residual liquid from around the cell, and so giving slight depressions on a cast surface that outlines the cell boundaries (Figure 5.43). The cellular structure of the front does lead on other occasions to some problem to feed Sr-modified castings, as contrasted with those cast from Na-modified alloy. Surface-initiated porosity is also favoured by modification with Sr whereas it is discouraged by modification with Na.

The choice between the use of sodium or strontium depends on the circumstances. Sodium is usually lost from a melt within a time of the

order of 15 to 30 minutes, as a result of evaporation. Strontium, on the other hand, is a normal, stable, alloy in liquid Al alloys. Although it is slowly lost by oxidation at the surface, and sometimes can soak away into refractory furnace linings until they saturate, the alloy will usually survive several remeltings.

The addition of strontium to the melt is an expensive option, but is taken in an effort to improve the ductility of the alloy. However, the results are not always straightforward to understand, and are accompanied by a number of problematical factors.

Much confusion has existed over whether strontium can be successfully added without the deleterious effect of hydrogen pick-up. Amid the confusion, the hydrogen already in solution in the strontium master alloy addition has often been blamed for this problem. However, we can, with

Figure 5.43 *Thin unfed casting of Al–7Si alloy modified with Sr, showing the cellular growth form.*

some certainty, dismiss this minute source as negligible. What alternative possibilities remain?

The enhanced rate of oxidation of the strontium means that any moisture in the environment is quickly converted to surface oxide, and the hydrogen is released into the melt (Equation 1.2). In furnaces open to the air therefore the addition of strontium is usually accompanied by an increase of hydrogen porosity. The porosity naturally increases with increasing strontium content, temperature, time and environmental water. This makes it practically impossible to add strontium without an increase in porosity in most foundry melting systems. In the experience of the author a single 0.05 weight per cent addition to a 1000 kg holding furnace caused the gas level to rise to such a high level that gas porosity caused the castings to swell, solidifying oversize, and had to be scrapped. It took three days of waiting for the melt to return to a castable quality. (The arrival of rotary degassing has, thankfully, eliminated such difficulties nowadays.) The absorption of hydrogen can, of course, be reduced by reducing the time available for absorption, for instance, by the casting of the whole melt immediately after the strontium treatment. Treating the strontium as a late addition in this way has been adopted successfully (Valtierra 2001).

Strontium is, however, generally used with success in low-pressure casting furnaces where the melt is transferred immediately after treatment into an enclosed furnace that excludes any environmental moisture. It may then be held indefinitely, provided that the pressurizing gas that is introduced into the furnace from time to time is dry or inert, and that the furnace lining is already saturated with strontium from previous additions.

A factor that may be important is the amount of disturbance of the melt. Gruzleski has carried out careful laboratory measurements on Al–7Si–0.36Mg alloy (Dimayuga et al. 1988; Mulazimoglu et al. 1989) and found no increase in the rate of absorption of hydrogen after additions of strontium. Thus when undisturbed, the melt may continue to be protected from the environment by its oxide. The stabilization of the oxide by moisture as seen in Figure 5.34 is a further factor to strengthen this conclusion. In an industrial furnace the stirring and ladling actions may fracture the protective oxide and so allow further reaction, encouraging the ingress of hydrogen in a furnace open to the atmosphere.

A recent observation by Liu and co-workers (2002) of polished sections of an Al–Si alloy modified with Sr has discovered SrO_2 particles in pores. One side of the pores is seen to be formed by a beta-iron platelet, indicating the pore was almost certainly initiated on a bifilm. Thus ordinary alumina or spinel films are probably present in addition to the SrO_2 particles. A little reckless speculation may be forgiven. Perhaps therefore the alumina films

contain strontium oxide particles as islands, resembling icebergs or boulders, so that on entrainment, the film will be unable to close so easily, thus assisting the easier opening and easier initiation of pores. These thoughts indicate the complexity of the subject. It will take years to sort out the real answers.

Turning now from the question of gas pick-up and aided nucleation to other factors. It is possible that the finer structure of the silicon phase in the Al–Si eutectic may have better properties than the coarse unmodified structure. There are many results of published work that cite the benefits to strength and ductility following modification. However, some researchers have reported only mediocre improvements. Some have reported no benefit. A few have reported reductions in properties. This confusion of experience requires some examination.

Recently, the Cosworth casting foundry operated by Ford/Nemak in Windsor, Ontario, has reported that mechanical properties of A319 alloy are actually reduced by Sr modification (Byczynski and Cusinato 2001). This is an important result. It is often not easy to control small batches of Al alloys so as to carry out reproducible experiments in laboratory conditions. However, in a continuously operating plant handling tons of material each hour the melt quality can be assessed accurately, and is closely reproducible because of the process controls that are in place. Furthermore, most experimenters will have cast under gravity, mostly using relatively poor filling system designs, and thus their material will be expected to be impaired with new oxide films. On the other hand, the Cosworth process will have very few new films because of the relatively quiescent handling, and the counter-gravity filling under the control of a pump. (However, it is known that the metal delivered by the process contains old films, mostly spinels.)

A reasonable, but unproven, interpretation of these observations is that strontium acts to speed up the deactivation of newly entrained films. Any benefit from the refinement of the eutectic may also be present, but seems practically negligible. However, the problem from the increase of hydrogen leading to additional porosity is a severe disadvantage.

Strontium will be expected to aid the rebonding of bifilms by the conversion of the newly formed alumina films into spinels. The accompanying reorganization of the crystal lattice will encourage the diffusion bonding between the two sides of the bifilm. This may be sufficiently rapid that it may occur in the short time available between pouring and solidification of the casting.

Alternatively, the apparent pairing of βFe platelets, like chromosomes, on a polished section of an Al–Si alloy modified with Sr has been observed occasionally in the author's laboratory, but not yet been investigated. Perhaps these are indications that Sr aids the liquid metal to wet the dry inner sides of the bifilm, and thus creeps between the films, separating the two halves. The separation is only apparent in those cases where iron-rich particles are attached. If there is any truth in this suggestion, such twin intermetallic phases should not, of course, contain a central crack, and both sides on the intermetallic would be expected to be well bonded to the matrix. The question also arises, what happens to the entrained air between the films? This would be pushed ahead of the advancing liquid, and might be exuded as a bubble. Alternatively it may react and be consumed by reaction with the advancing fresh interface.

In summary, the most likely explanation of the action of Sr is that most operators using poorly designed gravity filling systems will benefit because the new bifilm defects introduced by pouring are partially healed. Some will enjoy a useful net benefit if the increase of hydrogen can be controlled. In contrast, those operators using a process such as the Cosworth process will have few new films and so cannot benefit from any healing process, and will only suffer from the extra porosity as a result of any increase in hydrogen. The refinement of the eutectic, much sought-after by metallurgists and assumed to be the main mechanism of property enhancement, appears to have little effect either way.

In passing, it is noted that even with the Cosworth process, where gas content of the continuously processed liquid metal is nicely controlled, the hydrogen level will rise during its passage into the mould because of the reaction with the organic sand binder. In heavy sections, there will be several minutes for the hydrogen to diffuse into the casting sections. From Figure 1.6 the coefficient of diffusion of hydrogen is seen to be close to 10^{-6} m^{2}s^{-1}. From Equation 5.21, for a typical time around 100 s the average diffusion distance for hydrogen in aluminium is close to 10 mm. Thus the gas will easily penetrate the thickest sections of castings such as automotive cylinder blocks.

This interesting observation calls into question the self-imposed task that the foundry adopts to reduce hydrogen levels. Clearly, if sand casting, or even semi-permanent mould casting (i.e. a metal mould with sand cores), the degradation of the sand binder will always raise the hydrogen level at the worst moment, as the melt enters the mould. Extremely low hydrogen content of the melt therefore is not feasible.

It seems inescapable therefore that the really important quality requirement should perhaps be the absence of nuclei for pores, i.e. absence of bifilms. This means really clean metal and excellent designs of melt handling. In such a case, the gas content probably need not be controlled.

5.5 Cast irons

5.5.1 Films on liquid cast iron

When cast iron is held at a high temperature (e.g. 1500°C) in a furnace or ladle lined with a traditional refractory material such as ganister, a fascinating sight can be witnessed. The surface of the liquid iron is seen to be continuously punctuated by the silent and mysterious arrival of bright patches. These suddenly appear and spread from nothing to their full size of several centimetres within about a second. The patches drift around, coalesce with other patches, and finally attach themselves to the wall of the vessel where they cool and add to the solidified rim of slag. These patches are droplets of liquid refractory, melted from the walls and bottom of the vessel. As the vessel is tipped and emptied, upstanding 'stalactites' on the base, and upward runs and drips on the walls, can usually be clearly seen, marking the sites where the drops detached.

In common with all the components of molten metal systems, the slag will be changing its composition rapidly as it interacts with the molten metal. At high temperature its contents of iron, manganese and silicon will be reduced from their respective oxides and taken into solution in the liquid iron, whereas the remaining stable oxides, such as those of aluminium and calcium, will remain to accumulate as a dry slag, sometimes called a dross. These reactions will be explained further in the next section.

Such layers of slag on the surface of molten iron can be anything from 0.1 mm thickness upwards. (In the cupola, of course, the thickness is often around 100 mm or more.) It is not intended to consider such macroscopic surface-layer problems in this section. Practical solutions to deal with these large-scale surface problems will be dealt with in Volume II in the sections on running systems and slag traps. This section considers only the microscopically thin surface film that, under certain conditions, will form automatically on the surface of the melt (no matter how good is the melting resistance of the lining material of the holding vessel).

5.5.1.1 Oxide films

Work by Heine and Loper at the University of Wisconsin, dating from 1951, has done much to explain the complex film formation in cast irons. A slightly later study by Merz and Marincek (1954) is also illuminating. Based on these studies we can explain the changes that occur as the temperature falls.

When the iron is at a high temperature, 1550°C, the Ellingham diagram indicates that CO is a more stable oxide than SiO_2. Thus carbon oxidizes preferentially, and is therefore lost at a higher rate than silicon, as is seen in Figure 5.44. Here the blowing of air on to the surface of a small crucible of molten metal serves to accentuate the effect. Silicon is observed to fall only after all the carbon has been used up. At this high temperature no film is present on the melt – any silicon oxide, SiO_2, would be immediately reduced to silicon metal, which would be dissolved in the melt, simultaneously forming CO, which would escape to atmosphere.

At around 1420°C the stability of the carbon and silicon oxides is reversed. The exact temperature of this inversion seems to be dependent on the composition of the iron; de Sy (1967) reports a range of 1410–1450°C for the irons that he investigated, whereas the Ellingham diagram (Figure 1.5) predicts an inversion temperature for pure Fe–C alloys of about 1500°C. The agreement is, perhaps, as good as can be expected because Merz and Marincek point out that the inversion temperature is sensitive to composition. Below approximately 1400°C, therefore, SiO_2 appears on the surface as

a dry, solid film, rather grey in colour. This film cannot be removed by wiping the surface, since it constantly re-forms.

At a temperature of 1300°C, and in alloys that contain some manganese, it is clear from the Ellingham diagram that MnO is the least stable, SiO_2 is intermediate and CO the most stable. Thus manganese is oxidized away preferentially, followed by silicon and finally by carbon. The contribution of MnO to the film at this stage may reduce the melting point of the film, causing it to become liquid.

At around 1200°C, iron oxide, FeO, contributes to the further lowering of the melting point at the ternary eutectic between FeO, MnO and SiO_2. If sulphur is also present in the iron then MnS will contribute to a complex eutectic of melting point 1066°C (Heine and Loper 1966a).

The author finds that, in general terms, the above considerations nicely explain his observations in an iron foundry where he once worked. For a common grade of grey iron, the surface of the iron was seen to be clear at 1420°C. As the temperature

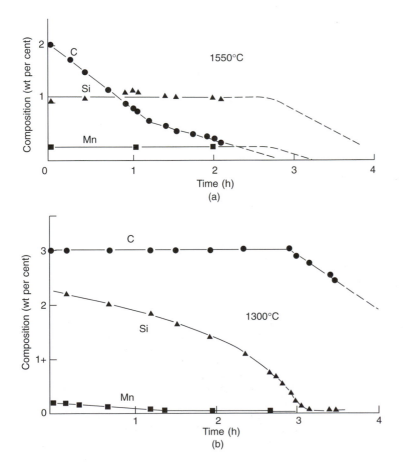

Figure 5.44 *Change in composition of 3.6 kg of molten grey iron held in a silica crucible, while air was directed over its surface at the rate of 22 ml/s: (a) melt at 1550°C; and (b) melt at 1300°C (data from Heine 1951).*

fell, patches of solid grey film were first observed at about 1390°C. These grew to cover the surface completely at 1350°C. The grey film remained in place until about 1280°C, at which temperature it started to break up by melting, finally becoming completely liquid at 1150°C.

When casting grey iron in an oxidizing environment, the falling temperature during the pour and the filling of the mould will ensure that the surface film will be liquid at this critical late stage. If it becomes entrained in the molten metal it will therefore quickly spherodize into compact droplets. The droplets are of much lower density than the iron, and so will float out rapidly. On meeting the surface of the casting they will mutually assimilate, and be assimilated by, the existing liquid film, and so spread over the casting surface. The glassy sheen of some grey iron castings may be this solidified skin. The harmless dispersal of the oxide film in this way is the reason for the good natured behaviour of cast iron when cast into greensand moulds; it is one of the very few metal/mould combinations that is tolerant of surface turbulence.

Only on one occasion has the accumulation of liquid oxide at a casting surface given the author some problems. This was in a grey iron casting where a small amount of surface turbulence was known to be present just inside the ingate, because it was not easy to lower the velocity below 0.4 ms^{-1} at this point, and was judged to be a negligible risk of any kind of internal defect. However, so much liquid surface was entrained, and so much floated out at a point just downstream, that the layer of surface slag accumulated at this location exceeded the machining allowance, scrapping the casting.

In ductile irons the entrainment of the surface is nearly always a serious matter. The small percentage of magnesium that is required to convert the iron from flake to the spheroidal graphite type dramatically alters the nature of the oxide film.

Above 1454°C Heine and Loper (1966b) find that the surface of liquid ductile iron remains clear of any film. Below this temperature a film starts to form, increasing in thickness to 1350°C, at which point the surface exhibits solidified crusty particles. By the time the temperature has reached 1290°C the entire surface is covered with a dry dross. Magnesium vapour distils off through dross since the molten iron is above the boiling point of magnesium. Presumably the oxidation of the vapour to powdery MgO at the upper surface of the dross is a major contributor, causing the dross to grow quickly and copiously. The dross makes life difficult for the ductile iron foundryman, forming films and agglomerating into dry, non-wetting heaps, which, if entrained, spoil otherwise excellent castings. Ductile iron is renowned for being difficult to cast cleanly, without unsightly dross defects.

5.5.1.2 Graphite films (lustrous carbon)

The liquid film present on cast iron at low temperature in an oxidizing environment has made iron easy to cast free from serious defects. This marvellous natural benefit of cast iron when cast into moulds made from sand bonded with clay and water must have played an important part in the success of the industrial revolution. In general, apart from a few infamous and tragic exceptions, the bridges did not fall into the river and the steam engines continued to power machinery. Later this benefit was to be extended to moulds made using one of the first widely used chemical binders: sodium silicate. This environmentally friendly chemical is still widely used today as a low-cost sand binder for the production of strong moulds (despite a number of significant disadvantages that some foundries are prepared to live with).

However, it is one of those ironies of history that the arrival of modern chemical binders was to change all this.

Binders based on various kinds of resins – furane, phenolic, acrylic, polyurethane, etc. – were heralded as the breakthrough of the twentieth century. Indeed, the new binders had many desirable properties, making accurate and stable moulds, with excellent surface finish, good breakdown after casting and at good rates of production from simple low-cost equipment.

However, when iron was poured into some of these early resin-bonded moulds, especially those based on polyurethane, a new defect was discovered. It became known as lustrous carbon. Although it had been occasionally seen, especially if the volatile additions to greensand had been high, it was never so common nor so damaging. This shiny, black material resulted in casting skins wrinkled like elephant hide. Studies concluded that it was pyrolytic carbon (a microcrystalline form) that was deposited from the gas phase on to hot surfaces in the temperature range of at least 650–1000°C (Draper 1976).

The hot surface was always assumed to be the sand grains of the mould, and somehow the deposit was pushed ahead of the advancing liquid front, to become incorporated into the surface as folds (Naro and Tanaglio 1977). The explanation is clearly problematical on several fronts: in most instances where lustrous carbon causes problems the sand surface is rather cold, and thus incapable of chemically 'cracking' (i.e. breaking down) the polymeric gases to precipitate graphite. Also, it is difficult to imagine how a film deposited on the complex and rough sand grains could be detached from its grip on these three-dimensional shapes before the arrival of the liquid metal. After the arrival of the metal the film would be assisted in keeping its place by being held against the surface of the sand grains by the pressure of the liquid.

The only explanation that fits all the facts is that the graphitic film forms on the surface of the molten metal itself. Photographs of lustrous carbon defects, particularly those seen on the fracture surfaces of parts that have suffered brittle failure, beautifully reveal their origin as the surface film on the liquid metal (Bindernagel *et al.* 1975; Naro and Tanaglio 1977). The caption to the photograph of the oxide skin on an aluminium alloy (Figure 5.45) could be changed to read that it was a graphitic skin on a grey iron; to the unaided eye the appearance of the two types of film is practically identical.

Part of the confusion that has surrounded the lustrous carbon film, and that has claimed that it deposited on sand grains, dates from a misreading of the brilliant original work by Petrzela (1968). This Czech foundry researcher devised a test in which he demonstrated that the vapours released from coal tar and other hydrocarbon additions to moulding sands would decompose, depositing carbon as a shiny, silvery film on a metal strip, resistance heated to at least 1300°C. In his test, it happened that the sand was also heated to this temperature. Thus he observed carbon to be deposited directly on to the sand grains in addition to that deposited on the heated metal strip. The mistake of subsequent generations of researchers has been to assume that lustrous carbon always deposits on to sand grains, even though, at the instant that the metal is filling the mould, the sand is usually nowhere near the temperature at which a hydrocarbon vapour could be decomposed.

The reader is recommended to Petrzela's engaging chatty and candid account. He was clearly one of our great foundry characters. His writing contains other fascinating asides to some of his observations on the release of carbon from hydrocarbons. On one occasion he recorded a sooty deposit that had a fibrous, woolly appearance among (hot?) sand grains.

Other work has studied the generation of lustrous carbon in greensand moulds. It is clear that the mould atmosphere can provide a hydrocarbon environment for the liquid metal if sufficiently high concentrations of hydrocarbons are added to the sand mixture. Such additives help the mould to resist wetting by the metal, and so improve surface finish, as appreciated in the original work of Petrzela (1968), and later by Bindernagel *et al.* (1975). Excess additions have sometimes been claimed to give lustrous carbon defects. However, it is certain that the defects form only if the surface turbulence can cause the film to be entrained.

The mechanism for the improvement of surface finish by the addition of hydrocarbons to the mould repays examination in some detail. The carbon film forms on the front of the advancing liquid. It becomes trapped between the melt and the mould, and is held there by friction. Thus, as the meniscus advances, it is forced to tear, splitting apart. The film is therefore continuously formed and laid down between the melt and the mould by the advancing metal, as though the advancing metal were rolling out its own track like a track-laying vehicle. The film forms a mechanical barrier between the metal and the mould. It seems most likely that it is the mechanical rigidity of this barrier, helping to bridge

Figure 5.45 *(a) Oxide skin on liquid Al–9Si–4Mg alloy wrinkled by repeated disturbance of the surface: and (b) a cross-section of the solidified metal. The appearance in both cases is extremely similar to graphite films on grey iron (courtesy of Agema and Fray 1990).*

the sand grains that confers the improved smoothness to the cast surface. The action is that shown in Figure 2.2 for all film-forming alloys.

History now appears to have turned full circle because some resin binders for sands have recently been developed to yield iron castings with reduced incidence of lustrous carbon defects. It is not clear whether the surface finish of the castings has suffered as a result.

Lustrous carbon films cause troublesome defects in lost-foam castings where the foam consists of polystyrene. In this situation the vaporization of the polystyrene to styrene, and the subsequent decomposition of the styrene to lower hydrocarbons, deposits thick carbonaceous films on the advancing surface of the iron (Figure 4.9 shows the decomposition products). Gallois et al (1987) found the film to consist of three main layers: (1) an upper lustrous multilayered structure of amorphous carbon; (2) an intermediate layer of sooty fibres consisting of strings of crystallites; and (3) a layer adhering strongly to the surface of the iron consisting of polycrystalline graphite enriched in manganese, silicon and sulphur. Clearly there has been some exchange of solutes from the iron into the film.

It is probably worth bearing in mind that the visible evidence of the thermal decomposition of hydrocarbons C_xH_y on the surface of the melt is the surface film of remaining carbon. But there is also an invisible legacy: the hydrogen that will have dissolved in the liquid.

Graphite films that have grown on molten iron have been studied in the form of crystals formed on the surface of Fe–C alloys held in graphite crucibles, and so saturated with carbon before being allowed to cool. These shiny black sheets that float on the surface are analogous to the kish graphite that separates from hypereutectic cast irons during cooling. (For the scientifically minded, the graphite films in this research had interesting features. They were single crystals with numerous cracks along certain crystal directions, and hexagonal growth steps on the underside that showed how the film grew by gradual deposition of carbon atoms, probably on to ledges from emergent screw dislocations.)

The growth of graphite on melts saturated with carbon, as above, is easy to understand; but how does graphite grow in the case of lustrous carbon films where the composition of the iron is far from saturated? Such films should go into solution in the iron! Every student of metallurgy knows that at the eutectic temperature the carbon in solution in iron has to exceed 4.3 weight per cent before free (kish) graphite will precipitate. At higher temperatures the carbon concentration for saturation increases, following the liquidus line for the solidification of kish graphite in hypereutectic irons, as is clear from the Fe–C phase diagram.

In an atmosphere containing hydrocarbons, if the rate of arrival of reactants at the free liquid surface is low, then both carbon and hydrogen can diffuse away from the surface into the bulk liquid.

However, in a highly concentrated environment of hydrocarbon gases the rate of arrival of reactants may exceed the rate of diffusion away into the bulk. Thus carbon will become concentrated on the surface (hydrogen less so, because its rate of diffusion is much higher) and may exceed saturation, allowing carbon to build up at the surface as a solid in equilibrium with the local high levels of carbon. Once formed, it would then take time to go into solution again, even if the conditions for growth and stability were removed. Thus it would appear to have a pseudo stability, with a life just long enough so that in some conditions the film could be frozen into the casting if a chance event of surface turbulence were to enfold the surface into the melt.

Recent research has indicated that the conditions for the growth of the graphite film on liquid metals are similar to the conditions required for the growth of diamond films. Reviews by Bachmann and Messier (1984) and Yarborough and Messier (1990) list conditions for the growth of diamond as the breakdown of hydrocarbons and the presence of hydrogen. In the case of iron, the temperature is a little too high, and would tend to stabilize the growth of graphite films. But for metals such as aluminium in a hydrocarbon environment, the conditions seem optimum for the creation of diamond on the metal surface. Prospectors and investors will be disappointed to note, however, that the rate of growth is slow, only one micrometre per hour. Thus in the time that most liquid metal fronts exist, the diamond layers, if any, will be so thin as to be a disappointing investment.

5.5.2 Entrained films

Once folded into the bulk melt by surface turbulence, the newly created bifilm may take a number of forms in cast irons. Since very little research has been carried out in this field, this section promises to be short, and regrettably, at this stage, speculative.

The folding in of the graphitic film is known to result in the familiar lustrous carbon defect. This is, of course, simply a crack lined with the lustrous carbon films. In heavy section such defects are not seen, probably because they will have time to go into solution before being frozen into the casting. In thinner sections of a ferritic matrix, or mixed ferrite/ pearlite matrix, the films can be seen to be partly dissolved as indicated by the layers of the higher carbon content pearlite on either side of the defect.

More speculatively, many irons are poured so turbulently that it is to be expected that huge numbers of graphitic films will be entrained. This raises a distinct possibility that many of the graphite flakes seen on a polished section of grey iron will not be formed as a result of a metallurgical precipitation reaction, but may be the remnants of entrained graphitic bifilms. The occasional appearance on microsections of what seem to be isolated large flakes amid uniform smaller flakes is suggestive of the bi-modal distribution to be expected if such a mixed source of graphite were present.

5.5.2.1 Nitrogen fissures in grey iron

Nitrogen fissures in grey iron castings are large cracks, often measured in centimetres, that appear to have been associated with the use of sand binders that contain high levels of nitrogen. They are an enigma that has never been satisfactorily explained. The high-nitrogen binders that are blamed for these features usually contain amines, whose breakdown probably contributes both nitrogen and hydrogen to the liquid iron. However, although such binders have been associated with fissure defects, their use does not always result in fissures. Perhaps entrained bifilms, perhaps consisting of nitride films, are also required, so that the filling system may also be highly influential. Any involvement of the filling system has not previously been suspected, but would explain the confusion in results. Once entrained, the high hydrogen and nitrogen pressure in the iron might be sufficient to open any bifilms to some extent, revealing their presence as crack-like features.

5.5.2.2 Bifilms in ductile iron

The ductile casting industry has referred to entrained surface films as 'dross stringers'. This name, based on their one-dimensional appearance on a polished section, has led to a comforting self-deception, concealing their obvious real nature as extensive two-dimensional defects in the form of films. The occasional appearance of clusters of graphite nodules that have floated up and been trapped under such 'stringers' corroborates their real nature as films. Also, as we are now aware, if the film is solid, the entrainment process will fold them in dry side to dry side, thus forming a crack.

The films appear only at low temperature as we have seen, and seem to be mainly magnesium silicate, $MgO.SiO_2$ (alternatively written as $MgSiO_3$) probably with a thick upper layer of solid MgO as discussed earlier. If the ductile iron is cast at a low temperature, and if the surface is entrained, the creation of seriously damaging bifilms is guaranteed. Naturally, as the iron cools during its passage through the running system it is likely to cool to the temperature at which the solid film can form, so that defects will be expected in most filling systems in which surface turbulence is not controlled. Once entrained, the defects can, of

course, lead to a variety of additional problems. One potential problem is discussed below.

5.5.2.3 Plate fracture defect in ductile iron

Like nitrogen fissures in grey irons, plate fracture in ductile irons has also never been satisfactorily explained. Ductile irons, should, of course, always exhibit a ductile mode of failure. Sometimes, however, a casting will exhibit poor strength and poor elongation to failure, with the fracture surface appearing to consist of large embrittled grains. These unpredictable events give rise to serious concern that the material is not under the proper control that either the foundry or the customer would like to see. Everyone's faith is shaken. The question naturally arises, 'Is ductile iron a reliable engineering material?' This is a question that should never arise, and that no one wishes to hear.

Following the description given by Karsay (1980) and Gagne and Goller (1983), the features of the plate fracture are large, flat, apparently brittle fracture planes, in ductile irons that should exhibit only ductile failure (Figures 5.46 and 5.47). When viewed closely, the planes are seen to be studded with small, irregularly shaped graphite spheroids, that are arranged with an accuracy almost resembling a crystal lattice (Figure 5.47b). The planes are nearly always close to a right angle with the cast surface, and grow mainly vertically. The plane is in a matrix that is somewhat lighter than the rest of the casting after etching. Karsay suggests that the colour

Figure 5.46 *Plate fracture in the feeder neck of a ductile iron casting (Karsay 1980).*

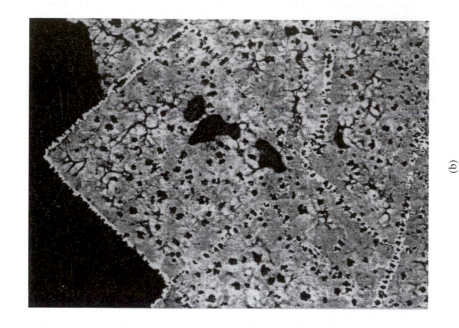

(b)

(a)

Figure 5.47 *(a) Fracture surface showing plate fracture in about 50 per cent of grains; (b) polished section through the fracture (Barton 1985) (courtesy Casting Technology International).*

difference may be the result of a higher Si content in this region. Finally, in this region, there is a high incidence of small inclusions that appear to be mainly magnesium silicates.

All these features are consistent with the defect being an oxide bifilm, probably a magnesium silicate, explaining the high Si content and the higher inclusion content, and possibly malformed spheroids as a result of local loss of Mg. The planar form arises from the bifilm being pushed by the raft of austenite dendrites and organized into an interdendritic sheet, similar to that commonly seen in other alloy systems (Figures 2.41–2.44). The vertical orientation is explained by the greater rate of heat transfer from the base of the casting where gravity retains the contact with the mould, so that these grains grow fastest and furthest. In addition, the buoyancy of the magnesium silicate bifilm will encourage its vertical orientation, and so assist the advancing dendrite to straighten the film. Spheroids in interdendritic regions would then be revealed at the regular spacing dictated by the dendrite arm size (normally, a section at a random angle to the dendrite growth directions would obscure this natural regularity that is almost certainly present in all ductile iron structures. Thus it should not be looked upon as a defective structure in itself, as has occasionally been assumed.) The bifilm probably disintegrates to some extent because of its surface energy tending to spheroidize it; the high temperature also assisting this effect. What remains are the changes in chemistry and numerous silicate fragments as inclusions to encourage the direction of growth of the crack that finally causes failure.

Other features of plate fracture are its occurrence in slowly cooled regions, such as in a feeder neck. This may be the result of the lower rate of growth allowing the dendrites to straighten films more successfully (at high growth velocity, the drag resistance of films would resist dendrite growth, and resist film straightening).

The less common appearance of plate fracture in irons of higher carbon equivalent value (above 2.9 per cent CEV), and its reduction in resin-bonded sand moulds reported by Barton (1985) is probably not so much the result of a more rigid mould but an indication that the entrainment of the oxide film is less damaging in this more carbonaceous environment.

5.5.3 Nucleation and growth of graphite

The properties of most graphitic cast irons are dominated by the graphite form; the shape, size and distribution. The over-riding effect on strength and especially ductility is simply the result of the graphite having practically zero strength and/or poor bonding with the matrix, and thus behaving like a crack in the iron matrix. The shape, size and

distribution of the cracks control the properties. (The analogy with light alloys containing a high density of bifilms is compelling! In the case of spheroidal graphite iron the spheroids are analogous to the convoluted form of the bifilms, whereas grey irons are analogous to the aluminium alloys with unfurled bifilm cracks.)

In the past, little attention has been paid to the structure of the iron dendrites, nor the as-cast grain size of the iron matrix. Despite the scientific interest of such questions, the approach seems actually sound and pragmatic and, in general, is adopted here. This is a case where the as-cast matrix structure is accepted as relatively unimportant. The important features are (i) the high density of defects (the graphite particles acting as cracks) that dominate properties like elongation and ductility, and (ii) the room temperature structure of the metallic matrix, whether ferritic or pearlitic, etc., that dominates strength and hardness.

In view of the massive research effort devoted to cast iron, and the many books written on the subject, it may seem unnecessary to add to this impressive literature. Certainly, a review of cast iron properties is not intended. Nevertheless, recent thinking is assisting to clarify some of the traditional mysteries such as inoculation. Thus it is worthwhile to outline some of these new concepts.

The nucleation of graphite in cast irons by the deliberate addition of foreign nuclei is called inoculation. Inoculation of cast irons is beneficial to achieve a reproducible type and distribution of graphite, so important for the achievement of reproducible mechanical properties and good machinability.

Successful inoculants include ferrosilicon (an alloy of Fe and Si, usually denoted FeSi, and usually containing approximately 75 weight per cent silicon), calcium silicide and graphite. These are added to the melt as late additions, just prior to casting. Additions designed to work over a period of 15 to 20 minutes are used in a granular form, of size around 5 mm diameter, whereas very late additions (made to the pouring stream) are generally close to 1 mm. Late inoculation is carried out because the inoculation effect gradually disappears; a process called 'fade.'

Ferrosilicon is the normally preferred addition, and is known as a 'clean' inoculant. Calcium silicide is known to be a rather 'dirty' addition, almost certainly because the calcium will react with air to give solid CaO surface films (in contrast to FeSi that will cause liquid silicate films). The CaSi addition would probably be much more acceptable with better-designed filling systems that reduce surface turbulence, as is the case of ductile iron spheroidized with magnesium.

It is immediately clear that the common inoculant FeSi does not perform any nucleating role itself.

This is because liquid iron at its casting temperature is above the melting point of the FeSi intermetallic compound, so that the whole FeSi particle melts.

The evidence now suggests that the molten inoculant continues to exist as a high Si region in the liquid iron. Although the Si-rich region is liquid, and the iron is liquid, and the two liquids are completely miscible, the two nevertheless take time to inter-diffuse. This time is probably the fade time. The Si-rich region slowly dissipates in the melt, eventually disappearing completely. However, in the meantime it provides a local environment with a highly effective carbon equivalent value (CEV). To get some idea of the scale and importance of this effect it is instructive (although admittedly not really justified as we shall see) to calculate the carbon equivalent in one of these regions. For an iron of carbon content about 3 per cent, assuming CEV = (per cent C) + (per cent Si/3) we have CEV = 3 + 75/3 = 28 per cent C. Extrapolating the carbon liquidus line on the equilibrium diagram to an iron alloy with 28 per cent C predicts a liquidus temperature in the region of several thousand degrees Celsius. (This is actually not surprising because graphite itself has an effective melting point of over 10 000°C.) Clearly therefore there seems good reason for believing that the carbon in solution in the Si-rich regions is, in effect, enormously undercooled. It is a form of artificial constitutional undercooling (because the graphite is effectively undercooled as a result of a change in the constitution of the alloy).

Now, in reality, it is not appropriate to extrapolate the CEV beyond the eutectic value of 4.3 per cent

C. In fact, when this part of the equilibrium phase diagram is calculated, the liquidus surface is nothing like linear, as seen in Figure 5.48 (Harding et al. 1997). Even so, this figure shows the liquidus in the hypereutectic region to be very high, so that the essential concept is not far wrong. The path of the dissolving particle is marked on the figure, confirming its progress though high constitutional undercoolings through high Si regions, where it will experience large driving forces for the precipitation of graphite.

The size of the driving force is almost certainly the reason why, over the years, so many different nuclei have been identified for the initiation of graphite. It seems that even nuclei that would hardly be expected to work at all are still coaxed into effectiveness by the extraordinary undercooling conditions that it experiences. Studies have shown that many particles that are found in the centres of graphite spherules, and thus appear to have acted as nuclei, are also seen to be floating freely in the melt of the same casting, having nucleated nothing (Harding et al. 1997). This is understandable if the nuclei are not particularly effective. They will only be forced to act as nuclei if they happen to float through a region that is highly constitutionally undercooled.

Studies by quenching irons just after inoculation have revealed a complex series of shells around the dissolving FeSi particle. Although FeSi itself contains almost no carbon, the carbon in the cast iron diffuses into the liquid FeSi region quickly. Data from Figure 1.8 and Equation 5.21 indicate a time of 1 s for an average diffusion distance d =

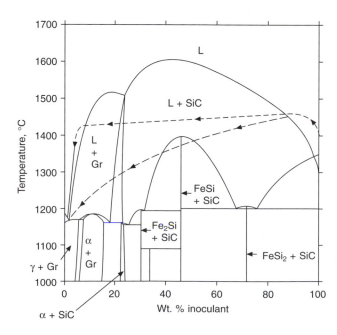

Figure 5.48 *The Fe–FeSi phase diagram showing possible melting and mixing routes for a dissolving FeSi inoculant particle (Harding et al. 1997).*

0.1 mm, and 100 s for $d = 1$ mm. The flow resulting from the buoyancy of the high Si melt, and the internal flows of metal in the mould cavity, will smear the liquid Si-rich region into streamers, reducing the diffusion distance to give the shorter estimated times of homogenization of carbon. Thus the shell of SiC particles around a dissolving FeSi particle (Figure 5.49) appears logical as a result of the high undercooling in the part of the phase diagram where SiC should be stable (Figure 5.48). It seems likely that the SiC nucleates homogeneously because of the high constitutional undercooling. In a shell further out from the centre of the dissolving inoculant particle, graphite starts to form. It seems that graphite may not simply nucleate homogeneously by the generous undercooling but can also form in this region by the decomposition of some of the SiC particles.

If all this were not already complicated enough, there is even more complexity. In addition to the local solute enrichment from the dissolving particle there will also be a release of sundry complex inclusions including oxides and sulphides. Commercially available inoculants contain various impurities, and various deliberate additions that supplement the natural nucleating action in this way. At least some of these may be good heterogeneous nuclei for the formation of new graphite crystals (or perhaps new SiC crystals that will subsequently transform to graphite particles). Also, of course, these particles are provided exactly where they are needed, in the heart of the highly undercooled region.

This action of the inoculating material in providing a combination of good growth conditions and copious heterogeneous nuclei explains the action of graphitizers such as ferrosilicon, and the importance of the traces of impurities such as aluminium and rare earths that raise the efficiency of inoculation.

Ferrosilicon and calcium silicide are not, of course, the only materials that can act as inoculants. Silicon carbide (SiC) is also effective, as is graphite itself. Both of these materials can be seen to provide in a similar way the transient conditions of high constitutional undercooling that are needed for the nucleation of graphite in cast irons.

Jacobs *et al.* (1974) were probably the first to carry out some elegant electron microscopy to demonstrate that within graphite nodules there is a central seed of a mixed (Ca,Mg) sulphide, surrounded by a mixed (Mg, Al, Si, Ti) oxide spinel. There are matching crystal planes between the central sulphide, the spinel shell, and the graphite nodule, indicating a succession of nucleating reactions. This exemplary work has been confirmed a number of times, most recently by Solberg and Onsoien (2001).

However, because the undercooling is high at this initial time, once nucleated, graphite will be expected to grow dendritically as thin flakes (analogous to metal growth at high undercooling

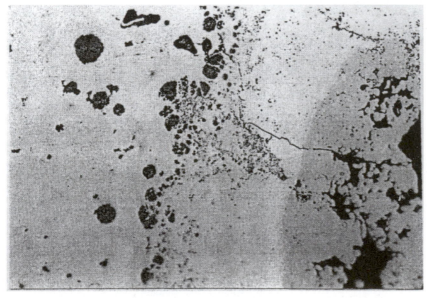

Graphite   SiC   Fe–Si phases

100 µm

Figure 5.49 *Microsection of a dissolving FeSi particle in a ductile iron, quenched from the liquid state (Bachelot 1997)*

as thin dendrites). It seems unlikely therefore that the initial form of graphite is spheroidal as has often been supposed. Later, at the edges of the supercooled region, the thin dendritic form will start to coarsen, its form becoming more bulbous (Figure 5.50). As the embryonic particles of graphite move further out into the open liquid, growth conditions will reverse; the particles will become unstable and start to dissolve. Even so, of course, many will be expected to survive to approach the solidification front of the austenite, where their instability will be reduced. They will become fully stable when the eutectic is reached, and finally grow once again as further cooling takes the metal below the equilibrium eutectic temperature.

This complex chain of nucleating effects has the outcome that graphite particles exist in the melt at temperatures well above the eutectic. The prior existence of graphite particles in the liquid at high temperature, well above the temperature at which austenite starts to form, is quite contrary to normal expectations based on the equilibrium phase diagram, but explains many features of cast iron solidification. The expansion of graphitic irons prior to freezing (the so-called 'pre-shrinkage expansion') has in the past always been difficult to explain (Girshovich *et al.* 1963). The existence of graphite spheroids growing freely in the melt above the eutectic temperature has been a similar problem, seemingly widely known, and seemingly widely ignored, but now provided with an explanation, despite the desirability of much confirmatory effort over future years.

Whether the subsequent growth of graphite occurs in the form of flakes or spheroids is a completely separate issue, unrelated to the nucleation/inoculation treatment. This is a growth problem. The separate nature of the problem can

be appreciated from a close look at the graphite structure around some central nucleating particles. The structure in graphite spheroids close to the nucleating particle is usually seen to be highly irregular (Figure 5.51). The graphite form in this region appears almost turbulent. Clearly, after a very short growth distance, the crystallographic orientation is not under any influence of the nucleating particle. However, after a small further distance, the graphite organizes itself, and develops its nicely ordered radial grains typical of a good spheroid. Thus the organization of the growth takes time to develop, and is a macroscopic phenomenon. The analogy with the planar growth condition of a metal under conditions of low constitutional undercooling is striking. The spheroidal growth has been widely proposed to be the result of a detailed atomic mechanism. For an elegant exposition the reader is recommended to the classic paper by Double and Hellawell (1974). However, in addition, if not actually dominant, the growth form almost certainly has at least some contribution from macroscopic influences. To influence the roundness of the growth form, a mechanism must act on the scale of the spheroid itself. Such mechanisms might include (i) a low constitutional undercooling condition in the surrounding liquid when in the free-floating state, or (ii) a mechanical constraint imposed on the expanding sphere when surrounded by solid, but plastically deforming, austenite. It is just possible that (iii) some adsorption on the surfaces of the growing crystal may be important. There are no shortages of theories on this issue, and facts are hard to establish.

5.5.4 Nucleation and growth of the matrix

The nucleation of the austenitic matrix of cast irons has, to the author's knowledge, never been researched. Furthermore, it is not especially clear that the problem is at all important. For instance, if a fine austenite grain size could be obtained, would it be beneficial? The answer to this question appears to be not known. Moreover, in the section on steels the grain refinement of austenite is seen to be unsolved. Thus in all this disappointing ignorance, we shall turn to other matters about which at least something is known.

Only recently, two different teams of researchers have revealed for the first time the growth morphology of the austenite matrix in which the graphite spherulites are embedded. Ruxanda *et al.* (2001) studied dendrites that they found in a shrinkage cavity, finding them to be irregular, each dendrite being locally swollen and misshapen from many spherulites beneath their surfaces. Rivera *et al.* (2002) developed an austempering treatment directly from the as-cast state that revealed the austenite grains clearly. The grains were large, about

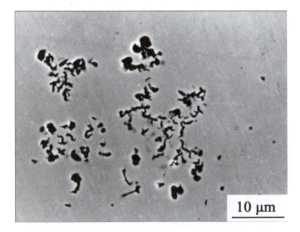

Figure 5.50 *Coarsening of graphite particles on emerging from the undercooled FeSi region (Benaily 1998)*

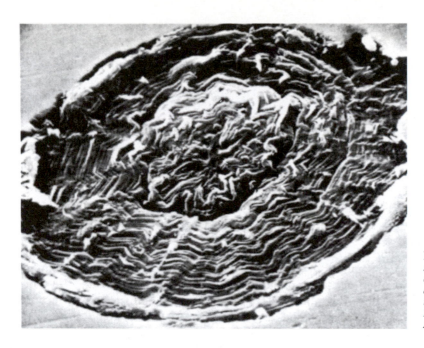

Figure 5.51 *The chaotic growth structure of a graphite spherule, cathodically etched in vacuum, and viewed at a tilt of 45 degrees in the SEM (Karsay 1985, 1992). Reprinted with permission of the American Foundry Society.*

1 mm across, clearly composed of many irregular dendrites, several hundred eutectic cells, and tens of thousands of spherulites. The dendrites from both these studies are not unlike the aluminium dendrite shown in Figure 5.20.

It seems fairly certain, therefore, that the growth of the austenite dendrites occurs into the melt in which there exists a suspension of graphitic particles. The particles hover almost non-buoyant because of their small size, having such a low Stoke's velocity that they are carried about by the flow of the liquid. Using the Stoke's relation it is quickly shown that a 1 µm diameter particle has a rate of flotation of only about 1 µms^{-1}, corresponding to a movement of the order of one dendrite arm spacing in a minute. Particles of 10 µm diameter would have a dendritic form (Figure 5.50), reducing their overall average density difference, and increasing their viscous drag, so their flotation rate would hardly be higher, despite their larger size, thus still allowing plenty of time for incorporation into the dendrite structure.

Once trapped, the surrounding dendrite would be expanded and distorted by the continued growth of the graphite particle, since, at these temperatures, the surrounding solid will be no barrier to the rapid diffusion of carbon to feed its growth. This micro-expansion of the dendrites translates of course to the macroscopic expansion of the whole casting, the expansion of the mould, and even the expansion of the surrounding steel moulding box, if any. Submicroscopic rearrangements of atoms can accumulate to irresistible forces in the macroscopic world.

5.6 Steels

5.6.1 Inclusions in steels; general background

Svoboda *et al.* (1987) report on a large programme carried out in the USA, in which over 500 macroinclusions were analysed from 14 steel foundries. This valuable piece of work appears to have given a definitive description of the types of inclusions to be found in cast steel, and the ways in which they can be identified. A summary of the findings is presented in Figure 5.52 and is discussed below.

Each inclusion type can be identified by (i) its appearance under the microscope, and (ii) its composition.

1. Acid slags can be identified by their high FeO content (typically 10–25 per cent), and glass-like microstructure.
2. Basic slags and furnace slags from high-alloy melts can be traced by the calcia (lime), alumina, and/or magnesia that they contain.
3. Refractories from furnace walls and/or ladles have characteristic layering, flow lines, and a pressed and sintered appearance including sintered microporosity. Their compositions are reminiscent of those of the refractories from which they originated (e.g. pure alumina, pure magnesia, phosphate bonded materials, etc.).
4. Moulding sand is identified from the shape of residual sand grains and from its composition high in silica.
5. Mould coat material is normally easily

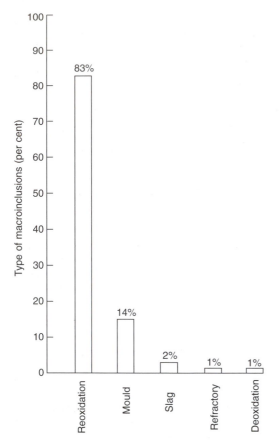

Figure 5.52 *Distribution of types of macroinclusions in carbon and low alloy steel castings, from a sample of 500 inclusions in castings from 14 foundries (Svoboda et al. 1987).*

distinguished because of its composition (e.g. alumina or zircon).

6. Deoxidation products are always extremely small in size (typically less than 15 μm) and are composed of the strongest deoxidizers. These inclusions are likely to have formed at two distinct stages: (1) during the initial addition of strong deoxidizer to the liquid steel, when small inclusions will be nucleated in large numbers as a result of the high supersaturation of reactive elements in that locality of the melt – any larger inclusions will have some opportunity to float out at this time, (ii) during solidification and cooling. These stages will be discussed further below.

7. Reoxidation products are large in size, usually 5–10 μm in diameter, and consist of a complex mixture of weak and strong deoxidizers. In carbon steels the mixture contains aluminium, manganese and silicon oxides. In high-alloy steels the mixture often contains a dark silica-rich

phase, and a lighter coloured Mn + Cr oxide-rich phase. Entrapped metal shot is found inside most of these inclusions. At the present time it seems uncertain whether the shot is incorporated by turbulence or by chemical reduction of the FeO by the strong deoxidizers. (These larger inclusions have been previously known as ceroxide defects, as a result of their content of cerium and other powerful rare earth deoxidizers. The rare earth deoxidizers are used in an attempt to control the shape of sulphide inclusions.)

5.6.2 Entrained inclusions

Previously, most inclusions introduced from outside sources have been called exogenous inclusions, but this name, besides being ugly, is unhelpful because it is not descriptive. 'Entrained' indicates the mechanism of incorporation. Also, the word 'entrained' draws attention to the fact that as a necessary consequence of their introduction to the melt, such inclusions have passed through its surface, and so will be wrapped in a film of its surface oxide. Depending on the dry or sticky qualities of the oxide, and the rate at which the wrapping may react with the particle, the fragment can act later as an initiation site for porosity or cracks. Metal, too, may become entrapped in the entraining action, and thus form the observed shot-like particles.

Svoboda has determined the distribution of types of macroinclusion in carbon and low-alloy cast steels from the survey. The results are surprising. He finds that reoxidation defects comprise nearly 83 per cent of the total macroinclusions (Figure 5.52). These are our familiar bifilms created by the surface turbulence during the transfer of the melt from the furnace into the ladle, and from the ladle, through the filling system into the mould. In addition, he found nearly 14 per cent of macroinclusions were found to be mould materials. Since we know that mould materials are also introduced to the melt as part of an entrainment process, it follows that approximately 96 per cent of all inclusions in this exercise were entrainment defects from pouring actions.

Only approximately 4 per cent of inclusions were due to truly extraneous sources; the carry-over of slag, refractory particles and deoxidation products.

This sobering result underlines the importance of the reaction of the metal with its environment after it leaves the furnace or ladle. The pouring and the journey through the running system and into the mould are opportunities for reaction of those elements that were added to reduce the original oxygen content of the steel in the furnace. The unreacted, residual deoxidizer remains to react with the air and mould gases. Such observations confirm the overwhelming influence of reactions during pouring or in the running system as a result of surface turbulence; these effects are capable of ruining the quality of the casting.

However, good running systems are not usually a problem for small steel castings. Large steel castings are another matter because of the high velocities that the melt necessarily suffers. This is partly a consequence of the use of bottom-pour ladles, and partly the result of the fall down tall sprues.

The historical use of rather poor filling system designs has given steel the reputation for a high rate of attack on the mould refractories. Unfortunately, the solution has resulted in the use of pre-formed refractory tubes and corners for the running system. The joining of these standard pipe shapes means that nicely tapered sprues cannot easily be provided, with the result that much air goes through the running system with the metal. The chaos of surface turbulence in the runner, and the splashing and foaming of bursting bubbles rising through the metal in the mould cavity, will mean that reoxidation product problems are an automatic penalty.

It follows that a common feature of steel foundries is that the foundry often employs more welders in the 'upgrading' department repairing castings than people in the foundry making castings. This regrettable fact follows from the surface turbulence caused during pouring. Even so, it has to be admitted that this conclusion is probably more easily reached than acted on. It has not been easy to provide steel castings with a good filling system. The difficulties are addressed in Volume II, in which better-moulded systems are recommended, returning to sand moulding for the front end of the filling system (a good pouring basin and tapered sprue combination is not usually harmed by the steel) while ceramic tubing might be acceptable and convenient for the remainder of the system. In the meantime, we shall examine the problems caused by the current poor filling systems.

Some liquid steels have strong, solid oxide films covering their surface. The high melting point of these oxides ensures that they behave as though they are quite dry films. They occur on chromium- and molybdenum-rich stainless steels, especially the super duplex stainless steels. In casting above about 250 kg in weight the filling systems are sufficiently large to pass bifilms up to 100 mm across or more. Entrained air bubbles and surface turbulence in the mould cavity create even more films *in situ*. These are found to be arranged in clusters, often near the ingate, or just under the cope. They are identified on radiographs as resembling faint, dispersed microshrinkage porosity. When grinding into such areas, and checking periodically with the red penetrant dye, the bifilms appear as an irregular spider's web. The bifilms

are mainly at grain boundaries, of course, and are often somewhat opened up by cooling strains. When viewed under the optical microscope they have given rise to the description 'loose grain effect' in some stainless steel foundries. This maze of thin, deep cracks often has to be excavated completely through walls of 100 mm thickness and greater section before these regions can be rebuilt by welding.

However, in small castings of these particular steels, there now seems to be evidence that the ingates can be sufficiently narrow so that the strong, rigid plates of oxide cannot pass through (Cox *et al.* 1999). Thus, paradoxically, this notoriously difficult material can be used to make small castings that are relatively free from defects.

Low-carbon/manganese and low-alloy steels are typically deoxidized with Si, Mn and Al in that order. They can suffer from a stable alumina film on the liquid if the final deoxidation with Al has been carried out too enthusiastically. This causes similar problems to those described above.

An aluminium addition has been recommended to liquid steel to reduce MnO and FeO, which contribute to slag-type defects (Rouse 1987). However, the resulting solid alumina film on the liquid will give rise to its own type of defect problems in the form of internal films that could be even more serious if the level of addition were not carefully controlled.

However, for the usual level of final deoxidation with Al, at approximately only 1 kg or less Al per 1000 kg steel, most low-carbon/manganese and low-alloy steels do not usually suffer such severe internal defects. Because of the high melting temperatures of such steels, the surface oxides contain a mix of SiO_2, MnO and Al_2O_3, among other oxide components. The mix is usually partially molten. On being entrained during pouring the internal turbulence in the melt tumbles the films into sticky agglomerates. Because of the presence of the liquid phases that act as an adhesive, the bifilms cannot reopen, and grow by agglomeration. The matrix becomes therefore relatively free from defects in this way. Also, the oxide is now rather compact and can float out rapidly, gluing itself to the surface of the cope as a ceroxide defect, so called because of the presence of cerium oxide as one of the more noticeable of the many phases in the inclusion. Cope defects are common *surface defects* in these steels. In castings weighing 1000 kg or more the defects can easily grow to the size of a fist. They are, of course, labour intensive to dig out and repair by welding. However, their compact form makes this job somewhat easier, and not quite in the league of the extensive webs of bifilms presented by the super duplex stainless.

Over recent years it has become popular to give a final deoxidation treatment with calcium in the form of calcium silicide (CaSi) or ferro-silicon-calcium because the steel has been found to be much cleaner. This is quickly understood. Alumina and calcia form a low melting point eutectic. Thus the dry Al_2O_3 surface oxide is converted into a liquid oxide of approximate composition $Al_2O_3.CaO$ that has a melting point near 1400°C. Any folding-in of the liquid film will quickly be followed by agglomeration of the film into droplets. The compact form and low density of the droplets will ensure that they float out quickly and will be assimilated into the original liquid eutectic film at the surface, leaving the steel without defects.

In passing, it seems worth mentioning a class of defect that has been the subject of huge amounts of research, but which has never been satisfactorily explained. A tentative explanation is presented here. The phenomenon was the so-called 'rock candy fracture' appearance of some cast steel. This type of defect was seen when the ductility of the casting was especially low, despite the metal appearing to have precisely the correct chemistry and heat treatment. The fracture surface was characterized by intergranular facets that on examination in the scanning electron microscope were found to contain aluminium nitride. Naturally, the aluminium nitride was concluded to be brittle.

This defect seems most likely to be an entrained surface film. The film would probably originally consist of alumina, but would also contain some enfolded air. The nitrogen in the entrained air would be gradually consumed to form aluminium nitride as a facing to the crack. The defect would, of course, be pushed by the growing dendrites into the interdendritic spaces, particularly to grain boundaries. The central crack in the bifilm would give the appearance of the nitride being brittle. On examination, only the nitrogen is likely to be detected, constituting four-fifths of the air, and the oxygen being in any case not easily analysed. The defect is analogous to the plate fracture defect in ductile irons, and the planar fracture seen in Al alloys and other alloy systems (Figures 2.41–2.44).

Thus, despite the chemistry of the steel being maintained perfectly within specification, the defect could come and go depending on chance entrainment effects. Such chance effects could arise because of slight changes in the running system, or the state of fullness of the bottom-pour ladle, or the skill of the caster, etc. It is not surprising that the defect remained baffling to metallurgists and casters for so long.

5.6.3 Primary inclusions

When the liquid alloy is cooling, new phases may appear in the liquid that precede the appearance of the bulk alloy. We have already dealt with the formation of the primary phase in section 5.2.2. Whether any newly forming dense phase gets called

a phase or an inclusion largely depends on whether it is wanted or not: keen gardeners will appreciate the similar distinction between plants and weeds!

New phases that precede the appearance of the bulk alloy are especially likely following the additions to the melt of such materials as deoxidizers or grain refiners, but may also occur because of the presence of other impurities or dilute alloying elements.

For instance, in the case of steel that has a sufficiently high content of vanadium and nitrogen, vanadium nitride, VN, may be precipitated according to the simple equation:

$$V + N \Leftrightarrow VN$$

Whether the VN phase will be able to exist or not depends on whether the concentrations of V and N exceed the solubility product for the formation of VN. To a reasonable approximation the solubility product is defined as:

$$K = [\%V].[\%N]$$

where the concentrations of V and N are written as their weight per cent. More accurately, a general relation is given by using, instead of weight per cent, the activities a_V and a_N, in the form of a product of activities:

$$K' = a_V.a_N$$

It is clear then that VN may be precipitated when V and N are present, where sometimes V is high and N low, and vice versa, providing that the product $\%V \times \%N$ (or more accurately, $a_V \times a_N$) exceeds the critical value K (or K'). It is interesting to speculate that [N] may be high very close to the surface where the melt may be dissolving air. Thus the formation of a surface film of VN may be more likely.

In the case of the deoxidation of steel with aluminium, the reaction is somewhat more complicated:

$$2Al + 3O \rightleftharpoons Al_2O_3$$

and the solubility product now takes the form:

$$K'' = [a_{Al}]^2 \cdot [a_O]^3$$

where the value of K'' increases with temperature. Again, the surface conditions are likely to be different from those in the bulk, with the result that a surface film of AlN or Al_2O_3 is to be expected, even if concentrations for precipitation in the bulk are not met.

These examples only relate to the case where the newly formed phase is in equilibrium with the melt. In practice higher concentrations of the individual constituents of the phases will be required to overcome the problem of nucleation of the new phase.

Turpin and Elliot (1966) were among the first to study the problem of the nucleation of new dense phases from the melt. Using the approach of classical nucleation theory as illustrated in Equation 5.15, these authors used the standard free energy changes for the formation of oxides, which they took from the literature on thermodynamics, to find the energy for formation of a nucleus of the new material. We shall not follow their argument in detail, but merely quote their result in Figure 5.53 for the Fe–O–Si system. In this example two oxides are considered. The first is from the reaction:

$$Si + 2O \rightleftharpoons SiO_2$$

so that the equilibrium constant is now approximately:

$$K''' = [\%Si] \cdot [\%O]^2$$

Figure 5.53 shows this equilibrium threshold with its slope of 2 (i.e. an increase of a factor of 10 in oxygen concentration together with a decrease of a factor of 100 in silicon concentration will still result in the nucleation condition being satisfied). The higher threshold shown in Figure 5.53 corresponds to the concentrations required for nucleation, assuming a surface energy of the interface of 1.3 Nm^{-1}. (In fact, the threshold required to nucleate silica can be shown to lie at increasing concentrations as the assumed value for the surface energy is raised.) (We shall continue to use Nm^{-1} in uniformity with the rest of this book. Otherwise, it would have been logical to quote surface energy in the identical units Jm^{-2}.)

Turning now to the possibility of forming FeO in this system, the equation is:

$$Fe + O \rightleftharpoons FeO$$

This simple equation becomes even more simplified in its solubility product form, because the concentration of iron is very closely 100 per cent (i.e. unity in the above equation). Thus the FeO can exist in equilibrium in an iron melt only if the oxygen concentration is high enough (since the iron concentration is already fixed at its maximum). Thus in Figure 5.53 the threshold for the formation of FeO is very nearly a vertical line. The parallel line denoting the threshold to overcome the resistance to the nucleation of FeO is quite close; this is because the surface energy of the interface is low, in the region of only 0.25 Nm^{-1}.

Turpin and Elliott take their analysis further to show that a melt that has been allowed to come into equilibrium at a high temperature may reach a sufficient supersaturation to cause nucleation as the melt is cooled. They effectively work their analysis backwards, aiming for a nucleation at the freezing point of iron, 1536°C, and calculating what equilibrating temperature would have been required to achieve this. Their results are summarized in Figure 5.54.

These results demonstrate that it is possible, in principle, to predict the arrival and stability of particles in melts, as a function of temperature and composition. Turpin and Elliot were not able to confirm their theoretical predictions for this system because of experimental limitations. However, much work on the grain refinement of metals would surely benefit from a careful, formal approach of this kind.

All this work so far has neglected the problem of the nucleation of the inclusion. We have considered examples of nucleation at various points in the book, especially in section 5.2.2. At this stage we shall simply note that any primary

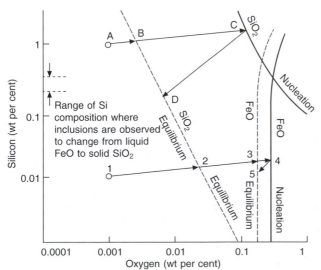

Figure 5.53 Equilibrium and nucleation thresholds for silica and iron oxide inclusions in solidifying iron. Data on thresholds from Turpin and Elliot (1966).

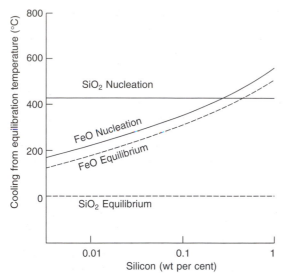

Figure 5.54 *The cooling required, from a temperature where the system was allowed to come into equilibrium, down to the freezing point of iron (1536°C), to nucleate oxides in the Fe–O–Si system (from Turpin and Elliot 1966).*

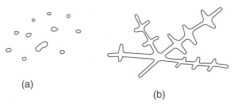

(a)   (b)

Figure 5.55 *Alumina inclusion in an aluminium-killed low-carbon steel, showing: (a) a two-dimensional section; and (b) a three-dimensional view (from Rege et al. 1970).*

inclusions form prior to the arrival of the matrix primary phase. Thus they appear in a sea of liquid. During this 'free-swimming' phase, primary inclusions are thought to grow by collision and agglomeration (Iyengar and Philbrook 1972).

For liquid inclusions this is expected to result in large spherical inclusions whose compact shape will enable them to float rapidly to the surface and become incorporated into a slag or dross layer which can be removed by mechanically raking off, or can be diverted from incorporation into the casting by the use of bottom-pouring ladles, or teapot spout ladles.

For solid inclusions, the agglomeration process may form loosely adhering aggregates or clouds. For instance, alumina inclusions in aluminium-killed and rolled steel appear to be fine clouds of dispersed particles, arranged in stringers, on a polished section. There seems to be more than one potential explanation of this appearance: (i) when revealed by deep etching the inclusion is sometimes seen to have a three-dimensional dendritic shape (Figure 5.55) – it is easy to see how the spindly dendrite arms of these alumina inclusions could align, elongate and fracture to form the long stringers observed in longitudinal sections of rolled steels, (ii) alternatively, the entrained and ravelled alumina films may condense into arrays of compact particles, analogous to the way in which sheets of liquid metal break up into droplets (a spectacular example is given in Figure 2.13), an effect driven by the

reduction of surface energy. The rolling out of these clouds of discrete particles will again explain the observed stringers. Work to clarify these possibilities would be welcome.

Hutchinson and Sutherland (1965) have studied the formation of open-structured solids. They find that flocs can form by the random addition of particles. If these particles are spherical and adhere precisely at the point at which they first happen to encounter the floc, then the floc builds up as a roughly spherical assembly, with maximum radius R, and about half the number of spheres within a region R/2 from the centroid. The central core has an almost constant density of 64 per cent by volume of spheres. Occasional added spheres will penetrate right into the heart of the floc. Graphite spheres in ductile iron appear to be a good example of this kind of flocculation. Melts of hypereutectic ductile irons suffer a loss of graphite by the floating out of loose flocs of spherulites (Rauch *et al.* 1959).

We have only touched on examples of oxides and nitrides as inclusions in cast metals. Other inclusions are expected to follow similar rules and include borides, carbides, sulphides and many complex mixtures of many of these materials. Carbonitrides are common, as are oxy-sulphides. In C–Mn steels the oxide inclusions are typically mixtures of MnO, SiO_2, and Al_2O_3 (Franklin *et al.* 1969) and in more complex steels deoxidized with ever more complex deoxidizers the inclusions similarly grow more complex (Kiessling 1978).

Kiessling points out that steel that contains only as little as 1 ppm oxygen and sulphur will contain over 1000 inclusions/g. Thus it is necessary to keep in mind that steel is a composite product, and probably better named 'steel with inclusions'. Even so, steels are often much cleaner than light alloy castings, that might contain 10 or 100 times more inclusions, partly helping to explain the relatively poor ductility of Al-based casting alloys compared to steel casting alloys.

Not all of these inclusions will be formed during the liquid phase. Many, if not most, will be formed later as the metal freezes. These are termed secondary inclusions, or second phases, and are dealt with in the following section.

5.6.4 Secondary inclusions and second phases

After the primary alloy phase has started to freeze, usually in the form of an array of dendrites, the remaining liquid trapped between the dendrite arms progressively concentrates in various solutes as these are rejected by the advancing solid. Because the concentration ahead of the front is increased by a factor $1/k$, where k is the partition coefficient, the number of inclusions can be greatly increased compared to those that occurred in the free-floating stage in the liquid. However, the size population is usually different, being somewhat finer and more uniform as a result of the more uniform growth conditions.

The secondary inclusions or second phases form at the freezing front. One of the most common and important second phases is a eutectic. We have already seen how microsegregation can lead to the formation of eutectic at bulk compositions that are much below those expected from the equilibrium phase diagram.

For the remainder of this short section, we shall consider the arrival of other phases. Oxide inclusions in the Fe–O–Si system are taken as an example.

Take, for instance, a melt that contains 1 per cent silicon and 0.001 per cent oxygen, shown as point A on Figure 5.53. As freezing progresses and the concentration of the residual liquid region increases in both silicon and oxygen, the composition moves to B. This is the point at which silica is in stable equilibrium with the melt. Thus silica could form if there were pre-existing silica or some other favourable nucleus present. However, in the effectively isolated pockets of liquid trapped between the dendrite arms the probability of a suitable substrate is low. Thus the liquid continues to supersaturate along the line BC. At C the concentration is sufficiently high to allow silica to nucleate without any assistance. It is said to nucleate homogeneously. Forward and Elliot (1967) calculated that the supersaturation required to nucleate silica occurs at about 98 per cent solidification.

Once the new phase has been nucleated, the surrounding residual liquid will be quickly depleted of solutes. These will diffuse to the growing phase, causing the local concentrations to fall until they meet the equilibrium threshold at D. The concentration of those solutes will then remain stable in the local region, the inclusion only growing to take up any excess solute as it is comes available because of rejection from the advancing dendrites.

If now we take a second example containing, for instance, only 0.01 per cent silicon and 0.001 per cent oxygen, we start at point 1 in Figure 5.53. At and beyond point 2, silica could form if there were any favourable nucleus or pre-existing silica.

In the absence of this, on passing point 3, FeO could form if a favourable foreign substrate happened to be present. However, in the absence of pre-existing FeO or SiO_2 or any favourable nuclei for either of these phases, then point 4 will be reached. At this point FeO will nucleate spontaneously. Its subsequent growth will cause the supersaturation to fall until the local melt is once again in equilibrium with the new phase at point 5.

(Because the path 4 to 5 on the composition map lies within the regime in which silica is stable, it is possible in principle that some silica may dissolve in the growing FeO particle, or might nucleate on it. In fact it is likely that neither will occur: FeO and SiO_2 are relatively immiscible, and liquid FeO is unlikely to constitute a favourable nucleus for solid SiO_2.)

Turpin and Elliott (1966) go on to examine the Fe–O–Al system that contains, in addition to FeO and Al_2O_3, the mixed oxide hercynite $FeO.Al_2O_3$. Thus the succession of phases that can appear becomes more complicated. In real systems, of course, the situation is vastly more complex still, with very many alloying elements being concentrated in interdendritic regions, and all able to react with a number of fellow concentrates.

However, the nucleation of a first phase is likely to prevent the subsequent nucleation of any other phase that might also require one of the same elements for its composition. The availability of solute is clearly limited by a naturally occurring 'first come first served' principle.

In the subsequent observation of inclusions in cast steels, those that have formed in the melt prior to any solidification are, in general, rather larger than those formed on solidification within the dendrite mesh. The possible exceptions to this pattern are those inclusions that have formed in channel segregates, where their growth has been fed by the flow of solute-enriched liquid. Similarly, in the cone of negative segregation in the base of ingots the flow of liquid through the mesh of crystals would be expected to feed the growth of inclusions trapped in the mesh, like sponges growing on a coral reef feeding on material carried by in the current. In Figure 5.31b the peak in inclusions in the zone of negative segregation is composed of macro-inclusions that may have grown by such a mechanism. Elsewhere, particularly in the region of dendritic segregation around the edge of the ingot, there are only fine alumina inclusions.

It would not be right to leave the subject of inclusions without mentioning the special importance of the role of sulphide inclusions in cast steel. The ductility of plain carbon steel castings is sensitive to the type of sulphide inclusions that form.

Type 1   sulphides have a globular form. They are produced by deoxidation with silicon.

Type 2   sulphides take the form of thin grain boundary films that seriously embrittle the steel. They usually form when deoxidizing with aluminium, zirconium or titanium.

Type 3   sulphides have a compact form, and do not seriously impair the properties of the steel. They form when an excess of aluminium or zirconium (but not apparently titanium!) is used for deoxidation.

Mohla and Beech (1968) investigated the relation between these sulphide types, and concluded that the change from type 1 to type 2 is brought about by a lowering of the oxygen content. Additionally, it seems that the new mixed sulphide/oxide phase has a low interfacial energy with the solid, allowing it to spread along the grain boundaries. Also, it might constitute a eutectic phase. Type 3 sulphides were thought by Mohla and Beech to be a primary phase.

However, type 2 inclusions have all the hallmarks of an entrained film defect. It is significant that this type of inclusion forms only when the melt is deoxidized with Al or other powerful deoxidizers that are known to create solid films on the melt. The surface film might originally have been enriched with the other highly surface-active element, sulphur. The entrainment of an oxide film would in any case be expected to form a favourable substrate for the precipitation of sulphides. The film would naturally be pushed into the interdendritic regions by the growing dendrites, so that it would automatically sit at grain boundaries.

Even so, an explanation of type 3 sulphides remains elusive. These results illustrate the complexity of the form of inclusions, and the problems to understand their formation. Much additional research is required to elucidate the mechanism of formation of these defects.

A final question we should ask is 'How do inclusions in the liquid become incorporated into the freezing solid?'

It seems that for small inclusions, especially those that are in the relatively quiet region of the dendrite mesh, the particles are pushed ahead of the front, concentrating in interdendritic spaces.

For larger inclusions, generally above about 10 μm diameter, trapping between dendrite arms is only likely if the inclusion is carried directly into the mesh by an inward-flowing current. This may be the mechanism by which large inclusions are originally trapped within the cone of negative segregation, where they subsequently grow to large size (Figure 5.31b).

Where the front is relatively planar and strong currents stir the melt, the larger inclusions are not frozen into the advancing solid as a consequence of the velocity gradient at the front. Delamore *et al.* (1971) found that those particles which do approach the interface cannot be totally contained within the boundary layer, and as a result spin or roll along it because of the torque produced in the velocity gradient. In this way the larger particles finally come to rest in the centre of castings. For the same reason rimming steels benefited from an absence of large inclusions in their pure rim.

Now for an absolutely final point about inclusions. Take care not to confuse those inclusions that arise from the melt or from freezing with those which occur as a result of later solid-state precipitation. Precipitation within the solid is usually on a scale at least a factor of 10 finer than anything that occurs when liquid phases are still present. This is the direct result of the considerably smaller rate of diffusion in the solid compared to the liquid.

For those readers who are interested in checking whether nitrides or other inclusions might occur in the solid, and particularly because of the problems of embrittlement when such precipitation occurs on grain boundaries, the logic of the approach is broadly the same as that presented above for the nucleation in the liquid phase. In fact more accurate predictions are often possible because better data are usually available for reactions within solid metals.

5.6.5 Nucleation and growth of the solid

During the cooling of the liquid steel, a number of particles may pre-exist in suspension, or may precipitate as primary inclusions. The primary iron-rich dendrites will nucleate in turn on some of these particles. The work by Bramfitt (1970) illustrates how only specific inclusions act as nuclei for delta-iron (δFe).

Bramfitt carried out a series of elegant experiments to investigate the effect of a variety of nitrides and carbides on the nucleation of solid pure iron from the liquid state (in this case, of course, the solid phase is delta-iron). In his work he found that his particular sample of iron froze at approximately 39°C undercooling (i.e. 39°C below the equilibrium freezing point). Of the 20 carbides and nitrides that were investigated, 14 had no effect, and the remaining six had varying degrees of success in reducing the undercooling required for nucleation.

The results are shown in Figure 5.56. They give clear evidence that the best nuclei are those with a lattice plane giving a good atomic match with a lattice plane in the nucleating solid. Extrapolating Bramfitt's theoretical curve to the value for the supercooling of his pure liquid iron indicates that any disregistry between the lattices beyond approximately 23 per cent means that the foreign material is of no help in nucleating solid iron from liquid iron.

Another interesting detail of Bramfitt's work

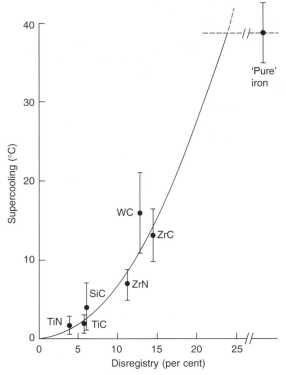

Figure 5.56 *Supercooling of liquid iron in the presence of various nucleating agents. Data from Bramfitt (1970).*

was that a number of additions were ineffective because they either melted or dissolved in the liquid iron prior to it being cooled to promote freezing. This consequent lack of effectiveness was despite, in some cases, quite low values of disregistry. This underlines the perhaps self-evident point (but often forgotten) that any addition has to be present in solid form for it to nucleate another solid.

All of Bramfitt's work was concerned with the nucleation of δFe, the body-centred-cubic form of iron.

The face-centred-cubic form of iron, γFe, or austenite, has been considerably more resistant to past attempts to nucleate it. No one has yet succeeded to identify any prior-existing solid that may act as a nucleus for the γFe phase. The grain refinement of austenitic stainless steels remains a challenge to future metallurgists. It seems likely that only those cast austenitic stainless steels that solidify first to δFe, prior to their subsequent transformation in the solid state to γFe, are able to benefit from grain refinement.

Jackson (1972) lists a large number of additives that were unsuccessful in attempts to refine austenitic steels. His first success was the addition of FeCr powder together with floor sweepings! This impressively economic but hardly commendable

formula caused him to persevere, searching for more scientifically chosen additions. He later found that calcium cyanamide ($CaCN_2$) was quite useful, but required the nitrogen content in the alloy to be raised to 0.3 per cent to be successful in 18:8 stainless steels. At this level of gas content severe nitrogen porosity is the unwelcome product. Jackson was able to define satisfactory conditions for 18/10/3Mo stainless steel, again providing that the nitrogen content was above 0.08 weight per cent. The modest improvement that he reports in mechanical properties he attributes not so much to the reduction in grain size as to the increase in the alloying effect of nitrogen! This work is further complicated by the expectation that the compound $CaCN_2$ will probably have decomposed at steel casting temperatures. Thus a viable grain-refining agent for austenitic steel remains a challenge for future researchers.

Suutala (1983) proposes a factor that allows the prediction of whether the steel will solidify to austenite or primary ferrite; this is the ratio of the chromium equivalent to the nickel equivalent, Cr_{eq}/Ni_{eq}, where the chromium and nickel equivalents are calculated from (elements in weight per cent):

$$Cr_{eq} = \%Cr + 1.37Mo + 1.5Si + 2Nb + 3Ti$$

$$Ni_{eq} = \%Ni + 0.31Mn + 22C + 14.2N + Cu$$

$$Cr_{eq}/Ni_{eq} = 1.55$$

The ratio 1.55 is the critical value at which solidification changes from primary austenite to ferrite. (This value applies for shaped castings and ingots. The equivalent value for welds is 1.43.)

Barbe and co-workers (2002) find that a ratio below 3.5 is useful for mainly ferritic steels. The retention of some austenite limits the grain growth of ferrite, and may reduce the susceptibility that ferritic steels have to 'clinking', i.e. cracking during continuous casting. When slabs are cooled to room temperature cracks are found on the slab edges. Sometimes the slabs crack into two. This problem is reminiscent of problems in continuously cast direct chill (DC) aluminium alloys, particularly the strong 7000 series alloys. These appear to suffer from oxide films that cause the ingots to crack catastrophically, often weeks after being cast. It is dangerous to be near such an event. The high Cr ferritic steels might be expected to behave analogously as a result of entrained Cr-rich films.

It is interesting that the 18/8 stainless steel that was resistant to refinement in Jackson's work had a ratio of around 2, indicating freezing to ferrite. In the presence of 0.3 per cent nitrogen the Cr_{eq}/Ni_{eq} ratio fell to 1.42, indicating solidification to austenite, and suggesting that Jackson may indeed have been successful to refine this fcc structure with $CaCN_2$ additions. The success is repeated for 18/10/3 stainless steel, where the nitrogen acts once

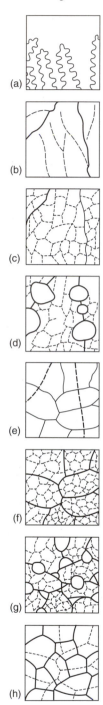

Figure 5.57 *Successive stages of grain evolution in a low-carbon steel, from its freezing point to room temperature (see text for full explanation).*

again to reduce the ratio to around 1.4, indicating solidification to austenite, and a successful grain refinement with $CaCN_2$.

Roberts *et al.* (1979) confirm that only ferritic material was refinable with titanium additions, and confirm that TiC and TiN have lattices that are good fits with ferrite, but poor fits with austenite. Baliktay and Nickel (1988) report that titanium additions can also refine the grain size of the widely used high-strength stainless steel 17-4-PH. However, Equation 7.2 gives a ratio of about 2 for this material, confirming that it solidifies to ferrite, in agreement with the finding by Roberts.

Such grain refinement does not seem to be necessarily beneficial, as inferred from the work by Campbell and Bannister (1975) on the ferritic alloy Fe–3Si. They showed that the best refinement was obtained by the addition of TiB_2 to the melt. However, on metallographic examination the grain boundaries were found to be surrounded by a phase that appeared to be iron boride, which, like iron carbide, is brittle. The mechanical properties were not tested, but were likely to have been impaired. It would be valuable to explore further whether conditions can be found in which grain refinement would improve properties.

Encouragement that such useful results might be gained is given by Church *et al.* (1966). This work on a high-strength steel, 0.33C–0.7Mn–0.3Si–0.8Cr–1.8Ni–0.25Mo–0.040S–0.040P, revealed that although grain refinement was successfully accomplished with 0.60Ti, the benefit was negated by the presence of interdendritic films of titanium sulphide, causing severe embrittlement. However, toughness and ductility could be improved by smaller additions of titanium in the range 0.1–0.2 per cent, which was still successful in achieving grain refinement. The doubt remains that much of this research was undermined by poor casting technique, introducing quantities of deleterious bifilms, particularly at grain boundaries. Sulphur would be expected to precipitate preferentially on such substrates, giving the impression of sulphide films at these locations.

5.6.6 Structure development in the solid

The grain structure that forms on solidification may turn out to be the same as that seen in the finished casting. However, this would be somewhat unusual. It happens only in those cases where the metal is a single phase from the freezing point down to room temperature. Examples include some austenitic stainless, and some ferritic stainless. However, even the ferritic stainless may undergo a transformation to martensite or bainite depending mainly on its carbon content. The transformer steel, Fe–3.25 per cent Si, is a common steel that, on a polished and etched section, clearly displays at room temperature a structure that is not too dissimilar to that originating during solidification. For that reason it is a useful research model alloy.

Even in these single-phase materials there is opportunity for grain boundary migration, possibly grain growth, and possibly recrystallization. Complete recrystallization would be expected in those parts of castings that had been subject to considerable plastic deformation during cooling. This would be expected, for instance, at junctions of flanges that restrain the contraction of the casting.

In materials that change phase during cooling to room temperature the situation can be very much more complicated. Low-carbon and low-alloy steels are a good example, illustrating the problems of understanding a structure that after freezing has undergone at least two further phase changes during cooling to room temperature. Figure 5.57 lists the changes.

(a) The liquid solidifies to delta iron dendrites.
(b) When solidification is complete, the principal grain boundaries have their positions delineated and to some extent fixed in position by segregates, particulate inclusions and bifilms. The slight misalignments between parts of the dendrite raft result in a network of less important subgrain boundaries.
(c) During cooling and differential contraction of the casting, the plastic strains will create dislocations that will migrate to form an additional network of new subgrain boundaries. These are, of course, all low-angle boundaries and may not be readily visible.
(d) On reaching the temperature for the formation of the gamma-iron phase, austenite grains will nucleate on the original grain boundaries or other discontinuities. Their growth into the delta-grains will sweep away most traces of the subgrain network.
(e) When the conversion to austenite is complete the original delta-grain boundaries will still usually be discernible as ghost boundaries because of the fragmentary lines of segregates.
(f) Further cooling strains will generate a new subgrain structure.
(g) Austenite will start to convert to ferrite, usually nucleating at grain corners and boundaries, sweeping away the substructure once again.
(h) The final ferrite grains will again show the ghost boundaries of the previous austenite grains because these will have experienced sufficient time at temperature to have gathered a certain amount of segregates by diffusion to the boundary.
(i) Subsequently a further series of subgrains may be created, although by now the temperature is sufficiently low that any strains will generate fewer dislocations, and that such dislocations will not be sufficiently mobile at lower temperature to migrate into low energy positions, forming low-angle boundaries. Thus the alloy will have become sufficiently strong to retain any further strain as elastic strain. The structure of the alloy will no longer be affected during further cooling.

The final structure on a polished section will be a grain size that has been refined by two successive phase changes (but possibly coarsened a little by intervening grain growth) and still retains ghost boundaries of delta-iron and austenite. The underlying structure of the original delta-iron dendrites will probably still be present, as can be revealed by etching to highlight the differences in chemical composition.

For a formal review of the development of structure in castings, see Rappaz (1989). Further detailed work on cast structures has been carried out during extensive work on the structures of welds in steels. For a review of this work, see Sugden and Bhadeshia (1987). This work draws attention to the complicating effects of the formation of Widmanstatten and acicular ferritic structures, and the presence of martensite, bainite, pearlite and retained austenite. The solidification morphology of the steel in this work seems to be principally cellular, or possibly cellular/dendritic (i.e. dendrites without side branches). Also, of course, in successive weld deposits there are the additional effects of the subsequent heat treatment of the previous runs in the laying down of the subsequent deposits.

Chapter 6

Gas porosity

6.1 Nucleation of gas porosity

Although the problem of the nucleation of cavities and bubbles is in principle similar to that of the nucleation of dense phases such as inclusions, there are differences that make it worthwhile to look at non-condensed phases separately in some detail. In particular, we shall find that there are special difficulties with the nucleation of void and gas phases, forcing us to adopt new concepts.

6.1.1 Homogeneous nucleation

Following the beautifully elegant approach by Fisher (1948), we can quantify the conditions required for the formation of porosity in liquid metals. A quantity of work is associated with the reversible formation of a bubble in a liquid. If the local pressure in the liquid is P_e, we need to carry out an amount of work $P_e V$ to push back the liquid far enough to create a bubble of volume V.

The formation and stretching out of the new liquid/gas interface of area A requires work γA, where γ is the interfacial energy per unit area.

The work required to fill the bubble with vapour or gas at pressure P_i is negative and equal to $-P_i V$. (The negative sign arises because the pressure inside the bubble clearly helps the formation of the bubble, as opposed to the other work requirements, which tend to oppose bubble formation.) Thus the total work is:

$$\Delta G = \gamma A + P_e - P_i$$
$$= 4\gamma \pi r^2 + (4/3)\pi r^3 (P_e - P_i)$$

where clearly $(P_e - P_i)$ is the pressure difference between the exterior and the interior of the bubble, which we may write as ΔP for convenience. Similarly to dense phase nucleation, a plot of ΔG versus bubble radius r shows a maximum that

constitutes an energy barrier to nucleation, as in Figure 5.15. The critical radius r^* in this case is:

$$r^* = 2\gamma/\Delta P^* \tag{6.1}$$

Since bubbles growing from the bulk liquid will grow an atom at a time as the result of statistical thermal fluctuations, it is evident that small bubbles with radii less than r^* will tend to disappear. Only exceptionally will a long chain of favourable energy fluctuations produce a bubble exceeding the critical radius r^*. When this rare event does happen, the microscopic bubble will then have the potential to grow to an observable size.

Fisher goes on to apply some delightfully elegant rate theory to derive values for the critical pressure difference AP^* at which nucleation will occur. The reader is strongly recommended to consult Fisher's original paper. However, for our purposes we can obtain a sufficiently good estimate very easily and quickly using Equation 6.1. Using experimentally determined values of atomic sizes and surface energy γ for liquid metals and assuming that the critical radius is perhaps in the region of an atomic diameter, we obtain Table 6.1.

The reasonable agreement between the calculated critical pressures is corroboration that the critical embryo is actually about two atoms across, and therefore occupies the volume of approximately eight atoms. However, whether or not these figures really are accurate is a detail that need not concern us here. The important message is that the pressures that are required for nucleation are *extremely* high, and reflect the real difficulty of homogeneous nucleation of pores in liquid metals. It is clear that the strengths of liquid metals are almost as high as those of solid metals (for liquid iron the fracture strength corresponds to approximately 7 GPa). This is hardly surprising since the atomic structure is similar, liquid metals being close-packed random structures, compared with solid metals being close-

Table 6.1 Fracture pressures of liquids

Liquid	Surface tension (Nm⁻¹)	Atomic diameter (nm)	ΔP*		
			From Equation 6.1 (atm)	From Fisher (atm)	Complex inclusion (atm)
Water	0.072	–	–	1 320	16
Mercury	0.5	0.30	16 700	22 300	200
Aluminium	0.9	0.29	31 000	30 000	360
Copper	1.3	0.26	50 000	50 000	600
Iron	1.9	0.25	76 000	70 000	850

packed regular structures. In either case, the atoms are about the same distance apart, and it is similarly difficult to separate them, forcing them apart to create a void.

It is certain that in practice the problem of nucleation is reduced by the presence of surface-active impurities in the melt. The non-metals oxygen, sulphur and phosphorus are particularly active in iron melts: the presence of only 0.2 weight per cent of oxygen reduces the surface tension of liquid iron from 1.9 to approximately 1.0 Nm⁻¹. This approximately halves the estimates of the pressure required for nucleation as shown by Equation 6.1. Similar reductions in surface tension (and therefore in fracture pressures) are to be found in liquid copper when contaminated with high levels of the non-metals, O, S and P.

Such high concentrations of oxygen (and the other non-metals) are probably often found on solidification because of the concentration of solutes ahead of the freezing front. For the case of liquid iron once again, the partition coefficient for oxygen is approximately 0.05, giving a factor of 20 increase in concentration at the advancing front. Thus an average of only 0.01 weight per cent oxygen in the bulk melt can produce 0.2 weight per cent at the front.

If the levels of oxygen rise sufficiently to precipitate FeO as a liquid inclusion at the front, then the nucleation problem is reduced yet further because FeO has a surface tension of between 0.6 and 0.5 Nm⁻¹, depending on the oxygen content (Popel and Esin 1956). Thus a gas pore will preferentially nucleate in such a liquid inclusion, where the critical pressure is easily shown to be reduced to around 17 000 atmospheres. Effectively, this is still homogeneous nucleation in a pure liquid where in this case the pure liquid in the liquid Fe is in the form of regions, possibly minute droplets, of FeO.

Even this pressure is still so high as to be probably unattainable. What other possibilities are there?

It is possible that nucleation might occur on a solid impurity particle. A solid foreign substrate, if a poorly wetted surface, might make a location for nucleation. This is known as heterogeneous

nucleation. If this poorly wetted solid surface happened to be inside the liquid FeO inclusion, we shall see how we can reduce the 17 000 atmospheres yet further in the following section.

6.1.2 Heterogeneous nucleation

Fisher considers the case of the nucleation of a bubble against the surface of a solid substrate. The liquid is considered to make an angle θ with the solid. This contact angle defines the extent of wetting; θ = 0 degrees means complete wetting, whereas θ = 180 degrees is complete non-wetting. The geometry is shown in Figure 6.1. Fisher shows that nucleation is easier by a factor:

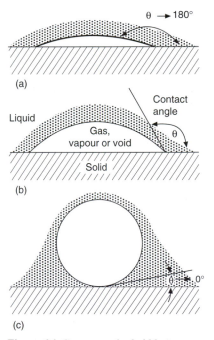

(a)

(b)

(c)

Figure 6.1 *Geometry of a bubble in contact with a solid, showing: (a) poor wetting and easy decohesion of liquid from the solid; (b) medium wetting; and (c) good wetting, where cohesion of the liquid to the solid is high, and the bubble is therefore displaced to a more energy-favoured position in the bulk liquid, out of contact with the solid.*

$$P^*_{het}/P^*_{hom} = 1.12\{(2 - \cos\theta)(1 + \cos\theta)^2/4\}^{1/2}$$
$$(6.2)$$

The factor 1.12 arises because of the fewer sites for nucleation on a plane surface of atoms compared to the greatly increased number of possibilities in the bulk liquid. This factor makes the nucleation of pores on wetted surfaces unfavourable. In fact, nucleation on solid surfaces does not become favourable until the contact angle exceeds 60 or 70 degrees, as shown in Figure 6.2. For this reason the nucleation of pores against the growing solid such as a dendrite is *not* favoured (since a melt wets the solid formed from itself).

This factor is contrary to those other factors that favour the nucleation of a pore close to a front. These favourable factors include the high gas contents and low surface tension usually present in the highly segregated liquid. Additionally there are likely to be inclusions present, pushed and concentrated by the advancing front into the residual liquid. Again, the inclusions that are pushed by the front are likely to be the non-wetted variety, and so will constitute good nuclei for pores. Contrary, therefore, to many published opinions elsewhere, the interface is theoretically *not* a favoured site. Yet, paradoxically, pores will in practice nucleate there because of all the other favourable conditions that prevail adjacent to an advancing solidification interface. The received wisdom turns out to be correct for the wrong reasons!

It is also important to note that not all inclusions are good nucleation sites for porosity. Those that are well wetted will not be favoured. These include the rather more metallic inclusions such as borides, carbides and nitrides. (However, being well wetted, they are mostly good nuclei for the solid phase, and so can assist with grain refinement, as we have seen at several points in Chapter 5.)

The wetting requirements for the nucleation of pores are completely opposite: good nuclei must not be wetted. Such substrates include the non-metals such as oxides. However, the situation is especially bad for *entrained* oxides as will be described later in this chapter.

The reader should be aware that there is widespread misunderstanding of the important fact outlined above, that the nuclei required for the formation of solid (i.e. for grain refinement) are quite different to the nuclei required for the formation of pores and voids.

A nice experiment was carried out by Gernez in 1867 in which he demonstrated that crystalline solids which had been grown in the liquid, and which had never been allowed to come into contact with air, were incapable of inducing effervescence in a liquid supersaturated with gas. Otherwise identical solids which had surfaces which had been allowed to dry always caused effervescence.

Oelsen describes a related experiment that he carried out in 1936, in which he isolated a sample of liquid iron on all sides by a liquid slag. Since the iron contained 2 per cent carbon and 0.035 per cent oxygen it was supersaturated considerably in excess of equilibrium. In fact Oelsen estimated that the internal pressure of carbon monoxide would be approximately 40 atm. When an iron rod was immersed in the melt to destroy the perfection of the containment, a violent eruption of gas immediately occurred, which ceased once again when the rod was withdrawn.

Figure 6.2 indicates that as the contact angle increases to 180 degrees any difficulty of heterogeneous nucleation should fall to zero. In fact there are good reasons to believe that such perfect non-wetting is probably not possible, and that the maximum contact angle attainable in practice is perhaps close to 160 degrees. Certainly no contact angle greater than this appears ever to have been observed (see, for instance, the work by Livingston

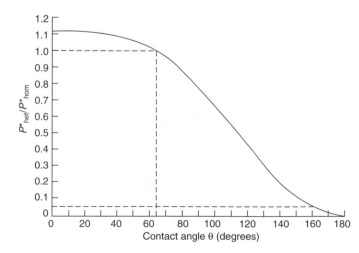

Figure 6.2 *Relative difficulty of nucleating a pore as the contact angle with the solid changes from wetting to non-wetting. Only when the angle exceeds about 65 degrees does heterogeneous nucleation on the solid become favourable.*

and Swingley (1971)). Assuming this to be true, then from Figure 6.2 we see that heterogeneous nucleation on the most non-wetted solid known requires only about one-twentieth of the pressure required for nucleation in the bulk liquid.

Returning to our liquid FeO inclusion in solidifying iron; if a highly non-wetting inclusion were present inside the liquid FeO inclusion, then the lowest pressure for nucleation of a pore in this complex inclusion would be approximately 17 000/20 = 850 atmospheres. Although this pressure is still high, it might now (just) be attainable in iron and steel castings. In Al and Mg alloys such a nucleation condition from an equivalent complex inclusion seems unlikely to be attained since these weaker materials would collapse plastically under the internal tension. The problem of nucleating voids under conditions of high hydrostatic tensions in castings is dealt with in Chapter 7.

Similar reasoning can almost certainly be applied to other alloy systems: complex inclusions are likely to be present in all alloys. In fact it seems likely that most if not all inclusions are complex. The apparent simplicity of many inclusions may be an illusion; microscopic regions of impurities, perhaps only an atom or so thick, may be distributed in patches. It would be difficult to find such patches, and, conversely, difficult to prove that they did not exist.

Perhaps therefore these large fracture pressures predicted by the classical nucleation theory are to be expected to be always reduced by the presence of low surface energy liquid inclusions that in turn contain very poorly wetted solid particles. If this is the case, then assuming that all the values are reduced by the same total factor, approximately 83 found for pure iron, the final column in Table 6.1 is estimated.

There is a natural concern that although these stresses are greatly reduced from those required for nucleation in the pure liquid, they are still potentially unattainable, possibly in all cases. Thus, in a nutshell, even fairly dirty liquid metals do not meet a reasonable criterion for fracture considering the classical nucleation theory.

Thus although the formation of pores by random, thermal atomic fluctuations within the liquid, and against plane solid surfaces, has been a problem in classical physics that has fascinated many scientists since the first attempt by Fisher in 1948, all of the solutions that have been found so far (see, for instance, the short review by Campbell (1968)) have shown that pores are difficult, and actually probably impossible, to nucleate, despite invoking the most active of heterogeneous substrates.

If liquid metals do not fracture, the internal initiation of porosity in castings is impossible. Nevertheless, the fact is that pores in castings are the norm rather than the exception. Clearly, there is a major mismatch between theory and fact. The failure of classical nucleation theory to account for porosity is not widely appreciated. In this work, the fundamental inability of the classical theory has driven the search for other pore initiation processes as outlined below.

This conclusion is so important and so surprising that it is worth emphasizing by stating it another way. The student will usually have been persuaded that because classical homogeneous nucleation of pores in liquid metals is difficult, classical heterogeneous nucleation on a solid particle must therefore occur. The key message here is that this is almost certainly untrue. Classical nucleation of a cap-shaped bubble cannot occur even on the most non-wetted solids. This alarming fact forces a re-evaluation of initiation processes for porosity in castings.

Two potential processes are examined below in 6.1.3.1 and 6.1.3.2. However, it will become clear that the most likely candidate mechanism for pore initiation are entrainment defects, as will be described in section 6.1.3.3.

6.1.3 Non-classical initiation of pores

6.1.3.1 High-energy radiation

The radioactive decay of naturally occurring isotopes, and, unfortunately, contaminating radioactive substances, occurs around us all at every point of our lives. Naturally occurring radioactive materials are relatively common in metals and alloys, and general radioactive contamination is presently increasing annually, arising from industrial and medical sources, and general fallout since the first nuclear explosions in 1945. This is a sad outcome, but even if all future contamination of the environment were to be prevented, we will still have to come to terms with accepting the historical legacy of a radioactive environment as a fact of life.

Liquid metals are, in common with all other present-day materials, subjected to a constant barrage of high-energy particles from these internal radioactive decay processes. The passage of these high-energy particles through the liquid causes thermal or displacement spikes, the name given to regions of intense heating, or actual displacement of atoms, effectively raising the local temperature of the liquid to well above its boiling point. It is possible, therefore, that these transient heated regions might become vapour bubbles sufficiently large to satisfy Equation 6.1, and thus constituting effective nucleation sites for gas or shrinkage pores.

Johnson and Orlov in their review (1986) describe defect regions in solid metals of up to 100 atoms in diameter. Energy can be channelled away from such events along crystallographic directions in ordered solids, reducing the local damage. The lack of any long-range order in liquids means that no such safety valve is possible, so that energy deposition is much more localized. It follows that the production of a bubble of 100 atoms diameter in liquid iron should be easy, giving an equivalent fracture pressure of approximately 1500 atm, much lower than that required for classical homogeneous nucleation. In liquid FeO the fracture pressure would be 400 atm. If the event occurred close to a non-wetted surface, then the fracture pressure would be only 20 atm. Such pressures are much more easily met in castings.

Analogous events are actually observed directly in the bubble chamber, a device full of a transparent liquid that can be vaporized by a high-energy particle, and so thus define its path. It is sobering to note that a bubble chamber can only be constructed using steel made prior to 1945, commonly sourced from the German battleships on the seabed at Scapa Flow. Later steel introduces too much spurious background radiation.

Claxton (1967) has carried out a detailed study of the nucleation of vapour bubbles in liquid sodium subjected to a wide variety of different high-energy particles, including photons, electrons, protons, neutrons, alpha-particles, xenon and strontium as fission fragments, and alpha-recoils. His preliminary analysis suggests that the only interactions capable of initiating nucleation are 'knocked-on' atoms of the liquid produced by fast neutrons. Claxton (1969) suggests that for heavy recoils arising from alpha-particle decay the rate of energy transfer in liquid aluminium would be about three times higher than that in sodium, and in liquid iron should be about ten times, giving rise to the possibility of nucleation, depending on the isotope responsible.

Significantly, the microbubble spectrum in water is seen to be augmented when the water is irradiated with neutrons (Figure 6.3).

Additional factors that are likely to enhance any effects of the presence of radioactive isotopes in metals will be the concentration of such elements ahead of the freezing front. This is precisely where such events can be most effective. The region has high gas content, low surface tension, and high density of assorted solid debris, some of which may be effective nuclei.

Furthermore, if a particularly troublesome radioactive isotope happened to be present in the melt, as an alloy in solution, it would be fundamentally different in character from suspensions of bubbles or inclusions, in that it could not simply be eliminated by filtration. These uncertainties have never been investigated.

Differences in the levels of contamination by trace isotopes may be just one of many possible reasons for the occasional different behaviour from one batch of metal to the next that is often experienced in the foundry, and that often seems inexplicable. Kato (1999) is one of the first to record a check for the presence of alpha-emitters in his high purity copper castings. It is a concern that one day such checks may have to become a routine.

6.1.3.2 Pre-existing suspension of bubbles

A number of studies have indicated that there may be a microbubble spectrum in most liquids.

Studies on tap water have demonstrated that there seem to be approximately 300 bubbles of around 5 μm diameter in each ml (Figure 6.3). Hammitt (1973) has carried out work that implies similar distributions of bubbles in liquid sodium circulating as coolant in atomic reactors. Even higher densities

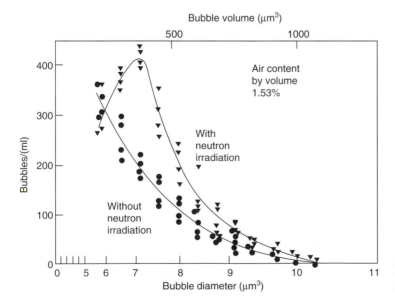

Figure 6.3 *Microbubble spectrum in tap water with and without neutron irradiation (Hammitt, 1973).*

of bubbles have been measured by Outlaw *et al.* (1981) in vacuum-cast pure aluminium (Figure 6.4). Although, of course, we have to be careful not to assume that this pore distribution in the solid reflects that originally present in the liquid, the result does underline the fact that there are distributions of fine pores in circumstances where we may not have any reason to suspect their presence.

Chen and Engler (1994) examine the very old proposal that pores exist in irregular crevices of solid inclusions. They propose that for a conical cavity a gas pocket would have an indefinite existence (they overlook the complications of chemical reactions and dissolution that we shall consider below). In the case of old oxides, from the surfaces of melting or holding furnaces, that have been relatively recently entrained, this model is probably accurate. Even so, it seems likely that in most current situations, this source of inclusions will be of less importance than new films from more recent entrainment events. Bifilms are expected to exist in many liquid metals and alloys. They are not envisaged as a distribution of spherical pores, but as a wide spectrum of sizes and shapes of films of air trapped between folded entrained surface films (the crack-like pores illustrated by Chen and Engler are good examples of bifilms in the early stages of opening). The films will usually be oxides, but may be a number of other non-metallic phases such as graphite or nitrides, etc. The bifilms are, of course, a kind of crevice, so that, to some extent, the theory elaborated by Chen and Engler remains appropriate. Even so, we shall see how the bifilm model as a somewhat flexible folded film, exhibiting some rigidity, and enclosing films and bubbles of entrained gas, in general, fits the facts more closely than that of a model of a solid particle with a gas-filled crevice.

For an entrained oxide film, the oxygen in the air entrapped in the folded film will be quickly consumed by reaction with the metal to form more oxide. The nitrogen may subsequently be consumed more slowly to form nitrides. Ultimately, however, there will be a tiny residue of only about 1 per cent of the original volume of entrapped air, consisting mainly of argon. The inert gases are practically insoluble in liquid metals (Boom *et al.* 2000). Thus a spectrum of very fine volumes of inert gas, trapped within oxide fragments, will be expected to be rather stable over long periods of time.

Figure 6.5 shows the order of magnitude relation showing the size and number of bubbles equivalent to that volume percentage of porosity. For instance, 1 per cent porosity can correspond to either a mere 10 pores per ml when the pores are 1 mm diameter, or 10 million pores per ml when the pores are only 10 µm diameter. (In Figure 6.5 the scale of gas content of the melt, assuming the melt to be aluminium, is of course only accurate at larger bubble sizes. It becomes increasingly inaccurate as sizes fall below 0.1 mm diameter because the internal gas becomes increasingly compressed by the action of surface tension.)

In other more reactive metals, such as liquid titanium, both oxygen and nitrogen have high solubility, leaving only the inert gases that may be insoluble. A microbubble distribution introduced on the surface oxide (if any) may be less stable in this material after the oxide has dissolved, since the bubbles would be more free to float out.

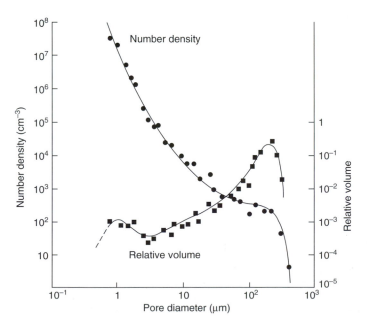

Figure 6.4 *Hydrogen porosity in vacuum melted and cast 99. 995A1. Total porosity is 0.71 per cent (Outlaw et al. 1981).*

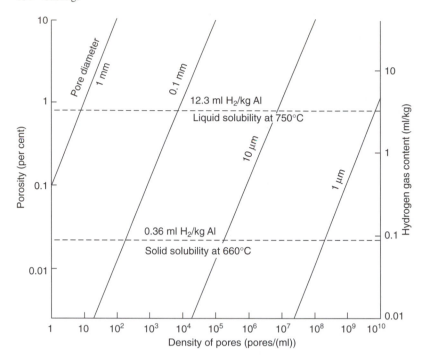

Figure 6.5 *General relation between the volume percentage of porosity and the number and size of pores. The solubility of hydrogen in aluminium is superimposed.*

Similarly, in other high-temperature liquids such as irons and steels the greater speed of reactions, the higher density of the melt and its higher surface tension will all tend to limit the lifetime and spread of sizes of any microbubble population introduced from this source. Nevertheless, enough of the population may still survive for long enough to cause problems in the short time needed to achieve solidification.

6.1.3.3 Pore initiation on bifilms

In contrast with all other pore initiation mechanisms, apart from the pre-existing population of pores with which this section has much in common, the bifilm is seen to possess the potential to initiate pores with negligible difficulty. It simply opens by the separation of its unbonded halves. Surface tension is not involved (as is usually assumed when nucleating a pore in a liquid). This is simply a mechanical action between parallel films of (effectively) vanishingly thin oxide separated by a vanishingly thin layer of gas. The acid test of any theory of pore initiation is its capability to explain the experimental data described below. These findings have so far proved inexplicable.

The definitive research on the growth of porosity in aluminium alloy castings was that carried out at Alcan Kingston Laboratories by a team led by Fred Major (Tynelius 1993). In this exemplary work small tapered plates and end-chilled plates of Al–7Si–0.4Mg alloy were cast under varying conditions to separate the effects of gas content, alloy composition,

freezing time and solidus velocity on the growth of porosity in the castings. The reader is recommended to consult this impressively logical piece of work; the first of its kind, and not since repeated at the time of writing.

Normally of course, the effects of solidification time, temperature gradient and solidification rate are all so closely linked that they are effectively inseparable in most practical casting experiments. However, the experiment was designed so that the effect of casting geometry was cleverly separated out, revealing that the most appropriate thermal parameters to predict porosity were the solidification time and solidus velocity. These gave better results than any of the various temperature gradient terms. They also quantified the dominant effect of hydrogen, and the important contribution of strontium.

The outstanding mystery from this exemplary work was the parameter 'areal pore density' (i.e. the number of pores per unit area). The results are shown in Figure 6.6. It was found that at short solidification times the pore density *increased* with increasing hydrogen content. However, at long solidification times the pore density *decreased* as hydrogen content increased. The authors correctly surmised that this curious and baffling result must depend somehow on the nucleation processes at work. However, the finding has remained unexplained ever since. We shall see how the action of bifilms explains this enigma in a natural way.

We assume that conditions at the lower left corner of Figure 6.6 will be characterized by convoluted

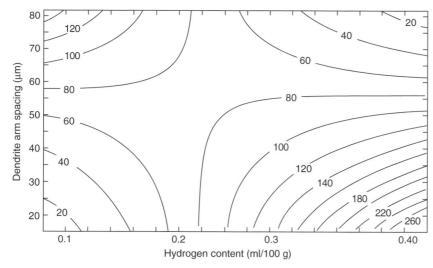

Figure 6.6 *Experimental results of the phenomenological study by Tynelius et al. (1994) showing that pore density per cm² increases at fast freezing rates (small DAS) but decreases at slow freezing rates (large DAS).*

bifilms of which only the longest and most outside fold is inflated with sufficient gas to be seen as a pore on a polished section. With time (going vertically on the figure), or with additional concentration of gas (travelling along the bottom towards the right side of the figure) the bifilm will further inflate additional sections. Eventually, as the remaining sides of Figure 6.6 are traversed towards the top right corner, the separate sections will become sufficiently inflated that the whole bifilm will unfurl, blowing up like a balloon.

We can gain some idea of the rate of unfurling of bifilms from a simple mechanical model as illustrated in Figure 6.7. We shall assume that the unfolding of the bifilm is resisted by a force F of the same type as that resisting the motion of a sphere in a viscous liquid (as in the derivation of Stoke's law). Thus the force would be $3\pi\eta RV$ if it were evenly distributed over the square face of area $R \times R$. Since the velocity V is that at the tip of the bifilm, and the pivot of the bifilm is fixed, on average the resisting force is $3\pi\eta RV/2$. The opening force is that due to the pressure P in the gas phase of the bifilm, acting over the area hR. Equating moments we have

$$2PRh \cdot (h/2) = 3\pi\eta RV \cdot (R/2) \qquad (6.3)$$

so that we can find the opening time t from the speed V and the distance travelled πR:

$$t = (3\pi^2/2(\eta/P)(R/h)^2$$
$$= 15(\eta/P)(R/h)^2 \qquad (6.4)$$

For viscosity $\eta = 1.4 \times 10^{-3}$ Nsm⁻², and reasonable figures for P of about 0.2 atmosphere (0.2×10^5 Pa) above ambient, and for $R = 5$ mm and $h =$

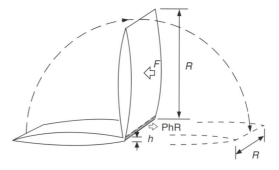

Figure 6.7 *Schematic model of the unfolding of a bifilm.*

5 μm, t is only approximately 1 second. Whereas if $R = 10$ mm and $h = 1$ μm t is approximately 2 minutes. Thus the rate of opening is seen to be highly dependent on the geometry of the bifilm, as might be expected. Nevertheless, the rates are of the correct order of magnitude to explain the rate of loss of properties in most ordinary castings as the freezing time increases. (This wide scatter in the performance of bifilms supports the interpretation of variable nucleation conditions surmised as a result of the subsurface porosity described in section 6.2.)

Returning to our interpretation of Figure 6.6, Figure 6.8 illustrates the general scheme. In the lower left-hand corner, gas content is so low, and time available so short, that any bifilms will mostly still be closed as a result of the action of bulk turbulence. If anything, only the longest section of the bifilm at the outside of the compacted inclusion will be observable as a pore on a polished section

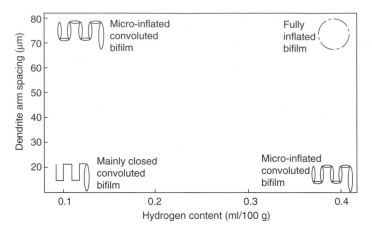

Figure 6.8 *Interpretation of Figure 6.6 according to a bifilm model.*

because this part of the bifilm will open first, having the largest area to gain gas from solution, and being unshielded from the arrival of supplies of diffusing gas. (Internal folds within the convoluted film will be shielded from outside supply of gas until the defect unfolds.)

As we move from the lower left-hand corner up to the top left-hand corner, the increased solidification time will have allowed additional gas to diffuse to the bifilm, and will begin to open many of the folds. We can call this the micro-inflated stage. A polished section through the bifilm will give the appearance of about ten times the number of pores, in agreement with observations (Figure 6.6). The move from the lower left corner directly to the lower right corner will similarly micro-inflate the bifilm. Although the time is still short, the amount of gas available is now sufficient to achieve this.

Finally, moving to the top right gives sufficient time and sufficient gas to power the complete inflation of the film. This is expected to occur by successive unfolding actions, as fold after fold opens out. Finally, the bifilm is fully inflated like a balloon. Thus on a polished section only one pore is seen. The main features of Figure 6.6 are thereby explained.

Keen readers can check the observed pore densities from the original publication, allowing a rough calculation of the number of bifilms in the original cast metal and the size distribution of the final pores. A further check is the total fraction of porosity. Such checks confirm the consistency of the whole scenario.

(Most treatments for the growth of gas pores assume that the rate of growth is limited by the rate at which gas can diffuse to the pore, as shall be assumed in section 6.3. This is probably an accurate model for those liquids such as molten carbon steel in which hydrogen, nitrogen, or carbon monoxide act to grow the pore, and solid folded films do not exist, or are unimportant. In the case of many alloys, however, in which substantial bifilms

are present, as is probably most often the case, the classical diffusion growth model is unlikely to be accurate. The area of the bifilm is already large, so that diffusion distances to the film are reduced, and the rate of arrival of gas by diffusion may no longer impede the growth. Effectively, the gas in the bifilm can be considered to be in equilibrium with the gas in the liquid. In those cases where the bifilms are smaller or fewer in number, the growth of pores may exhibit mixed diffusion and mechanical limitations. The transition from diffusion-limited growth to the mechanics of the unfurling of bifilms would be expected when $(Dt)^{1/2} < R$ where D is the diffusion coefficient, t is the time available and R is the radius of the unfurling bifilm as shown in Figure 6.7.)

The assumption that bifilms are present allows a description of initiation and growth of considerable sophistication and complexity, and that can explain experimental observations that have so far been inexplicable.

6.2 Subsurface porosity

A widely accepted theory of the origin of subsurface porosity has been summarized by Turkdogan (1986). He describes how subsurface porosity occurring in cast irons and steels poured into greensand moulds is a consequence of metal/mould interaction. Gas bubbles form in crevices of the mould in contact with the metal and bubble into the metal, where they become trapped during the early stages of solidification. The action of alloying elements on the process is discussed in terms of their effect on the surface tension of the liquid metal; a lower surface tension allows bubbles to enter the metal more easily, thereby increasing the subsurface porosity.

The theory is similar to the microblow theory outlined in section 6.4.1. However, microblows are probably effectively suppressed by the presence of a strong surface film, such as is normally found on many Al alloys, and irons and steels. Thus there are a number of difficulties with the theory as put forward by Turkdogan.

1. The metal does not in general enter the crevices of the mould. The high surface tension causes the liquid to bridge between high spots, leaving the crevices empty. Mould washes that promote the wetting of the mould by the metal, such as those that are based on sodium silicate, actually reduce subsurface porosity.
2. The pressure required to force small bubbles (radius of 1 mm or less) into the liquid metal against the resistance of surface tension is high. Conversely, it is known that the pressures attainable at the surface of a mould, from which gas can easily migrate away through the mould to the atmosphere, are very low. Thus it seems unlikely that small bubbles could be forced into the metal in this way. (Large bubbles, with radius measured in centimetres, and therefore small pressure requirements, can be forced into the metal from outgassing cores, as we shall see in section 6.4).

Features (1) and (2) may not always be relevant, because the microblow conditions may apply as described in section 4.5.1. However, features (3) and (4) below remain important contrary evidence.

3. Tellurium additions to cast iron reduce surface tension and should, according to the penetration theory, increase porosity. In fact tellurium additions are found to decrease porosity.
4. The theory is not capable of explaining the occurrence of subsurface porosity in inert moulds such as investment moulds, which are free from gas-forming materials such as moisture and hydrocarbons.

We shall therefore assume that in general the formation of subsurface porosity does not occur by mechanical penetration of the liquid surface by bubbles from the mould. We shall see how it is to be expected as the consequence of normal segregation ahead of the solidification front, and the normal processes of nucleation and growth of pores from gases in solution in the metal. A supply of gas diffusing from the mould into the casting surface will enhance the effect. Also, assuming the presence of bifilms completes an attractive model as we shall see.

We have seen in section 5.3.1 how the early growth of the freezing front from the mould wall is planar because of the high temperature gradient. Thus the 'snow plough' build-up of solute occurs, starting from the original solute content C_0, and

increasing over the first millimetre or so of travel of the front, until the solute reaches a level at which a pore may nucleate. A typical form of this porosity is shown in Figure 6.9. It should perhaps be better named surface-layer-free porosity, since, if the gas content of the melt is sufficiently high, the porosity in such cases is rarely simply subsurface, but is distributed everywhere *except* in the surface layer.

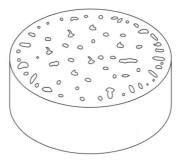

Figure 6.9 *Subsurface porosity revealed on a cut section of a bar casting.*

This microsegregation process is common in many alloy systems where the solute has a low partition coefficient k, resulting in a high concentration of solute ahead of the advancing front. (Recall from section 5.3.1 that the maximum concentration of solute is C_0/k.) For these systems subsurface porosity is the standard form of gas porosity. It may or may not be the result of the release of gas from a metal/mould reaction that can subsequently diffuse into the casting. Such a source will certainly increase the gas available, and may in fact be the sole source of gas. However, it must not be forgotten that subsurface porosity is the normal appearance of gas porosity whether metal/mould reactions contribute or not.

For instance, Klemp (1989) describes subsurface porosity in a low-alloy steel cast into an investment mould. Such moulds are fired at high temperature (commonly about 1000°C) and can be safely assumed to be dry and free from outgassing materials. In this case the subsurface porosity is definitely *not* the result of surface reaction; it was simply in solution in the cast metal.

In contrast, but not in conflict, Turkdogan (1986) reports subsurface porosity in cast irons cast in greensand, but had no reports of any such defects in irons cast into iron moulds. In this case the porosity in the greensand moulds *is* the result of surface reaction; little gas was in solution in the cast metal.

It is not easy to make a clear separation in this account of the various gases, since all cooperate in some alloy systems, and many systems are analogous. However, a rough division will be attempted, the reader being requested to overlook the necessarily ragged edges between sections.

6.2.1 Hydrogen porosity

It is also important to remember that both water and hydrocarbons (that are available in abundance in most sand castings) can decompose at the metal surface, both releasing hydrogen. The surface will therefore have no shortage of hydrogen; in fact, from section 4.3, it is seen that in general the mould atmosphere often contains up to 50 per cent hydrogen, and may be practically 100 per cent hydrogen in many cases.

What happens to this hydrogen?

Although much is clearly lost by convection to the general atmosphere in the mould, some will diffuse into the metal if not prevented by some kind of barrier (see later). If the hydrogen does manage to penetrate the surface of the casting, how far will it diffuse?

We can quickly estimate an average diffusion distance d from the useful approximate relation:

$$d = (Dt)^{1/2} \qquad (6.5)$$

Some researchers increase the right-hand side of this relation by a factor of 2 in a token attempt to achieve a little more accuracy. We shall neglect such niceties, and treat this equation merely as an order of magnitude estimate. Taking the diffusion rate D of hydrogen as approximately 10^{-7} m^2 s^{-1} for all three liquid metals, aluminium, copper and iron (see Figures 1.6 to 1.8), then for a time of 10 seconds d works out to be approximately 1 mm. For a time of 10 minutes d grows to approximately 10 mm.

Clearly, hydrogen from a surface reaction can diffuse sufficiently far in the time available during the solidification of an average casting to contribute to the formation and growth of subsurface porosity.

The distance that the front has to travel before the solute peak reaches its maximum is actually identical to the figures we have just derived, as explained in section 5.3.1. Thus conditions are exactly optimum for the creation of the maximum gas pressure in the melt at a point a millimetre or so under the surface of the casting. The high peak will favour conditions for the nucleation of pores while the closeness to the surface will favour the transport of additional gas, if present or if required, from the surface reaction. Naturally, if there is enough gas already present in the melt, then contributions from any surface reaction will only add to the already existing porosity.

In aluminium alloys where hydrogen gas in solution in the melt segregates strongly on freezing: the partition coefficient is approximately 0.05, corresponding to a concentrating effect of 20 times. Figure 6.10 shows subsurface porosity in an Al–

7Si–0.4Mg alloy solidified against a sand core bonded with a phenolic urethane resin. The general gas content of the casting is low, so that pores are only seen close to the surface, within reach by diffusion of the hydrogen from the breakdown of the resin binder. Some of the pores in Figure 6.10a are clearly crack-like, and seem likely therefore to have formed on bifilms.

Close-up views of subsurface porosity in the same alloy and same bonded sand (Figure 6.11a) confirms that the pores are of widely different form, some being perfectly round (6.11b), some dendritic (6.11c) and some of intermediate form (6.11d). It seems reasonable to assume that all the pores experienced the same environment, consisting of a uniform field of hydrogen diffusing into the melt from the degeneration of the core binder. Their growth conditions would therefore have been expected to be identical. Their very different forms cannot therefore be the result of growth effects (i.e. some are not shrinkage and others gas pores). The differences therefore must be a result of differences in ease of nucleation. The simple explanation is that the round pores nucleated early because of easy nucleation, and thus grew freely in the liquid. The dendrite-lined pores are assumed to nucleate late, as a result of a greater difficulty to nucleate, so that they expanded when the dendrites were already well advanced. The differences in ease of nucleation can be easily understood in terms of the randomly different conditions in which the nuclei, bifilms, are created. Some will come apart easily, whereas others will be ravelled tightly, or may be partially bonded as a result of being older or being contaminated with traces of liquid salts from the surface of the melt. Additional evidence for the differences in the rate of opening of bifilms was highlighted in section 6.1.3.3.

The case of some subsurface pores initiating their growth very late, when freezing must have been 80 per cent or more complete, raises an interesting extrapolation. If the bifilms had been even more difficult to open, or if not even present at all, then *no* pores would have nucleated. This situation may explain the well-known industrial experience, in which subsurface porosity comes and goes, is present one day, but not the next, and is more typical of some foundries than others. It is a metal quality problem.

Note that both round and dendritic pores can be both gas pores. They could also both be shrinkage pores, or both (having some combined gas + shrinkage contribution). Whether grown by gas or shrinkage, or both, the shape difference merely happens because of the *timing* of the pore growth in relation to the dendrite growth. Although it is common for gas pores to form early and so be rounded, and shrinkage pores to form late and so take on an interdendritic morphology, it is not

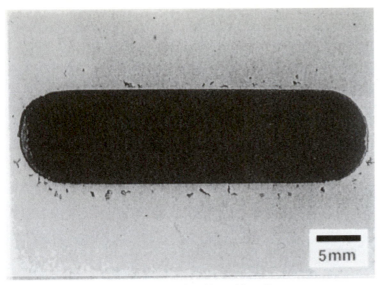

(a)   Macrosection (Aqua blasted)

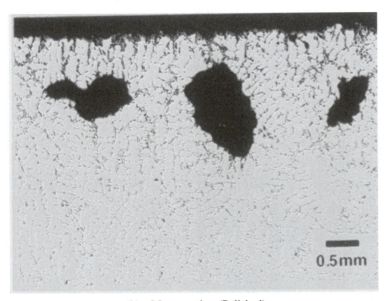

(b)   Macrosection (Polished)

Figure 6.10 *(a) Polished section, lightly blasted with fine grit, showing subsurface porosity around a sand core in an Al–7Si–0.4Mg alloy casting of low overall gas content; and (b) an enlarged view of some pores on a polished section.*

necessary. It is very important not to fall into the standard trap of assuming that round pores result from gas and dendritic forms result from shrinkage.

The work by Anson and Gruzleski (1999) describes particularly careful work in an attempt to distinguish between gas and shrinkage pores. Their study concentrated on the appearance and spacing of pores. They pointed out that on a polished section, groups of apparently separate, small interdendritic pores were almost certainly a single pore of irregular shape (Figure 6.12). Despite apparently clear differences in shape and spacing, it is finally evident in their case that all pores were gas pores, since they all grow at the same rate as hydrogen is increased. In this case the pores that were assumed to be shrinkage pores were almost certainly partially opened and/or late opened bifilms. Their irregular cuspoid outlines probably derived

(a)

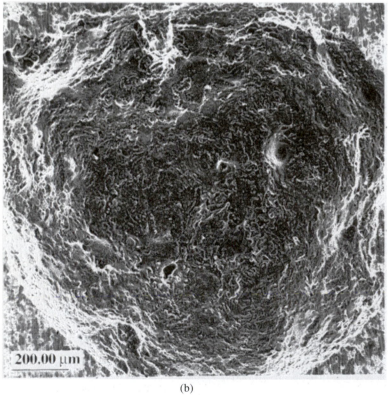

(b)

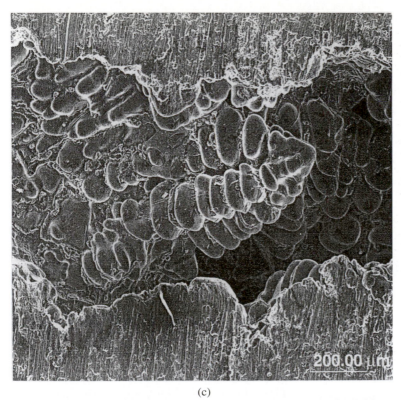

(c)

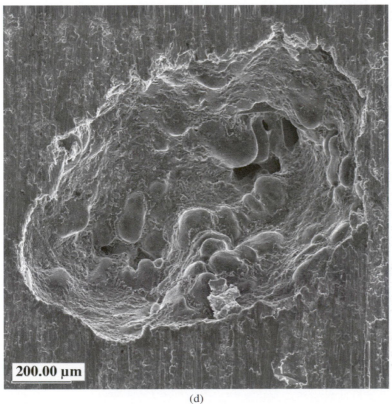

(d)

Figure 6.11 *Thin slice of an Al–7Si–0.4Mg alloy casting taken from around a phenolic urethane bonded sand core: (a) a general view, showing the sand-cast surface made by the core and several subsurface pores; and close-ups of (b) a spherical pore; (c) a dendritic pore; and (d) a mixed pore.* (Courtesy S. Fox and Ashland Chemical Co.)

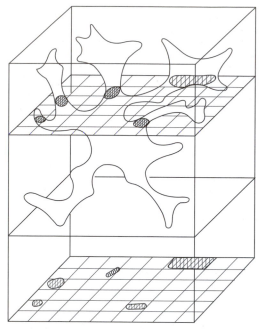

Figure 6.12 *Complex interdendritic pore, appearing as a group of pores on a polished section (after Anson and Gruzleski 1999).*

partly from the irregular, crumpled form of the bifilms, together with their late opening in the interdendritic spaces. Such misidentification of pore shapes is easily understood, and is common.

6.2.2 Carbon, oxygen and nitrogen

For the case of the casting of aluminium alloys we have only to concern ourselves about hydrogen, the only known gas in solution.

For the case of copper-based alloys a number of additional gases complicate this simple picture. The rate of diffusion of oxygen in the liquid is not known, but is probably not less than 10^{-8} m^2 s^{-1}. In the case of liquid iron-based alloys, oxygen and nitrogen diffuse at similar rates (Figure 1.8). Thus for all of these liquids the average diffusion distance d is 1 to 2.5 mm for the time span of 1–10 minutes. It seems therefore that all these gases can enter and travel sufficiently far into castings of all of these alloy systems to contribute to the formation of porosity.

In copper-based alloys the effect is widely seen and attributed to the so-called 'steam reaction':

$$2H + O \leftrightarrows H_2O \qquad (6.6)$$

which is practically equivalent to the alternative statement in Equation 1.6. It seems certain, however, that SO_2 and CO will also contribute to the total pressure available for nucleation in copper alloys

that contain the impurities sulphur and carbon. Carbon is an important impurity in Cu–Ni alloys such as the monels. Zinc vapour is also an important contributing gas in the many varieties of brasses and gunmetals.

From the point of view of nucleation the action of oxygen is likely to be central. This is because it is probably the most strongly segregating of all these solutes (with the possible exception of sulphur). Thus deoxidation practice for copper-based alloys is critical.

When gunmetal has been deoxidized with phosphorus, Townsend (1984) reports that an optimum rate of addition is required, as illustrated in Figure 6.13. Too little phosphorus allows too much oxygen to remain in solution in the melt, to be concentrated to a level at which precipitation of water vapour will occur as freezing progresses. Too much phosphorus will reduce the internal oxygen to negligible levels, suppressing this source of porosity. However, the melt will then have enhanced reactivity with its environment, the excess phosphorus picking up oxygen and hydrogen from a reaction at the metal surface with water vapour from the mould. The porosity in the cast metal is the result of the sum of the internal and external reactions. This has a minimum at approximately 0.015 per cent phosphorus for the case of this particular sample of alloy as seen in Figure 6.13.

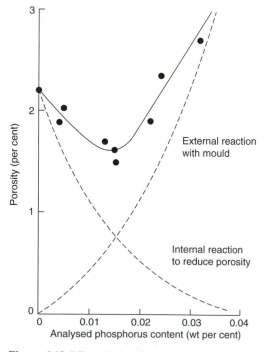

Figure 6.13 *Effect of phosphorus on the porosity in 75 mm thick plates of leaded gunmetal LG2 cast at 1100°C. Data from Townsend (1984).*

A similar reaction occurs in the presence of 0.005–0.02 per cent aluminium or 0.04 per cent titanium in grey iron. The reaction is characterized by subsurface pores that have a shiny internal surface covered with a continuous graphite film. (The graphite film is present simply because the free surface provided by the pore allows the graphite to accommodate its volume expansion on precipitation most easily. Similarly, these pores are often seen to be filled with a frozen droplet of iron, again simply because the pore is an available volume into which liquid can be exuded during the period when the graphite expansion is occurring. Such droplets would be expected to be more common in castings made in rigid moulds, where the expansion could not be easily accommodated by the expansion of the mould.)

Carter *et al.* (1979) describe the analogous problem caused by the presence of magnesium in ductile iron. Clearly, this double effect of the addition of a strong deoxidizer, resulting in an optimum concentration of the addition, is a general phenomenon.

In much of the work on *subsurface* pores in irons and steels the phenomenon is called *surface* pinhole porosity. This is almost certainly the result of the loss of the surface of the casting by a combination of oxidation and/or severe grit blasting. The surface pinholes almost certainly originated as subsurface pinholes.

In the case of low-carbon equivalent irons it is found that small surface pinholes occur that have an internal surface lined with iron oxide, and whose surrounding metal is decarburized, as witnessed by a reduction in the carbide content of the metal. Although Dawson *et al.* (1965) and others make out a case for these defects to be the result of a reaction with slag, it seems more reasonable to suppose that once again the pores were originally subsurface, but the high oxygen content of the metal promoted early nucleation, with the result that the pores were extremely close to the surface of the casting. The thin skin of metal quickly oxidized, opening the pore to the air at an early stage, and allowing plenty of time for oxidation and decarburization while the casting was still at a high temperature. Tests to check whether the pores have been connected to the atmosphere do not appear to have ever been carried out.

Dawson reports that an addition of 0.02 per cent aluminium usually eliminates the problem. This relatively high addition of aluminium is probably to be expected because the oxygen in solution in these low-carbon equivalent irons will be higher than that found in normal grey irons. However, if even higher levels of aluminium were added, the problem would be expected to return because of the increased rate of reaction with moisture in the mould, as shown in the similar example in Figure 6.13.

Oxygen and carbon are important when CO is the major contributing gas, although, of course, in cast irons, where carbon is present in excess, the CO pressure is effectively controlled solely by the amount of oxygen present. This deduction is nicely confirmed for malleable cast irons by the Italian workers Molaroni and Pozzesi (1963), who found a strong correlation between their proposed '*oxidation index*', I, defined as:

$$I = C + 4Mn + 1.5Si - 0.42FeO - 5.3$$

where the symbols for the elements carbon C, manganese Mn, etc., represent the weight percentage of the alloying elements in the iron. Compositions of irons that gave a positive index were largely free from pores, whereas those with an increasingly negative index were, on average, more highly porous.

In steels there are several gases that can be important in different circumstances. The most important are CO (Equation 1.8), N_2 (Equation 1.10) and H_2 (Equation 1.3). Since, at the melting point of iron, hydrogen has a solubility in the liquid of approximately 245 $mlkg^{-1}$ and in the solid of 69 $mlkg^{-1}$ (extrapolating slightly from Brandes (1983)), its partition coefficient is $69/245 = 0.28$, with the result that it is concentrated ahead of the solidification front by a factor of $1/0.28 = 3.55$. It therefore makes a modest contribution to the gas pressure for nucleation of pores in iron alloys.

Nitrogen seems to have a similar importance in nucleation. Its solubility at the melting point of iron is 0.0129 weight per cent in the solid and 0.044 weight per cent in the liquid (Brandes 1983) giving a partition coefficient 0.29, and a concentration effect for nitrogen ahead of the freezing front of approximately 3.4 times. Subsurface porosity is common when steels are cast into moulds bonded with urea formaldehyde resin (Middleton 1970), or bonded with other amines that release ammonia, NH_3, on heating. These include hexamine in Croning shell moulds (Middleton and Canwood 1967). The ammonia breaks down at casting temperatures to release both nitrogen and hydrogen. This situation has already been discussed in section 4.5.2 on metal/mould reactions.

Since $k = 0.05$ for oxygen in iron, and $k = 0.2$ for carbon in iron, the concentration factors are 20 and 5 respectively, so that when combined, the equilibrium CO pressure at the solidification front is $20 \times 5 = 100$ times higher than in the bulk melt (this is before activities are allowed for, which will increase this factor further). The distribution coefficients refer to bcc delta-iron; those for fcc gamma-iron would be nearer to unity, implying much less concentration ahead of the solidification front for solidification to austenite. Because of the multiplying factor 100, oxygen in solution in the iron is the major contributing gas in the nucleation of CO gas pores during the solidification of most irons and steels.

In an investigation of a wide variety of different binders for the moulding sand, Fischer (1988) finds that subsurface porosity in copper-based castings is highly sensitive to the type of binder, although degassing and deoxidizing of the metal did help to reduce the problem. These observations are all in line with our expectations based on the model described above.

In a more detailed earlier study (Jones and Grim 1959) it was found that different clays used in greensands release their moisture at different temperatures. This might have a significant effect on the creation of subsurface pores.

6.2.3 Nitrogen porosity

There has been a massive effort to understand the metal/mould reactions in which nitrogen is released. This gives problems in both iron and steel castings as subsurface pores. A review for steel castings is given by Middleton (1970).

The nitrogen problem in ferrous castings has resulted in the production of a whole new class of sand binders known as 'low nitrogen' binders. However, later work (Graham *et al.* 1987) investigating the relation between total nitrogen content of the binder and the subsurface porosity and fissures in iron castings found no direct correlation. However, Graham did find a good correlation with the ammonia content of the binder. Ammonia is released during the pyrolysis of important components of many binders, such as urea, amines (including hexamine used in shell moulds) and ammonium salts. The ammonia in turn will decompose at high temperature as follows:

$$NH_3 \leftrightarrows N + 3H$$

thus nascent nitrogen and nascent hydrogen are released (the word *nascent* meaning *in the act of being born*). Both will contribute to the formation of pores in the metal. Both nitrogen and hydrogen will have a similar influence in the nucleation of a pore, concentrating strongly ahead of the freezing front. For the subsequent growth, however, hydrogen will be the major influence because of its much faster rate of diffusion. The fact that both gases are released simultaneously by ammonia explains the extreme effectiveness of ammonia in creating porosity. Nitrogen alone would not have been particularly effective. Even if it may have been successful in nucleating pores, without the additional help from hydrogen any subsequent growth would have been limited. The high rate of diffusion of hydrogen ensures that hydrogen dominates the feeding of the growth of the pore. It is supplied by gas in solution in the liquid that drains from the surrounding casting, and any fresh supply through the surface from a surface reaction.

It seems that ammonia can build up in greensand systems as the clays and carbons absorb the decomposition products of cores. Lee (1987) confirms that an ammoniacal nitrogen test on the moulding sand was found to be a useful indicator of the pore-forming potential of the sand, even though the test was not a measure of total nitrogen.

The action of gases working in combination is illustrated by the work of Naro (1974) in his work on phenolic urethane–isocyanate binders. He showed that, from a range of irons, ductile iron (high carbon and low oxygen) was least susceptible and low-carbon equivalent irons (high oxygen) were most susceptible to porosity from the binder. Once again, it seems logical that the oxygen remaining in solution in the iron plays a key role encouraging nucleation, and hydrogen and nitrogen from the binder encourage growth.

6.2.4 Barriers to diffusion

In some unusual conditions, hydrogen appears to be prevented from diffusing into some metals.

For magnesium alloys, potassium borofluoride, KBF_4, has been known for many years to be an effective suppressant of metal/mould reactions for Mg alloys. In fact, if not added to the sand moulds of some Mg castings both mould and casting will be consumed by fire – the ultimate metal/mould reaction! However, Al–5Mg and Al–10Mg casting alloys, and even Al–7Si–0.4Mg alloy, also benefit from KBF_4 or K_2TiF_6 additions to suppress reactions with the mould. We might speculate that liquid oxyfluorides, produced by the dissolution of the alumina film in the flux, assist to seal the surface of the liquid metal.

The Al–7Si–0.4Mg alloy similarly benefits from Sr additions to the metal. This effect may be associated with a more impermeable oxide of the modified alloy.

Naro (1974) confirms the widely reported fact that the addition of 0.25 per cent iron oxide to phenolic urethane–isocyanate-bonded sands reduces subsurface pores in a wide range of cast irons. This is a curious fact, and difficult to explain at this time. One suggestion is that the oxide creates a surface flux, possibly an iron silicate. This glassy liquid phase is likely to reduce the rate at which gases can diffuse into the casting.

The rate of uptake of nitrogen in stainless steel is inhibited by the presence of silicon in the steel that, at certain oxidation potentials, forms SiO_2 on the surface in preference to Cr_2O_3 (Kirner *et al.* 1988).

Even when the surface film consists only of a layer or so of adsorbed surface-active atoms, the presence of the layer reduces the rate at which gases can transfer across the surface. This happens, for instance, in the case of carbon steels: sulphur and other surface-active impurities hinder the rate at which nitrogen can be transferred. An excellent review of this phenomenon is given by Hua and Parlee (1982).

However, the precise mechanisms of many of these inhibition reactions are not clear at this time.

6.3 Growth of gas pores

The presence of bifilms, and their action to initiate porosity in liquid metals, has been discussed in section 6.1.3.3. The interesting feature of the mechanical model for the opening of bifilms in relation to the growth of pores is illustrated in Equation 6.4. If the gas in solution in the liquid is approximately in equilibrium with the entrapped gas in the bifilm, the internal pressure will be proportional to $[H]^2$, assuming for a moment that the gas involved is hydrogen (other diatomic gases will act similarly of course, although their approach to equilibrium may be slower). The rate of unfurling is therefore especially sensitive to the amount of gas in solution in the alloy.

In the case of iron and steel where an important contributor to the internal pressure will be expected to be carbon monoxide, CO, the internal pressure will approach that dictated by the product of the activities of carbon and oxygen in the melt, approximately $[C].[O]$. In addition, of course, nitrogen and hydrogen will also contribute to the total pressure.

In one of the most exciting pieces of research published in this field, Tiberg (1960) describes the growth of carbon monoxide bubbles in liquid steel while actually observing the inside surface of the growing bubbles. He achieves this miracle by using high-speed cine film to record the nucleation and growth of bubbles on the inside wall of a fused silica tube that contained the steel. The classical theories of pore growth assume that the geometry of the pore and its collection volume are spherical, and that growth is steady. This seems to be far from true in the experiment in which Tiberg tested these assumptions.

At high rates of growth he found that the speed of expansion of the bubble surface dr/dt was indeed constant from the time the bubble was first observed at a size of 30 μm. However, after the addition of the deoxidizers, aluminium or silicon, the growth rate was slower and varied considerably from one bubble to another. In some bubbles growth suddenly halted and then continued at a slower rate. In fast-growing bubbles a small bright spot was observed.

The observation of the bright spot is interesting. It is most likely to have been an inclusion of alumina or silica (possibly actually in the form of a bifilm?)

since this behaviour was only observed after the addition of the corresponding deoxidizer. The transparency or translucency (or even its hollowness if in the form of a partially opened bifilm) of the inclusion would have allowed the interior of the inclusion to be visible, giving an observer a view into the interior of the melt. This would appear as a bright enclosure, the classical 'black body cavity' of the physicist, radiating a full spectrum corresponding to the temperature of the interior of the steel, and therefore appearing as a bright spot. (The remainder of the bubble surface radiating its heat away to the outside world via the transparent silica vessel, and partially reflecting the cooler outside environment from its surface, and therefore appearing cooler.) We may speculate that the enhanced rate of transfer of gas into the bubble may have resulted from either (i) the short-circuiting of a surface layer that was hindering the transfer of gas into the bubble, or (ii) the attached inclusion having a large surface area and a high rate of diffusion for gas. Its surface area would then act as a collecting zone, funnelling the gas into the growing bubble through the small window of contact. A bifilm would have been expected to be especially effective in this way.

Such complicated growth effects apply to 'dirty' (i.e. 'real') liquids.

In what remains of this section we shall consider the classical mechanisms by which gas pores can grow in clean liquid metals.

In general the growth of gas pores in clean liquid metals appears to be controlled mainly by the rate of diffusion of gases through the liquid metal. There are many data in support of this, especially in simple systems such as the Al–H system. Apart from the bifilm effects, usually in this book we shall make the assumption that the rate of growth of pores is controlled by diffusion through the bulk liquid or solid phase.

Usually, therefore, it follows that the rate of growth is dominated by the rate of arrival of the fastest diffusing gas. From Figure 1.8 it is clear that in liquid iron, hydrogen has a diffusion coefficient approximately ten times higher than that of any other element in solution. Thus the average diffusion distance d is approximately $(Dt)^{1/2}$ so that in comparison with other diffusing species, the radius over which hydrogen can diffuse into the bubble is $(10/1)^{1/2} = 3$ times greater. Thus the volume over which hydrogen can be collected by the bubble, in comparison with other diffusing species, is therefore $3^3 = 30$ times greater. Thus it is clear that hydrogen has a dominant influence over the *growth* of the bubble.

It should be remembered that hydrogen makes a comparatively small contribution to the *nucleation* of the bubble, because it concentrates relatively little ahead of the advancing freezing front, in comparison with the combined effects of oxygen and carbon to form CO in liquid iron and steel. The situation is closely paralleled in liquid copper alloys, where oxygen controls the nucleation of pores because of the snow plough mechanism, whereas hydrogen contributes disproportionately to growth because of its greater rate of diffusion.

This clarification of the different roles of oxygen and hydrogen in copper and steel explains much early confusion in the literature concerning which of these two gases was responsible for subsurface pores. Zuithoff (1964, 1965) published the first evidence that confirmed the present hypothesis for steels. He succeeded in showing that aluminium deoxidation would control the appearance of pores. Clearly, if the oxygen was high, then pores could nucleate, but they would not necessarily grow unless sufficient hydrogen was present. Conversely, if hydrogen was high, pores might not form at all if no oxygen was present to facilitate nucleation. The hydrogen would therefore simply remain in solution in the casting. The same arguments apply, of course, to the roles of hydrogen and oxygen in copper-based alloys.

A useful simple test for steels which deserves wider use is proposed by Denisov and Manakin (1965): a sample test piece was developed 110 mm high, and 30×15 mm at the top, tapering down to 25×12 mm at the base. A metal pattern of the sample quickly creates the shaped cavity in the sand, into which the metal is poured. Immediately after casting, the sample is knocked out and quenched in water. It is then broken into three pieces in a special tup. The entire process takes 1 to 2 minutes. It was found that the tapered test piece gave an accurate prediction of the risk of subsurface porosity; if such problems were seen in the sample they were seen in the castings and vice versa. The test therefore warned of danger, and avoiding action could be taken, such as the addition of extra deoxidizer to the ladle. This test for steel castings cast in greensand moulds should be applicable to other alloy and sand systems prone to this problem. Perhaps the 'look and see' test by the author, described in section 6.4 might be even simpler and quicker. Quick, reliable tests are very much needed. The reader is recommended to try these techniques.

In some alloy systems the rate of growth of pores is not expected to be simply dependent on the rate of diffusion. The rate can also be limited by a surface film as we have seen in section 6.2 in which barriers are discussed.

Ultimately, however, the maximum amount of gas porosity in a casting depends partly on simple mechanics, as illustrated by the well-known general gas law. The use of this law assumes that the gas in the pore behaves as a perfect gas, which is an excellent approximation for our purposes. We shall also assume that all the gas precipitates (which is a less good approximation of course).

$$PV = nRT \qquad (6.7)$$

where n is the amount of gas in gram.moles (in most use of this equation, n is somewhat misleadingly assumed to be unity), R is the gas constant $8.314 \, \text{JK}^{-1} \, \text{mol}^{-1}$, and P is the applied pressure.

The equation can be restated to give the volume V explicitly as:

$$V = nRT/P \qquad (6.8)$$

It follows as a piece of rather obvious logic that the volume of the porosity is directly proportional to n, the amount of gas present in solution. This is graphically shown in Figure 6.14.

The illustration shows sections of the small sample that is cast into a metal cup about the size of an egg cup, and is then solidified in vacuum. The test is sometimes known as the reduced pressure test (RPT), or the Straube–Pfeiffer test. The solidification under reduced pressure expands the pores, making the test more sensitive and easier to use than the old foundry trick of pouring a small pancake of liquid on to a metal plate, and watching closely for the evolution of tiny bubbles.

The general gas equation also shows that the volume of a gas is inversely proportional to the pressure applied to it. For instance, in the RPT to determine the amount of hydrogen in a liquid aluminium alloy, the percentage porosity is commonly expanded by a factor of 10 by freezing at 0.1 atm (76 mmHg) residual pressure rather than at normal atmospheric pressure (760 mmHg).

This sensitivity to pressure needs to be kept in mind when using the test. For instance, if the vacuum pump is overhauled and starts to apply not 76 mmHg but only 38 mmHg (0.05 atm) as a residual pressure, then the porosity in the test samples will be doubled, although, of course, the gas content of the liquid metal will be unchanged. Rooy and Fischer (1968) recommend that for the most sensitive tests the applied pressure should be reduced to 2 to 5 mmHg (approximately 0.003 to 0.006 atm). Clearly this will yield about a further tenfold increase in porosity in the sample for any given gas content. However, care needs to be taken because these simple numerical factors are reduced by the additional loss of hydrogen from the surface of the test sample during the extra time taken to pump down to these especially low pressures.

As has been mentioned before in the case of vacuum casting, the effect of pressure on pore growth is an excellent reason to melt and pour under vacuum, but to solidify under atmospheric pressure. It makes no sense to solidify under vacuum because pore expansion will act to negate the benefits of lower gas content. In terms of the general gas law,

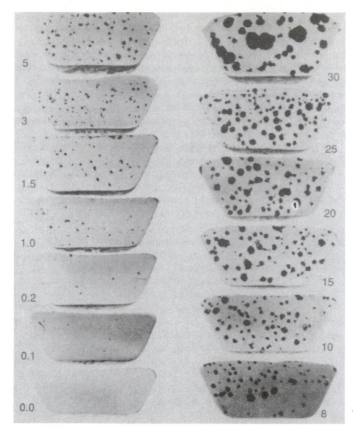

Figure 6.14 *Gas porosity at various percentage levels in sectioned samples from the reduced pressure test (courtesy Stahl Speciality Co. 1990).*

the pore volume *V* will be decreased by lower *n*, but increased correspondingly by low *P*. Whether the effects will exactly cancel will depend, among other things, on whether the melt has had time to equilibrate with the applied vacuum so as to reduce its gas content *n*. Taylor (1960) gives a further reason for not freezing under vacuum: For a nickel-based alloy containing 6 per cent aluminium, the vapour pressure of aluminium at 1230°C is sufficient to form vapour bubbles at the working pressure of the vacuum chamber. He correctly concludes that the only remedy is to increase the pressure in the chamber immediately after casting. During the melting of TiAl intermetallic alloys at temperatures close to 1600°C, the evaporation of Al causes a loss of Al from the alloy, and a messy build-up of deposits in the vacuum chamber. Melting under an atmosphere of argon greatly reduces these problems. (However, pouring under argon cannot be recommmended if the pouring is turbulent because of the danger of the entrainment of argon bubbles; another reason for the adoption of counter-gravity.)

If the rate of diffusion of the gas in the casting is slow, the volume of the final pore will be less than that indicated by the general gas law, and will be controlled by the time available for gas to diffuse into the pore. In Figure 6.15 the benefits of increasing

feeder size are seen to be enjoyed up to a critical size. After that any further increase in the feeder merely delays solidification of the casting so that gas porosity increases. The complete curve is therefore seen to be the sum of the effects of two separate curves. The first curve decreases linearly from about 7 to 0 per cent porosity as shrinkage is countered by good feeding; and the second increasing parabolically from zero as more time is available for the diffusion of gas into pores as solidification time increases.

Although these general laws for the volume of a gas-filled cavity are well known and nicely applied in various models of pore growth (see, for instance, the elegant work by Kubo and Pehlke (1985), Poirier (1987) and Atwood and Lee (2000)) some researchers have shown that the detailed mechanism of the growth of pores can be very different in some cases.

A direct observation of pore growth has been carried out for air bubbles in ice. At a growth rate of 40 μms⁻¹, Carte (1960) found that the concentration of gas built up to form a concentrated layer approximately 0.1 mm thick. He deduced this from observing the impingement of freezing fronts. When the bubbles nucleated in this layer, their subsequent rapid growth so much depleted the

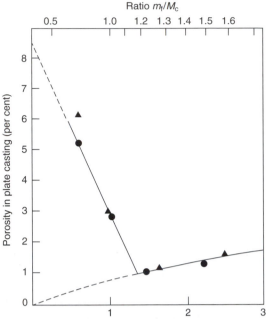

Ratio m_f/M_c

Porosity in plate casting (per cent)

Ratio of freezing time of separately cast feeder to
freezing time of separately cast plate casting

Figure 6.15 *Effect of increasing feeder solidification time on the soundness of a plate casting in Al–12Si alloy. Data from Rao et al. (1975).*

solution in the vicinity of the front that growth stopped and clear ice followed. The concentration of gas built up again and the pattern was repeated, forming alternate layers of opaque and clear ice.

On examination of the front under the microscope, Carte saw that the bubbles seemed to originate behind the front; the first 0.1 mm deep layer of solid appeared to be in constant activity; threads of air approximately 10 μm in diameter spurted along what seemed to be water-filled channels, and were squeezed out of the ice. Sometimes bubbles arrived in quick succession, the first being pushed away and floating to the surface. Those bubbles that remained attached to the front would then expand, but finally be overtaken and frozen into the solid. It seems that pore growth might involve more turmoil than we first thought! Much of this activity arises, of course, from the expansion of the ice on freezing, and so forcing liquid back out of interdendritic channels. The opposite motion will occur in most metallic alloys as a result of the contraction on freezing. Also, it is to be expected that the movement in metals will be somewhat less frenetic. Nevertheless, no matter how the pore might grow in detail, we can reach some conclusions about the final limits to its growth.

Poirier (1987) uses the fact that the pore deep in a dendrite mesh will grow until it impinges on the surrounding dendrites. The radius of curvature of

the pore is therefore defined by the remaining space between the dendrite arms. However, of course, although the smallest radius defining the internal pressure is now limited, the pore can continue to grow, forcing its way between the dendrite arms. Again, as has been mentioned before, an interdendritic morphology should not be taken as the definition of a shrinkage pore. Whether grown by gas or shrinkage, its morphology of spheroidal or interdendritic is merely an indication of the timing of its growth relative to the timing of the growth of the dendrites. Thus round pores are those that have grown early. Interdendritic pores are those that have grown late.

Fang and Granger (1989) found that hydrogen porosity in Al–7Si–0.3Mg alloy was reduced in size and volume percentage, and was more uniformly distributed when the alloy was grain refined. In this case the growth of bubbles will be limited by their impingement on grains. It may be that a certain amount of mass feeding may also occur, compressing the mass of grains and pores. In later work Poirier and co-workers (2001) confirm by a theoretical model that finer grains do reduce porosity to some degree.

Another limitation to growth occurs when the bubble can escape from the freezing front. This will normally happen when the front is relatively planar, and is typified by the example of the rimming steel ingot. In general, however, escape from a mesh of dendrites is likely to be rare.

A final limitation to growth is seen in those cases where the porosity reaches such levels that it cannot be contained by the casting. This occurs at around 20 to 30 per cent porosity. During the freezing of the sample in the reduced pressure test, gas can be seen to escape by the bursting of bubbles at the surface of the sample. The effect is clearly seen in Figure 6.14. Also, Figure 6.16 shows that as gas in increased, measurements of porosity above 20 per cent in such samples show a lower rate of increase of porosity than would be expected because of this loss from bubbles bursting at the surface, releasing their gas to the environment. (The theoretical curve in Figure 6.16 is based on 1 per cent porosity in the solid being equal to 10 ml hydrogen in 1000 ml aluminium; this is equivalent to 10 ml hydrogen in 2.76 kg aluminium, or 3.62 ml hydrogen in 1 kg aluminium. Incidentally, Figure 6.16 also shows the reduction in porosity as a result of increasing the difficulty of nucleation, because of the removal of nuclei by filtration.)

Finally, it is worth mentioning the considerable volume of work over many years in which people have drilled into steel castings immersed in mercury or oil and have collected and analysed the gases in pores. In almost every case the dominant gas was hydrogen. This led early workers to conclude that the bubbles were caused by hydrogen (see, for

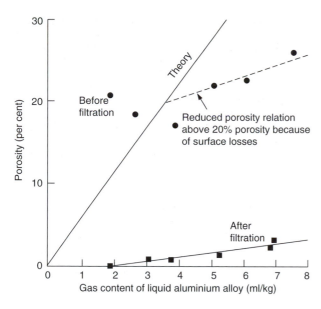

Figure 6.16 *Porosity of reduced pressure test samples frozen at 0.005 atm as a function of gas content. Data from Rooy and Fischer (1968).*

instance, the review of early work by Hultgren and Phragmen (1939)). This was despite calculations by Muller in 1879 that the CO pressure in pores in steel castings was up to 40 atmosphere, and the consequent correct deduction by Ledebur in 1882 that the hydrogen content of pores in steel castings at room temperature was the result of the continued accumulation of hydrogen after solidification was complete.

We can see that, in summary, the high final hydrogen content of pores at room temperature is a natural consequence of the high rate of diffusion of hydrogen in both the liquid and solid states. Thus hydrogen is the dominant gas contributing to the growth of pores, and continuing to contribute additional gas during the cooling to room temperature. (Conversely, remember, in copper-based and ferrous alloys, oxygen is the dominating gas contributing to the nucleation of pores, as we have seen earlier.)

Even after the casting reaches room temperature the growth of gas pores may still not be complete! Talbot and Granger (1962) showed that hydrogen in cast aluminium could continue to diffuse into pores in the solid state during a heat treatment at 550°C. Porosity was found to increase during this treatment, with pores becoming larger and fewer. With very long annealing in vacuum the hydrogen could be removed from the sample and the porosity could be observed to fall or even disappear.

Beech (1974) was perhaps the first to point out that the environment of a long bubble is not necessarily homogeneous. In other words, what may be happening at one end of the bubble may be quite different from what is happening at the other. This is almost certainly the case with the kind of subsurface porosity that continues to grow into an

array of wormholes. The metal/mould reaction will continue to feed the base of the bubble, which of course remains within the diffusion distance of the surface. If the gas content of the melt is high the front of the bubble may also be gaining gas from the melt. However, if the melt has a low gas content, the front part of the bubble may lose gas. The bubble effectively acts as a diffusion short-circuit for the transfer of gas from the surface reaction to the centre of the casting. The effect is shown in Figure 6.17.

Depending on the relative rates of gain and loss, the pore may grow, or even give off bubbles from the growing front. Alternatively it may stop growing and thus be overtaken by the dendrites and frozen into the thickening solid shell. The swellings and narrowings of these long bubbles probably reflect the variation in growth conditions, such as sudden variations in gap between the casting and the mould, or variations from time to time of convection currents within the casting.

The long CO bubbles in rimming steel ingots, cast into cast iron moulds, were a clear case where the growing bubble was fed from the growing front, and not from the mould surface.

Conversely, in aluminium bronze castings made in greensand moulds Matsubara *et al.* (1972) provide an elegant and clear demonstration of the feeding of the pore from the mould, together with the combined effect of residual gas in the metal. Radiographs show wormhole porosity approximately 100 mm long. The amount of porosity was shown to increase as the gas content of the metal was increased, and as the water content of the moulds was increased from zero to 8 per cent.

Halvaee (1997) illustrates a similar structure for aluminium bronze cast into a sand mould bonded

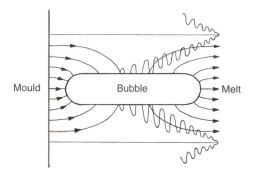

Figure 6.17 *Subsurface pore growing in competition with dendrites, into a melt of low gas content. The pore gains gas by diffusion from the surface reaction, and loses it from its growing front (after Beech 1974).*

with a phenolic urethane resin (Figure 6.18). A proper melting procedure to give a low gas content in the melt prior to casting eliminated the porosity completely.

6.4 Blowholes

The reader needs to be aware that the name 'blow-hole' is, unfortunately, widely misused to describe almost any kind of hole in a casting. In this work the name 'blow-hole, blowhole', or simply 'blow',

is strictly reserved solely for those defects which are forced (actually 'blown') into the liquid metal via the mechanical penetration of the liquid surface. The term is therefore quite specific and accurately descriptive. (The term excludes pores nucleated internally by the precipitation of gas dissolved in the liquid, or diffusing in from the surface, and also excludes bubbles entrained by surface turbulence.) The reader is encouraged to use the name 'blowhole' with accuracy.

If the gas pressure in a core increases to the point at which it exceeds the pressure in the surrounding liquid, gas will force its way out of the core and into the liquid. (The contribution of surface tension to increasing the pressure required is practically negligible in the case of *core blows* because of the large size of the bubbles that are formed in this process).

Campbell (1950) (not the author) devised a useful test that assessed the pressure in cores compared to the pressure in the liquid metal. A modified version of his test is shown in Figure 6.19 and the following is a modification and further development of his explanation.

If the mould is filled quickly then the hydrostatic pressure due to depth in the liquid metal is built up more rapidly than the pressure of gas in the core. This dominance of metallostatic pressure can persist throughout filling and solidification. The higher metal pressure effectively suppresses any bubbling

Figure 6.18 *Radiograph of a Cu–10Al casting 200 × 100 × 10 mm with a high hydrogen content poured at 1285°C with gate velocity 0.85 ms^{-1} into a sand mould (Halvaee 1997).*

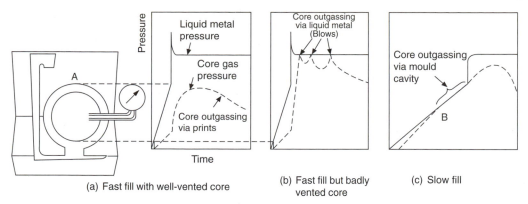

(a) Fast fill with well-vented core   (b) Fast fill but badly vented core   (c) Slow fill

Figure 6.19 *Effect of fill rate on core outgassing (adapted from Campbell 1950)*

of gas through the core at point A as illustrated in Figure 6.19a. Bubbles will form at A only if the core pressure reaches the metal pressure because either the venting of the core is poor (Figure 6.19b) or the core is covered too slowly (Figure 6.19c).

If the mould is filled slowly, the gas pressure in the core exceeds the metal pressure at B and remains higher during most of the filling of the mould. Gas escapes through the top of the core during the late stages of the filling, so that, although the metal pressure in Figure 6.19c finally exceeds the gas pressure in this case, it is doubtful whether the gas will stop flowing. This is because, in practice, when the liquid metal reaches the top of the casting at A it is prevented from closing and joining together by the constant evolution of gas. The migration of volatiles ahead of the slowly rising metal front also concentrates the gas source near A, and the creation of a well-established, well-oxidized bubble trail will further hinder the welding and closing together of the meniscus at this point. Gas issuing from the top of a submerged core can be seen to cause the melt to tremble and flutter against the surface of the core, the melt attempting to close together to stop the flow, but repeatedly failing in this attempt as drossy surface films thicken.

Figure 6.19b illustrates the case in which the pressure in the core rises only just above the metal pressure, with the result that a limited number of bubbles escape. If the casting has started to solidify, then although the bubble may be forced into the melt, it is unlikely to be able to escape from the upper surface of the casting. Medvedev and Kuzovkov (1966) summarize the situation aptly: 'One can postulate that the thinnest of continuous solidified crust will present an insuperable obstacle to a bubble seeking to escape from the casting. There is more probability of fresh bubbles entering the casting through the crust than of existing bubbles escaping to the surface again. It follows therefore that gas will continue to enter the casting from the core for a longer period than it can escape again.'

This behaviour is the result of the core gas pressures considerably exceeding the pressure that the bubble can exert arising from its own buoyancy. The bubbles from cores rise through the casting and generally become trapped under the top solidified skin of the casting as illustrated in Figure 6.20.

The bubbles preferentially detach from upward pointing features of cores (Figure 6.21), as droplets

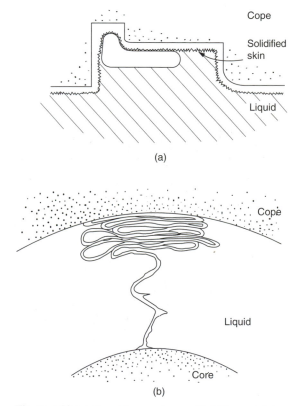

Figure 6.20 *(a) Core blow – a trapped bubble containing core gases. (b) A bubble trail, ending in an exfoliated dross defect as the result of the passage of copious volumes of core gas (after Frawley et al. 1974).*

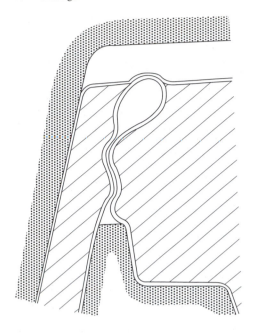

Figure 6.21 *Detachment of a bubble from the top of a core, bequeathing a bubble trail as a permanent legacy of its journey. This bubble, unusually, may be early enough to escape at the free surface of the rising metal.*

of water will detach from the downwardly pointing tip of a stalactite. The theory of droplets and bubbles indicates (Campbell 1970) that the size of the detaching bubble is linked to the size of the source. The results of calculated bubble diameters are shown in Figure 6.22, and reveal that even though a bubble might detach from a horizontal surface, and so be of maximum size, its size is limited to 10 to 15 mm or so, depending on the liquid.

However, final sizes of core blow defects vary typically from 10 to 100 mm in diameter, showing that they are usually the accumulation of many bubbles, or possibly even the result of a constant stream of gas funnelled along a bubble trail, arriving and coalescing under the solidifying skins of castings.

Bubbles blown from cores contrast with bubbles introduced by turbulent filling of the casting. The bubbles entrained during pouring arrive quickly at the top surface of the casting usually prior to any solidification. Providing that they have sufficient buoyancy to break the double film at this point (the film on the casting surface and their own film on the bubble surface) the entrained gases can escape. Bubbles blown from cores arrive much later, and often do not escape from the casting. This is because a core takes time to build up its internal pressure, as heat slowly diffuses into its interior. At this later

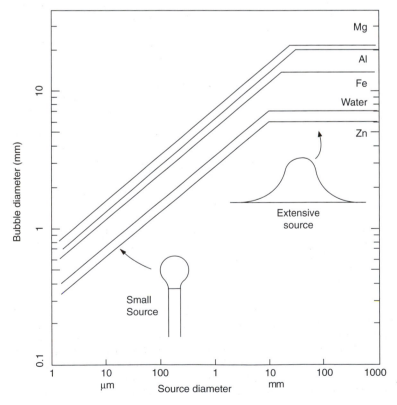

Figure 6.22 *Size of bubbles detaching from a range of sizes of source in a variety of liquids.*

stage a layer of solid has usually already formed against the mould surface, so that any subsequent bubbles are trapped. The small buoyancy force that the bubble possesses is unable to force the bubble through the dendrite mesh, even though the mesh may be relatively open, with liquid still present through to the mould surface. The trapped bubble(s) spread under the dendrites, as, inversely, a large limp balloon would float down to sag on the tips of conifer trees.

These trapped bubbles are recognizable in castings by their size (often measured in centimetres or even decimetres!), their characteristic flattened shape, and their position several millimetres under the uppermost surfaces of local regions of the casting (Figure 6.20a). In addition, of course, they will faithfully follow the contours of the upper surface of the casting. Like the bubble trail mentioned earlier, the core blow defect is sometimes so impressively extensive as to be difficult to perceive on a radiograph.

As an example, the author remembers making a delightful compact aluminium alloy cylinder block for a small pump engine. At first sight the first casting appeared excellent, and unusually light. It seemed that the designer had done a good job. A close study of the radiograph revealed it to be clear of any defects. However, on standing back to take a more general view it was noticed that a curious line effect, as though from a wall thickness, could not be found on the designer's drawing. It took some time to realize the casting was completely hollow, the smooth and extensive form of the core blow faithfully followed the contours of the casting. The prolific outgassing of the water jacket core around the bores of the block were found to be the source. The water jacket core was stood upright on its core prints (the only vents) that located in the drag. The problem arose because of the very slow filling velocity that had been chosen. Vapours were driven ahead of the slowly rising metal, concentrating in the tip of the water jacket core (unvented at its top). Thus by the time the melt arrived at the top, it was unable to cover the core because of the rate of gas evolution. One unhelpful suggestion was that the technique seemed an excellent method for making ultra-lightweight hollow castings. The problem was solved by simply increasing the fill rate, covering the core by the melt before outgassing started. Core gases were then effectively sealed into the core, and were vented in the normal way via the prints at the base of the core.

Where the bubbles are not trapped, but have repeatedly escaped, then oxides from bubble trails and the breaking of the surface of the casting build up to create a spectacular defect sometimes known as an exfoliated dross defect (Figure 6.20b). Exfoliated structures are those that have been expanded and delaminated, like puff pastry, or like the conversion of slate rock into vermiculite, or like leaves from a book.

Exfoliated dross defects are recorded for steels containing 0.13C, 1Cr, 0.5Mo and 0.4A1 (Frawley *et al.* 1974), and for ductile iron (Loper and Saig 1976). The content of film-forming elements in these alloys is of central significance. For ductile iron (containing magnesium) the defect is described as surrounded by a wrinkled or pastry-like area, and when chipped out appears to be a mass of tangled flakes resembling mica that can be easily peeled away. Polished microsections of the defect confirmed the structure of tangled films, and contrasted with the structure of the rest of the casting, which was reported as excellent.

The various actions that can be taken to eliminate blowholes are:

1. Use a core binder that has minimal outgassing (i.e. minimize the binder so far as possible) or delayed outgassing (change to a late outgassing material).
2. If possible, vent the core from its top-most point. Otherwise
3. Redesign the core to avoid upward-pointing features (or invert the casting).
4. Improve the permeability or artificial venting of the core to reduce its internal gas pressure.
5. Fill to cover the core with liquid metal as quickly as possible, before its internal pressure rises high enough to force a bubble into the melt.
6. Use a high hydrostatic head of liquid metal to suppress the evolution of gas from the surface of the core.
7. Raise the casting temperature. This commonly applied technique has the effect of giving more time for the core to outgas through the liquid metal before a solidified shell can form, so that bubbles will not be trapped. The approach can only be recommended with some reluctance, perhaps justifiable in an emergency, or perhaps only for grey iron cast with dry sand cores. In most situations the passage of the bubbles through the melt will, even if the bubbles escape, result in damage to the casting in the form of the creation of bubble trails, leading possibly to both leakage and dross defects as illustrated in Figures 6.20 and 6.21.

6.4.1 Microblows

It is conceivable that small pockets of volatiles on the surface of moulds might cause a small localized explosive release of vapour that would cause gas to be forced through the (oxidized) liquid surface of the casting to form an internal bubble. Poorly mixed sand containing pockets of pure resin binder might perform this. Alternatively, sand particles with small cavities filled with a volatile binder material might act as cannons to blast pockets of vapour into the metal. Figure 6.22 indicates that for the blowing of bubbles of less than 1 mm diameter in some common liquid metals the source of the high-pressure gas needs to be only a few micrometres diameter. The pressures inside such minute bubbles are easily shown to be high, of the order of 1 MPa (10 atm).

There is no direct evidence for this mechanical mechanism of subsurface pore formation. The mechanism for the creation of subsurface porosity remains obscure. It awaits creative and discriminating experiments to separate the problem from classical processes such as diffusion, in which the gases travel from the mould, across the mould/metal interface into the casting as separate atoms, albeit in swarms, the interface being mechanically undisturbed by this inward flow of elements. Such diffusive transfer is assumed in most of the remainder of this work. However, the reader should be aware of the possibility of mechanical transfer of gases, possibly as eruptive events, disrupting the casting interface as minute explosions.

An observation of a surface interaction that is worth reporting, but whose cause is not yet known, is described below. It may therefore have been allocated to the wrong section of the book. It is hoped that future research will provide an answer.

In the study of interactions between chemically bonded sand moulds and liquid metals, a novel and simple test was devised in the author's laboratory (Mazed and Campbell 1992). It was prosaically named the 'wait and see' test. It consisted simply of a flat horizontal surface of a bonded aggregate on to which a small quantity of metal was poured to give a puddle of 50 to 100 mm diameter. The investigator then waited to see what happened. If anything was going to happen it was not necessary to wait long.

When Al–7Si alloy was poured on to the flat surface of either a phenolic–urethane (PF) or furane–sulphonic (FS) acid bonded silica sand the liquid alloy maintained its mirror smooth upper surface. The same was true when Al–7Si–0.4Mg alloy was poured on PF-bonded sand. When the puddles of liquid metal had finally solidified they had retained their top mirror-like surfaces and after sectioning were found to be sound.

However, when the 0.4Mg-containing alloy was poured on the FS-bonded sand the effect was startlingly different. After several seconds a few bubbles not exceeding 1 mm diameter were observed to arrive at the surface, underneath the surface oxide film, raising up minute bumps on the mirror surface. After a few more seconds the number increased, finally becoming a storm of arriving and bursting bubbles, destroying the mirror. The final sessile drop was found to be quite porous throughout its depth, corroborating the expectation that the bubbles had entered the drop at its base because of a reaction with the sand binder.

The fact that this reaction was only observed with certain binders and with certain alloys originally suggested to the author a chemical mechanism. At the time of the work this was thought unlikely because the diffusive transfer of gas into the melt would result in gas in solution. Having an equilibrium pressure of only about 1 atmosphere (since the pressure at the mould surface was originally thought to be limited to this level) it would have been difficult to nucleate small bubbles. However, it has been more recently realized that the presence of bifilms in the alloy practically eliminates the nucleation problem, and so might make the process feasible. This will be especially true since some melts will, partly by accident and partly as a result of alloy susceptibility, have different quantities and different qualities of bifilms. The 0.4Mg alloy would have been expected to contain a different concentration and a different type of bifilm.

Even so, a mechanical process cannot be ruled out at this stage. For instance, the presence of Mg in the alloy may cause MgO formation at the metal/mould interface leading to a convoluted and microscopically fractured surface oxide film that may be more easily penetrated by high-pressure gas. High local pressures might arise particularly if the FS binder was not so well mixed as the PF binder, leaving minute pockets of pure binder in the sand mixture of the plate core. Thus the quality of mixing may be important.

Clearly, there is no shortage of important factors to be researched. Probably it will always be so. In the meantime, the best we can do is be aware of the possible effects and possible dangers, and have patience to live with uncertainty until the truth is finally uncovered.

Chapter 7

Solidification shrinkage

7.1 General shrinkage behaviour

The molten metal in the furnace occupies considerably more volume than the solidified castings that are eventually produced, giving rise to a number of problems for the founder.

There are three quite different contractions to be dealt with when cooling from the liquid state to room temperature, as Figure 7.1 illustrates.

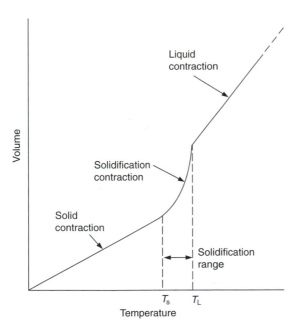

Figure 7.1 *Schematic illustration of three shrinkage regimes: in the liquid; during freezing; and in the solid.*

1. As the temperature reduces, the first contraction to be experienced is that in the liquid state. This is the normal thermal contraction observed by

everyone as a mercury thermometer cools; the volume of the liquid metal reduces almost exactly linearly with falling temperature.

In the casting situation the shrinkage of the liquid metal is usually not troublesome; the extra liquid metal required to compensate for this small reduction in volume is provided without difficulty. It is usually not even noticed, being merely a slight extension to the pouring time if freezing is occurring while the mould is being filled. Alternatively, it is met by a slight fall in level in the feeder.

2. The contraction on solidification is quite another matter, however. This contraction occurs at the freezing point, because (in general) of the greater density of the solid compared to that of the liquid. Contractions associated with freezing for a number of metals are given in Table 7.1. The contraction causes a number of problems. These include (i) The requirement for 'feeding', which is defined here as any process that will allow for the compensation of solidification contraction by the movement of either liquid or solid, and (ii) 'shrinkage porosity', which is the result of failure of feeding to operate effectively. These issues are dealt with at length in this chapter.

3. The final stages of shrinkage in the solid state can cause a separate series of problems. As cooling progresses, and the casting attempts to reduce its size in consequence, it is rarely free to contract as it wishes. It usually finds itself constrained to some extent either by the mould, or often by other parts of the casting that have solidified and cooled already. These constraints always lead to the casting being somewhat larger than would be expected from free contraction alone. This is because of a certain amount of plastic stretching that the casting necessarily suffers. It leads to difficulties in predicting the size of the pattern since the degree to which the

Table 7.1 Solidification shrinkage for some metals

Metal	Crystal structure	Melting point C	Liquid density (kgm^{-3})	Solid density (kgm^{-3})	Volume change (%)	Ref.
Al	fcc	660	2368	2550	7.14	1
Au	fcc	1063	17 380	18 280	5.47	1
Co	fcc	1495	7750	8180	5.26	1
Cu	fcc	1083	7938	8382	5.30	1
Ni	fcc	1453	7790	8210	5.11	1
Pb	fcc	327	10 665	11 020	3.22	1
Fe	bcc	1536	7035	7265	3.16	1
Li	bcc	181	528	–	2.74	4, 5
Na	bcc	97	927	–	2.6	4, 5
K	bcc	64	827	–	2.54	4, 5
Rb	bcc	303	11 200	–	2.2	2
Cd	bcp	321	7998	–	4.00	2
Mg	bcp	651	1590	1655	4.10	3
Zn	scp	420	6577	–	4.08	2
Ce	bcp	787	6668	6646	–0.33	1
In	tcl	156	7017	–	1.98	2
Sn	tetrag	232	6986	7166	2.51	1
Bi	rbomb	271	10 034	9701	–3.32	1
Sb	rhomb	631	6493	6535	0.64	1
Si	diam	1410	2525	–	–2.9	2

References: 1 Wray (1976); 2 Lucas (quoted by Wray 1976): 3 This book; 4 Lida and Guthrie (1988); 5 Critical review by J. Campbell in Brandes (1983).

pattern is made oversize (the 'contraction allowance' or 'patternmaker's allowance') is not easy to quantify. The mould constraint during the solid-state contraction can also lead to more localized problems such as hot tearing or cracking of the casting. The conditions for the generation of these defects are discussed in Chapter 8.

7.2 Solidification shrinkage

In general, liquids contract on freezing because of the rearrangement of atoms from a rather open 'random close-packed' arrangement to a regular crystalline array of significantly denser packing.

The densest solids are those that have cubic close-packed (face-centred-cubic, fcc, and hexagonal close-packed, hcp) symmetry. Thus the greatest values for contraction on solidification are seen for these metals. Table 7.1 shows the contractions to be in the range 3.2–7.2 per cent. The solidification shrinkage for the less closely packed body-centred-cubic (bcc) lattice is in the range 2–3.2 per cent. Other materials that are less dense in the solid state contract by even smaller amounts on freezing.

The exceptions to this general pattern are those materials that expand on freezing. These include water, silicon and bismuth (Table 7.1), and of course, perhaps the most important alloy of all, cast iron.

Remember the milk left on the doorstep, which, in freezing weather, expands sufficiently on solidification to burst the glass bottle (this story reveals my age!). For the same reason the ice cubes

are always stuck fast, having expanded in the ice tray. Analogously, the success of 'type metal', a bismuth alloy used for casting type for printing, derives some of its ability to take up fine detail required for lettering because of its expansion on freezing. Graphitic cast irons with carbon equivalent above approximately 3.6 (Figure 7.2) similarly expand because of the precipitation of the low density phase, graphite. The non-feeding requirements of grey irons are in sharp contrast to other cast irons that solidify 'white', i.e. containing the non-equilibrium carbide phase in place of graphite. In fact the white irons in general have solidification contractions similar to those of steels.

For the majority of materials that do contract on freezing it is important to have a clear idea of what happens in a poorly fed casting. As an ideal case of an unfed casting, it is instructive to consider the freezing of a sphere. We shall assume that the sphere has been fed via an ingate of negligibly small size up to the stage at which a solid shell has formed of thickness x (Figure 7.3). The source of supply of feed metal is then frozen off. Now as solidification continues with the freezing of the following onion layer of thickness dx, the reduced volume occupied by the layer dx compared to that of the original liquid means that either a pore has to form, or the liquid has to expand a little, and the surrounding solid correspondingly has to contract a little. If we assume for the moment that there is no favourable nucleus available for the creation of a pore, then the liquid has to accommodate this by expanding, creating a state of tension, or negative pressure. At

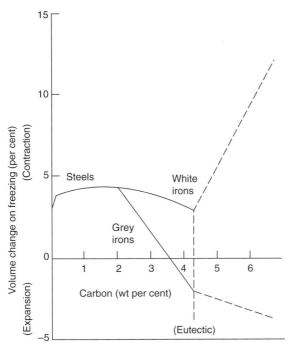

Figure 7.2 *Volume change on freezing of Fe–C alloys. The relations up to 4.3 per cent carbon are due to Wray (1976); the relations for higher carbon have been calculated by the author.*

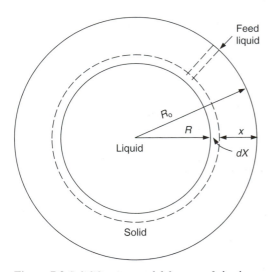

Figure 7.3 *Solidification model for an unfed sphere.*

the same time, of course, the liquid is in mechanical equilibrium with the enclosing solid shell, effectively sucking it inwards. As more onion layers form, so the tension in the liquid increases, the liquid expands, and the solid shell is drawn inwards by plastic collapse.

The reader may find the solidifying sphere model

easier to visualize by thinking of the exact converse situation of a material that expands on solidification: in this case the formation of solid squeezes the remaining liquid into a smaller volume. The liquid therefore experiences a positive pressure. The internal positive pressure then clearly acts on the liquid to compress it. In turn, the liquid acts on the solid shell to expand it.

Further examples to clarify the concept of negative pressures in liquids are presented in section 7.2.1.

Calculations based on a solidifying sphere model were carried out by the author (Campbell 1967, 1968) that show the high tensile stresses in the residual liquid cause the solid to collapse plastically by a creep process. This inward movement of the solid greatly reduces the tension in the liquid. Even so it seems that negative pressures of −100 to −1000 atm may be expected under ideal conditions. The liquid itself is easily able to withstand such stresses since the tensile strength of liquid metals is in the range of −10 000 to −100 000 atmospheres. Incidentally, the liquid/solid interface (whether planar or dendritic is immaterial) is also easily able to withstand the stress, because it is not a favourable site for failure by the nucleation of a cavity (see Chapter 6).

It can also be shown that the resistance to viscous flow of the residual liquid through a pasty zone can cause the pressure to fall, eventually becoming large enough for the casting to collapse by plastic or creep flow. This condition is described more fully later in the sections on interdendritic feeding and shrinkage porosity.

Thus there are a number of situations in which the pressure in the solidifying casting can fall to low or negative values. This hydrostatic stress is a driving force for the formation of shrinkage porosity. However, at the same time, of course, the pressure gradient between the outside and inside of the casting is also the driving force for the various feeding mechanisms that help to reduce porosity.

Whether the driving force for pore formation wins over the driving force for feeding will depend on whether nuclei for pore formation exist. If not (i.e. the metal is clean), then no pores will nucleate, and feeding is forced to continue until the casting is completely frozen. If favourable nuclei are present, then pores will be created at an early stage before the development of any significant hydrostatic stress, with the result that little feeding will occur, and the casting will develop its full percentage of porosity as defined by the physics of the phase change (as given, for instance, in Table 7.1). In most cases the real situation is somewhere between these extremes, with castings displaying some evidence for partially successful liquid feeding in the form of feeder heads which have drawn down somewhat, and parts of the casting displaying evidence of some solid feeding

(Figure 7.4) because of some surface sinks, but the interior exhibiting some porosity. Clearly in such cases feeding has continued under increasing pressure differentials, until the development of a critical internal stress at which some particular nuclei, or surface puncture, can be activated at one or more points in the casting. Feeding is then stopped at such a locality, and pore growth starts.

In those cases where a pore does form at an early stage of solidification, a freezing contraction of perhaps 3 per cent does not at first sight appear to be of any great significance. However, the reader can easily confirm that in a 100 mm diameter sphere the corresponding cavity will be 31 mm in diameter. For an aluminium casting at 7 per cent contraction, the cavity would be 41 mm in diameter. These considerable cavities require a dedicated effort to ensure that they do not appear in castings. This is especially difficult when it is realized that on occasions a substantial casting might be scrapped if it contains a defect of only approximately 1 mm in size. (The scrapping of castings because of the imposition of unreasonably high inspection standards is a widespread injustice that would benefit from the injection of common sense.)

For the vast majority of cast materials, therefore, shrinkage porosity is one of the most important and common defects in castings. Paradoxically, this even includes grey cast irons because of the effect of mould dilation. This problem occurs if the mould is weak, allowing the benefits of the expansion of the graphite phase to be wasted on outwards expansion of the casting, rather than being used on inwards compression of cavities.

The serious consequences of the porosity occurring on the inside of the casting, where it causes a disproportionately large hole, contrasts with the occurrence of porosity on the outside of the casting. If our 100 mm diameter Al sphere had no internal nuclei for the initiation of porosity, no pores would form here, and the solid would therefore be forced to collapse. The final casting would be slightly smaller by just over 1 mm over its whole surface. This 2.5 per cent reduction in diameter is sufficiently small to be not noticeable in most situations. If it were to be important for a casting requiring special accuracy, this small adjustment could be added to the patternmaker's shrinkage allowance. Thus the elimination of internal porosity and its displacement to the outside of the casting is a powerful technique that is strongly recommended.

Although shrinkage porosity can be reduced by

improvements to the cleanness of the metal as mentioned above, this section deals with the other major approach to this problem, the design and provision of good feeding, ensuring good conditions for easy accommodation of the volume change during freezing. (However, the special case of provision of good liquid feeding, the usual way to deal with a feeding problem, is dealt with as a 'methoding' issue in Volume II.)

7.2.1 Hydrostatic tensions in liquids

In the original book *Castings* this section was to have been an appendix. However, it proved to be so interesting, and so central to the understanding of defect generation in castings, that the subject was included as part of the main text. The reader will understand why the subject is included here even though few of the examples described in this section are metal castings. Although most relate to water or other organic systems, the principles will be seen to apply perfectly to metals.

Liquids have been known to be able to withstand tension for many years: the mercury in a clean glass barometer would 'hang up', supporting the weight of a long length of the liquid like a rope in tension, until the glass was tapped, breaking the liquid away from the glass, and allowing the liquid to fall to the height at which it could be supported by the atmospheric pressure.

In 1850 Berthelot cooled various organic liquids in sealed glass tubes, watching until cavitation occurred as a shower of small bubbles. The number of bubbles was subsequently explained by Lewis (1961), who observed that the rupture was always associated with the sudden appearance and collapse of a single bubble, which was immediately followed by the formation of a large number of tiny bubbles along the length of the tube. The asymmetric collapse of the primary bubble would form an impacting jet and create a series of shock waves. These would be ideal conditions for the formation of additional bubbles. This simple experiment is of special interest, since it is analogous to the solidifying casting, with the internal solidifying liquid progressively occupying less volume, and with the surrounding solid gradually being pulled inwards, its elastic and plastic resistance to deformation steadily increasing, building up the tensile stress in the liquid, to the point at which the liquid fractures.

Berthelot's experiment has been repeated many times in different ways. For instance, Vincent and

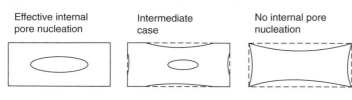

Figure 7.4 *The three regimes of shrinkage porosity: (a) internal; (b) mixed; and (c) external shrinkage porosity.*

Simmons (1943) sealed their tube by freezing the liquid in the mouth of the capillary. They found that water would withstand −157 atm and a mineral oil −119 atm. This is a fascinating observation, since it indicates that the freezing interface will withstand high tensions; it is not necessarily a good nucleation site for pores, as has been indicated theoretically (section 6.1) This conclusion contrasts with the assumptions widely held in the solidification literature.

Scott *et al.* (1948) calculated the build-up of tensile stress ΔP as the liquid cools an amount ΔT, using the coefficient of thermal expansion of the liquid α and the glass σ, together with the modulus of rigidity G and the coefficient of compressibility β of the glass tube:

$$\Delta P = [(\alpha - \sigma)/(\beta - (1/G)]\Delta T$$

They also checked their results by measuring the length of the liquid columns after fracture. This work emphasizes the elastic strains built up in the liquid and the surrounding solid during the experiment.

Lewis (1961) found tensile strengths in water and carbon tetrachloride in excess of −60 atm. Also, when water and carbon tetrachloride were introduced into the same tube the two liquids separated at a clear interface. Fracture was not observed to occur at the interface. The liquid/liquid interface seems as poor a nucleation site as the liquid/solid interface above. These observations confirm the expectations from classical nucleation theory as presented in section 6.1.

Tyndall (1872) carried out experiments on the internal melting of ice by shining light through the ice. The light was absorbed by particles frozen inside. The heated particles then melted a small surrounding region of the shape of a snowflake (known nowadays as a Tyndall figure, or, occasionally, a negative crystal). However, because ice reduces in volume on melting, the liquid volume created in this way was under tension. As melting progressed, the tensile stress became so great that the liquid fractured with an audible click. The melted region was then observed to contain a small bubble. Kaiser (1966) demonstrates that the critical pressure for fracture is in excess of −150 atm. Tyndall proved that the bubble was a vapour cavity containing practically no air by carefully melting the ice towards the figure, using warm water. As soon as a link was established with the bubble it collapsed and disappeared; no air bubble rose to the surface. The Tyndall figures are analogous to the isolated regions of metal castings, where vapour cavities are expected to be formed at critical stresses during solidification.

Briggs (1950) centrifuged water in a capillary tube, spinning it at increasing speed to stretch the water until it was observed to snap at a tensile stress of −280 atm.

The early work on the determination of the tensile strengths of liquids held in various containers such as a Berthelot tube gave notoriously scattered results. This was probably to be expected from the different degrees of cleanness of the container, and any microscopic imperfections it may have had on its walls. Experiments to test the liquid away from the influence of walls have been carried out using ultrasonic waves focused in the centre of a volume of liquid. The rarefaction (tensile) half of the vibration cycle can cause cavitation in the liquid. These experiments have been moderately successful in increasing the limits, but impurities in the liquid have tended to obscure results.

More recently, Carlson (1975) has reflected shock waves through a thin film of mercury and recorded strengths of up to −30 000 atm. This result is one of the first to give a strength value of the same order as that predicted by various theoretical approaches. Despite the dynamic nature of the test method, the result is not particularly sensitive to the length of the duration of the stress (varying only from −25 000 to −30 000 atm as the time decreased from 1 s to 10^{-7} s). This again is to be expected, since nucleation is a process involving the rearrangement of atoms, where the atoms are vibrating about their mean positions at a rate of approximately 10^{13} times per second.

An attempt to measure negative pressures in solidifying aluminium ingots directly was made by Ohsasa *et al.* (1988a, b). They immersed a stainless steel disc into the top of a solidifying metal, so as to almost fill the top of the crucible in which the metal was held. When the solidification front reached the disc, a volume of liquid was effectively trapped underneath. As this volume solidified, the stress in the liquid was measured by a transducer connected to the disc. Stresses needed to cause pore formation were typically only about −0.1 atm. However, occasionally stresses of up to −2.5 atm were recorded. On these occasions the walls of the ingot were observed to suffer some inward collapse. Clearly, in this work the stresses were limited by the nucleation of cavities against the surface of the stainless steel disc. This would be poorly wetted as a result of the presence of the Cr-rich oxide on the steel. At the higher stresses that were measured, these were in turn limited by the inward plastic collapse of the solidified casting.

In general, therefore, it seems that the various attempts to measure the strengths of mercury and the organic liquids have achieved results intermediate between the minimum predicted for the complex inclusion, and the maximum set for the prediction of homogeneous nucleation (Table 6.1). This is probably as good a result as can be expected. However, the one high result by Carlson is reassuring that the physics seems basically correct, and really does apply to liquid metals.

7.3 Feeding criteria

Since this book is concerned with the properties of cast metals, the practical aspects of feeding cannot be covered in detail here. It will be an important section in the next volume *Castings II – Practice*. However, it is desirable to indicate some central concepts, and some small amount of repetition will do no harm.

To allow for the fact that extra metal needs to be fed to the solidifying casting to compensate for the contraction on freezing, it is normal to provide a separate reservoir of metal. We shall call this reservoir a *feeder*, since its action is to *feed* the casting (i.e. to compensate for the solidification shrinkage). The American term *riser* is not recommended. It is not descriptive in a helpful way, and could lead to confusion with other features of the *rigging* such as vents, up which metal also rises. However, the American term *rigging* is helpful. It is a general word for the various appendages of runners, gates and feeders, etc. No equivalent term exists in UK English.

The provision of a feeder can be complicated to get right. There are seven rules that the author has used to help in the systematic approach to a solution. Readers of the first book *Castings* will note this is one more rule than before. This arises because of the introduction of Rule 1. This was cited in *Castings* but not elevated to the status of a rule. The reader will appreciate that it has a valuable place in the new listing.

Even more recently, Tiryakioglu (2001) has further simplified the seven rules as is noted in more detail at the end of this section.

7.3.1 The seven feeding rules

Rule 1. The main question relating to the provision of a feeder on a casting is 'Should we have a feeder at all?' This is a question well worth asking, and constitutes Rule 1 '*Do not feed (unless absolutely necessary)*'.

The avoidance of feeding is to be greatly encouraged, in contrast to the teaching in most traditional foundry texts. There are several reasons to avoid the placing of feeders on castings. The obvious one is cost. They cost money to put on, and money to take off. In addition, many castings are actually impaired by the inappropriate placing of a feeder. This is especially true for thin-walled castings, where the filling of the feeder diverts metal from the filling of the casting, with the creation of a misrun casting. Probably half of the small and medium-sized castings made today do not need to be fed. Sometimes the casting suffers delayed cooling, impaired properties and even segregation problems as a result of the presence of a feeder. Finally, it is easy to make an error in the estimation of the appropriate feeder size, with the result that the casting can be more defective than if no feeder were used at all. The aspects will be covered in detail in *Castings II – Practice*.

However, there is one potential major problem if a feeder is not used, but where a feeding problem remains in the casting. The feeding problem will show itself as a region of reduced pressure in the casting at a late stage of freezing. The reduced pressure will act to open bifilms as in the reduced pressure test (RPT) used for non-ferrous alloys. Thus in an aluminum alloy casting with medium thickness sections, enhanced size and number of internal microshrinkage pores will be experienced if the sections are unfed. Whether this is important or not depends on the specification of microstructure and properties that the casting is required to meet. If the casting is required to have good elongation properties in a tensile test it is likely that feeding to maintain pressure during solidification will be needed to achieve this.

Just for the moment we shall assume that a feeder is required. The next question is 'How large should it be?'

There is of course an optimum size. Figure 6.15 shows the results of Rao *et al.* (1975), who investigated an increasing feeder size on the feeding of a plate casting in Al–12Si alloy. Interestingly, when the data are extrapolated to zero feeder size the porosity is indicated to be approximately 8 per cent, which is close to the theoretical 7.14 per cent solidification shrinkage for pure aluminium (Table 7.1). At a feeder modulus of around 1.2 times the modulus of the casting, the casting is at its most sound. The residual 1 per cent porosity is probably dispersed gas porosity. As the feeder size is increased further the solidification of the casting is now progressively delayed by the nearby mass of metal in the feeder. Thus while this excessive feeder is no disadvantage in itself, the delay to solidification of the whole casting increases the time available for further precipitation of hydrogen as gas porosity and the unfolding of bifilms. However, it is clear from this work that an undersized feeder will result in very serious porosity, while an oversize feeder causes less of a problem (although, of course, it does adversely influence the economics!).

There is a vast literature on the subject of providing adequate-sized feeders for the feeding of castings. It is mainly concerned with two feeding rules. Numbering onwards from Rule 1 above, the conventional rules are:

Rule 2. The feeder must solidify at the same time as, or later than, the casting. This is the heat-transfer criterion, attributed to Chvorinov.

Rule 3. The feeder must contain sufficient liquid to meet the volume-contraction requirements of the casting. This is usually known as the volume criterion.

However, there are additional rules which are also often overlooked, but which define additional thermal, geometrical and pressure criteria that are absolutely necessary conditions for the casting to freeze soundly:

Rule 4. The junction between the casting and the feeder should not create a hot spot, i.e. have a freezing time greater than either the feeder or the casting. This is a problem that, if not avoided, leads to '*under-feeder shrinkage porosity*'. The junction problem is a widely overlooked requirement. The use of Chvorinov's equation systematically gives the wrong answer for this reason, so the junction requirement is often found to override Chvorinov.

Rule 5. There must be a path to allow feed metal to reach those regions that require it. The reader can see why this criterion has been often overlooked as a separate rule; the communication criterion appears self-evident! Nevertheless it does have a number of geometrical implications which are not self-evident, and which will be discussed.

Rule 6. There must be sufficient pressure differential to cause the feed material to flow, and the flow needs to be in the correct direction.

Rule 7. There must be sufficient pressure at all points in the casting to suppress the formation and growth of porosity. In section 9.2.5 we shall see that the action of pressure also retains the good mechanical properties of the casting by suppressing the opening of (usually invisible) bifilms. The bifilms effectively constitute a low background level of porosity (that may or may not be just visible) that is highly effective in reducing strength and ductility of castings.

It is essential to understand that all the rules must be fulfilled if truly sound castings are to be produced. The reader must not underestimate the scale of this problem. The breaking of only one of the rules may result in ineffective feeding and a porous casting. The wide prevalence of porosity in castings is a sobering reminder that solutions are often not straightforward.

Because the calculation of the optimum feeder size is therefore so fraught with complications, is dangerous if calculated wrongly, and costs money and effort, the casting engineer is strongly recommended to consider whether a feeder is really necessary at all. This is Rule 1. You can see how valuable it is.

Of the remainder of castings that do suffer feeding demand, many could avoid the use of a feeder by the judicious application of chills or cooling fins.

This leaves only those castings that have heavy sections, isolated heavy bosses or other features that *do* need to be fed with the correct size of feeder.

Finally, in a recent development of the concepts of feeding, it is worth drawing attention to the work of Tiryakioglu (2001). He has demonstrated that Rules 2, 3 and 4 can be gathered together under only one new criterion 'The thermal centre of the total casting (i.e. the casting + feeder combination) should be in the feeder.' This deceptively straightforward statement has been shown to allow the calculation of optimum feeders with unprecedented accuracy. Discussion of this approach has to await Volume II.

7.3.2 Criteria functions

The development of computer software to predict the solidification of castings is not yet developed to predict the occurrence of porosity from first principles, i.e. calculating the pressure drop in the various parts of the casting, and thereby assessing the potential for nucleation and growth of cavities. This represents a Herculean task. As a useful short cut, therefore, many workers have searched for parameters that can be relatively easily calculated and enable an assessment of the potential for pore formation.

Niyama and co-workers (1982) were among the first to develop a useful criterion function. Based on a simple model assuming Darcy's law for the flow of residual liquid through the dendrite mesh they proposed the parameter $G/V_S^{1/2}$ to assess the difficulty of providing feed liquid, where G is the temperature gradient at the solidus temperature and V_S the velocity of the solidus isotherm. This is probably the most widely used criteria function. It has been found useful to predict the propensity for porosity in steels.

In their study of the formation of porosity in steel plates of thickness 5–50 mm, with and without end chills, Minakawa *et al.* (1985) confirmed that usefulness of the Niyama criterion in assessing the conditions for the onset of porosity in their castings. (They also looked at G, the temperature gradient along the centreline of the casting at the solidification front, and the fraction solid along the centreline. Neither of these alone was satisfactory.)

In another theoretical study Hansen and Sahm (1988) support the usefulness of $G/V_S^{1/2}$ for steel castings. However, in addition they go on to argue the case for the use of a more complex function $G/V_S^{1/4} V_L^{1/2}$ where V_L is the velocity of flow of the residual liquid. They proposed this relation because they noticed that the velocity of flow in bars was five to ten times the velocity in plates of the same thickness, which, they suggest, contributes to the additional feeding difficulty of bars compared to plates. (A further contributor will be the comparatively high resistance to the plastic collapse of bars, compared to the efficiency of solid feeding in plates as will be discussed later.)

They found that $G/V_S^{1/4} V_L^{1/2}$ < a critical value, which for steel plates and bars is approximately 1 Ks$^{3/4}$ mm$^{-7/4}$. Their parameter is, of course, less easy to use than that due to Niyama, because it needs flow velocities. The Niyama approach only requires data obtainable from temperature measurement in the casting.

Unfortunately, however, there are many limitations of criteria functions that are widely overlooked. They include

1. Strictly, they assess only the difficulty of interdendritic feeding.
2. They apply best to strong materials such as steels in cold moulds. For steels in hot investment moulds and for light alloys in cold moulds where significant solid feeding can occur the criteria become inaccurate. Interdendritic feeding can become impossibly difficult, generating high internal tension in the casting, a condition that the computer would interpret as leading to porosity. However, the softer solid can yield plastically, collapsing slightly to give an internally sound casting.

They cannot predict conditions for porosity arising as a result of the many other mechanisms for the production of porosity in castings. These include the major shrinkage pores as a result of the isolation of major liquid regions; the creation of surface-initiated porosity; and the mechanically entrained porosity originating from bubbles of air and mould gases introduced by poor filling systems. Experience shows that in general entrainment defects cause most of the porosity found in castings.

7.4 Feeding – the five mechanisms

During the solidification of a casting, the gradual spread and growth of the solid, often in the form of a tangled mass of dendrites, presents increasing difficulties for the passage of feeding liquid. In fact, as the freezing liquid contracts to form the solid, the pressure in the liquid falls, causing an increasing pressure difference between the inside and the outside of the casting. The internal pressure might actually fall far enough to become negative, as a hydrostatic tension.

The generation of such negative pressure differentials, and even actual hydrostatic tension, is undesirable in castings. Such phenomena provide the driving force for the initiation and growth of volume defects such as porosity and surface sinks on castings, and the lowering of properties because of the opening of bifilms.

However, there appear to be at least five mechanisms by which hydrostatic tension can be reduced in a solidifying material, although, of course, not all five processes are likely to operate in any single case. Adequate feeding by one or more of these processes will relieve the stress in the solidifying liquid and thus reduce the possibility of the formation of defects.

The five feeding mechanisms as set out in Figure 7.5 were first identified and described by the author in 1969. In the original publication, solid-state diffusion was added as a sixth effect that would cause pore shape to change somewhat after solidification was complete. Shape changes in pores would occur because the forces of surface tension and mechanical stress are sufficient to cause material flow at temperatures near the melting point of the solid. Such changes to the pore shape and size are detectable under a microscope. However, these

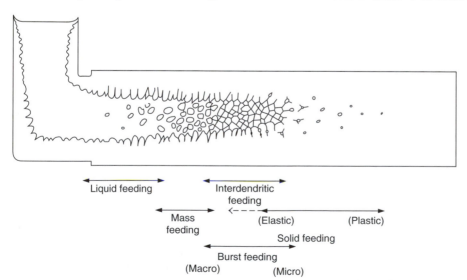

Figure 7.5 *Schematic representation of the five feeding mechanisms in a solidifying casting (Campbell 1969).*

considerations are the reserve of the research scientist, and reflect the author's early interests, having been trained as a physicist. Nowadays, as a somewhat more practical foundryman trying to make good castings, the first five mechanisms are all that matter.

The mechanisms are dealt with in the order in which they might occur during freezing. The order coincides with a progressive but ill-defined transition from what might usefully be termed 'open' to 'closed' feeding systems.

7.4.1 Liquid feeding

Liquid feeding is the most 'open' feeding mechanism and generally precedes other forms of feeding (Figure 7.5). It should be noted that in skin-freezing materials it is normally the only method of feeding. The liquid has low viscosity, and for most of the freezing process the feed path is wide, so that the pressure difference required to cause the process to operate is negligibly small. Results of theoretical model of a cylindrical casting only 20 mm diameter (Figure 7.6) indicate that pressures of the order of only 1 Pa are generated in the early stages. By the time the 10 mm radius casting has a liquid core of radius 1 mm (i.e. is 99 per cent solid) the pressure

difference has increased only to 100 Pa (approximately one-thousandth of an atmosphere; smaller than about one-tenth of the hydrostatic pressure due to depth). (It is worth emphasizing that the theoretical model represented in Figure 7.6 and elsewhere in this book represents a worst case. This is because the temperature gradient in the solidified shell has been neglected. The lower temperature of the outer layers of the shell will cause the shell to contract, compressing the internal layers of the casting, and thus reducing the internal hydrostatic tension. In some cases the effect is so large that the internal pressure can become positive, as shown in the excellent treatment by Forgac et al. (1979).)

For all practical purposes, therefore, liquid feeding occurs at pressure gradients that are so low that these gentle stresses will never lead to problems.

The rules for adequate liquid feeding are the seven feeding rules listed in section 7.3.

Inadequate liquid feeding is often seen to occur when the feeder has inadequate volume. Thus liquid flow from the feeder terminates early, and subsequently only air is drawn into the casting. Depending on the mode of solidification of the casting, the resulting porosity can take two forms:

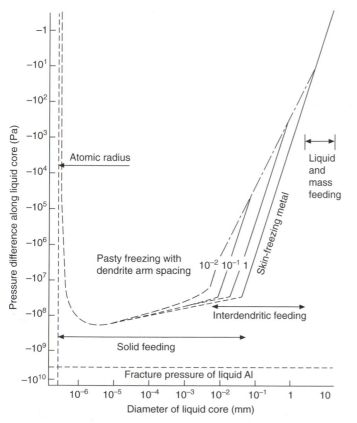

Figure 7.6 *Hydrostatic tensions in the residual liquid calculated for the various feeding regimes during the freezing of a 20 mm diameter aluminium alloy cylinder (Campbell 1969).*

1. Skin-freezing alloys will have a smooth solidification front that will therefore result in a smooth shrinkage pipe extending from the feeder into the castings as a long funnel-shaped hole. In very short-freezing-range metals the surface of the pipe can be as smooth and silvery as a mirror.
2. Long-freezing-range alloys will be filled with a mesh of dendrites in a sea of residual liquid. In this case liquid feeding effectively becomes interdendritic feeding, of course. In the case of an inadequate supply of liquid in the feeder, the liquid level falls, draining out to feed more distant regions of the casting and sucking in air to replace it. The progressively falling level of liquid will define the spread of the porosity, decreasing as it advances because of the decreasing volume fraction of residual liquid as freezing proceeds. The resulting effect is that of a partially drained sponge, as shown in the tin bronze casting in Figure 7.7. Sponge porosity is a good name for this defect.

Figure 7.7 *Porosity in the long-freezing-range alloy Cu–10Sn bronze, cast with an inadequate feeder, resulting in a spongy shrinkage pipe.*

When sectioned, the porosity resembles a mass of separate pores in regions separated by dendrites. It is therefore often mistaken for isolated interdendritic porosity. However, it is, of course, only another form of a primary shrinkage pipe, practically every part of which is connected to the atmosphere through the feeder. It is a particularly injurious form of porosity, therefore, in castings that are required to be leak-tight, especially since it can be extensive throughout the casting, as Figure 7.7 illustrates. Furthermore this type of porosity is commonly found. It is an indictment of our feeding practice.

The author recalls an investigation into porosity in the centre of a balanced steel ingot, to ascertain whether the so-called secondary porosity was connected to the atmosphere via the shrinkage cavity in the top of the ingot. Water was poured on to the top of the ingot, creating a never-forgotten drenching from the shower that issued from the so-called secondary pores. The lesson that the pores were perfectly well connected was also not forgotten.

7.4.2 Mass feeding

Mass feeding is the term coined by Baker (1945) to denote the movement of a slurry of solidified metal and residual liquid. This movement is arrested when the volume fraction of solid reaches anywhere between 0 and 50 per cent, depending on the pressure differential driving the flow, and depending on what percentage of dendrites are free from points of attachment to the wall of the casting. However, it seems that smaller amounts of movement can continue to occur up to about 68 per cent solid, which is the level at which the dendrites start to become a coherent network, like a plastic three-dimensional space frame (Campbell 1969).

In thin sections, where there may be only two or three grains across the wall section, mass feeding will not be able to occur. The grains are pinned in place by their contacts with the wall. However, as the number of grains across the section increases to between five and ten the central grains are definitely free to move to some extent. In larger sections, or where grains have been refined, there may be 20 to 100 grains or more, so that the flow of the slurry can become an important mechanism to reduce the pressure differential along the flow direction. Clearly, the important criterion to assess whether mass flow will occur is the ratio (casting section thickness)/(average grain diameter). This is probably one of the main reasons why grain refinement is useful in reducing porosity in castings (the other main reason being the greater dispersion of gases in solution and their reduced segregation).

At the point at which the grains finally impinge strongly and stop is the point at which feeding starts to become appreciably more difficult. This is the regime of the next feeding mechanism, interdendritic feeding.

In passing, we may note that in some instances mass feeding may cause difficulties. There is some evidence that the flow of the liquid/solid mass into the entrance of a narrow section can lead to the premature blocking of the entrance with the solid phase. Thus the feed path to more distant regions of the casting may become choked.

7.4.3 Interdendritic feeding

Allen (1932) was one of the first to use the term 'interdendritic feeding' to describe the flow of residual liquid through the pasty zone. He also made the first serious attempt to provide a quantitative theory. However, we can obtain an improved estimate of the pressure gradient involved simply by use of the famous equation by Poisseuille that describes the pressure gradient dP/dx required to cause a fluid to flow along a capillary:

$$\frac{dP}{dx} = \frac{8v\eta}{\pi R^4} \tag{7.1}$$

where v is the volume flowing per second, η is the viscosity, and R is the radius of the capillary. It is clear without going further that the resistance to flow is critically dependent on the size of the capillary. For a bunch of N capillaries, which we can take as a rough model of the pasty zone, the problem is reduced somewhat:

$$\frac{dP}{dx} = \frac{8v\eta}{\pi R^4 N} \qquad (7.2)$$

For the sake of completeness it is worth developing this relation to evaluate a more realistic channel that includes the effect of simultaneous solidification so as to close it by slow degrees. The treatment is based on that by Piwonka and Flemings (1966) (Figure 7.8). Given that the average velocity V is $v/\pi R^2$, and, by conservation of volume, equating the volume flow through element dx with the volume deficit as a result of solidification on the surface of the tube beyond element dx, we have, all in unit time:

$$\pi R^2 V = 2\pi R(L-x)\left(\frac{\alpha}{1-\alpha}\right)\frac{dR}{dt} \qquad (7.3)$$

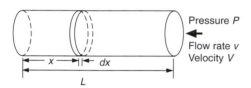

Pressure P

Flow rate v
Velocity V

Figure 7.8 *A tube of liquid, solidifying inwards, while being fed with extra liquid from the right.*

By substituting and integrating, it follows directly that:

$$\Delta P = \frac{16\eta}{N}\left(\frac{\alpha}{1-\alpha}\right)\frac{dR}{dt}\cdot\frac{1}{R^3}\left(Lx-\frac{x^2}{2}\right) \qquad (7.4)$$

We can find the maximum pressure drop ΔP at the far end of the pasty zone by substituting $x = 0$. At the same time we can substitute the relation for freezing rate used by Piwonka and Flemings, $dR/dt = -4\lambda^2/R$ approximately, where λ is their heat-flow constant. Also using the relation $Nd^2 = D^2$ where d is the dendrite arm spacing and D^2 is the area of the pasty zone of interest, we obtain at last:

$$\Delta P = 32\eta\left(\frac{\alpha}{1-\alpha}\right)\frac{\lambda^2 L^2 d^2}{R^4 D^2} \qquad (7.5)$$

This final solution reveals that the pressure drop by viscous flow through the pasty zone is controlled by a number of important factors such as viscosity, solidification shrinkage, the rate of freezing, the dendrite arm spacing and the length of the pasty zone. However, in confirmation of our original conclusion, the pressure drop is most sensitive to the size of the flow channels.

Additional refinements to this equation, such as the inclusion of a tortuosity factor to allow for the non-straightness of the flow, do not affect the result significantly. However, more recent improvements have resulted in an allowance for the different resistance to flow depending on whether the flow direction is aligned with or across the main dendrite stems (Poirier 1987).

The overriding effect of the radius of the flow channel leads to ΔP becoming extremely high as R diminishes. In fact, in the absence of nuclei that would allow pore formation to release the stress, the high hydrostatic stress near the end of freezing will be limited by the inward collapse of the solidified outer parts of the casting, as indicated in Figure 7.6. This plastic flow of the solid denotes the onset of 'solid feeding', the last of the feeding mechanisms. The natural progression of inter-dendritic feeding followed by solid feeding is confirmed by more recent models (Ohsasa *et al.* 1988a, b).

7.4.3.1 Effect of the presence of eutectic

The rapid increase of stress as R becomes very small explains the profound effect of a small percentage of eutectic in reducing the stress by orders of magnitude (Campbell 1969). This is because the eutectic freezes at a specific temperature, and progress of this specific isothermal plane through the mesh corresponds to a specific planar freezing front for the eutectic. The front occurs ahead of the roots of the dendrites, so that the interdendritic flow paths no longer continue to taper to zero, but finish, abruptly truncated as shown in Figure 7.9. Thus the most difficult part of the dendrite mesh to feed is eliminated.

Larger amounts of eutectic liquid in the alloy reduce AP even further, because of the increased size of channel at the point of final solidification. As the percentage eutectic increases towards 100 per cent the alloy feeds only by liquid feeding, of course, which makes such materials easy to feed to complete soundness.

Since most long-freezing-range alloys exhibit poor pressure tightness, the use of the extremely long-freezing-range alloy 85Cu–5Sn–5Zn–5Pb for valves and pipe fittings seems inexplicable. However, the 5 per cent lead is practically insoluble in the remainder of the alloy, and thus freezes as practically pure lead at 326°C, considerably easing feeding, as discussed above.

The appearance of non-equilibrium eutectic in pure Fe–C alloys is predicted to be rather close to the equilibrium condition of 2 per cent C (Clyne and Kurz 1981) because carbon is an interstitial atom in iron, and therefore diffuses rapidly, reducing the effect of segregation during freezing. However, in the presence of carbide-stabilizing alloys such

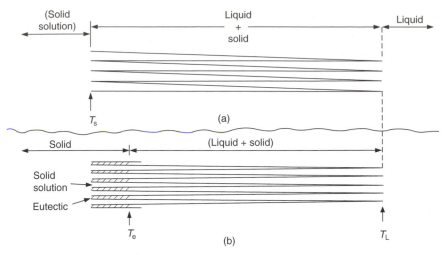

Figure 7.9 *A diagrammatic illustration of (a) how the tapering interdendritic path increases the difficulty of the final stage of interdendritic feeding, and (b) how a small percentage of eutectic will eliminate this final and narrowest portion of the path, thereby greatly easing the last stages of feeding.*

as manganese, the segregation of carbon is retained to some extent, causing eutectic to appear only in the region of 1.0 per cent C as seen in Figure 5.28.

In Al–Mg alloys, layer porosity is observed in increasing amounts as magnesium is increased, illustrating the growing problem of interdendritic feeding as the freezing range increases. However, at a critical composition close to 10.5 per cent Mg the porosity suddenly disappears, and the eutectic beta-phase is first seen in the microstructure (Jay and Cibula 1956) The actual arrival of eutectic at 10.5 per cent Mg confirms the non-equilibrium conditions, and compares with the prediction of 17.5 per cent Mg for equilibrium. Lagowski and Meier (1964) found a similar transition in Mg–Zn alloys as zinc is progressively increased. Their results are presented later in Figure 9.6.

However, one of the most spectacular displays of segregation of a solute element in a common alloy system is that of copper in aluminium. In the equilibrium condition, eutectic would not appear unless the copper content exceeded 5.7 per cent. However, in experimental castings of increasing copper content, eutectic has been found to occur at concentrations as low as approximately 0.5 to 0.8 per cent. This concentration corresponds to a peak in porosity, and the predicted peak in hydrostatic tension in the pasty zone (Figure 7.10).

Many property-composition curves are of the cuspoid, sharp-peaked type (note that they are not merely a rounded, hump-like maximum). Examples are to be found throughout the foundry research literature (although the results are most often interpreted as mere humps!). For instance, the porosity in the series of bronzes of increasing tin

content exhibits a peak in porosity at 5 per cent Sn, not 14 per cent as expected from the equilibrium phase diagram. Pell-Walpole (1946) was probably the first to conclude that this is the result of the maximum in the effective freezing range. Spittle and Cushway (1983) find a sharp maximum in the hot-tearing behaviour of Al–Cu alloys at approximately 0.5–0.8 per cent Cu (Figure 8.21). The analogous results by Warrington and McCartney (1989) can be extrapolated to show that their peak is nearer 0.5 per cent Cu (Figure 8.18), close to the peak in porosity as described above.

7.4.4 Burst feeding

Where hydrostatic tension is increasing in a poorly fed region of the casting, it seems reasonable to expect that any barrier might suddenly yield, like a dam bursting, allowing feed metal to flood into the poorly fed region. This feed mechanism was proposed by the author simply as a logical possibility based on such straightforward reasoning (Campbell 1969). As solidification proceeds, both the stress and the strength of the barrier will be increasing together, but at different rates. Failure will be expected if the stress grows to exceed the strength of the barrier. The barrier may be only a partial barrier, i.e. a restriction to flow, and failure may or may not be sudden.

In terms of Figure 7.11, the nucleation threshold diagram, the threshold for burst feeding will be unique for each poorly fed region of the casting. For small or intermediate barriers, bursts will reduce the internal stress and allow the casting to remain free from porosity. It is possible that repeated bursts

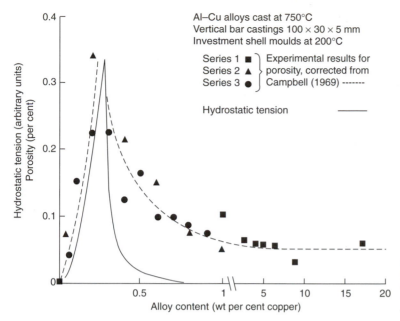

Figure 7.10 *Predicted peak in hydrostatic tension in the pasty zone, and the measured porosity in test bars, as a function of composition in Al–Cu alloys (after Campbell 1969).*

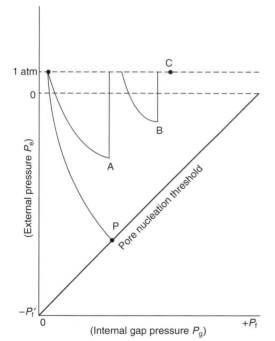

Figure 7.11 *Gas-shrinkage map showing the path of development to early pore nucleation at P. In a contrasting case, slow mechanical collapse of the casting delays the build-up of internal tension, culminating in complete plastic collapse in the form of burst feeding processes at A and B. This delay is successful in avoiding pore nucleation, since freezing is complete at C.*

the feeding barrier is substantial then it may never burst, causing the resulting stress to rise and eventually exceed the nucleation threshold. This time the release of stress corresponds to the creation and growth of a pore. There can be no further feeding of any kind in that region of the casting after this event; the driving force for feeding is suddenly eliminated.

Previously, the author has quoted the following observation as a possible instance of a kind of microscopic type of burst feeding. During observation of the late stages of solidification of the feeder head of many aluminium alloy castings it is clearly seen that the level of the last portion of interdendritic liquid sinks into the dendrite mesh not smoothly, but in a series of abrupt, discontinuous jumps. It was thought that the jumps may be bursts of feeding into interdendritic regions. However, it now seems more likely that the jumps are the result of the repeated, sudden, brittle failure of the surface oxide film, caught up and stretched between supporting dendrites at the surface. The liquid draining down into the dendrite mesh will attempt to drag down its surface film, which will repeatedly burst and repair, resisting failure again for a time. The phenomenon is an illustration of the strength of the film, its capacity for stretching to some extent elastically, and the capacity of the solidified material at its freezing point to exhibit a certain amount of elastic recoil behaviour.

A macroscopic type of barrier can be envisaged for those parts of castings where mass flow has occurred, causing equiaxed crystals to block the entrance to a section of casting.

might help to maintain the casting interior at a low stress until the casting has solidified. However, if

Macroscopic blockages have been observed directly in waxes, where the flow of liquid wax along a glass tube was seen to be halted by the formation of a solidified plug, only to be restarted as the plug was burst. This behaviour was repeated several times along the length of the channel (Scott and Smith 1985).

In iron castings such behaviour was intentionally encouraged in the early twentieth century. Nearly all large castings were subjected to 'rodding' – one or two men would stand on the mould and ram an iron rod up and down through the feeder top. Extra feed metal might be called for and topped up from time to time. This procedure would last for many hours until the casting had solidified. Nowadays it is more common to provide a feeder of adequate size so that feeding occurs automatically without such strenuous human intervention!

On a microscale, a type of burst feeding is the rupture of the casting skin, allowing an inrush of air or mould gases. However, this is, of course, a gaseous burst that corresponds to the growth of a cavity, not a feeding process. Pellini (1953) drew attention to this possibility in bronze castings. It is expected to be relatively common in castings of many alloys.

In conclusion, it has to be admitted that while burst feeding might be an important feeding mechanism, it is not easy to quantify its effects by modelling. Despite some interest in using the concept of burst feeding as an explanation of some casting experiments, these uses remain speculative. The existence of burst feeding has never been unambiguously demonstrated. It therefore seems difficult to understand it and difficult to control it. At this stage we have to be content with the conclusion that logic suggests that it does exist in metal castings.

7.4.5 Solid feeding

At a late stage in freezing it is possible that sections of the casting may become isolated from feed liquid by premature solidification of an intervening region.

In this condition the solidification of the isolated region will be accompanied by the development of high hydrostatic stress in the trapped liquid; sometimes high enough to cause the surrounding solidified shell to deform, sucking it inwards by plastic or creep flow. This inward flow of the solid relieves the internal tension, like any other feeding mechanism. In analogy with 'liquid feeding', the author called it 'solid feeding'. An equally good name would have been 'self feeding'.

When solid feeding starts to operate, the stress in the liquid becomes limited by the plastic yielding of the solid, and so is a function of the yield stress Y and the geometrical shape of the solid. The yield stress Y is, of course, a function of the strain rate at these temperatures when assuming an elastic/plastic model. The procedure is practically equivalent to the assumption of a creep stress model, and results in similar order-of-magnitude predictions for stress (Campbell 1968a, b). For instance, for a sphere of radius R_0, with internal liquid radius R (Figure 7.3):

$$P = 2Y \ln(R_0/R)$$

which is curiously independent of the solidification shrinkage α. Mechanical engineers will recognize this relation as the classical formula for the failure of a thick-shell pressure vessel stressed by internal pressure to the point at which it is in a completely plastic state. This equation is expected to give maximum estimates of the hydrostatic tensions in castings because: (i) the shape is the most difficult to collapse inwardly; and (ii) the equation neglects the opposing contribution of the thermal contraction of the solidified shell which will tend to reduce internal tension (Forgac et al. 1979). Nevertheless it is still interesting to set an upper bound to the hydrostatic tensions that might arise in castings.

This early model (Campbell 1967) used the concept that the liquid radius R had to be expanded to some intermediate radius R', and the solid had to be shrunk inwards from its original internal radius $R + dR$ to the new common radius R'. At this new radius the stress in the liquid equals the stress applied at the inner surface of the solid.

The working out of this simple model indicated that for a solidifying iron sphere of diameter 20 mm, the elastic limit at the inner surface of the shell was reached at an internal stress of about -40 atm; and by the time the residual volume of liquid was only 0.5 mm in diameter a plastic zone had spread out from the centre to encompass the whole shell. At this point the internal pressure was in the range of approximately -200 to -400 atm and the casting was 99.998 per cent solid. Solidification of the remaining drop of liquid increased the pressure in the liquid to approximately -1000 atm. Later estimates using a creep model and cylindrical geometry confirmed similar figures for iron, nickel, copper and aluminium (Campbell 1968a, b).

A minute theoretical point of interest to those of a scientific disposition is the effect of the solid/liquid interfacial tension. Although this is small, it starts to become important when the liquid region is only a few hundred atoms in diameter. The interfacial tension causes an inward pressure $2\gamma_{LS}/R$ that starts to compress the residual liquid. This is the explanation for the theoretical curves to take an upward turn in Figure 7.6 as freezing nears completion, creating a limit to the maximum internal tension.

We have to bear in mind that these estimates of the internal tension are upper bounds, likely to be reduced by thermal contraction of the shell, and reduced by geometries that are easier to collapse, such as a cylinder or a plate. Also the predictions are in any case lower for smaller trapped volumes of liquid, as might occur, for instance, in inter-dendritic spaces. Figure 7.12 shows the effect of plastic zones spreading from isolated unfed regions of the casting.

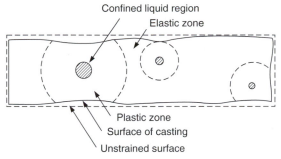

Figure 7.12 *Plastic zones spreading from isolated volumes of residual liquid in a casting, showing localized solid feeding in action (Campbell 1969).*

For an infinite, flat plate-shaped casting in a skin-freezing metal, the internal stress developed is zero, which is an obvious solution, since there can be no restraint to the inward movement of infinite flat plates separated by a solidifying liquid, the plates simply move closer together to follow the reduction in volume. For real plates, their surfaces are held apart to some extent by the rigidity of the edges of the casting, so the development of internal stress would be expected to be intermediate between the two extreme cases. The ease of collapse of the central regions of flat plates emphasizes the importance of geometry.

Figures 7.13 and 7.14 show results of measurements of porosity in small plates of an investment-cast nickel-based alloy. This is an excellent example of solid feeding in action. At low mould temperatures the solid gains strength rapidly during freezing and therefore retains the rectangular outer shape of the casting, and the steep temperature gradient concentrates the porosity in the centre of the casting. As mould temperature is increased, the falling yield stress of the solidified metal allows progressively more collapse of the centre, reducing the total level of porosity by solid feeding. However, some residual porosity remains noticeable nearer the side walls, where geometrical constraint prevents full collapse. Note that these results were obtained in vacuum, with zero contribution from exterior positive atmospheric pressure. It follows, therefore, that all of the solid feeding in this case is the result of internal negative pressure. In fact, surface sinks are commonly seen in vacuum casting. They are not therefore solely

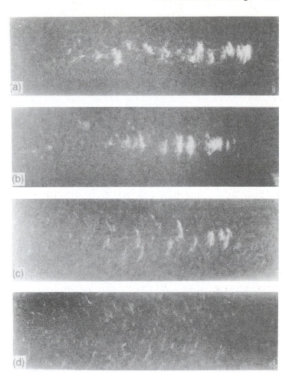

Figure 7.13 *(a) Radiographs of bar castings 100 × 30 × 5 mm in nickel-based alloy cast at 1620°C in vacuum 15 μmHg into moulds at: (a) 250°C; (b) 500°C; (c) 800°C; and (d) 1000°C (Campbell 1969). Centreline macroporosity is seen to blend into layer porosity, and finally into dispersed microporosity.*

the consequence of the action of atmospheric pressure, as generally supposed.

Figure 7.15 shows solid-feeding behaviour in wax castings. The example is interesting because it is evident that sound castings can, in principle, be produced without any feeding in the classical sense. In this case feeding has been successfully accomplished by skilful choice of mould temperature to facilitate uniform solid feeding.

Figure 7.16 shows a similar effect in unfed Al–12Si alloy as a function of increasing casting temperature. The full 6 or 7 per cent of internal shrinkage porosity is gradually replaced by external collapse of the casting as casting temperature increases (Harinath *et al.* 1979).

If solid feeding is controlled so that it spreads itself uniformly in this way, then the accompanying movement of the outer surface of the casting becomes negligible for most purposes. For instance, the high-volume shrinkage of about 6 per cent suffered by Al–Si alloys corresponds to a linear shrinkage of only 2 per cent in each of the three perpendicular directions (i.e. 6 per cent in 3-D corresponds to 2 per cent in 2-D). For a datum in

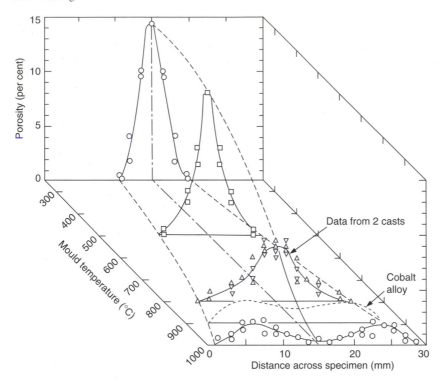

Figure 7.14 *Porosity across an average transverse section of vacuum-cast nickel-based alloy as a function of mould temperature, quantifying the effect shown in Figure 7.13 (Campbell 1969). The effect of solid feeding by the plastic collapse of the section is clear from the shape of the porosity distribution at high mould temperatures.*

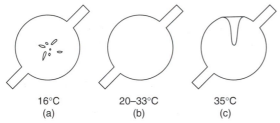

Figure 7.15 *Cross-section of 25 mm diameter wax castings injected into an aluminium die at various temperatures.*

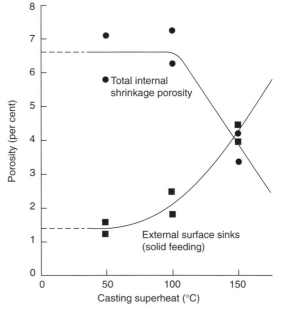

Figure 7.16 *Al–12Si alloy cast into unfed shell moulds showing the full 6.6 per cent internal shrinkage porosity at low casting temperature, giving way to solid feeding at higher casting temperature. Data from Harinath et al. (1979).*

the centre of the casting this means an inward wall movement of only 1 per cent from each of the opposite surfaces. Thus a 25 mm diameter boss would be 0.25 mm small on radius if it were entirely unfed by liquid. In practice, of the 6 per cent volume contraction in aluminium alloy castings, usually at least 4 per cent is relatively easily fed by liquid and interdendritic modes, leaving only 2 per cent or less for solid feeding. Thus dimensional errors resulting from solid feeding reduce to the point at which they are not measurable.

In contrast to the 0.25 mm worst case reduction in radius for the 25 mm diameter feature, if all the shrinkage were concentrated at the centre of the casting, the internal pore would have a diameter of 10 mm. The difference between the extreme seriousness of internal porosity, compared to its

harmless dispersion over the exterior surfaces of the casting, is a key factor to encourage the development of casting processes that would automatically yield such benefits.

It is also worth emphasizing that solid feeding will occur at a late stage of freezing even if the liquid is not entirely isolated. The case has been discussed in the section on interdendritic feeding, and is summarized in Figure 7.6. It is also seen in Figures 7.13 and 7.14. The effect is the result of the gradual build-up of tension along the length of the pasty zone because of viscous resistance to flow. At the point at which the tension reaches a level where it starts to cause the collapse of the casting the region is effectively isolated from the feeder. Although liquid channels still connect this region to the feeder they are by this time too small to be effective to feed.

An experimental result by Jackson (1956) illustrates an attempt to reduce solid feeding by increasing the internal pressure within the casting by raising the height of the feeder. Jackson was casting vertical cylinders 100 mm in diameter and 150 mm high in Cu 85-5-5-5 alloy in greensand. He employed a plaster-lined feeder of only 50 mm diameter (incidentally, failing feeding Rules 2 and 3, which explains why he observed such high porosity in the castings). Nevertheless the beneficial effect of increasing the feeder height is clear in Figure 7.17. His data indicate that, despite the unfavourable geometry, if he had raised his feeder height to 250 mm, all exterior shrinkage would have been eliminated. The interior porosity would have fallen to about 2.0 per cent, almost certainly being the residual effects from the combination of gas porosity, and the residual shrinkage from his poorly sized feeder.

In a study of two small shaped castings in three different Al–Si alloys, of short, medium and long freezing ranges, Li et al. (1998) measured the internal porosity of the castings by density, and the external porosity (the total surface sink effect) by measuring the volume of the casting in water. They found that the internal porosity in the castings in all three alloys was about the same at approximately 1 volume per cent. However, the external sinks grew from an average of 3.1, to 6.4 to 7.5 volume per cent for the short, medium- and long-freezing-range alloys. This significant increase in solid feeding for the long-freezing-range material probably reflects the easier collapsibility of the thinner solidified shell and its internal mesh of dendrites. The more severe internal stress because of the greater difficulty in interdendritic feeding may also be a significant contributor. Conversely, of course, the absence of any corresponding increase in internal porosity confirms that feeding of the castings in the shorter-freezing-range alloys occurred by the simpler and easier more open liquid feeding mechanisms.

A reminder of the possible dangers accompanying solid feeding is probably worth summarizing. Clearly, if the liquid is free from bifilms, the casting will not contain internally initiated pores. However, it may generate:

1. Surface-initiated pores or even
2. Surface sinks.

In the presence of one or more easily opened bifilms, the situation changes significantly:

3. A large interior shrinkage pore in the presence of a bifilm in the stressed region, if the hydrostatic stress becomes sufficiently high and if the stressed volume is large.
4. A population of internal microscopic cracks. This is the subtle danger arising from the usual presence of a population of bifilms in the stressed liquid. In this situation the compact bifilms are subjected to a strong driving force to unfurl. The mechanical properties, especially the ductility and strength, of the casting are thereby impaired in this region. In a nearby region of the casting that had enjoyed better feeding the

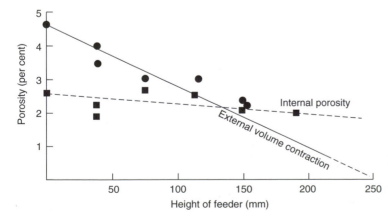

Figure 7.17 Gunmetal casting showing the reduction in solid feeding as liquid feeding is enhanced by extra height and volume of feed metal. Data from Jackson (1956).

ductility and UTS would be significantly improved.

A final personal remark concerning solid feeding that is a source of mystery to the author is the widespread inability of many to comprehend that it is a fact. This lack of comprehension is not easy to understand, in view of the obvious evidence for all to see as surface sinks (even in castings solidified in vacuum) and the fact that isolated bosses can be cast sound provided that the metal quality is good (i.e. few nuclei for pores). Foundries that convert poor filling systems to well-designed filling systems suddenly find that internal porosity and hot tears vanish, but the castings now require extra feeding to counter surface sinks (Tiryakioglu 2001). The increased solid feeding at higher mould temperatures is widely seen in investment castings. The easy collapse of flat plates, especially of alloys weak at their freezing points like Al alloys, explaining their long and difficult-to-define feeding distances. The better-defined feeding distances of steels are the consequence of their better-defined resistance to collapse; their greater strength resisting solid feeding. Additionally, of course, hot isostatic pressing (hipping) is a good analogy of an enforced

plastic collapse of the casting, as is also direct squeeze casting.

7.5 Initiation of shrinkage porosity

In the absence of gas, and if feeding is adequate, then no porosity will be found in the casting.

Unfortunately, however, in the real world, many castings are sufficiently complex that one or more regions of the casting are not well fed, with the result that the internal hydrostatic tension will increase, reaching a level at which an internal pore may form in a number of ways. Conversely, if the internal tension is kept sufficiently low by effective solid feeding, the mechanisms for internal pore formation are not triggered; the solidification shrinkage appears on the outside of the casting. All this is discussed in more detail below.

7.5.1 Internal porosity by surface initiation

If the pressure inside the casting falls, then liquid that is still connected to the outside surface may be drawn from the surface, causing the growth of porosity connected to the surface (Figure 7.18).

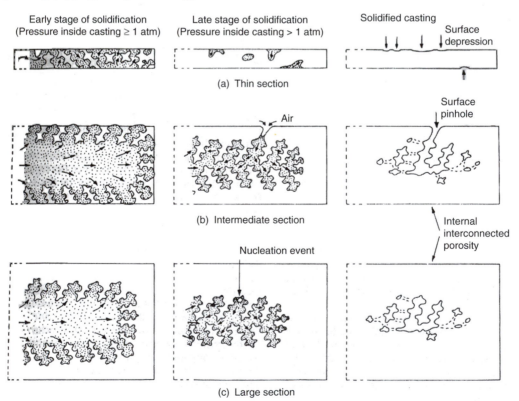

Figure 7.18 *Schematic representation of the origin of porosity as section thickness is increased. The thin sections contain negligible porosity, intermediate sections surface-linked porosity, and thick sections internally nucleated porosity (Campbell 1969).*

The sucking of liquid from the surface in this way naturally draws in air, following interdendritic channels, spreading along these routes into the interior of the casting. The phenomenon is a kind of feeding by a fluid, where the fluid in this case is air. The porosity in the interior of the casting is usually indistinguishable from microporosity caused in other ways: on a polished section it appears to be a series of separate interdendritic pores, whereas in reality it is a single highly complex shaped interconnected pore, linked to the surface.

Figure 7.18 illustrates how the withdrawal of surface liquid is negligible in thin-section castings, that explains why thin sections require little feeding, or even no apparent feeding, but automatically exhibit good soundness. The effect is easily seen in gravity die castings because of their shiny surface when lifted directly from the die. In a section of intermediate thickness the experienced caster will often notice a local frosting of the surface. This dull patch is a warning that interdendritic liquid is being drained away from the surface indicating an internal feeding problem that requires attention.

Pericleous (1997) was the first to predict this form of porosity using a computer model of the freezing of a long-freezing-range alloy. His result is shown in Figure 7.19.

This pore-formation mechanism seems to be much more common than is generally recognized. It is especially likely to occur in long-freezing-range alloys at a late stage in freezing, when the development of the dendrite mesh means that drawing liquid from the nearby surface becomes easier than drawing liquid from the more distant feeder. The point at which liquid may be drawn from the surface may be anywhere for an alloy of sufficiently wide freezing range.

However, in an alloy of intermediate freezing range, the initiation site is often a hot spot such as an internal corner or re-entrant angle. As has been mentioned before, the gravity die caster pouring an Al–Si alloy looks for such defects on each casting as it is taken from the die. If such a 'draw' or cavity is noticed in a re-entrant angle, he immediately doses the melt with sodium. The straightening of the solidification front (Figure 5.42) strengthens the alloy at the corner so that it can better resist local collapse. The outcome is a pore hidden inside the casting if the melt quality is poor so that nucleation is easy. Alternatively, if the melt quality is good, no internal pore can easily form, so that the rise in internal tension will cause more general collapse of the casting. Solid feeding will have been encouraged.

The connection of two opposite surfaces of the casting by pores that are extensively connected internally is one of the major reasons why long-freezing-range alloys cannot easily be used for pressure-tight applications such as hydraulic valves or automobile cylinder heads. In such complex castings it is often difficult to meet the essential requirement that the interior of the casting has a positive pressure at all locations so as to prevent surface-connected internal porosity.

7.5.2 Internal porosity by nucleation

Short-freezing-range alloys, such as aluminium bronze and Al–Si eutectic, do not normally exhibit surface-connected porosity. They form a sound, solid skin at an early stage of freezing, and liquid feeding continues unhindered through widely open channels. Any final lowering of the internal pressure due to poor feeding towards the end of freezing may then

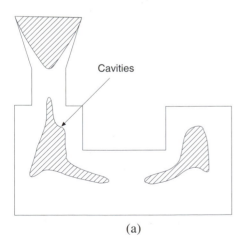
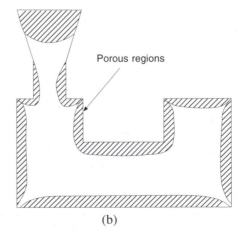

(a)   (b)

Figure 7.19 *Regions of computer-simulated shrinkage porosity: (a) internally in a short-freezing-range alloy; and (b) externally (surface-initiated) in a long-freezing-range alloy. The latter was the first prediction of surface-initiated porosity by computer simulation (after Bounds et al. 1998).*

create a pore by nucleation in the interior liquid. In this case there is clearly no connection to the outside surface of the casting, as illustrated in the larger section shown in Figure 7.18. After nucleation, further solidification will provide the driving force for growth of the pore, which, on sectioning, may be more or less indistinguishable from the surface-initiated type.

In alloys of short freezing range, therefore, porosity is probably normally nucleated, and is concentrated near the centre of the casting, usually well clear of the casting surface. In castings of large length to thickness ratio this is widely referred to as centreline porosity. Thus unless subsequent machining operations cut into the porosity, castings in such alloys are normally leak-tight. (The leak paths commonly provided by folded oxide films or bubble trails generated during a turbulent fill are a separate problem requiring solution by other means such as improved filling, and/or the use of filters.)

Tiwari et al. (1985) has suggested a way of initiating internal porosity in specific regions of castings by the addition of nuclei in the form of fragments of refractory. These foreign particles contain much porosity, so that the growth of pores from such sites proceeds without difficulty. The result is a large internal pore, which, to some extent, can be sited in a chosen location in the casting. Additional feeders or chills are therefore not required, and internal porosity in an unwanted location is avoided. External porosity is also successfully avoided because internal pressure is prevented from falling to negative values. However, as inventive as this technique is, for the majority of castings that are required to be sound throughout, and are required to be free of pieces of refractory, it is, unfortunately, only of academic interest.

7.5.2.1 The nucleation of shrinkage pores

The problem of the nucleation of shrinkage cavities is widely overlooked. Somehow it is assumed that they are fundamentally different to gas pores, and that they 'just arrive'. After all, it is argued, they *must* occur in an unfed isolated volume of liquid, because the concept of shrinkage means that there is a volume deficit. It is assumed that this volume deficit must result in a cavity.

However, we shall go on to show in this section that there really is a difficulty in the initiation of a cavity in a liquid, as we have seen for various analogous systems described in section 7.2.1. If we accept this, then it follows that the liquid is stretched elastically, and the surrounding solid drawn inwards, first elastically, then plastically as the stress in the liquid increases (Campbell 1967). These predictions explain many common observations in the foundry, as will be referred to repeatedly in this work. Only when the stress in the liquid reaches

some critical value, referred to here as a fracture pressure P_f, will a pore appear, growing in milliseconds to a size which will dispel the stress, as a crack would flash across a tensile test specimen as it failed under load.

Interestingly, the analysis by Fisher for the nucleation of gas pores described in section 6.1 applies exactly for the case of the nucleation of cavities. Instead of the diffusion of gas atoms to the embryonic pore we now consider the diffusion of vacancies. Also, for a pore of volume V having zero internal pressure, but in a liquid providing an external pressure P_e, the work required, P_eV, for the formation of the pore is negative. Thus, as before, most embryos shrink and disappear, until a chance chain of additions of vacancies causes it to exceed the critical size. At this point it grows explosively, releasing the tension in the liquid. The fracture pressures calculated from Fisher's theory are identical to those calculated assuming that the diffusing species was a gas. Thus the answers are already given in Table 6.1.

The high tensile strengths of liquids reflect the difficulty of separating atoms that are bonded by strong interatomic forces. Solid metals have similar high theoretical strengths because both the interatomic forces and spacings are similar.

Fracture strengths are, of course, reduced by the presence of weakly bonded surfaces in the liquid. Thus section 6.1.2 on heterogeneous nucleation also applies. Shrinkage cavities are therefore expected to nucleate only on non-wetted interfaces.

Good nuclei for shrinkage cavities include oxides. Complex inclusions that consist of low-surface-tension liquid phases containing non-wetted solids might be especially efficient nuclei, as discussed in section 6.1.2.

Unfavourable nuclei on which the initiation of a shrinkage cavity will *not* occur include wetted surfaces such as carbides, nitrides and borides, and other metal surfaces such as the dendrites that constitute the solidification front. Readers need to be aware that many authors assume, incorrectly, that dendrites are good nuclei for pores (although the reader is referred to other complicating effects listed in Chapter 6). All these substrates are unfavourable for decohesion simply because the bonding between the atoms across the interface is so strong. This is reflected in the good wetting (i.e. small contact angle) of the liquid on these solids.

Interestingly, although oxides are included above as good potential nuclei for pores, this is only true of their non-wetted surfaces. Those surfaces that have grown off the melt, and are thereby in perfect atomic contact with the melt, are not expected to be good nuclei. This illustrates the important distinction between wetting defined by contact angle, and wetting defined as being in perfect atomic contact with the liquid. Ultimately, it is the atom-to-atom contact, and the strong interatomic bonding that is important.

7.5.2.2 Nucleation of combined gas and shrinkage pores

Following Fisher's analysis through once again, considering now that there is gas at pressure Pg on the inside of the pore effectively pushing, and a negative pressure Ps in the bulk liquid effectively pulling, the final result is:

$$P_f = P_g + P_s \qquad (7.6)$$

This equation illustrates how gas and shrinkage cooperate to exceed the critical pressure for nucleation. The significance of this equation can perhaps be better appreciated by deriving an analogous relation as follows.

If we write the condition for simple mechanical equilibrium of a bubble of radius r in a liquid of surface tension γ, where the bubble has internal pressure P_i and external pressure in the liquid P_e, we have:

$$2\gamma/r = P_i - P_e \qquad (7.7)$$

When the pore is of critical size, radius r^*, the internal pressure is the pressure of gas P_g in equilibrium with the liquid, and the external pressure P_e is the (negative) pressure due to shrinkage $-P_S$, then Equation 7.8 becomes the analogue of Fisher's equation:

$$2\gamma/r^* = P_g + P_s \qquad (7.8)$$

The fracture pressures of various liquid metals can then be estimated from this relation assuming that r is approximately one or two atomic diameters, giving the values presented in Table 6.1.

The cooperative action of gas and shrinkage quantified above in Equations 7.6 and 7.8 was predicted by Whittenberger and Rhines (1952). Their ground-breaking concept was enshrined by them in a nucleation diagram as shown in Figure 7.20. We shall devote some space here to a consideration of this insightful map and show how it can be developed to a fascinating degree of sophistication, greatly assisting the description of pore-forming conditions within a casting.

Considering Figure 7.20, for a well-fed casting, $P_s = 0$, and as freezing proceeds the gas is progressively concentrated in the residual liquid, raising the equilibrium gas pressure. Thus conditions in the casting progress along the line ADCE. At point E the conditions for heterogeneous nucleation of a gas pore on nucleus 1 are met, so that a gas pore will pop into existence at that instant. The initial rapid growth of the gas bubble will deplete its surroundings of excess gas in solution, so that conditions in the locality of the bubble will reverse

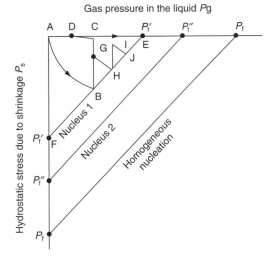

Figure 7.20 *Gas-shrinkage map showing the path of conditions within the residual liquid in the casting in relation to the nucleation threshold for pore formation.*

off the nucleation threshold, back towards D. Thus a second pore will be unable to nucleate in the immediate neighbourhood of the first pore. Other gas pores may nucleate elsewhere, beyond the diffusion catchment area of pore number 1.

Notice also that other heterogeneous nuclei are also present in this situation (threshold 2 in Figure 7.20) but, being less favourable, are not activated.

We turn now to the quite different situation where the casting is free from gas, but is poorly fed. The internal pressure in the casting falls, progressing along the line AF. At F the fracture pressure for nucleation on heterogeneous nucleus 1 is met, and a cavity forms. The hydrostatic tension is explosively released and conditions in the casting shoot back to point A.

In practice, both gas and shrinkage will be present to some degree in the average casting, and both will cooperate, causing the conditions to progress along a curve AB. In the absence of any foreign nuclei it is unlikely that the condition P_f would be met before the casting had solidified completely. This is because the homogeneous nucleation threshold is so far away. (Figure 7.20 is not to scale. If it had been, then the heterogeneous threshold would have been a minute triangle only a millimetre or two in size up at the top left-hand corner of the page, making all the action undecipherable!) Thus the casting would have been sound if no favourable inclusions had been present. However, in the presence of nuclei 1 and 2, the combined gas and shrinkage pore will form at B on nucleus 1. Nucleus 2 will never be needed.

On the formation of a mixed pore at B, the pressure in the liquid immediately reverts to point C. Subsequent slower diffusion of gas into the pore will deplete the immediate surroundings of the pore, causing the local environment to progress to D. Outside this diffusion distance, conditions elsewhere in the casting continue to be free of stress, and may progress to point E at which many more gas pores may nucleate. These will add to the original mixed pore that will in the meantime have continued to grow under the combined action of shrinkage and gas.

The area of influence of the release of the hydrostatic tension due to shrinkage has been assumed in the above discussion to apply to the whole casting. In practice it is true that its influence is vastly greater than that of the depletion of gas. The distance over which shrinkage is depleted is probably a factor of 1000–10 000 times greater than that of gas. This is evident from the fact that diffusion distances of solutes in the liquid are usually measured in micrometres, whereas the distances between layers in typical layer porosity can be measured in millimetres or centimetres. As will be explained in section 7.7.2, each layer in layer porosity represents a separate nucleation event, effectively isolated from its neighbouring layer, because the threads of liquid that connect the two have become so fine as to be practically impassable due to viscous restraint.

Thus in terms of our nucleation diagram, Figure 7.20, the nucleation event at B may cause the reduction in tension in its own locality to drop to nearly zero at point C. However, at a distant location elsewhere, it may drop only to about a quarter of the original tension at G (as is explained in the case of the formation of layer porosity to be discussed in section 7.7.2). A second nucleation event may then occur at H, leading via I to a third pore formation event at J, and so on.

The nucleation diagram is useful for visualizing the effect of such variables as casting under an applied pressure, so that the starting point A is raised, effectively making the nucleation threshold more distant. If it is sufficiently distant then there is a chance that it may not be reached before the casting has solidified.

The diagram also illustrates how it is of no benefit to the soundness of a casting to melt in vacuum and cast in vacuum. This is because both the gas pressure and the pressure in the liquid are both shifted by one atmosphere, in the same direction, meaning that the pore nucleation threshold remains the same distance away, cancelling any advantage. Conversely, melting under vacuum but solidifying under atmospheric pressure is seen to be a benefit, pushing the threshold away by 1 atmosphere pressure.

7.5.3 External porosity

If internal porosity is not formed (either by surface-linked initiation, or by nucleation events) then the lowering of the internal pressure will lead to an inward movement of the external surface of the casting. If the movement is severe and localized, then it constitutes a defect known as a 'sink' or a 'draw'. The feeding of the internal shrinkage by the inward flow of solid is, of course, 'solid feeding' or 'self feeding'.

Adequate internal pressure within the casting will reduce or eliminate solid feeding, so maintaining the shape of the casting and keeping it sound. In such favourable feeding conditions neither internal nor external porosity will occur.

If the remedy of the application of internal pressure is carried out too enthusiastically, the natural inward movement of solid feeding will be reversed, causing the casting to swell. Such growth is common in grey iron castings, and castings that have a high head of metal. Swelling of cast metal is also commonly seen in pressure die casting if the casting is removed from the die before it is fully solid. This is because the gases entrained by the extreme turbulence are under extremely high pressure. The technique is useful for identifying hot spots in the casting (i.e. regions last to freeze, and which therefore require additional local cooling in the die to raise productivity). Gas entrained in hot spots can cause the casting to locally explode when released early from the constraint of the die.

Returning to conventional gravity castings, in real situations it sometimes happens that a certain amount of external collapse of the casting will occur before the internal pressure falls to the level required to nucleate an internal pore. Once the pore is formed, the internal stress will be eliminated so that further solid feeding is arrested. The action of the remaining solidification shrinkage is simply to grow the pore. The final effect of solidification shrinkage on the casting is usually found to be partly external and partly internal as illustrated in Figure 7.4. The balance between external and internal porosity can be widely seen in foundries. Examples are Al–12Si (Figure 7.16) and aluminium-based MMCs (Emamy and Campbell 1996).

7.6 Growth of shrinkage pores

For internal pores that are nucleated within a stressed liquid, the initial growth is extremely fast; in fact it is explosive! The elastic stress in the liquid and the surrounding solid can be dissipated at the speed of sound. The tensile failure of a liquid is like the tensile failure of a strong solid; it goes with a bang.

The audible clicks that are heard as pores nucleate within the melted Tyndall volumes in ice are an exact analogy, and a reminder of the considerable tensile stress that is supported in the liquid water prior to a cavity being nucleated (see section 7.2.1).

In the direct observation of the freezing of Pb–Sn alloys under a glass cover, Davies (1963) recorded the solidification of isolated regions of alloy on cine film at 16 frames per second. It was observed that as these regions shrank, a pore would suddenly appear in between frames. Thus the growth time for this initial growth phase had to be shorter than approximately 60 ms. Almost certainly it was very much faster than this. If in fact the rate of expansion is close to the speed of sound, close to 6 kms^{-1} in liquid aluminium, a pore 1 mm diameter will form a thousand times faster, in 0.1 μs.

After this explosive growth phase the subsequent growth of the pore observed by Davies was more leisurely, occurring at a rate assumed to be controlled by the rate of solidification. Thus in general it seems safe to conclude that the second phase of growth is controlled by the rate of heat extraction by the mould.

For pores that are surface initiated, the initial stress is probably lower, and the puncture of the surface will occur relatively slowly as the surface collapses plastically into the forming hole. Thus the initial rapid growth phase will be less dramatic.

For shrinkage porosity that grows like a pipe from the free surface of the melt there is no initial fast growth phase at all. The cavity grows at all stages simply in response to the solidification shrinkage, the rate being dictated by the rate of extraction of heat from the casting.

7.7 Final forms of shrinkage porosity

7.7.1 Shrinkage cavity or pipe

During liquid feeding, the gradual progress of the solidification front towards the centre of the casting is accompanied by the steady fall in liquid level in the feeder. These linked advances by the solid and liquid fronts generate a smooth conical funnel as shown in Figure 7.21. This is a shrinkage pipe.

It is sometimes called a primary shrinkage pipe to distinguish it from so-called secondary shrinkage

seen on sections cut through the casting. Secondary shrinkage appears to take the form of a scattering of disconnected pores below the primary pipe. Generally, however, it is easy to demonstrate that water, or a dye, poured in the top of the primary pipe will find its way into most, if not all, of the so-called secondary cavities. Thus in reality the secondary shrinkage cavities are only an extension of the primary pipe, which penetrates further into the casting than might be expected at first sight, as Figure 7.21 makes clear.

In the situation where the shrinkage problem is in an isolated central region of the casting, a narrow-freezing-range material will give a smooth single cavity. This is occasionally called a macropore to distinguish it from microporosity. There has been much written to emphasize the differences between these two forms of porosity. However, as will be clear from evidence presented in the next section, there seems to be no fundamental difference between them; one gradually changes into the other as conditions change from skin freezing to pasty freezing.

In the case of the single isolated area of macroporosity, it is important to note that its final location will not be in the thermal centre of the isolated region, as might at first be supposed. This is because the pore will float to the top of the isolated liquid region. Conversely, of course, it can be viewed from the point of view of the liquid. This phase, being heavier, will sink, finding its own level in the bottom of the isolated region. The final position of the shrinkage cavity in this case will be as shown in Figure 7.22. Also, of course, the shape and the position of the porosity can be altered by changing the angle of the casting, since the pore floats (or the residual liquid finds its lowest level). An analogy is a stoppered bottle, partly filled with liquid, which is turned through different angles. The bottle corresponds to the outline of the isolated liquid volume in Figure 7.22.

Note once more that the long parallel walls of the casting give a corresponding long tapering extension of the shrinkage cavity. On a cut section, any slight out-of-straightness of this tubular cavity

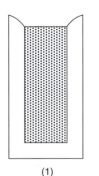 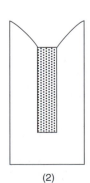 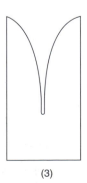 

(1) (2) (3) (4)

Figure 7.21 *Stages in the development of a primary shrinkage pipe. Stage (4) is the appearance of stage (3) on a planar cut section if the central pipe is not exactly straight (i.e. it is not a series of separate pores).*

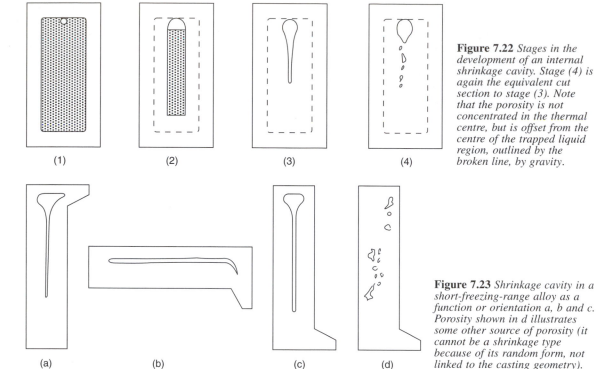

Figure 7.22 *Stages in the development of an internal shrinkage cavity. Stage (4) is again the equivalent cut section to stage (3). Note that the porosity is not concentrated in the thermal centre, but is offset from the centre of the trapped liquid region, outlined by the broken line, by gravity.*

Figure 7.23 *Shrinkage cavity in a short-freezing-range alloy as a function or orientation a, b and c. Porosity shown in d illustrates some other source of porosity (it cannot be a shrinkage type because of its random form, not linked to the casting geometry).*

can once again be easily misinterpreted as dispersed porosity, so-called 'secondary pipe', as it weaves its way in and out of the plane of sectioning. In a casting of more complicated shape the shape of the shrinkage pore will take on a corresponding complexity.

The effect of gravity on the final form and distribution of porosity is illustrated in Figure 7.23. Clearly, the porosity can be moved from one end of the casting to the other simply by making the casting in different orientations, a, b and c. (Figure 7.23d is simply slipped in to emphasize that porosity having no relation to the casting geometry cannot be shrinkage porosity. Perhaps the pores are air bubbles entrained during the pour.)

7.7.2 Layer porosity

Alloys of long freezing range are particularly susceptible to a type of porosity that is observed to form in layers parallel to the supposed positions of the isotherms in the solidifying casting. It is known as layer porosity.

Conditions favourable to the formation of layer porosity appear to be a wide pasty zone arising from long freezing range and/or poor temperature gradients. Poor gradients are typical in alloys of high thermal conductivity (such as the light alloys and copper-based alloys), and in moulds of low rate of heat extraction either because of their high temperature (such as in investment casting) or low

thermal conductivity (as in sand or plaster castings).

Given these favourable conditions, layer porosity has been observed in practically all types of casting alloys, including those based on magnesium (Lagowski and Meier 1964), aluminium (Pollard 1964), copper (Ruddle 1960), steels and high-temperature alloys based on nickel and cobalt (Campbell 1969). An example in steel is shown in Figure 7.24, and in a nickel-based vacuum cast alloy in Figure 7.13.

It has been argued (Liddiard and Baker 1945; Cibula 1955) that layer porosity is the result of thermal contraction, in a manner analogous to hot tearing. Briefly, the theory goes that after the establishment of a coherent dendrite network by the impingement of the dendrite tips, subsequent cooling causes the structure to shrink, imposing tensile stresses on the network, that tears perpendicularly to the stress, i.e. parallel to the isotherms. If the tears are not filled by the inflow of residual liquid, then layers of porosity are frozen into the structure. Baker (1945) attempted to test whether the porosity was the result of thermal contraction on cooling by casting test pieces in a mould designed to accentuate hot tearing. However, no significant increase in porosity was found.

The negative result of this critical experiment by Baker is substantiated by numerous observations, particularly the observations by Lagowski and Meier (1964) on Mg–Zn alloys covering the range of zinc contents up to 30 per cent zinc. These clearly reveal

Figure 7.24 *Radiograph of interdendritic porosity in a carbon steel (Campbell 1969).*

that hot tearing peaks at 1 per cent Zn, well separated from layer porosity that peaks at 6 per cent Zn (Figures 8.20 and 9.6). Furthermore the thermal contraction maximum occurs at 1 per cent Zn, indicating that hot tearing, and not porosity, is associated with the contraction in the solid state.

Finally, the thermal contraction model is seen in any case to be fundamentally flawed; it is not easy to envisage how such differential cooling could arise to pull apart the centre, when the whole casting is solidifying in the absence of significant temperature differences.

A new explanation of layer porosity was put forward by Campbell (1968c). It avoids the difficulties mentioned above because it is based not on thermal contraction in the solid as a driving force, but on the contraction of the liquid on solidification. The mechanism of formation of this defect is easily understood from a consideration of Equations 7.4 and 7.5. The sequence of events in the solidifying casting is shown in Figure 7.25, and is discussed below.

From Equation 7.4 it is clear that the hydrostatic tension increases parabolically with distance x through the pasty zone of length L, as shown in stage 1, Figure 7.25. The stresses continue to

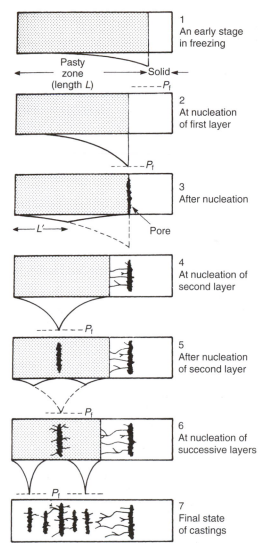

Figure 7.25 *Schematic representation of the formation of layer porosity (Campbell 1968a, b).*

increase with advancing solidification (as R continues to decrease) until the local stress at some point along the parabola exceeds the threshold at which a pore will form (by either nucleation or surface puncture). This threshold is labelled the fracture pressure P_f in stage 2.

As soon as a pore is created by some mechanism, it will immediately spread along the isobaric surface (this surface of constant pressure also probably coincides with an isothermal surface and an isosolid surface), forming a layer and instantly dissipating the local hydrostatic tension. The elastic energy that is available for the initial explosive growth stage is proportional to the difference in areas under the pressure–distance curves before and after this growth that, as we discussed in the previous section, would last probably only microseconds. The energy is clearly proportional to the area $L \times P_f$ under the curve.

As solidification proceeds further, the solidification contraction in the centre of the remaining liquid region is now fed both from the feeder and by fluid (whether residual liquid, gas or vapour) from the newly created pore. This is a slower growth phase for the new pore, extending via channels towards the region requiring feed metal. The new layer-shaped pore effectively provides a free liquid surface, adjacent to which no large stresses can occur in the liquid.

The maximum stress in the liquid has at this stage fallen by approximately a factor of 4, since the effective length L has now approximately halved. However, because of the progressive decrease of R, stress once again gradually increases with time until another pore-formation event occurs as at stage 4.

Further nucleation and growth events produce successive layers until the whole casting is solidified. The final state consists of layers of porosity that have considerable interlinking.

Although these arguments have been presented for the case of porosity being formed only by the action of solidification shrinkage, the reader is reminded that the action of gas and shrinkage in combination also fits the facts well as discussed in Section 7.5.2.2 and illustrated in Figure 7.20.

It is important to observe that layer porosity is quite different to a hot tear. A hot tear is formed by the linear contraction of the casting pulling the grains and/or dendrites apart. When they are sufficiently separated there is insufficient residual liquid to fill the increasing volume, so that a true crack opens up. The crack, of course, supports no load, and represents a serious defect. In contrast, layer porosity is formed by the nucleation of a pore in the residual liquid. The liquid is in a state of hydrostatic tension, so that the pore spreads along the surface of maximum hydrostatic tension, through the liquid phase. The dendrites stay fixed in place. The final defect is a layer-like pore threaded through with dendrites. It is akin to a crack spot-welded at closely spaced spots, and so has considerable tensile strength. On a polished microsection, therefore, a hot tear is clear, whereas porosity as part of a layer is often not easily identified. It only becomes clear on a radiograph when the radiation is aligned with the plane of the layers.

Figures 7.13 and 7.14 show layer porosity in an investment casting as a function of conditions that vary progressively from skin freezing to pasty freezing. It reveals that macroscopic centreline porosity, layer porosity and microscopic dispersed porosity transform imperceptibly from one to the other as mould temperature alone is increased, effectively reducing the temperature gradient during solidification. It is clear, therefore, contrary to many widely held views, that there are no fundamental differences between the various types of macroporosity and microporosity. They are simply different growth forms of shrinkage porosity under different solidification conditions.

Similarly, Jay and Cibula (1956) carried out interesting work on Al–Mg alloys, in which they showed that as the gas content of the alloy was increased, the porosity changed gradually from layer porosity to dispersed pinhole porosity (see later, in Figure 9.21). Thus these two extreme categories of shrinkage and gas microporosity were demonstrated to be capable of being mixed, allowing a complete spectrum of possibilities from pure shrinkage layer type to pure gas-dispersed type.

This merging and overlap of all the different types of porosity makes diagnosis of the cause of porosity sometimes difficult in any particular case. However, it is better to know of this difficulty and thus be better able to guard against falling into the trap of being dogmatic. Section 9.3.2 gives some guidance on the diagnosis of the various types of gas porosity.

A final note of complexity. The above classical description of the formation of layer porosity by the nucleation of pores by the accumulated build up tension because of the difficulty of interdendritic flow, and their subsequent growth along isobaric surfaces as originally proposed, seems to the author to be still probably largely correct. However, now that the presence of bifilms has been realized, it is necessary to take account of the effect they may have in such situations where interdendritic feeding is difficult. It seems likely to suppose that bifilms would interfere with the flow of residual liquid through the dendrite mesh. The close spacing of dendrites, providing support for the films, would ensure that they would be capable of resisting large pressure differences across their surface. Thus bifilms transverse to the flow would halt the upstream flow, but be sucked open by the downstream demand, creating a series of shrinkage cavities arranged generally transverse to the flow direction. Such a scenario is not greatly different from that of the classical theory, and might be difficult to separate experimentally, since the nucleation events in both explanations would be assumed to be bifilms. Ultimately, cleaning the melt will be a certain technique to eliminate layer porosity irrespective of mechanism.

7.7.3 Summary of shrinkage cavity morphologies

The various forms of shrinkage cavity are summarized in Figure 7.26. Without exception, all

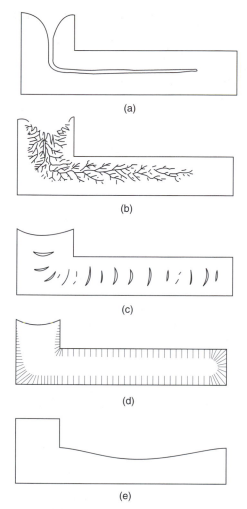

(a)

(b)

(c)

(d)

(e)

Figure 7.26 *Summary of the various types of shrinkage porosity.*

the morphologies are dictated by (i) the *geometry* of the casting, and (ii) *gravity*. These two key features allow shrinkage-dominated porosity to be clearly

differentiated from other sources of porosity. The various forms that shrinkage porosity can take include

1. Centreline porosity is formed in a skin-freezing alloy that has suffered an inadequate supply of liquid from the feeder. The geometry dictates that the pore is closely parallel to the thermal axis of the casting.
2. Sponge porosity. Formed in a long-freezing-range alloy, with adequate temperature gradient, but inadequate feed liquid from the feeder.
3. Layer porosity; the result of inadequate inter-dendritic feeding in a poor temperature gradient. The nucleation of internal porosity indicates a poor cleanness of the liquid metal. Geometry dictates that the pores are closely at right angles to the axis of the casting.
4. Surface-initiated porosity generated in a long-freezing-range alloy in conditions of poor temperature gradient, but good cleanness of the melt.
5. Surface sink (external shrinkage porosity) formed in conditions of no liquid available from the feeder, but good cleanness of melt, resulting in good solid feeding. Notice gravity dictates that the sink is usually sited on the cope surface of the casting.

If the porosity is clearly *not* strongly influenced by the shape of the casting and by gravity (for instance, porosity in random corners, or well away from the thermal axis of the casting as in Figure 7.23d) we can conclude it is *not* shrinkage porosity. (Most often it will be bubble damage.)

Notice that the shrinkage pores are always influenced by the geometry of the casting, whether sometimes parallel to the axis of the casting as in centreline shrinkage, or (perhaps seemingly perversely) sometimes at right angles to the axis of the casting, as in layer porosity. Fortunately, the other major influence to the distribution of shrinkage porosity, gravity, is kinder to our intuition, acting only downwards.

Chapter 8

Linear contraction

Following Figure 7.1, when the contraction in the liquid state and most of the contraction on freezing will have been completed, the casting has cooled sufficiently to develop some coherence as a solid. Further cooling in the solid state will cause the casting to contract as a whole.

The point at which the casting develops its solid-like character is different for long- and short-freezing-range alloys. For short-freezing-range material the point is reached when the casting has developed a solid skin. For the case of long-freezing-range alloys, the point is marked by the development of a coherent skeleton of solid dendrites.

At this stage of freezing, since the casting fits the mould rather well, having been poured in as a liquid (and therefore at that earlier stage fitting perfectly!), something has to give.

It is not the case that either the casting or the mould will yield. Both yield. The action of the casting on the mould causes an equal and opposite action of the mould on the casting. The degree of yielding of the casting and mould depends on the relative strengths of each. Naturally, this varies greatly from one casting/mould system to another.

This chapter is an examination of the various effects that follow from the strains and consequent stresses in the casting/mould system as a result of the linear contraction of the solid casting as it cools. Uniform contraction and distortion are considered separately. A distorted casting can usually be straightened, whereas a casting that is uniformly oversize or undersize is scrap.

8.1 Uniform contraction

The contraction of the casting from its freezing temperature to room temperature can cause the patternmaker sleepless nights. This is because the pattern must be made oversize by an amount known as the contraction allowance (or patternmaker's shrinkage allowance) so that the casting will finally finish at the correct size at room temperature. However, the patternmaker often does not know exactly what allowance to use when he starts to construct the pattern.

This was not so important in the 'bad old days' (perhaps the 'good old days'?) when castings were regarded only as 'rough castings' having plenty of machining allowance. However, now that greater accuracy is being sought in the quest for a 'near net shape' product, the problem has become serious; the patternmaker now finds that the wrong allowance was chosen only after the casting is made! This is the emergency scenario when the tooling has to be modified or remade, but the tooling budget has already been spent, and the deadline for delivery is about to be passed. The contents of this small chapter are recommended to the reader as the only procedure known to the author to avoid this disaster.

The Imperial System of measurements gave us a vast legacy of choice of presentation of this data. For instance, for simple and heavy aluminium alloy castings many are made using an addition of '5/32 inch per foot' to all the linear sizes. This corresponds to the widely used contraction allowance of '1 in 77'. The author has given up these various units and ratios in favour of a simple percentage, in this case 1.30 per cent.

For other aluminium alloy castings with larger internal cores, such as cylinder blocks, the allowance is only 1/8 inch per foot, or 1 in 96, or, as recommended here, 1.04 per cent.

Other aluminium alloy castings such as sumps (oil pans) and thin-walled pipes contract even less. A contraction of 0.60 per cent is common.

Whereas the patternmaker would originally have chosen a special wooden rule whose scale was expanded by the correct amount so that he could read the dimensions directly, without conversion problems, this clearly limited him to specific contraction values. Nowadays, the greater accuracy requires that intermediate values must be chosen, such as 1.15 or 1.20 per cent, etc. for different castings. These are now easy to apply with the use of electronic and digital measuring instruments, which can be programmed for any value of contraction. In fact, modern three-dimensional coordinate measuring machines should be capable of being programmed to read allowing for three different contractions along each of the three perpendicular axes.

The different contractions are the result of different degrees of constraint by the mould during cooling. For instance, in the case of zero constraint, a casting such as a straight bar will contract freely to its maximum extent. We can therefore calculate this rather easily, assuming an average linear contraction. For instance for Al–Si alloys the coefficient of thermal expansion is close to $20.5 \times 10^{-6} \, C^{-1}$ and the total cooling from 660°C to 25°C. From this we can predict the contraction as $20.5 \times 10^{-6} \times 635 = 0.0130$ or 1.30 per cent, in exact agreement with practice.

Turning now to the case of high mould constraint, it is possible to envisage an ideal case in which a large box casting with thin walls was cast around a large, rigid sand core. If the wall thickness of the casting is imagined to be vanishingly thin, like a sheet of paper, then its strength will be negligible and the mould not compressed at all. Thus the casting will not be allowed to contract; its paper-thin walls will be forced to stretch. We can therefore envisage in principle the case of infinite constraint in which the casting contraction is zero.

In practice, of course, the real world is filled with casting/mould combinations that lie intermediate between the case of zero and infinite constraint, that is part way between 0 and 1.30 per cent contraction in the case of aluminium alloy castings.

How can we obtain an estimate of the degree of constraint, so as to be able to predict the contraction allowance exactly? This is the patternmaker's problem. It is a difficult question, to which there is no accurate answer at this time. However, we can obtain a useful estimate by the following procedure, which, fortunately, is good enough for many purposes.

In the case of the straight parallel-sided bar casting made in a sand mould, the casting suffers no constraint. We can define this as being a fully dense metal casting, in the case of aluminium having a density of about $2700 \, kgm^{-3}$. This casting will contract the maximum amount, which for aluminium contracting from its melting point is 1.30 per cent. In contrast, our thin-walled box casting has maximum constraint, contains maximum sand and has (when cast and finally emptied of sand) a density

of practically $0 \, kgm^{-3}$. This simple theory gives us the two extreme points on our calibration curve given in Figure 8.1.

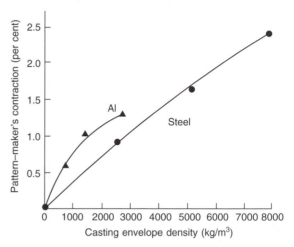

Figure 8.1 *Contraction allowance for aluminium and steel castings as a function of mould constraint.*

Intermediate points are found from measurements on actual castings, taking the weight of the casting divided by the overall volume occupied by the envelope of the casting (Figure 8.2). The envelope is that shape which a tight-fitting rubber balloon would make if stretched over the casting. This gives a measure of the amount of restraining sand it contains, compared with the amount of metal in the casting. Working this out for one or two castings quickly conveys the principle.

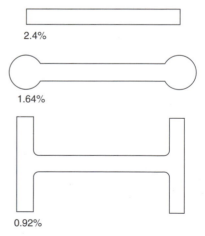

Figure 8.2 *Contraction of three different shapes cast in greensand from the same melt of steel (Steel Castings Handbook 1970).*

When this is carried out accurately, it is found that different varieties of casting are found to lie on a family of approximately parallel curves, all starting and finishing at our theoretical points. Thus the procedure is not absolute, it does not yield a single universal curve. Nevertheless it is a helpful guide in the absence of any better alternative at the present time.

In the case of steel castings the famous result shown in Figure 8.2 can be explained for the first time. Following the procedure that was outlined for aluminium: for the straight bar, the average thermal contraction of steel is around 16×10^{-6} C^{-1} and the cooling range to room temperature is close to 1500°C. Thus the contraction is $16 \times 10^{-6} \times 1500 \times 100$, which is 2.4 per cent, in agreement with the measured value (Figure 8.2). We can plot this at the full density of steel of approximately 7850 kgm^{-3} to define our theoretical point, coincident with our measured point, to define the zero constraint condition. The other theoretical point is, of course, the origin (zero contraction at zero envelope density) as before. Working out the area of the sand mould envelopes of the dumb-bell and H shapes in Figure 8.2, and dividing by the area of the casting, allows us to plot the two remaining

points, giving the nearly linear relation in Figure 8.1.

Work by Briggs and Gezelius (1934) also confirms the 2.4 per cent contraction of a medium-carbon steel, freely contracting in a sand mould. Their results in Figure 8.3 show how increasing constraint, provided in their case by increasingly powerful springs to resist the contraction of the casting, has resulted in lower final values: 1.65, 0.92 and finally down to only 0.47 per cent for the highest restraint which they applied. Clearly the increasing restraint has caused increasing stretching of the casting during cooling. We shall return to the detailed reasons for this later.

In the meantime, until better methods become available, it seems reasonable to suppose that each foundry will have to determine for itself an equivalent of Figure 8.1 for each of its processes. For instance, it is well known that the values of contraction allowance for greensand are dependent on the hardness of ramming. Similarly the percentage of binder in chemically bonded sands significantly affects the contraction of the casting. A standard trick to reduce the constraint provided

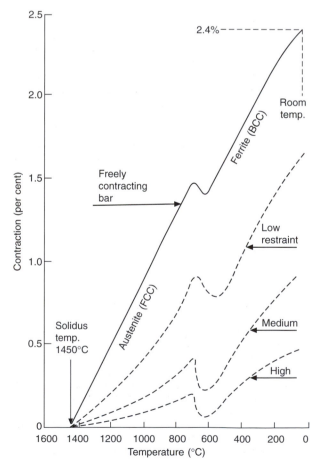

Figure 8.3 *Contraction of 0.35C steel cast into greensand moulds showing the effects of zero and increasing restraint. Data from Briggs and Gezelius (1934).*

by a central core is to reduce its binder level, or to make it hollow.

These relations for sand moulds and sand cores are not expected to apply accurately to metal dies. Here the casting is subject to high mould constraint up to the time of ejection. Clearly the casting contracts freely after this instant.

For gravity and low-pressure dies the contraction varies between 0.75 per cent for low-silicon alloys, and 0.5 per cent for eutectic (approximately Al–11Si) alloys (Street 1977), although much of the industry seems to work generally at 0.6 per cent. These low values reflect the high resistance of the die to the contraction of the casting. However, much lower contractions, close to zero, are occasionally found for thin boxes and window-frame-type castings.

For pressure die casting in magnesium the contraction allowance is 0.7 per cent, whereas for aluminium alloys it is close to 0.5 per cent. The value is at the lower end of the range for gravity and low-pressure dies, indicating the even greater constraint in high-pressure die design.

These figures for the contraction allowance of die castings are the result of the prior expansion of the die from room temperature to its working temperature, and the subsequent contraction of the casting after its ejection out of the die (we shall assume that its contraction while in the die is negligible). We can estimate this quantitatively, taking the die working temperature as roughly 350°C on average (the hot face will be nearer 450°C, but the interior of the die may be water cooled), ambient temperature as 25°C, and the temperature of the casting at ejection approximately 500°C, we have:

$$
\begin{aligned}
\text{Total casting contraction} &= \text{Die expansion} - \text{Casting contraction after ejection} \\
&= (350 - 25) \times 11.7 \times 10^{-6} \\
&\quad -(500 - 25) \times 20.5 \times 10^{-6} \\
&= 0.60 \text{ per cent}
\end{aligned}
$$

Using these rough assumptions and simple logic the answer is seen to be precisely correct. Furthermore, it is clear, as confirmed by experience, that the size of pressure die castings is controlled by the time of ejection of the casting from the die.

For lost-wax castings the problem is compounded by having to take account of the expansion of the aluminium die into which the hot wax is injected, the contraction of the wax pattern, the expansion of the ceramic shell, and the final contraction of the casting itself. This complicated equation is a major source of uncertainty in the accuracy of what is known as 'precision casting'. Regrettably, it is the reason why many so-called precision castings suffer dimensional out-of-tolerance problems.

It is important to note that the pattern contraction allowance can often be different in different directions. Thus each of the x, y and z axes may require a different value.

It is also important to remember that the casting contraction can be greatly affected by the precipitation of gas during solidification. Girshovich et al. (1963) draw attention to this problem in aluminium-, copper- and ferrous-based alloys. The author has experienced a major pick-up of hydrogen gas in a 1000 kg holding furnace after the addition of Sr to an Al–7Si–0.4Mg alloy. Over the next three days the castings suffered up to 3 volume per cent porosity, therefore growing linearly by 1 per cent. The castings were all outside their machining allowance and were consequently scrapped. Strontium addition was immediately discontinued at that time! The subsequent introduction of better degassing techniques has allowed the question of Sr addition to be revisited. In a related experience with steels, Schurmann (1965) describes how rimming steel ingots that fail to develop any significant rimming action tend to grow a solid crust over the ingot top. The internal pressure in the ingot cannot then be relieved by the escape of gas, so the ingot swells.

Finally, the reader would be forgiven for assuming that the size of the casting was fixed when at last it reached room temperature. However, this is rarely true. For instance, Figure 8.4 shows how the common zinc pressure die casting alloys continue to shrink in size for the first six months or so. Alloy A is then fairly stable for the next few decades. Alloy B, on the other hand, starts to reverse its shrinkage after about the first year. At 100°C these changes can be accelerated by about a factor of 250, effectively compressing years into days, as the scale for Figure 8.4 shows.

The zinc die casting alloy ZA27 (Zn–27 per cent Al) shrinks only about one-tenth of the amount of the lower-aluminium zinc alloys (8 and 12 per cent Al), but its expansion is greater, as seen in Figure 8.5. (Figures 8.4 and 8.5 are calculated assuming a factor of 2 increase in reaction rate for every 10°C rise in temperature. Thus the reader can quickly demonstrate that the peak expansion of the ZA27 alloy can be reached in roughly two weeks at 150°C – assuming that the extrapolation is valid.)

Aluminium alloy castings also show size changes. For instance, Al–7Si–0.5Mg alloy contracts by 0.1–0.2 per cent after solution treatment and quenching as alloying elements are taken into solution. The castings grow by 0.05–0.15 per cent during ageing as the alloying elements precipitate once again (Hunsicker 1980).

The Al–17Si alloy used for wear-resistant applications shows considerable growth at temperatures high enough to allow silicon to

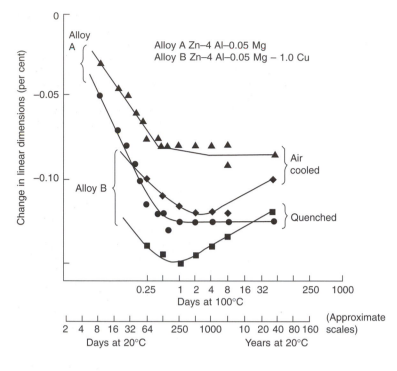

Figure 8.4 *Zinc pressure die casting alloys showing accelerated ageing at 100°C, or slow shrinkage followed by expansion in alloy B taking place over decades at 20°C. Data from Street (1977).*

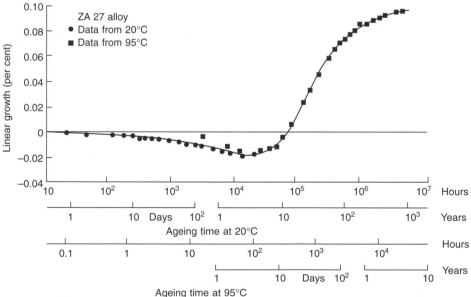

Figure 8.5 *Zn–27Al alloy dimensional changes with time. Data from Fakes and Wall (1982).*

precipitate. This growth in service can be reduced by a pre-age at a minimum of 230°C for 8 hours (Figure 8.6). A treatment of 260°C for 1 hour would be closely equivalent.

Even higher percentage growths are exhibited by white cast irons. At high temperatures, above approximately 900°C, the breakdown of the metastable cementite to stable graphite takes several days. During this period the casting will grow by up to 1.6 per cent (Johnson and Nohr 1970). Grey irons will also grow at temperatures down to 350°C, and growth can be catastrophic if the iron is cycled repeatedly through the ferrite/austenite phase change. Walton (1971) observed a growth of 3.5

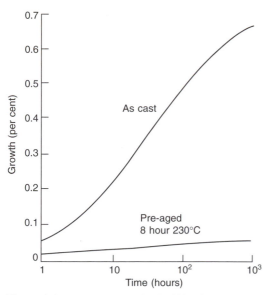

Figure 8.6 *Permanent growth of A390 alloy (Al–17Si) at 165°C with time (Jorstad 1971)*

per cent in only 500 hours in a grey iron subjected to cyclic heating to 800°C. Growth can be further enhanced by the internal oxidation of the material. Angus (1976) gives more details of the growth of cast iron at elevated temperatures, and the means by which it can be controlled by control of structure and chemical composition of the iron.

These are just a few examples of the growth and/or shrinkage of castings that can occur in the solid state because of microstructural changes occurring within the alloy. The casting engineer needs to be on guard against such problems.

The changes cause the patternmaker problems when attempting to decide what contraction allowance to use when constructing the pattern. The decision may be right or wrong, depending on whether the foundry check the dimensions of the casting before or (more correctly) after heat treatment, and will depend on the service conditions.

If the allowance was chosen wrongly, giving an undersized casting in grey cast iron, then a heat treatment may save the day by growing the casting by up to 1 per cent or so. However, such good fortune is rare. The choice of contraction allowance prior to making the casting will remain a difficult and risky decision, and will become more difficult as demands on casting accuracy increase.

8.2 Non-uniform contraction (distortion)

If the casting were to be cooled at a uniform rate and with a uniform constraint acting at all points

over its surface, then it would reach room temperature perfectly in proportion, perhaps a little large, or a little small, but not distorted.

In practice, of course, this utopia is never realized. Usually the casting is somewhat large, or somewhat small, and is not as accurate a shape as a discerning customer would prefer. Occasionally it may be very seriously distorted. We shall examine the reasons for these factors and see to what extent they can be controlled.

8.2.1 Mould constraint

Again, wishing ourselves into utopia, we can envisage that if the constraint by the mould were either zero or infinite, in both cases the casting would be of predictable size and correct shape.

In reality different parts of the casting experience different degrees of constraint by the mould. One of the most common examples of this problem is a simple five-sided box with its sixth side remaining open, as shown in Figure 8.7. The closed face wishes to contract as a straight length, whereas the vertical sides have maximum constraint. The result is a compromise, with the straight side shortening with an effective contraction allowance of perhaps 1.2 per cent in the case of an aluminium alloy casting. The vertical sides will be restrained from pulling inwards, and so have an effective contraction allowance of perhaps only 0.9 per cent, or even less if the walls are very thin. During cooling the casting therefore develops a bowed shape. For a hard, chemically bonded sand mould the camber will be approximately 1 mm in the centre of the long side of an open box 500 × 100 × 100 mm with walls 4 mm thick.

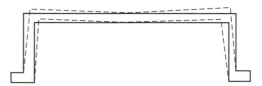

Figure 8.7 *Distortion of an open-sided box casting during cooling as a result of uneven mould constraint.*

The box casting may be cast somewhat straighter by a number of techniques well known to the foundry technologist:

1. The centre core can be made weaker, reducing its constraint on the contraction of the casting. This is achieved either by reducing the percentage addition of binder, or, usually more conveniently, by hollowing out the centre core. The thin shell of sand thereby becomes hotter, giving greater breakdown of the binder in the case of an organic

binder, and so allowing the core to collapse earlier.

2. Tie bars can be connected across the open side of the box, thereby holding the walls in place, and balancing the effect of the contraction of the closed side. The tie bar need not be a separate device. It can be the running system of the casting, carefully sized so as to carry out its two jobs effectively.

This raises the important issue of the influence of the running and feeding system. Unfortunately these appendages to the casting cannot be neglected. They can be used positively to resist casting distortion as above. Alternatively they can cause distortion and even tensile failure as shown in the simple case in Figure 8.8a. Alternatively, if this casting is fed at the flange end, leaving the plate free to contract along its length as shown in Figure 8.8b, the problem is solved. However, note the important point that whether or not casting 'a' has suffered any tensile failure, it will be somewhat longer than casting 'b'. Thus different pattern contraction allowances are appropriate for these two different constraint modes.

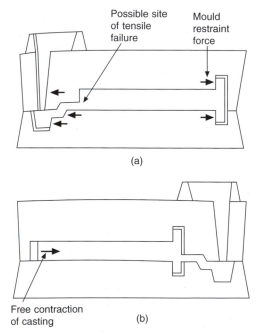

(a)

(b)

Figure 8.8 *(a) Effect of the running and feeding systems imposing constraint on the contraction of the casting. (b) Applying the running and feeding to the opposite end of the casting removes the problem.*

In more complex castings the effect of geometry can be hard to predict, and harder to rectify if the casting is particularly badly out of shape. Especially

for large, thin-walled castings requiring close dimensional tolerance it may be wise to include for a straightening jig in the tooling price.

8.2.2 Casting constraint

Even if the casting were subjected to no constraint at all from the mould, it would certainly suffer internally generated constraints as a result of uneven cooling. The famous example of this effect is the mixed-section casting shown in Figure 8.9a. If a failure occurs it always happens in the thicker section. This may at first sight be surprising. The explanation of this behaviour requires careful reasoning, as follows.

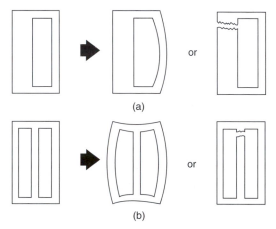

(a)

(b)

Figure 8.9 *(a) Thick/thin-section casting showing tensile stress in the thick section; (b) an even-walled casting showing internal tensile stress.*

First, the thin section solidifies and cools. Its contraction along its length is easily accommodated by the heavier section, which simply contracts under the compressive load since it is hot, and therefore plastic, if not actually still molten. Later, however, when the thin section has practically finished contracting, the heavier section starts to contract. It is now unable to squash the thin section significantly, which has by now become rigid and strong. The result is the bending of the thin section, or the failure in tension of the thick section, which, depending on its temperature, will experience a tensile load. It can therefore stretch plastically, or hot tear, or cold crack (these defect modes are discussed later).

The example shown in Figure 8.9b is another common failure mode. The internal walls of a casting remain hot for longest even though the casting may have been designed with even wall sections. This is, of course, simply the result of the internal sections being surrounded by other hot sections. The reasoning is therefore the same as that for the thick-

/thin-section casting above. The internal walls of the casting suffer tension at a late stage of cooling. This tension may be retained as a residual stress in the finished casting, or may be sufficiently high to cause catastrophic failure by tearing or cracking.

The same reasoning applies to the case of a single-component heavy-section casting such as an ingot, billet or slab, and especially when these are cast in steel, because of its poor thermal conductivity. The inner parts of the casting solidify and contract last, putting the internal parts of the casting into tension (notice it is always the inside of the casting that suffers the tension; the outside being in compression). Because of the low yield point of the hot metal, extensive plastic yielding occurs at high temperature. However, as the temperature falls, the stress cannot be relieved by plastic flow, so that increasing amounts of stress are built up and retained. There has been much experimental and theoretical work in this area. Some of this work will be touched on in section 8.5.

An example shown in Figure 8.10 shows the kind of distortion to be expected from a box section casting with uneven walls. The late contraction of the thicker walls collapses the box asymmetrically (the casting is at risk from tensile failure in the thicker walls, but we shall assume that neither tearing nor cracking occurs in this case). There is clearly some strong additional effect from mould constraint. If the central core were less rigid, then the casting would contract more evenly, remaining more square.

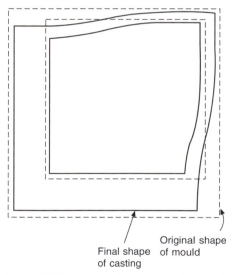

Final shape of casting · Original shape of mould

Figure 8.10 *Distortion in an uneven box section casting because of combined casting and core constraint.*

There is an important kind of distortion seen in plate-shaped castings which have heavy ribs adjoining the edges of the plate, or whose faces are reinforced by heavy-section ribs. It is often seen in thin-section boxes that have reinforcing ribs around the edges of the box faces. The general argument is the same as before: the thin, flat faces cool first, and the subsequent contraction of the heavier ribs causes the face to buckle, springing inwards or outwards. This is known as 'oil-can distortion'; an apt name which describes the exasperating nature of this defect, as any attempts to straighten the face cause it to buckle in the opposite direction, taking up its new reversed curvature. It can be flipped backwards and forwards indefinitely, but not straightened permanently. Once a casting exhibits oil-can distortion it is practically impossible to cure. The effect is more often seen after quenching from heat treatment, where, of course, the rate of cooling is greater than in the mould, and where the casting does not have the benefit of the support of the mould.

Oil-can distortion may be preventable by careful design of ribs, to ensure that their geometrical modulus (i.e. their cooling rate) is similar to, or less than, that of the thinner flat face. Alternatively, the cooling rate from the quench needs to be equalized better, possibly by the use of polymer quenchants and/or the masking of the more rapidly cooling areas.

In ductile iron castings that exhibit expansion on solidification due to graphite precipitation, the expansion can be used to good effect to reduce or eliminate the necessity for feeders, particularly if the casting cools uniformly. Tafazzoli and Kondic (1977) draw attention to the problem created if the cooling is not uniform. The freezing of sections that freeze first leads to mould dilation in those regions that solidify last. Although these authors attribute this behaviour to the mismatch between the timing of the graphite expansion and the austenite contraction, it seems more likely to be the result of the pressure within the casting being less easily withstood by those portions of the mould that contain the heavier sections of the casting. This follows from the effect of casting modulus on mould dilation; the lighter sections are cooler and stronger, and the thicker sections are hotter and thus more plastic. Any internal pressure will therefore transfer material from those sections able to withstand the pressure to those that cannot. The thinner sections will retain their size while the thicker sections will swell. Tafazzoli and Kondic recommend the use of chills or other devices to encourage uniformity.

In a classic series of papers Longden (1931–32, 1939–40, 1947–48, 1948) published the results of measurements that he carried out on grey iron lathe beds and other machine beds. The curvature of a casting up to 10 m long could result in a maximum out-of-line deviation (camber) of 50 mm or more. Longden summarized his findings in a nomogram that allowed him to make a prediction of the camber

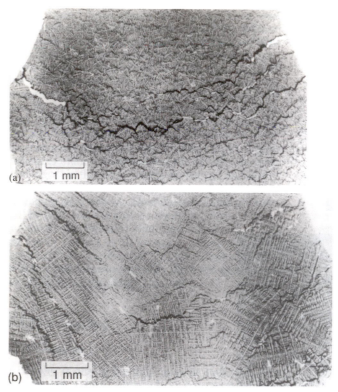

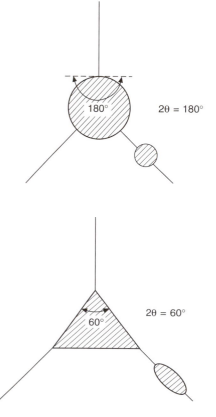

Figure 8.14 *(a) Radiograph of a hot tear in an Al–6.6Cu grain-refined alloy. Dark regions are copper-rich eutectic; white areas are open tears. (b) A radiograph of a hot tear in an Al–10Cu alloy, not grain refined (Rosenberg et al. 1960) (courtesy of Merton C. Flemings).*

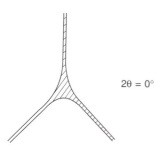

Figure 8.15 *Shapes of the liquid phase at the grain boundary as a function of dihedral angle (Smith 1948, 1952).*

is largely controlled by the relative surface energies of the grain-to-grain interface itself, γ_{gg} and the grain-to-liquid interface γ_{gL}. The balance of forces is:

$$\gamma_{gg} = \gamma_{gL} \cos \theta \qquad (8.1)$$

It is clear that for most values of the equilibrium dihedral angle 2θ the grain boundary liquid assumes compact shapes. However, it will, of course, occupy a greater area of the boundary as its volume fraction increases. The relation between (i) the area of the boundary which is occupied by liquid, (ii) the dihedral angle and (iii) the volume fraction of liquid present is a complicated geometrical calculation which was first tackled by the author (Campbell 1971), subsequently improved by Tucker and Hochgraf (1973), and, finally, comprehensively worked out by Wray (1976a). Hochgraf (1976) went on later to develop a fascinating study of the conditions for the spread of the liquid phase under non-equilibrium conditions, where the dihedral angle becomes effectively less than zero.

The importance of the dihedral angle being zero for complete wetting is illustrated in the work of Frederiksson and Lehtinen (1977). They observed the growth of hot tears in the scanning electron microscope. In Al–Sn alloys the liquid tin wetted the grain boundaries of the aluminium, leading to a brittle failure when subjected to tension. In Al–Cd alloys, the liquid cadmium at the grain boundaries did not wet and spread, but remained as compact pools. These alloys therefore failed by ductile fracture.

There have been a number of observations of failure by hot tearing where, on subsequent observation under the microscope, the fracture surface has been found to exhibit separate, nearly spherical droplets that appear to be non-wetting towards the fracture surface. This has been seen in systems as different as Al–Pb (Roth et al. 1980) and Fe–S (Brimacombe and Sorimachi 1977; Davies and Shin 1980). It seems certain that the liquid phase would wet a normal grain boundary. It is not clear therefore whether the observation is to be explained by the subsequent de-wetting of the liquid phase after the crack is exposed to the air, or because the boundary consists of a poorly wetted bifilm.

8.3.3 Pre-tear extension

While the casting is cooling under conditions in which liquid and mass feeding continue to operate, then clearly no tearing can occur. The problem starts when grains grow to the point at which they collide with each other, but are still largely surrounded by residual liquid.

Patterson and co-workers (1967) were among the first to consider a simple geometrical model of cubes. We shall develop this concept further as illustrated in Figure 8.16. It is clear that for grains of average diameter a separated at first by a liquid film of thickness b, the pre-tear extension ε is approximately:

$$\varepsilon = b/a \qquad (8.2)$$

For a three-dimensional cube model the reader can easily confirm the further relation:

$$b/a = f_L/3 \qquad (8.3)$$

where f_L is the volume fraction of liquid. (For either the cube or the hexagon models in two dimensions the relation becomes $b/a = f_L/2$) Thus for 3 to 6 per cent residual liquid phase we have 1 to 2 per cent extension prior to the impingement of the grains. The pre-tear extension being proportional to the amount of liquid present is an observation confirmed many times by experiment. Furthermore, those alloys with large amounts of eutectic liquid during freezing are usually free from hot tear problems probably for this reason, that there is plenty of extension that can be accommodated prior to any danger of the initiation of a tear.

Also, for a given amount of liquid present, the extension is inversely proportional to the grain size. Thus for finer grains, more strain can be accommodated by easy slip along the lubricated boundaries without the danger of cracking.

After the grains have impinged, a certain amount of grain boundary sliding may continue, as we shall discuss below, although this later phase may contribute only a limited amount of further extension.

Even in the case of the solidification of pure metals, the grain boundaries are known to have a freezing point well below that of the bulk crystalline material (see, for example, Ho et al. (1985) and Stoltze et al. (1988)). The presence of liquid at the grain boundaries even in pure metals, but perhaps only a few atoms thick, may help to explain why some workers have found tearing behaviour at temperatures apparently below the solidus temperature. However, many of the observations are also explainable simply by the presence of minute traces of impurities that have segregated to the grain boundaries. The two effects are clearly additive.

Because of the presence of the grain boundary film of liquid, bulk deformation of the solid will occur preferentially in the grain boundaries, so long as the strain is below a critical value (Burton and Greenwood 1970). This explains why the extension of the solid prior to fracture can be accounted for completely by the sum of the effects of grain boundary sliding plus extension due to the opening up of cracks (Williams and Singer 1968).

Later, during grain boundary sliding where the grains are now in contact, there has to be some deformation of the grains themselves. Novikov et

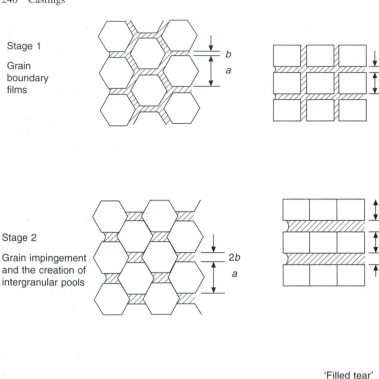

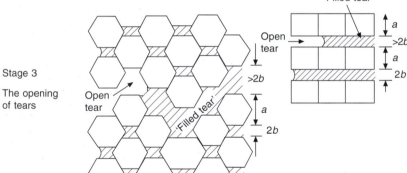

Figure 8.16 *Hexagonal and square models of grains, size a, surrounded by a liquid film of thickness b. The development of isolated regions of segregates is seen as tensile strain is applied. When this finally exceeds the ability of the liquid film to accommodate it, the action of the continued extension drains the liquid film, forming a tear. Compare the model with actual tears shown in Figures 8.13 and 8.14.*

al. (1966) found by careful X-ray investigation that the deformation is confined to the surface of the sliding grains. In addition, at a temperature close to the melting point, recovery of the grains is so fast that they do not work harden. Because they remain in a relatively soft and unstrained condition the general flow of the bulk material can continue relatively easily. Thus although the flow is actually controlled by bulk deformation of the grains, the appearance under the microscope is simply that of the sliding of the grains along their boundaries.

It is necessary to keep in mind that the total extension due to the various kinds of grain boundary sliding (whether 'lubricated' or not) amount to only perhaps 1 or 2 per cent strain. Further strain of at least this magnitude arises during the extension of the crack itself, as is discussed below.

8.3.4 Strain concentration

It was Pellini in 1952 that drew attention to the concentration of strain that could occur at a hot spot in a casting. It is instructive to quantify Pellini's theory by the following simple steps.

If the length of the casting is L, and if it has a coefficient of thermal expansion α, during its cooling by ΔT from the liquidus temperature it will contract by an amount $\alpha \Delta T L$. If all this contraction is concentrated in a hot spot of length l, then the strain in the hot spot is given by:

$$\varepsilon = \alpha \Delta T L / l \qquad (8.4)$$

Clearly, in the hot spot the casting contraction strain is increased by the factor L/l.

For a casting 300 mm long and a hot spot of

approximately 30 mm length at its end, the strain in the hot spot is concentrated ten times. This would be expected to be a fairly typical result – although it seems possible that strain concentrations of up to a hundred or more may sometimes occur.

It is interesting to note that the problem in the hot spot depends on the amount of strain concentrated in it, and this depends on the size of the adjacent casting and the temperature to which it has cooled while the hot spot is in a weak state.

We can clarify the size of the problem by evaluating an example of an aluminium casting. Assume that $\alpha = 20 \times 10^{-6}$ C^{-1} and that the casting has cooled 100°C. If its contraction is hindered, the strain that will result is, of course, $20 \times 10^{-6} \times 100 = 0.002 = 0.2$ per cent. This level of strain puts the material as a whole above the elastic limit. (In materials that do not show clear yield points, the yield stress is often approximated to the so-called proof stress, at which 0.1 or 0.2 per cent permanent strain remains after unloading.) In the hot spot, therefore, if the strain concentration factor lies between 10 and 100, then the strain will be between 2 and 20 per cent. These are strains giving an amount of permanent plastic extension that is relatively easily withstood by sound material. However, in material that is weakened by the presence of bifilms at the grain boundaries, and which can withstand typically only 1 or 2 per cent of strain prior to failure, as we shall see below, it is no wonder that the casting suffers problems.

In addition to the consideration of the amount of strain concentrated into the hot spot, it is also necessary to consider how many grain boundaries the hot spot will contain. If the grain size is coarse, the hot spot may contain only one boundary, with almost certain disastrous consequences, because all the strain will be concentrated in that one liquid film. If the hot spot contains fine grains, and thus many boundaries, then the strain is more widely distributed. We may quantify this, since the number of grains in the length l of the hot spot is l/a for grains of diameter a. Hence if we divide the strain in the hot spot (Equation 8.4) by the number of boundaries in it, then we have the strain per boundary ε_b

$$\varepsilon_b = \alpha \Delta T L a / l^2 \qquad \textbf{(8.5)}$$

It is clear that to reduce the strain that is trying to open up the individual grain boundaries, reduced temperature differences, smaller overall lengths between hot spots, and finer grain size all help. However, Equation 8.5 reveals for the first time that the most sensitive parameter is the length l of the hot spot; as this is halved, the grain boundary strain is increased four times.

8.3.5 Stress concentration

The problem of how sufficient stress arises during

cooling to initiate and grow the hot tear may not be relevant. This is because the forces available during cooling are massive, greatly exceeding what is necessary to create a failure in the rather weak casting. Thus we may consider the forces available as being irresistible, forcing the casting to deform. Since this deformation will always occur, the question as to whether a hot tear will arise is clearly not controlled by stress, but must depend on other factors, as we shall discuss in this section.

Nevertheless, although overwhelmed, the forces of resistance offered by the casting are not quite negligible. Guven and Hunt (1988) have measured the stress in solidifying Al–Cu alloys. Although the stresses are small, they are real, and show a release of stress each time a crack forms. The loads at which failure occur are approximately 50 N in a section 20 mm × 20 mm. Thus the stresses are approximately 0.1 MPa (compared to a strength of over 100 MPa at room temperature). Also, as an interesting detail, a simultaneous change in the rate of heat transfer across the casting/mould interface was detected each time the force holding the casting against the mould was relaxed.

In rough agreement with Guven and Hunts' results, Forest and Berovici, (1980) carried out careful tensile tests and found that an Al–4.2Cu alloy has a strength of over 200 MPa at 20°C, that falls to 12 MPa at 500°C, 2 MPa at the solidus temperature, and finally to zero at a liquid fraction of about 20 per cent.

As we have mentioned before, the other stress that may be present could be a hydrostatic tensile stress in the liquid phase. Although this may contribute to the nucleation of a pore, which in turn might assist the nucleation of a tear, the presence of a hydrostatic stress is clearly not a necessary condition for the formation of a tear, as we have discussed earlier. We need a uniaxial tensile stress to create a tear.

One final point should be emphasized about stresses at these high temperatures. Because of the creep of the solid at high temperature, any stress will depend on the rate of strain. The faster the solid is strained, the higher will be the stress with which it resists the deformation.

Zhao and colleagues (2000) have determined the rheological behaviour of Al–4.5Cu alloy, and thereby have determined the stress leading to the critical strain at which hot tearing will cause failure. This novel approach may require the densities of bifilms to be checked for similarity between their rheological sample and the hot tearing test piece, which is clearly poured rather badly.

8.3.6 Tear initiation

Probably the most important insight into the problem of tear initiation was provided by Hunt (1980) and

Durrans (1981). Until this time the nucleation of a tear was not widely appreciated as a problem. The tear was assumed simply to grow! The experiment by these authors is an education in profound insight using a simple technique.

These researchers constructed a transparent cell on a microscope slide that enabled them to study the solidification of a transparent analogue of a metal. The cell was shaped to provide a sharp corner around which the solidifying material could be stretched by the turning of a screw. The idea was to watch the formation of the hot tear at the sharp corner.

The amazing outcome of this study was that no matter how much the solidifying material was stretched against the corner, it was not possible to start a hot tear in clean material: the freezing mixture continued to stretch indefinitely, the dendrites continuing simply to move about and rearrange themselves.

However, on the rare occasion of the arrival of a small inclusion or bubble near the corner, then a tear opened up immediately, spreading away from the corner. In their system, therefore, hot tearing was demonstrated to be a process dependent on nucleation. In the absence of a nucleus a hot tear did not occur. This fact immediately explains much of the scattered nature of the results of hot-tearing work in castings: apparently identical conditions do not give identical tears, or at times even any tears at all.

It is necessary to remember that in Hunt and Durrans' work the liquid would wet the mould, adhering to the sharp corner, and so would require a volume defect to be nucleated; the defect would not be created easily. In the case of a casting, however, a sharp re-entrant corner may have liquid present at the casting surface, but the liquid will not be expected to wet the mould. In fact there is the complication that the liquid will be retained inside the surface oxide. The drawing of liquid away from this surface, analogous to the case of surface-initiated porosity, would represent the growth of the crack from the surface, and may involve a rather minimal nucleation difficulty. Thus in the case of cracks initiated at the surface, Hunt and Durrans' observation may not apply universally. However, the concept is still of value, as we shall discuss immediately below. In the case of internal hot tears their observation remains crucial. However, in the case of alloys that form strong oxide skins there may still be a difficulty in nucleating a tear on the underside of the surface film, making surface initiation difficult in most casting alloys.

However, even at the surface of a casting in an alloy that does not form a surface film, the initiation of a tear may not be straightforward. It is likely that the tear will only be able to start at grain boundaries, not within grains. This is because the dendrites composing the grain itself will be interconnected, all having grown from a single nucleation site. Dendrites from neighbouring grains will, however, have no such links, and in fact the growing together and touching of dendrite arms has not been observed in studies of the freezing of transparent models. The arms are seen to approach, but final contact seems to be prevented by the flow of residual liquid through the gap. Thus if a grain boundary is not sited conveniently at a hot spot, and where strain is concentrated, then a tear will be difficult to start. This will be more common in large-grained equiaxed castings, as suggested by Warrington and McCartney (1989).

If a grain boundary is favourably sited, it may open along its length. However, on meeting the next grain, that in general will have a different orientation, further progress may be restrained. Thus a tear may be limited to the depth of a single grain. The effect can be visualized as the first stage of the spread of the tear in the hexagon grain model shown in Figure 8.16. Considerably more strain will be required to assist the tear to overcome the sticking point in its further advance beyond the first grain.

For the case of columnar grains, the boundaries at right angles to the tensile stress direction will provide conditions for easy initiation of a tear along such favourably oriented grain boundaries. The effect is analogous to the rectangular grain model in Figure 8.16.

For fine-grained equiaxed material where the grain diameter can be as small as 0.1 to 0.2 mm, the dispersion of the problem as a large number of fine tears, all one grain deep, is effectively to say that the problem has been solved. This is because the crack depth would then be only approximately 0.1 mm. This is commensurate with the scale of surface roughness, because average foundry sands also have a grain size in the range 0.1–0.2 mm. The fine-scale cracking would have effectively disappeared into the surface roughness of the casting.

Nevertheless it is fair to emphasize that the problem of the nucleation of tears has been very much overlooked in most previous studies. Nucleation difficulties would help to explain much of the apparent scatter in the experimental observations. A chance positioning of a suitable grain boundary containing, by chance, a suitable nucleus, such as a folded oxide film, would allow a tear to open easily. Its chance absence from the hot spot would allow the casting to freeze without defect; the hot spot would simply deform, elongating to accommodate the imposed strain.

In practice there is much evidence to support the assertion that most hot tears initiate from entrained bifilms. The author has personally solved most hot-tearing problems he has encountered in foundries by simply improving the design of the casting filling system. The approach has generated almost universal disbelief. However, when implemented systematically in an aerospace foundry, all hot-tearing problems in the difficult Al–4.5Cu–0.7Ag (A201) alloy disappeared, to be replaced by surface sink problems. In comparison to the hot tears, the surface sinks were welcomed, and easily dealt with by improved feeding techniques (Tiryakioglu 2001).

The study by Chadwick and Campbell (1997) of A201 alloy poured by hand into a ring mould containing a central steel core, showed that failure by hot tearing in such a constrained mould was almost guaranteed. Conversely, when the metal was passed through a filter, and caused to enter the mould uphill at a speed of less than 0.5 ms^{-1} to ensure the avoidance of defects, no rings exhibited hot tears. (Bafflingly, because the experiment had led to a hot-tearing test in which no specimens had hot torn, Chadwick declared the test a failure.)

A similar study was repeated for Al–1 per cent Sn alloys by Chakrabarti (2000). Surfaces of hot torn alloys illustrate the brittle nature of the failure in this alloy (Figure 8.17). The alloy has such a large freezing range, close to 430°C (extending from close to pure Al at 660°C down to nearly pure Sn at 232°C) that in the hot tear ring test the alloy appears even more susceptible to failure by hot tearing than A201 alloy. When subjected to the uphill filled version of the ring test most castings continued to fail. However, about 10 per cent of the castings solidified without cracks. Once again, the existence of even one sound casting would be nothing short of amazing.

Further evidence can be cited from the work of Sadyappan *et al.* (2001) who demonstrated that

their as-melted Al alloy gave many and large hot tears, whereas after cleaning the metal by degassing they observed only a few small tears. Dion *et al.* (1995) found that in their very turbulently filled castings of yellow brass, the addition of aluminium to the alloy promoted hot tearing, as would be expected from the presence of the entrained alumina film.

8.3.7 Tear growth

We have touched on the problem of tear growth in the previous section. However, it bears some repeating in a section devoted solely to the issue of growth because (i) the birth of the hot tear and (ii) its growth, sometimes to awesome maturity, are really separate phenomena.

The evidence is growing that tears are closely associated with bifilms. It remains to clarify the nature of the link. For instance, (i) do tears initiate on bifilms and subsequently extend into the matrix alloy? or (ii) do tears grow along bifilms, or (iii) do bifilms constitute the tears, so that the growth of the tear is merely the opening of the bifilm so that the defect becomes obvious. The evidence is accumulating that the important mechanisms are (ii) and (iii) and that (ii) and (iii) are really the same mechanism.

The easy growth in columnar grains where the direction of tensile stress is at right angles to the grain boundaries has been mentioned. Spittle and Cushway (1983) observed that the linear boundary formed between columnar crystals growing together from two different directions was an especially easy growth route for a spreading crack. This is confirmed

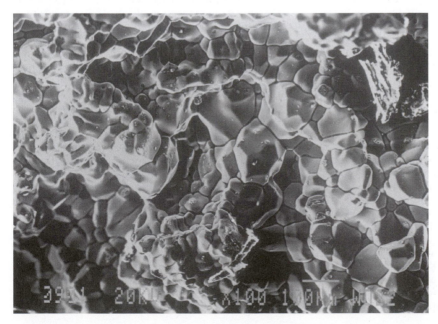

Figure 8.17 *Hot tear surface of an Al–1 per cent Sn alloy (courtesy of Charabarti 1999).*

by experience in the rolling industry, where the diagonal plane from the corners of rectangular ingots, defining the joint plane of the two sets of columnar grains from the two adjacent sides, is a common failure plane during the early reduction passes. The problem is reduced by rounding the corners or reducing the levels of critical impurities. In steel ingots the significant impurities are usually sulphur and the so-called tramp elements such as lead and tin.

The explanation in terms of bifilms for film-forming alloys (possibly applying to the steels cited above if they were deoxidized with aluminium) is that columnar grains push the bifilms into the intergranular spaces, so that failure along these surfaces is to be expected. The presence of an invisibly thin film organized into position between dendrite arms explains the fracture surface seen in Figure 8.13a. The fracture follows the film, the fracture surface exhibiting steps at integral numbers of dendrite arms, as explained in Figure 2.41. Naturally, the hot tear morphology is also seen in the room temperature fracture seen in Figures 2.42 and 2.43. Clearly, after being flattened out by the growth of dendrites, the bifilm is simply a hot tear waiting to be opened and so be revealed. If it is not opened during solidification, then it can wait and be opened later in a tensile test, or, more worryingly, opened in a failure in service.

X-ray radiographs by Fox and Campbell (2000) (Figure 2.46) of Al alloys in which the bifilms have been opened up by the action of reduced pressure show the films near to cast surfaces to be organized at right angles to the surface by the pushing action of the growing columnar grains.

In general, the bifilms and the consequential porosity are sited at grain boundaries. However, some films appear to traverse grains simply because the grains occasionally grow so as to impinge on a bifilm edge-on. In all cases, however, the individual dendrites that constitute a grain never cross a bifilm. They cannot grow through air.

Warrington and McCartney (1989) confirm the findings of Spittle and Cushway (1983) when they note that fine equiaxed grains also promote easy growth conditions for a hot tear. This seems to be because the tear can propagate intergranularly along a path that, because of the fine grain size, can remain almost perpendicular to the applied stress on a macroscopic scale. In terms of bifilms, the result is simply the observation of the opening of that bifilm favourably oriented with respect to the stress direction, of the many bifilms to be expected traversing the structure between small grains.

Conversely, coarse equiaxed grains gave increased resistance to the spread of the crack. In this case the bifilms would be segregated to planes sometimes well away from the stress direction, causing greater plastic deformation in the return of

the crack to its average growth direction. In addition, of course, the distance travelled by the crack would be significantly increased.

The question of the amount of plastic work that is expended during the propagation of a tear is interesting. The work of deformation is easily shown to be of the order of at least 10^4 times greater than the work required to create the newly formed surfaces of the tear. Thus arguments based on the effect of the surface energy of the crack limiting its growth (as in the case of a classical Griffiths crack in a brittle solid such as glass) are clearly not relevant in the case of the failure of plastic solids such as metals at their melting points.

Novikov and Portnoi (1966) draw attention to the fact that, despite the rather brittle appearance of the fracture, the hot torn surfaces cannot usually be fitted back together again, confirming the expectation that considerable plastic deformation occurs during hot tearing. Furthermore, they found that the gap between the poorly fitting surfaces corresponded almost to the total elongation, indicating that the elongation was associated almost entirely with crack propagation in their work. The further implication was that (i) nucleation of the crack was easy, since high stresses would have given high elongations prior to cracking, and (ii) the pre-tear extension due mainly to grain boundary sliding was limited in their experiments. These observations are again consistent with propagation along a bifilm. The brittle, intergranular appearance of the fracture surface is typical of a crack that has followed the central fold of a double oxide film. The pre-existing film has been automatically pushed into grain boundaries during the growth of the grains.

Although high stresses cannot be envisaged at the stage of the nucleation of a crack (because crack nucleation will find easy start sites such as the opening up of bifilms), when the crack has started to grow the sharpness of its tip ensures that high stress is available locally at this point. For large-grained material, therefore, the occasional absence of a favourably oriented grain boundary can be expected to result in further propagation by two means:

1. The continuation of the crack across the grain that blocks its path. Such occasional transgranular growth has been observed by Davies (1970) in models using low-melting-point Sn–Pb alloys. He saw that in such cases the crack followed the cell boundaries, which were clearly the next best route for the crack. Bifilms would be expected in both of these interfaces of course, and present a similarly cogent explanation.
2. The crack either renucleates a short distance ahead in a favourably oriented grain boundary, or more likely, travels around the grain by a path out of the plane of the section, appearing

once again a little ahead. Fredriksson and Lehtinen, (1977) have directly observed such behaviour by pulling specimens of Al–Sn alloys in the scanning electron microscope. As the crack continues to open, the intervening region of obstructing grain is seen to deform plastically, as might a collapsing bridge. The stepwise propagation of the crack, linked by plastic bridges at the steps, explains the irregular, branching appearance that is a characteristic feature of hot tears. The failure of plastic bridges between randomly sited bifilms explains the observations similarly.

Returning to our geometrical model, at the end of the pre-tear extension period the residual liquid is separated into pools between grains, as is seen in Figure 8.16. The corresponding pools in the real casting are seen in Figure 8.14. These compact segregates must pose a problem for subsequent solution heat treatments, even though their existence does not seem to have been previously recognized. At this point further strain must cause some concentration of failure at a weak point in the structure of the solidifying casting. If a grain near the surface is separated from its neighbour by the presence of a bifilm, the growth of the crack can then occur with little further hindrance. The necessarily irregular nature of its progress mirrors that of the real crack; once again, compare Figures 8.16 and 8.14. Straighter portions of the crack resemble the cube model in Figure 8.16.

A potentially important feature of the hot-tearing literature needs to be raised. From Figures 8.13, 8.14 and 8.16 it is quite clear that the portions of the cast structure in which the dendrites have separated but which still contain residual liquid have always contained residual liquid. This may seem self-evident. However, the casting literature is full of references to 'healed' hot tears, meaning tears containing residual liquid, but implying that the tears were once empty, and, fortuitously, were somehow filled by an inflow of liquid. Whether the term 'healed hot tears' is really intended to imply original emptiness and subsequent refilling is not clear but the term is misleading and would be better discontinued. The term 'filled tear' is more explicit. 'Un-emptied' tear would be even more accurate, but hardly an attractive name! If it solidifies while still full, as in Figure 8.13b, it will constitute a region of segregate, but still be relatively strong. Only if it becomes empty does it become a major defect meriting the name of hot tear.

Thus one type of hot tear is simply the separation of the grains to the point at which the residual liquid is no longer capable of keeping the tear filled as in our simple model (Figure 8.16), and as seems to be shown in Figure 8.14a.

However, there seems to be another important

possibility. It may be that the *filled tear* is simply the site of the residual liquid, now constituting a segregated region between grains that have been separated by uniaxial strain. This would contrast with the *hot tear* whose conditions for formation are identical, but the region between the grains now contains one or more bifilms that, on separation of the grains, also separates the halves of the bifilms.

Both of these mechanisms seem possible. However, the first is expected to be more difficult to grow, the grains deforming plastically as the crack attempts to propagate from grain to grain. The presence of the bifilm in the second mechanism is expected to create a defect of a serious size with ease.

8.3.8 Prediction of hot-tearing susceptibility

Over the years there have been many attempts to provide a useful working theory of hot tearing. Recently the attempts have narrowed to a few serious contenders. The exercise that has been found to be most useful to discriminate between them has been the attempt to predict susceptibility to hot tearing as a function of composition for binary alloys.

This is a useful test in alloy systems that display a eutectic. At zero solute content the theory has to contend with a pure metal; at low solute contents only solid solution dendrites are present; above a critical solute content eutectic liquid appears for the first time, steadily increasing to towards 100 per cent as the solute content increases towards the eutectic composition. The ability to deal with all of these aspects across a single alloy system constitutes a searching test of any theory, and covers the majority of solidification conditions in real castings.

A typical experimental result is shown in Figure 8.18. It reveals a steeply peaked curve that Feurer (1976) has called a lambda curve, after the shape of the Greek capital letter that takes the form of an upturned V. The problem is to find a theoretical description that will allow the lambda curves to be simulated for different alloy systems.

It is salutary to note how far wrong is any prediction from the equilibrium diagram. Here the maximum freezing range would be predicted to be at 5.7Cu, which might lead the unwary to believe that the maximum problem in porosity and hot tearing should be at this copper content. From Figure 8.18 the problem in hot tearing is clearly centred on a rather dilute alloy of approximately only 0.5Cu. Any problems of hot shortness have almost disappeared on reaching 5.7Cu!

It is interesting to look at the prediction by Campbell (1969) that deals with the analogous problem of porosity in a spectrum of binary alloy compositions. Here the relative hydrostatic tension developed by the flow of feed metal through the

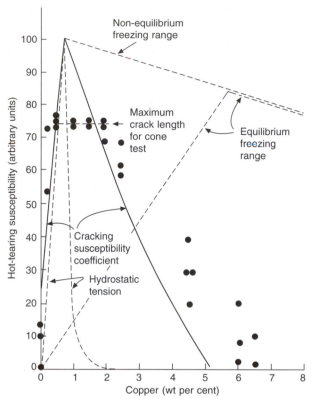

Figure 8.18 *Hot-tearing response of Al–Cu alloys, showing a peak (necessarily extrapolated somewhat) at approximately 0.7Cu from the conical ring die test of Warrington and McCartney (1989), compared to various theoretical models. Freezing ranges and hydrostatic tension by Campbell (1969); CSC by Clyne and Davies (1977).*

dendrite mesh was calculated. The form of the relationship calculated for the Al–Cu system is shown in the figure. This particular result is based on the assumption that the residual liquid is 1 per cent by volume. The peak is almost exactly in the correctly predicted location, confirming the fundamental importance of the arrival of eutectic liquid at that critical concentration of solute. The remainder of the curve follows the experimental data poorly. This is not surprising, since hydrostatic tension and hot tear formation are not expected to be closely related, as we have discussed above. The hydrostatic tension falls steeply with the arrival of eutectic liquid, dramatically reducing shrinkage porosity as seen in the original work, but not reducing hot tearing. Neither is hot tearing well related to the non-equilibrium freezing range, as is also clear in Figure 8.18.

The theoretical approach by Feurer (1976) that appears to explain the form of the lambda curves can be discounted because it is based on the modelling of liquid flow and hence the development of hydrostatic stress, not uniaxial tension (Campbell and Clyne 1991).

An alternative theoretical approach to hot tearing was proposed by Clyne and Davies (1979). They implicitly assume that the failure is the result of uniaxial tension, but point out that strain applied

during the stage of liquid and mass feeding is accommodated without problem by the casting. The problem of accommodating strain only occurs during the last stage of freezing, when the grains are no longer free to move easily. They define a cracking-susceptibility coefficient:

$$CSC = t_V/t_R \qquad (8.6)$$

Where t_R is the time available for stress-relaxation processes such as liquid and mass flow, and t_V is the vulnerable time period when cracks can propagate between grains.

Defining the limits of applicability of these various regimes is not easy. It is different for different alloy systems. However, as a first attempt the authors assume that the stress-relaxation period spans a fraction liquid f_L of approximately 0.6–0.1, and the vulnerable period spans f_V values 0.1–0.01. The scheme is shown in Figure 8.19. For the Al–Si system they correctly predict a lambda curve with the correct form and peak at approximately 0.3Si, as found by experiment. For the Al–Mg and Al–Zn systems the agreement is less good. The agreement for Al–Mg was improved later by Katgerman (1982), who modified the CSC limits. For magnesium-based alloys Clyne and Davies (1981) use their model to predict a peak at 2.0Zn for the Mg–Zn system and 3.0Al for the Mg–Al system. The poorer tearing

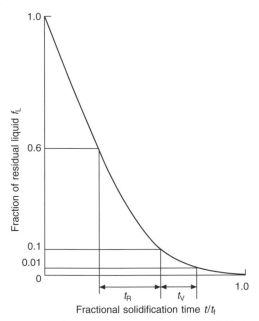

Figure 8.19 *Model by Clyne and Davies (1977) for the regimes during which either stress relaxation or vulnerability to hot tearing occur.*

resistance of the zinc-containing alloy is the result of its considerably greater freezing range. However, from the ring test results shown in Figure 8.20 the peaks are actually observed to be at approximately 1.0Zn and l.0Al.

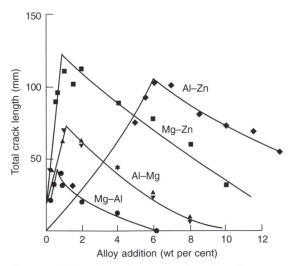

Figure 8.20 *Hot-tearing behaviour of various alloys subjected to the ring die test. Data from Dodd (1955), Dodd et al. (1957), Pumphrey and Lyons (1948) and Pumphrey and Moore (1949).*

Clyne *et al.* (1982) later extended the model to the cracking of steels. Here, as a consequence of

the complexity of iron-based alloys, the model is extremely useful in providing a framework to understand the phenomena. For instance, Rogberg (1980) found that for stainless steels that solidified to delta-iron the alloys were insensitive to the impurities As, Bi, Pb, Sn, P and Cu, whereas those that solidified to gamma-iron suffered serious loss of hot ductility. Kujanpaa and Moisio (1980) confirmed that S and P embrittled gamma-iron, but not delta-ferrite, but the best resistance to embrittlement was provided by a mixture of gamma- and delta-irons.

The approach therefore seems basically sound and useful, even though not always especially accurate in its predictions, as we have seen above. Thus, in common with most theories, it invites further development and refinement.

As a start it would seem useful to combine the cracking-susceptibility coefficient with Equation 8.5 derived above for the strain per boundary in the hot spot. This gives a modified CSC as:

$$CSC_b = \frac{\alpha \Delta TLa}{l^2} \cdot \frac{t_V}{t_R} \qquad (8.7)$$

This more general relation needs to be tested. Nevertheless there is already considerable evidence to suggest that it is at least approximately accurate. Figure 8.21 shows how grain size is significant. Also, several researchers with different diameter ring tests confirm that the cracking susceptibility is proportional to the circumference of the ring (Isobe *et al.* 1978). This proportionality to length is implicit in the design of the various I-beam tests using graded lengths of beam (see below).

8.3.9 Methods of testing

A complete survey of the methods used to assess hot tearing is beyond the scope of this short book. The interested reader is recommended to various reviews, including historical accounts by Middleton (1953), Dodd (1955) and Hansen (1975), and more recently a number of briefer accounts by Guven and Hunt (1988) and Warrington and McCartney (1989).

The methods fall into two main groups; the various I-beam (sometimes known as 'dog bone') tests and the ring mould test. Sundry other tests include the pulling of tensile test pieces during their solidification.

The I-beam has many variants. One of the most common is the arrangement of various lengths of rod castings from a single runner. Each of the rods has a T-shaped end to provide a restriction to its contraction. When metal is poured, the contraction of the rods will take place with various degrees of constraint, those with rods greater than a critical length failing by tearing at the hot spot, which is

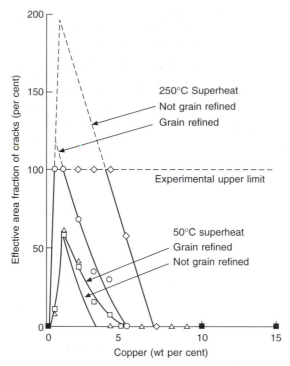

Figure 8.21 *Hot-tearing behaviour of Al–Cu alloys using an I-beam-type test showing the benefits of low casting temperature and grain refinement (Spittle and Cushway 1983). Peaks are extrapolated to illustrate the close agreement with Figure 8.18.*

the joint between the runner and the rod. From such a test, therefore, only one result is obtained, and its accuracy is limited by the increments by which the rods increase in length.

Other sorts of I-beam test gain a potentially more discriminating result by measuring the lengths of cracks in the hot spot region. However, even in this case the actual test volume of material is limited and the stress and strain distribution in the hot spot is far from uniform.

The ring test could hardly be simpler. It is normally carried out using a die. Metal is poured into the open annulus between the inner and outer parts of the die. As it cools it contracts on to the inner steel core. The core also expands slightly as it heats up. The resulting constraint on the casting is severe, opening up transverse tears all around the ring in a susceptible alloy. In this respect the technique is useful in that it tests a large volume of material, subjecting it all to a uniform constraint condition. The method of assessment is by measuring the total length of cracks. This gives a reasonably large number that can be assessed accurately, making the test more discriminating. Thus despite the criticisms normally levelled at the ring test, it has many advantages, one of the most important of which is simplicity. In general the

many different researches that have been carried out over the years by different workers using this approach are seen to agree tolerably well. Pumphrey (1955) carried out a series of repeated checks with different workers over a period of months. These showed an impressive consistency. This is more than can be said for most other tests for hot tearing.

Another aspect of the ring test die is worth comment: the test is unusual in that its rigid restraint is sufficient to cause hot tearing in the absence of any real hot spot and in the absence of strain concentration.

Gruznykh and Nekhendzi (1961) and DiSylvestro and Faist (1977) describe similar ring tests for steels using sand moulds.

The recent variant of the ring test by Warrington and McCartney (1989) provides a tapered centre core that gives a constraint condition increasing linearly along the axis of the cone. This seems a useful test, but is clearly limited in very hot short materials to a maximum tear of the length of the side of the cone. This artificial cut-off is seen in Figures 8.18 and 8.21. More work would be helpful to confirm the limitations of reproducibility of the test. For instance, from Equation 8.7 it is to be expected that the width of the hot spot that the authors place along the cone by painting a strip of insulating wash down its length will be critical. It is not clear whether such a hot spot is necessary or even desirable in the test.

8.3.10 Methods of control

The casting engineer can be reassured that there are a number of different approaches to tackling hot-tearing problems in castings, or even, preferably, preventing such defects by appropriate precautions.

8.3.10.1 Improved mould filling

In practice, the author has never failed to deal with hot tearing by simply upgrading the filling system. Whole foundries have been revolutionized in this way. Thus this is probably the single most important technique for dealing with hot-tearing problems. It is the most convincing evidence that hot tears are strongly linked to the presence of bifilms.

8.3.10.2 Design

Much can be achieved at the design stage of the casting. A publication by Kearney and Raffin (1987) concerns itself almost exclusively with the prevention of hot tears by adjustments to the casting design. In general it can be summarized by saying 'do not design sharp re-entrant corners; do not provide a straight join between two potential hot spots; curve such members; provide curves in gates so that deformation can be accommodated easily;

angle and off-set stiffeners and ribs to allow easier accommodation of strain; in gravity die (permanent mould) casting the rapid removal of any internal steel core is recommended to reduce constraint'. This severe condensation hardly does justice to the original.

It is also worth bearing in mind that flash on castings as the result of the poor fitting of moulds and cores can be a major source of casting constraint. To identify the real sources of constraint it is necessary, therefore, to check the casting straight out of the mould (not after it has been nicely dressed to appear on the manager's table!).

All this list of dire warnings needs to be set into context. In the experience of the author the provision of sharp corners and other so-called geometrical dangers have never given a problem providing the liquid metal is of good quality, and provided the metal has not been impaired by a poor filling system design. This point was reinforced in section 8.3.6.

In any case, it is usual to find that the design is fixed. The casting engineer needs to fall back on other options. These include the following.

8.3.10.3 Chilling

The chilling of the hot spot is a useful technique. This reduces the temperature locally, thus strengthening the metal by taking it outside of its susceptible temperature range before any significant strain and stress is applied. By reducing the temperature locally nearer to that of the casting as a whole, the temperature differential that drives the process is reduced, and any strain concentration is redistributed over a larger region of the casting. Local chilling is therefore usually extremely effective.

In addition to this conventional explanation, recent research has revealed that the action of chills is more complex. Chills cause the solidification front to move away from the chill in a rapid, unidirectional movement. Bifilms, with their enclosed layer of air, are the ultimate barrier to dendrites since the solidification cannot progress through the air. The result is that the bifilms are pushed ahead of the front, and so pushed away from the vulnerable hot spot.

This is the reason that many entrainment defects, as opposed to genuine shrinkage problems, appear to be cured by the placing of a chill. The usual (and usually quite wrong) interpretation being that the chill has cured the 'shrinkage' porosity.

Some films appear, however, to be attached to the casting surface (possibly at the point at which they were originally folded-in) so that they are not completely free to be pushed ahead. Also, some are oriented so that some of the tips of the dendrites in the advancing dendrite raft pass either side of the bifilm, as seen in Figure 2.41. Either way, the

bifilm is organized into a planar sheet, parallel to the dendrite growth direction, and then pinned in this position until solidified in place. The sheet exhibits characteristic steps, in integral numbers of DAS high. Such planes easily separate along the air layer as the result of a tensile strain. This may occur during freezing, in the form of a hot tear as shown in Figure 8.13. Alternatively, the planes and the characteristic steps can be seen in room temperature tensile fracture surfaces as in Figure 2.42.

8.3.10.4 Reduced constraint

A reduction in the stress on the contracting casting can be achieved by reducing the mould strength. As we have noted before, this is in principle relatively easily achieved. The options are: (i) reducing the level of binder in the sand core, although there is normally little scope for this, because those foundries not already operating at minimum strengths to reduce costs and ease shake-out are using strength as an insurance against core breakages and defects as a result of mishandling; and (ii) weakening the core by using it in a less dense form (such as produced by blowing rather than by hand ramming), or modifying its design by making it hollow. Earlier shake-out from the mould, and more rapid decoring, may also be useful. However, Twitty (1960) found the opposite effect; his white iron castings suffered more hot tears when shaken out earlier as a result of the extra stresses put on the casting during the removal of the mould!

El-Mahallawi and Beeley (1965) show how appropriate tests can be carried out on sands containing different binders. Their deformation/time test for sand under a gradient heating condition would be expected to provide a good assessment of the constraint imposed by different types of sand/binder systems. DiSylvestro and Faist (1977) use their sand moulded ring test to check the effect of different sand binders on the hot tearing in carbon steels. In later work on steel, SCRATA (1981) list the sand binders in increasing hot-tearing tendency for casting sections of less than 30 mm. These are:

Greensand (*least hot tearing*)
Dry sand (clay bonded)
Sodium silicate bonded (CO_2 and ester hardened types)
Resin-bonded shell sand
Alkyd resin/oil (perborate or isocyanate types)
Oil sand
Phenol formaldehyde resin/isocyanate
Furan resin (*worst tearing*)

In the case of the worst sand binder, the furan resin, it is difficult to believe that the thermal and mechanical behaviour of the binder is responsible for the increased incidence of hot tearing. It seems likely that the sulphur (or phosphorus) contained in the mixed binder will contaminate the surface of the steel casting, promoting grain boundary films of sulphides or phosphides, and thus rendering the metal more susceptible to tearing.

In thicker sections the greater amount of heat available causes more burning out of the binder, with better collapse of the sand mould. Organic binders benefit from this effect. Paradoxically the inorganic system based on sodium silicate in this list is also seen to benefit in a similar way. This is probably the result of the extra heat leading to greater general softening and/or melting of the binder at high temperature. As the binder cools, however, it is well known to become a fused, glassy mass, which is extremely difficult to remove from the finished casting.

For ferrous castings the inclusion in the sand of material that will burn away rapidly, leaving spaces into which the sand can move, allows faster and greater accommodation of the movement of the casting. Common additions are wood flour, cellulose and polystyrene granules. Slabs of polystyrene foam of 25 to 50 mm thickness have been inserted in the mould or core at approximately 6 to 12 mm away from the casting/mould interface, depending on the thickness of the casting.

Reinforcing rods or bars in sand moulds or cores greatly reduce the collapsibility of that part of the mould in which they are placed, and may cause local tearing.

In the case of gravity die casting, where the core may be made from cast iron or steel, it is common to construct the core so that it can be withdrawn or can be collapsed inwards as soon as possible after casting.

However, it has to be emphasized that removal of constraints after casting is not always a reliable technique. This is because the timing is difficult to control precisely; too early an action may result in a breakout of liquid metal, and too late will cause cracking. It is really better to rely on passive systems.

8.3.10.5 Brackets

The planting of brackets across a corner or hot spot can sometimes be useful. The brackets probably serve not only for strengthening but also as cooling fins. Perhaps predictably (because of the poorer conductivity of steel castings compared to those in aluminium, for instance), for steel castings they are generally less good at preventing tearing than are chills (SCRATA 1981). Even so, some large steel castings are sometimes seen to be covered in 'tear brackets', the casting bristling like a porcupine. Such features confirm the existence of their poor filling system designs.

8.3.10.6 Grain refinement

The grain refinement of the alloy is expected to be helpful in reducing crack initiation, as indicated by Equation 8.7. However, care is needed not to overlook the fact that if the conditions are good for initiation then the subsequent growth of the crack may be worse, as discussed above. Figure 8.21 shows the improved resistance to tearing by grain refinement of Al–Cu alloys. Twitty (1960) confirmed that his white cast iron alloyed with 30 per cent chromium was severely hot short when not grain-refined, whereas 0.10 to 0.25Ti addition reduced the grain size and eliminated hot tearing.

8.3.10.7 Reduced casting temperature

Reduction of the casting temperature can sometimes help, as is seen clearly for Al–Cu alloys (Figure 8.21). This effect is likely to be the result of the achievement of a finer grain size. If, however, the effect also relies on the reorganization of bifilms, then the effect may be sensitive to geometry: some directions may become more prone to failure if bifilms are caused to lie across directions of tensile strain.

8.3.10.8 Alloying

A variation of the alloy constituents, often within the limits of the chemical specification of the alloy, can sometimes help.

The addition of elements to increase the volume fraction of eutectic liquid can be seen to help by (i) increasing the pre-tear extension by lubricated grain boundary sliding, and (ii) decreasing the CSC. Couture and Edwards (1966) confirm that copper-based alloys benefit from increased amounts of liquid during the final stages of freezing.

Manganese in steels is well known for reacting with sulphur to form MnS, so that the formation of the deleterious FeS liquid at the grain boundary is reduced.

For other more complicated alloy systems the answers are not so straightforward. For instance, in early work on the Al–Cu–Mg system by Pumphrey and Lyons (1948), the relation between hot tearing and composition is complex, as was confirmed by Novikov (1962) for various Al–Cu–X systems. Ramseyer et al. (1982) investigated the Al–Cu–Fe system and found that for certain ranges of composition increasing levels of iron were desirable to control tearing. This is a most surprising conclusion, which most metallurgists would not have predicted; iron is usually an embrittling impurity in most high-strength aluminium alloys when assessed by room temperature tests of ductility. Later work by Chadwick (1991b) revealed that the effect of the iron is to provide a network of iron-rich crystals around the primary aluminium dendrites, as a scaffold framework that appears to support and reinforce the weaker dendrite array.

8.3.10.9 Reduced contracting length

Shortening the length over which the strain is accumulated is sometimes conveniently achieved by placing a feeder in the centre of the length. Such a large concentration of heat in the centre of the cooling section will allow the strain to be accommodated in the plastic region close to the feeder. Any opening up of intergranular pathways is likely to be easily fed from the nearby feeder. The careful siting of feeders in this way effectively splits up a casting into a series of short lengths. If each length is sufficiently short then strains that can cause a tear will be avoided. The siting of insulating coatings can also be used in the same way. In terms of Equation 8.5, the technique is equivalent to a method of reducing the strain concentration by multiplying the number of hot spots, and thus increasing the total length l of the hot spot while reducing the contracting length L.

8.3.11 Summary of the conditions for hot tearing and porosity

The findings of researches like those by Spittle and Cushway (1983) are summarized schematically in Figure 8.22a. The benefits of fine grain size are clear, but the benefits to be expected from clean metal are also emphasized as of overriding importance.

The differences between the conditions for the occurrence of porosity and the occurrence of hot tearing have been mentioned a number of times. Figure 8.22b represents a summary of these conditions.

Clearly, porosity forms in the hydrostatic tension as a result of poor feeding, especially in the kind of conditions found in a pasty freezing alloy. The interdendritic feeding leads to a reduction in pressure described by the so-called Darcy relation, in which laminar flow through interdendritic channels suffers viscous drag, causing the pressure to fall. Thus these conditions exist in those alloys that solidify as a solid solution, and thus form dendrites whose interdendritic channels taper down to nothing, therefore maximizing drag and maximizing the potential for the creation of porosity.

Increasing the solute content of the alloy sufficiently to form some eutectic, particularly in non-equilibrium freezing conditions, the hydrostatic (triaxial) tension is immediately greatly reduced because the dendrite channels now no longer taper to a point, but are now stopped off with a planar front of eutectic. The potential for porosity now falls precipitously as is clear in Figure 8.22b.

If any uniaxial tension exists, the potential for hot tearing at first increases similarly to the potential for pore formation, but does not drop steeply at the first appearance of the non-equilibrium eutectic liquid.

The different regimes of triaxial and uniaxial tension correspond to the regimes for the incidence of porosity and hot tearing in alloy systems as

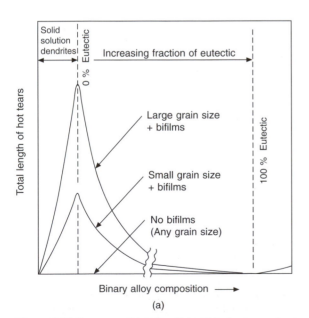

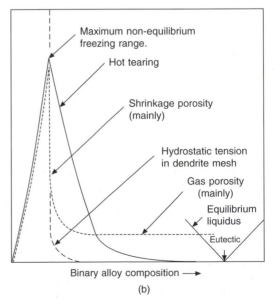

Figure 8.22 *Summary of (a) the effect of grain size and presence of bifilms on hot tearing, and (b) the relation between the conditions for the incidence of porosity or hot tearing.*

illustrated in Figure 8.22. The continuing low level of porosity at higher solute contents approaching the eutectic (based on the results shown in Figure 7.10) is the result of a residual amount of gas porosity. This too, of course, usually only occurs because of the presence of bifilms.

The maximum in the freezing range corresponds to the maximum potential for porosity and hot tearing (Campbell 1969). Interestingly, it also corresponds closely to the compositions of wrought alloys. Thus these alloys should suffer maximum problems. The only reason why such compositions are successful in practice is because of the great precautions taken by the wrought industry to cast metal as clean as possible, and under the highest temperature gradient possible provided by the water cooling. Even so, the continuous casting industry is troubled by cracking of ingots, particularly at the start of casting. This is because after all the cleaning of the melt, the cast is started by the melt being dropped several hundred millimetres into the mould where, of course, it re-creates huge bifilms. These cause the cracking of the cast material around the dovetail key at the top of the starter bar. Although, once the mould is filled, the remainder of the pour creates no further bifilms; the cracking troubles sometimes extend a long way up the length of the casting because of the mixing and progressive dilution of the original melt pool as the cast proceeds, spreading the original bifilms far up the length of the product.

8.4 Cold cracking

8.4.1 General

Cold cracking is a general term used to emphasize the different nature of the failure from that of hot tearing.

Whereas the word 'hot' in hot tearing implies a failure occurring at temperatures above the solidus, 'cold' simply means lower than the solidus temperature; in some cases, therefore, it can be rather warm! Solidus here means, of course, the real non-equilibrium solidus (not the value picked off an equilibrium phase diagram!).

The term 'cracking' is also to be contrasted with 'tearing'. Whereas a tear is a ragged failure in a weak material, a crack is more straight and smooth, and occurs in strong materials. Because it represents the failure of a strong material, the stress required to nucleate and propagate the failure is correspondingly more significant (stress was less significant in hot tearing, whereas strain was important).

Occasionally a failure appears to fall somewhere between the tear and crack categories. Such borderline problems include the cracks that form

in steels because of the presence of low melting point residual elements, such as copper-rich phases, at the grain boundaries. The copper-bearing phase is liquid between 1000 and 1100°C, with its dihedral angle falling to zero in this temperature range, thus wetting the grains. Over this range of temperature, therefore, the steel is particularly susceptible to tensile failure (Wieser 1983). The temperature is well below the solidus (in terms of the Fe–C system), and cracking will not occur easily at higher temperatures because the liquid phase does not wet so well, and therefore does not cover such a large proportion of the grain boundary area.

Returning to the 'cold' crack, the driving force for the nucleation and spread of a cold crack is stress. The various ways in which stress can arise and be concentrated in castings have been dealt with previously and are not repeated here. However, one mechanism not previously discussed is phase change in the solid state. In steels transforming from δ- to γ-structure, the large volume reduction has been suggested as a potential source of stress because of the large strains involved (Gelperin 1946). In fact the strain is so large that TRIP (transformation-induced plasticity) occurs, as discussed in the previous section, raising stress to over the yield point of the steel at that temperature.

The crack may be transgranular or intergranular, depending on the relative strength of the grains and their boundaries at the temperature at which the crack formed. We shall discuss examples of each below.

Depending on the temperature at which the crack formed, and whether it is open to the atmosphere, it may or may not be oxidized. Thus in steels the colour of a surface-connected crack is a useful guide to when it formed: an uncoloured metallic surface will indicate that the crack occurred at a temperature near to room temperature; the normal 'temper colours' ranging (the light interference colours reflecting the thickness of the oxide) from light straw, through yellow, blue and finally to brown indicate greater exposure to time at temperature, with temperatures probably approaching 600 or 700°C for the darker colours.

8.4.2 Crack initiation

Cracks start from stress raisers. A stress raiser can be an abrupt change of section in a casting. However, this problem is well known to designers, and in any case is not likely to cause increases of stress by much more than a factor of 2.

More severe stresses are raised by sharper features such as folded oxide films, which already constitute a kind of crack that has been cast into place at the time of the filling of the casting. These defects are all the more dangerous because they can occupy a large portion of the section of a casting, but at the same time are difficult to detect. Oxide films are probably the most important initiators of cracks in light alloys. This is clear because from daily foundry experience, castings made with good running and gating systems are usually not sensitive to problems of cracking.

However, during welding the stresses in the weld can be extremely high because of the constraint of the surrounding solid metal, which is near to room temperature and thus very strong. In such a case Dixon (1988) has observed that cracks start from such innocuous sites as micropores. However, it is probably unusual for cracks to start at such small and relatively rounded defects. They probably occur in this case partly because of the very high driving force. However, it is also most likely, of course, that the micropore is sited on a bifilm, and it is the bifilm that causes the initiation of the crack. Only very careful observation using the SEM would be capable of resolving such issues.

8.4.3 Crack growth

As the casting cools, stress-relaxation by creep becomes progressively slower, and eventually stops altogether. Thus from this stage, more contraction only builds up elastic strain and consequent stress. The accumulating stress is available to grow the crack with great speed to large size. Lees (1946) reports the casting of a high-strength aluminium alloy into a sand mould designed to produce hot tearing by constraining the ends of the cast bars. If the restraint was not released before the castings cooled to 200°C then a loud crack was heard, corresponding to the complete fracture of the bar. Similar failures during the cooling of steel castings are also well known, as was, for instance, reported by Steiger as long ago as 1913.

During the cooling of steel castings there is a succession of particularly vulnerable temperature ranges. The following list is not expected to be exhaustive.

1. Carbon steel castings will embrittle if they dwell for an excessive period between approximately 950 and 600°C (Harsem *et al.* 1968).
2. Impure low-alloy steels suffer similarly in the range 550–350°C (Low 1969).
3. Low-carbon steels are susceptible to brittle failure if deformation occurs in the range 300–150°C, the so-called blue brittleness, or temper embrittlement range (Sherby 1962).
4. Most body-centred-cubic iron alloys become brittle at subzero temperatures, −150 to −250°C. This is often known as the cold brittleness range.

Butakov *et al.* (1968) cites hydrogen, sulphur and phosphorus as increasing embrittlement in the 900–650°C range, and the intergranular fracture surface as exhibiting various forms of sulphides, particularly MnS and FeS. Harsem *et al.* (1968) adds carbides and AlN to this list. From our privileged gift of wisdom with hindsight, it seems probable that many of these researches were influenced by the presence of bifilms. New, clarifying researches are now needed.

In his review, Wieser (1983) lists the principal contributors to temper embrittlement as antimony, arsenic, tin and phosphorus. These elements have been found to segregate to the prior austenite grain boundaries. Since these boundaries will in general coincide with the ferrite boundaries, then the crack will usually be intergranular once again. However, the effect of these and other elements in the various kinds of steels is complicated, not being the same in every case; for instance, tin at concentrations up to 0.4 per cent embrittles Ni–Cr–Mo–V steel but not 2.25Cr–1Mo steel. Copper, manganese and silicon are also deleterious in some steels, whereas the effect of molybdenum is generally beneficial.

8.5 Residual stress

As a result of the contraction of the casting as it cools to room temperature, the strains and their consequential stresses that we have been considering at length in this chapter may have been released by the casting failing by slow tearing or sudden cracking. However, the casting may by now have survived these catastrophic failure modes. The stresses must therefore still be in the casting. What can be wrong with that?

Most foundries and machining operations have stories about the casting that flew into pieces with a bang when being machined, or even when simply standing on the floor! Benson (1946) describes one such event. The author and several colleagues narrowly missed the shrapnel from an aluminium alloy compressor housing that exploded without warning among them as it was being cut on a band saw.

It is easy to disregard such stories. However, they should be viewed as warnings. They warn that, in certain conditions, castings can have such high stresses locked inside that they are dangerous and unfit for service. In fact they can act like small bombs. We are unaware that the casting may be on the brink of catastrophic failure because, of course, the problem is invisible; the casting looks perfect.

Benson (1938) made the important commonsense observation that it is the last thermal treatment, and the rate of cooling from this final treatment, that is important so far as residual stresses are concerned. Thus the last treatment might simply be casting, or annealing, or quenching from a solution treatment. We consider all of these below.

8.5.1 Casting stress

There have been a number of test pieces that have been used over the years to help assess the parameters affecting the residual stress in castings. Most of these take the form of the three-bar frame casting shown in Figure 8.9.

In practice the stress remaining in the casting is usually assessed by scribing two lines on the central bar and accurately measuring their spacing. The bar, of length L, is then cut between the lines, and, usually, the cut ends spring apart as the cut is completed. The distance between the lines is then measured again and the difference ΔL is found. The strain ε is therefore $\Delta L/L$ and the stress σ is simply found, assuming the elastic (Young's) modulus E, by the definition:

$$E = \sigma/\varepsilon \qquad (8.8)$$

Such studies have revealed that the residual stress in castings is a function of the cooling rate in the mould, as shown for aluminium alloy castings from the effect of water content of the mould in Figure 8.23. Dodd (1950) cleverly illustrated that this effect is not the result of the change of mould strength by preparing greensand moulds with various water contents, then drying each carefully so that they all had the same water content. This gave a series of moulds with greatly differing strengths. When these were cast and tested there was found to be no difference in the residual stress in the castings. This result was further confirmed by testing castings

made in moulds rammed to various levels of hardness. Again, no significant difference in residual stress was found.

Dodd (1950) also checked the effect of casting temperature, and noted a small increase in residual stress as casting temperature was increased.

As with cases of the constraint of the casting by the mould, removing the casting from the mould at an early stage would be expected to be normally beneficial in reducing residual stress. Figure 8.24 shows the result for iron and a high-strength aluminium alloy. The higher residual stress for cast iron reflects its greater rigidity and strength. The effects of percentage water in the sand binder, and of stripping time and casting temperature, have been confirmed in other work on high-strength aluminium alloys and grey iron using a rather different three-bar frame (IBF Technical Subcommittee 1949, 1952).

All these observations appear to be explainable assuming that the main cause of the development of residual stress is the interaction of different members of the casting cooling at different rates. Beeley (1972) presents a neat solution to the problem.

The strain $\Delta L/L$ due to differential contraction is determined by the temperature difference ΔT and the coefficient of thermal expansion of the alloy α. We have:

$$\varepsilon = \Delta L/L = \alpha \Delta T \qquad (8.9)$$

and so from Equation 8.8:

$$\sigma = \alpha E \Delta T \qquad (8.10)$$

The stress therefore depends on the temperature difference between members. It is also worth noting that the stress is independent of casting length L.

Further influence of the geometry of the three-bar frame casting was found by Steiger (1913). He measured the increase of stress in the centre bar of grey iron castings by increasing the rigidity of the end cross members. Also he found that a centre bar of more than twice the diameter of the outer bars would suffer a residual stress of over 200 MPa, sufficient to fracture the bar during cooling. Working Equation 8.10 backwards, it is quickly shown that the temperature difference was only about 100°C to produce this failure stress. Clearly, such temperature differences will be common in castings and often exceeded. Thus high stresses are to be expected.

In ferrous castings that experience a γ/α phase change during cooling, any stresses that are built up prior to this event are probably reduced, their memory diluted by the plastic zone that the transformation causes. It seems probable therefore that the temperature differences and cooling rates applying below the gamma-alpha-transformation

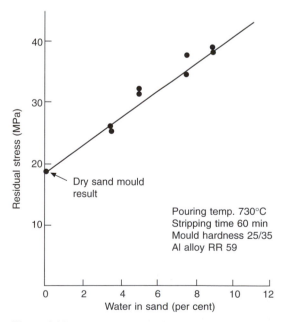

Figure 8.23 *Residual stress in the centre member of a three-bar frame as a function of water content of the greensand mould (Dodd 1950).*

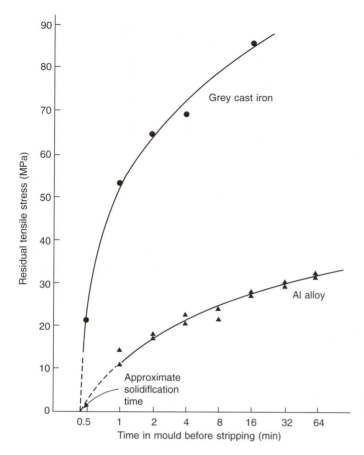

Figure 8.24 *Residual stress in aluminium alloy and grey iron castings as a function of stripping time. Data from Dodd (1950) and IBF Technical Subcommittee (1949).*

temperature that are the most important for the final remaining levels of stress.

This fact prompts Kotsyubinskii (1961–62) to recommend that heavy sections of ferrous castings be cooled by forced air or chills to equalize their cooling rates with those of the thinner sections, up to the point at which the pearlite reaction occurs. Below this temperature, little can be done to avoid the build up of stress. This is because the metal is largely elastic, and plastic relaxation, occurring only slowly by creep, becomes ineffective; thus cooling should at that stage be slow and even, so as to take advantage of as much natural stress relief as possible.

8.5.2 Quenching stress

Most of the reasons why stresses remain after the casting has cooled from its casting temperature apply also to the situation where the component has cooled from a high heat-treatment temperature. If the cooling was rapid, especially by quenching in water, then the stresses are likely to be even higher.

Following the same logic as in the previous section, the stress distribution in the casting after quenching is in general compressive in the outer features, which have the full benefit of the quench,

and tensile in the central members of the casting, which are somewhat shielded from the quenchant and so cool slowly. Stresses high enough to distort the casting are usual, and stresses even high enough to cause immediate fracture in the quench are not uncommon.

We can easily find how slowly we need to quench to avoid large temperature differences between the inside and the outside of the casting. From our familiar order of magnitude relations, we have the average distance x that heat will flow in time t in a material of thermal diffusivity D:

$$x = (Dt)^{1/2} \qquad (8.11)$$

where $D = K/\rho C_p$ and K is the thermal conductivity, about 200 $Wm^{-1} K^{-1}$ for aluminium, the density ρ is 2700 kgm^{-3}, and the specific heat C_p is approximately 1000 $Jkg^{-1} K^{-1}$. This gives the thermal diffusivity D as about $10^4 m^2 s^{-1}$.

For an aluminium bar of only 20 mm diameter, Figure 8.25 shows that quenching in water will cool the bar from 500 to 250°C within about 5 seconds. Below 250°C, stresses will start to accumulate because relaxation processes become slow. Substituting 5 s in Equation 8.11 shows that heat will on average have travelled 20 mm during

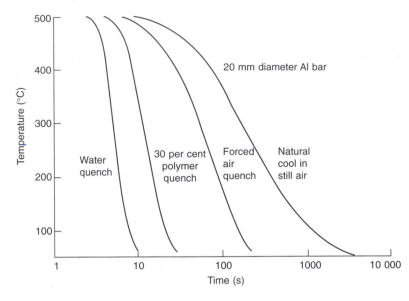

Figure 8.25 *Rates of cooling of a 20 mm diameter aluminium bar when quenched by various means from 500°C.*

this time. The bar will therefore be reasonably uniform in temperature, and large stresses will not be expected.

For a larger casting such as a cylinder block or head, however, the distance that heat would have to diffuse from the centre to the outside is now of the order of 100 mm. If it has walls of thickness 10 mm or less, then its outer walls will still cool at a similar rate to the 20 mm cylindrical bar (since they have similar modulus, defined as volume divided by the cooling area). Thus the time available is still only 5 s, so the heat will have travelled only 20 mm. The 100 mm distance in the cylinder block will therefore suffer extreme non-uniformity, and consequent high quenching stress.

To avoid high residual stress in the cylinder block the casting needs to be cooled at the rate at which it can equalize its temperature within tolerable limits. By using forced-air cooling, Figure 8.25 indicates that approximately 100 s is now available, sufficient time for the heat to diffuse the 100 mm distance, as Equation 8.11 confirms. Forced-air quenching would therefore be suitable for such a large component.

For steels the thermal diffusivity D is much lower, close to 10^{-5} m^2 s^{-1}. This results in average diffusion distances for heat of only 7 mm in 5 s, and 30 mm in 100 s. These limited distances highlight the problem of quenching steels without the generation of internal stress. They explain the reason for long cooling times from annealing temperatures, especially for large castings.

Even so, steel castings are not at the same risk of failure from a quenching stress as aluminium alloy castings. This is because steels in general enjoy an elongation to failure usually in the range 20 to 40 per cent. Thus 1 or 2 per cent quenching strain can easily be accommodated with a modest amount of plastic strain. In contrast, many aluminium castings are often found to have only 0.5 to 2.0 per cent elongation, so that the application of a 1 per cent quenching strain may result in failure. These differences in ductility are probably mainly the result of differences in the usual content of bifilms, the steel being relatively clean, compared to the aluminium that is expected to be full of bifilm defects.

Figure 8.25 shows that there are other intermediate quenching rate options available. The use of water, while being cheap and environmentally pleasant, causes problems for the quenching of most alloys, whether light alloys or ferrous. The rapidity of the quench is not suitable for larger parts, as we have seen above. In addition to this problem, water gives an uneven and non-reproducible quench because of its boiling action. When the parts are immersed in the water they are at a temperature hundreds of degrees above the boiling point. Thus the water in contact with the hot surface boils, coating it with a layer of vapour that conducts heat poorly. Thus that part of the casting is temporarily insulated while surrounding areas that happen to remain in contact with the water continue to cool rapidly. The pattern of contact varies rapidly and irregularly as the vapour film forms and collapses in the turbulent water. Thus the stress pattern in the casting is complex, and different from casting to casting.

To overcome the blanketing action of the vapour, liquids with higher boiling point such as oils have been used. However, the flammability hazard and the smoke and fumes have caused such quenchants to become increasingly unacceptable. Cleaning of the casting after the quench is also an environmental problem. Water-based solutions of polymers have therefore become widespread over the last decade or so. They are safer and somewhat less unpleasant in use. Fletcher (1989) reviews their action in detail. We shall simply consider a few general points.

Some polymers are used in solution in water and appear to act simply by the large molecular weight and length of their molecules increasing the viscosity and the boiling point of the water. Such viscous liquids are resistant to boiling and so provide a more even quench, with the quenchant remaining in better contact with the surface of the casting.

Sodium polyacrylate solution in water produces cooling rates similar to those of oils. However, its action is quite different. It seems to stabilize the vapour blanket stage by enclosing the casting in a gel-like casing. The fracture of this casing towards the later stages of the quench is said to be almost explosive.

Other polymers have a so-called reverse temperature coefficient of solubility. This long phrase means that the polymer becomes less soluble as the temperature of the water/polymer solution is raised. Many, but by no means all, of the polymers are based on glycol. One widely used polymer is polyalkylene glycol. This material becomes insoluble in water above about 70°C. The commercial mixtures are usually sold already diluted with water because the product in its pure form would be intractably sticky, like solid grease, and would therefore present practical difficulties on getting it into solution. It is usually available containing other chemicals such as antifoaming agents and corrosion inhibitors.

Such polymers have an active role during the quench. When the quenchant contacts the hot casting, the pure polymer becomes insoluble. It separates from the solution, and precipitates both on the surface of the casting and in the hot surrounding liquid, as clouds of immiscible droplets. The sticky, viscous layer on the casting, and the surrounding viscous mixture, inhibit boiling and aid the uniform cooling condition that is required. When the casting has cooled to below 70°C, the polymer becomes soluble once again in the bulk liquid, and can be taken back into solution. Re-solution is unfortunately rather slow, but the agitation of the quench tank with, for instance, bubbles of air rising from a submerged manifold, reduces the time required.

Polymer quenchants have been highly successful in reducing stresses in those castings that are required to be quenched as part of their heat treatment. The properties developed by the heat treatment are also found to be, in general, more reproducible. Capello and Carosso (1989) have shown that the elongation to failure of sand-cast Al–75I–0.5Si alloy, using 2.5 times the standard deviation to include 99 per cent of expected results, exhibits greater reliability as shown in Table 8.1. Thus the average properties that are achieved may be somewhat less than those that would have been achieved by a cold water quench, but the products have the following advantages:

Table 8.1 Elongation to failure results from different quenching media

	Elongation (%) Mean ± 2.5σ	Minimum
Hot-water quench (70°C)	4.73 ± 2.72	2.01
Cold-water quench	6.47 ± 1.67	4.80
Water–glycol quench	5.81 ± 0.96	4.85

1. The minimum values of the random distribution of results are raised.
2. With castings nearly free from stress the user has the confidence of knowing that all of the strength is available, and that an unknown level of stress is not detracting from the strength as indicated by a test bar.
3. The castings will have significantly reduced distortion.

Capello and Carosso (1989) carried out quenching tests on an aluminium plate $150 \times 100 \times 1.5$ mm, and found that, taking the distortion in cold water as 100 per cent, a quench with water temperature raised to 80°C reduced the distortion to 86 per cent of its previous value. Quenching in a water/20 per cent glycol mixture gave a distortion of only 3.5 per cent.

Other quenching routes to achieve a low stress casting have been developed involving the use of an intermediate quench into a molten salt at some intermediate temperature of approximately 300°C for approximately 20 s prior to the final quench into water (Maidment et al. 1984). Despite the advantages claimed by the authors, the expense and complexity of this double quench are likely to keep the technique reserved for aerospace components.

Not all residual stress need be bad.

Bean and Marsh (1969) describe a rare example, in which the stress remaining after quenching was used to enhance the service capability of a component. They were developing the air intake casing for the front of a turbojet engine. The casting has the general form of a wheel, with a centre hub, spokes and an outer shroud. In service the spokes reached 150°C and the shroud cooled to −40°C. With additional high loads from accelerations up to 7g and other forces, some casings were deformed out of round, and some even cracked. In order to counter this problem the casting was produced with tensile stress in the spokes and compressive loading in the shroud. This was achieved by wrapping the spokes in glass fibre insulation, while allowing the shroud to cool at the full quenching rate. By this means approximately 40 MPa tensile stress was introduced into the spokes. This was tested by cutting a spoke on each fifteenth casting, and measuring the gap opening of approximately 2 mm.

Another method of equalizing quenching rates in castings is by the clamping of shielding plates around thinner sections to effectively increase their section. The method is described by Avey *et al.* (1989) for a large circular clutch housing in a high-strength aluminium alloy. The technique improved the fatigue life of the part by over 400 per cent. It may be significant that both of these descriptions of the positive use of residual stress relate to rather simple circular-shaped castings.

The proper development of quenching techniques to give maximum properties with minimum residual stress is a technique known as quench factor analysis. It is also much used to optimize the corrosion behaviour of aluminium alloys. The method is based on the integration of the effects of precipitation of solute during the time of the quench. In this way any loss of properties caused by slow quenching or stepped quenching can be predicted accurately. The interested reader is recommended to the introduction by Staley (1981) and his later more advanced treatment (Staley 1986).

8.5.3 Stress relief

The original method of providing some stress relief in grey iron castings was simply to leave the castings in the foundry yard. Here, the long passage of time, of weeks or months, and the changeable weather, including rain, snow, frost and sun, would gradually do its work. It was well known that the natural ageing outdoors was more rapid and complete than ageing done indoors because of the more rapid and larger temperature changes.

Nowadays the more usual method of reducing internal stress is both faster and more reliable (although somewhat more energy intensive!). The casting is reheated to a temperature at which sufficient plastic flow can occur by creep to reduce the strain and hence reduce the stress. This is designed to take place within a reasonable time, of the order of an hour or so. As pointed out earlier, it is then most important that stress is not reintroduced by cooling too quickly from the stress-relieving treatment.

An apparently perverse and quite exasperating feature of internal tensile stress in castings is that the casting will often crack while it is being reheated as part of the stress-relieving process to avoid the danger of cracks! This happens if the reheating furnace is already at a high temperature when the

castings are loaded. The reason for this is that the casting may already have a high internal tensile stress. On placing the casting in a hot reheating furnace the outside will then be heated first and expand, before the centre becomes warm. Thus the centre, already suffering a tensile stress, will be placed under an additional tensile load, the total being perhaps sufficient to exceed the tensile strength.

The problem is avoided by reheating sufficiently slowly that the temperature in the centre is able to keep pace (within tolerable limits) with that at the outside. Consideration of the thermal diffusivity using Equation 8.11 will give some guidance of the times required.

Figure 8.26 shows the temperatures required for stress relief of various alloys. (Strictly, the figure shows results for 3 hours, but the results are fairly insensitive to time, a factor of 2 reduction in time corresponding to an increase in temperature of 10°C, hardly moving the curves on the scale used in the figure.) It indicates that nearly 100 per cent of the stress can be eliminated by an hour or so at the temperatures shown in Table 8.2.

There are numerous examples of the use of such heat treatments to effect a valuable degree of stress relief. One example is the work by Pope (1965) on cast iron diesel cylinder heads that were found to crack between the exhaust valve seats in service, despite a stress-relief treatment for 2 hours at 580°C. A modification of the treatment to 4 hours at 600°C cured the problem. From Figure 8.26 and Table 8.2 even 2 hours at 600°C would probably have been sufficient.

The work by Kotsyubinskii *et al.* (1968) highlights the fact that during the thermal stress relief the casting will distort. They carried out measurements on box-section castings in grey iron, intended as the beds of large machine tools, for which stress-relieving treatment is carried out after some machining of the top and base of the box section. He suggests that the degree of movement of the castings is approximately assessed by the factor $(w_1 - w_2)/w_c$ where w_1 and w_2 are the weights of metal machined from the top and base of the casting respectively, and w_c is the weight of the machined casting.

Moving on now from heat treatment, there are other methods of stress relief that are sometimes useful. In simple castings and welds it is sometimes possible to effect relief by mechanical overstrain as described in the excellent review by Spraragen and Claussen (1937).

Kotsyubinskii *et al.* (1962) describe a further related method for grey iron in which the castings are subjected to rapid heating and cooling between 300°C and room temperature at least three times. The differential rates of heating within the thick and thin sections produce the overstrain required for stress relief by plastic flow.

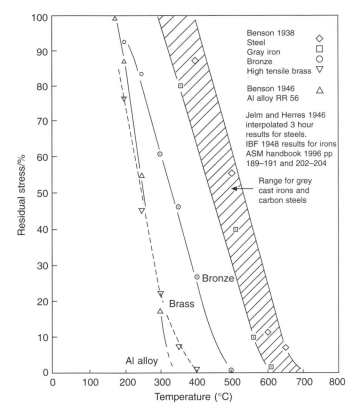

Figure 8.26 *Stress relief of a selection of alloys treated for three hours at temperature. Data from Benson (1938), Jelm and Herres (1946) and IBF (1948).*

Table 8.2 Approximate stress relieving temperature for some alloys

Alloy	Stress-relief temperature (°C)
Al–2.2C–1Ni–1Mg–1Fe–1Si–0.1Ti	300
Brass Cu–35Zn–1.5Fe–3.7Mn	400
Bronze Cu–10Sn–2Zn	500
Grey iron 3.4C–2Si–0.38Mn–0.1S–0.64P	600
Steel (C–Mn types)	700

More drastic heating rates are required to effect stress relief by differential heating in aluminium alloys because of the thermal smoothing effected by the high thermal conductivity. Hill *et al.* (1960) describe an 'up-quenching' technique in which the casting is taken from cryogenic temperatures, having been cooled in liquid nitrogen, and is reheated in jets of steam. This thermomechanical treatment introduces a pattern of stresses into the casting that are opposite to those introduced by normal quenching. One of the benefits of this method is that it is all carried out at temperatures below normal ageing temperatures, so that the effects of the final heat treatment and the resulting mechanical

properties are not affected. One of the possible disadvantages that the authors do not mention is the enhanced tensile stress in the centre of the casting during the early stages of the up-quench. Some castings would not be expected to survive this dangerous moment.

A variant of these approaches is stress relief by vibration. This is undoubtedly effective in some shapes, but it is difficult to see how the technique can apply to all parts of all shapes. This is particularly true if the component is treated at a resonant frequency. In this condition some parts of the casting will be at nodes (will not move) and some parts at antinodes (will vibrate with maximum amplitude). Thus the distribution of energy in the casting will be expected to be highly heterogeneous. Some investigators have reported the danger of fatigue cracks if vibrational stresses over the fatigue limit are employed (Kotsyubinskii, *et al.* 1961). The technique clearly requires some skill in its application, since null results are easily achieved (IBF Technical Subcommittee 1960a).

There may be greater certainty of a valid result with subresonant treatment. This technique has emerged only recently as a possible method of stress relief (Hebel 1989). In this technique the casting is vibrated not on the peak of the frequency – amplitude curve, but low on the flank of the curve. At this off-peak condition the casting is said to absorb energy more efficiently. Furthermore, it is claimed that the progress towards complete stress relief can be monitored by the gradual change in the resonant frequency of the casting. When the resonant frequency ceases to change, the casting is said to be fully stress relieved. If this technique could be verified, then it would deserve to be widely used.

Chapter 9

Structure, defects and properties of the finished casting

9.1 Grain size

9.1.1 General

The development of small grains during the solidification of the casting is generally an advantage. When the grain size is small, the area of grain boundary is large, leading to a lower concentration of impurities in the boundaries. The practical consequences that generally follow from a finer grain size are:

1. Improved resistance to hot tearing during solidification.
2. Improved resistance to cracking when welding or when removing feeders by flame cutting (for steel castings).
3. Reduced scattering of ultrasonic waves and X-rays, allowing better non-destructive inspection.
4. Improved resistance to grain boundary corrosion.
5. Higher yield strength (because of Hall–Petch relationship).
6. Higher ductility and toughness.
7. Improved fatigue resistance (including thermal fatigue resistance).
8. Reduced porosity and reduced size of pores. This effect has been shown by computer simulation by Conley *et al.* (1999). The effect is the consequence of the improved intergranular feeding and better distributed gas emerging from solution. Improved mass feeding will also help as described in section 7.4.2.
9. Improved hot workability of material cast as ingots.

However, it would not be wise to assume that all these benefits are true for all alloy systems. Some types of alloys are especially resistant to attempts to reduce their grain size, while others show impaired properties after grain refinement. Furthermore, this impressive list is perhaps not so impressive when the effects are quantified to assess their real importance. These apparent inconsistencies will be explained as we go.

In addition, important exceptions include the desirability of large grains in castings that require creep resistance at high temperature. Applications include, in particular, ferritic stainless steel for furnace furniture and high-temperature nickel-based alloy castings. Single-crystal turbine blades are, of course, an ultimate development of this concept. These applications, although important, are the exception, however. Because of the limitations of space, this section neglects those specialized applications that require large grains or single crystals, and is devoted to the more usual pursuit of fine grains.

Some of these benefits are explained satisfactorily by classical physical metallurgy. However, it is vital to take account of the presence of bifilms. These will be concentrated in the grain boundaries. The influence of bifilm defects is, on occasions, so important as to over-ride the conventional metallurgical considerations.

For instance, in the case of the propagation of ultrasonic waves through aluminium alloy castings, this was long thought to be impossible. Aluminium alloys were declared to be too difficult. They were thought to prevent ultrasonic inspection because of scatter of the waves from large as-cast grains.

No back-wall echo could be seen amid the fog of scattered reflections. However, in the early days of the Cosworth process, with long settling time of the liquid metal, and quiescent transfer into the mould, suddenly back-wall echoes could be seen without difficulty despite the absence of any grain refining action. It seems that the scatter was from the gas film between the oxide layers of the bifilms at the grain boundaries.

By extrapolation, it may be that the so-called 'diffraction mottle' that confuses the interpretation of X-ray radiographs, and usually attributed to the large grain size, is actually the result of the multitude of thin-section pores, or the glancing angle reflections from the air layer of bifilms at grain boundaries. It would be interesting to compare radiographs from material of similar grain size, but different content of bifilms to confirm this prediction.

The strong link between bifilms and microstructure, particularly grain size, is illustrated particularly well in Figure 2.42. The images (a) and (b) are the fracture surfaces of test bars taken from different parts of a single casting whose filling was observed by X-ray video radiography. The large grain size in the turbulently filled test bar (a) contrasts with the fine grain size in the quietly cast bar (b). The large grains are probably the result of reduced thermal convection in the casting because of the presence of the large obstructing bifilms, so that dendrites could grow without thermal and mechanical disturbance that is needed to melt off dendrite arms, and so lead to grain multiplication. In (b) the presence of numerous pockets of porosity suggests the presence of many smaller bifilms (that cannot be seen directly). These are older bifilms already present in the melt prior to pouring. The small bifilms will not be a hindrance to the flow of the melt, so that the small grain size is the result of grain multiplication because of convection during freezing.

Unfortunately, nearly all the experimental evidence that we shall cite regarding the structure and mechanical properties of castings is influenced by the necessary but unsuspected presence of bifilms. We shall do our best to sort out the effects so far as we can, although, clearly, it is not always possible. Only new, carefully controlled experiments will provide the certain answers. At this time we shall be compelled to make our best guess.

9.1.2 Grain refinement

As the grain size d of a metal is reduced, its yield strength σ_y increases. The widely quoted formula to explain this result is that due to Hall and Petch (see, for instance, the derivation by Cottrell 1964):

$$\sigma_y = a + bd^{-1/2} \qquad (9.1)$$

where a and b are constants. The equation is based on the assumption that a slip plane can operate with low resistance across a grain, allowing the two halves of the grain to shift, and so concentrating stress on the point where the slip plane impinges on the next grain. With the further spread of yielding temporarily blocked, the stress on the neighbouring grain increases until it exceeds a critical value. Slip then starts in the next grain, and so on. The process is analogous to the spreading of a crack, stepwise, halting at each grain boundary.

The Hall–Petch equation has been impressively successful in explaining the increase of the strength of rolled steels with a reduction in grain size, and has been the driving force behind the development of high-strength constructional steels based on manufacturing processes, especially controlled rolling, to control the grain size. This is cheaper than increasing strength by alloying, and has the further benefit that the steels are also tougher – an advantage not usually gained by alloying.

The development of higher-strength magnesium casting alloys with zirconium as the principal alloying element has also been driven by such thinking. The action of the zirconium is to refine the grain size, with a useful gain in strength and toughness. The zirconium is almost insoluble in both liquid and solid magnesium, so that any benefit from other alloying mechanisms (for instance, solid solution strengthening) is negligible. The test to ensure that the zirconium has successfully entered the alloy is simply a check of the grain size. Chadwick (1990) has used squeeze casting to demonstrate the impressive benefits of the grain refinement of magnesium castings. This is especially clear work that is not clouded by other effects such the influence of porosity. It appears that magnesium benefits significantly from the effect of small grain size because factor b in the Hall–Petch equation is high. This is the consequence of the grain boundaries being particularly effective in preventing slip, because in hexagonal close-packed lattices there are few slip systems, and only on the basal plane, so that slip is not easily activated in a randomly oriented neighbour.

This behaviour contrasts with that of face-centred-cubic materials such as aluminium, where the slip systems are numerous, so that there is always a slip system close to a favourable slip orientation in a neighbouring grain. Thus although grain refinement of aluminium alloys is useful, and is widely practised, Flemings (1974) draws attention to the fact that its effects are generally over-rated. However, little useful work on the problem had been carried out at that time. Recent measurements by Hayes and co-workers (2000) reveal the quantitative benefits of fine grain size in an Al–3Mg for the first time. They find huge increases in 0.2 proof stress of over 500 MPa when the grain size is only 0.1 μm (Figure 9.1). However, of course, such fine grain sizes are not normally obtainable in cast structures. For normally attainable fine grain sizes in the range reducing from 1 mm down to 100 μm the proof stress increases from about 55 to 65 MPa, confirming a useful, if modest, benefit. Figure 9.1 indicates that if the grain size could be reduced to 10 μm the proof strength would rise to 100 MPa. Such a valuable increase is usually beyond the scope of normal shaped casting processes.

The usual method of grain refinement of aluminium alloys is by the addition of titanium, or a mixture of titanium and boron. The effect has been discussed in section 5.4.3. The practical difficulties of controlling the addition of grain-refining materials are discussed by Loper and Kotschi (1974), who were among the first to draw attention to the problem of fade of the grain-refinement effect. Sicha and Boehm (1948) investigated the effect of grain size on Al–4.5Cu alloy, and Pan et al. (1989) duplicated this for Al–7Si–0.4Mg alloy. However, both these pieces of work confirm the useful but relatively unspectacular benefits of refinement. They appear to be complicated by the alloying effects of titanium, particularly the precipitation of large $TiAl_3$ crystals

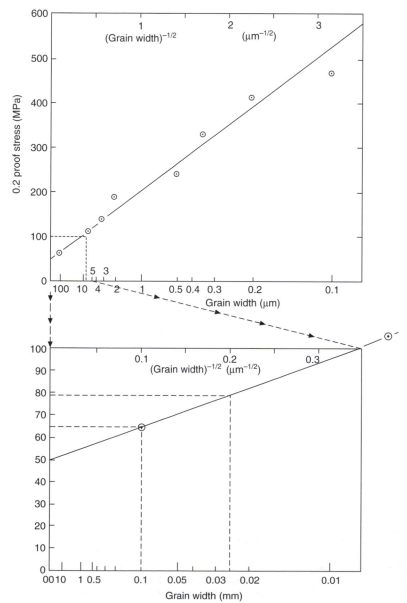

Figure 9.1 *Effect of grain size on yield strength of an Al–3Mg alloy (Hayes et al. 2000).*

at higher titanium levels. They do not therefore reveal the expected linear increase in yield strength with (grain diameter)$^{-1/2}$.

It should not be assumed that the advantages, even if small, of fine grain size in wrought steels and cast light alloys automatically extend to other alloy systems. It is worth devoting some space to the difficulties and imponderables elsewhere.

For instance, steels that solidify to the body-centred-cubic (bcc) form of the iron lattice are successfully grain refined by a number of additives, particularly compounds of titanium and similar metals, as discussed in section 5.4.3 (although not necessarily with benefits to the mechanical properties, as we shall see below). In contrast, steels that solidify to the face-centred-cubic (fcc) lattice do not appear to respond to attempts to grain refine with titanium, and are resistant to attempts to grain refine with most materials that have been tried to date.

The grain-refinement work carried out by Cibula (1955) on sand-cast bronzes and gunmetals showed that these alloys could be grain refined by the addition of 0.06 per cent of zirconium. This was found to reduce the tendency to open hot tears. However, this was the only benefit. The effect on strength was mixed, ductility was reduced, and although porosity was reduced in total, it was redistributed as layer porosity, leading to increased leakage in pressure-tightness tests. Although this was at the time viewed as a disappointing result, an examination of the tests that were employed makes the results less surprising. The test castings were grossly underfed, leading to greatly enhanced porosity. Had the castings been better poured and better fed, the result might have been greatly improved.

Remarkably similar results on a very different casting (but that also appears to have been underfed) were obtained by Couture and Edwards (1973). They found that various bronzes treated with 0.02Zr exhibited a nicely refined grain structure, and had improved density, hot tear resistance, yield and ultimate strengths. However, ductility and pressure-tightness were drastically reduced. It is possible to conjecture that if the alloy had been better supplied with feed metal during solidification then pressure-tightness would not have been such a problem. The presence of copious supplies of bifilm defects are to be expected to complicate the results as a consequence of poor casting technique.

The poor results by Cibula in 1955 have been repeatedly confirmed in Canadian research; Sahoo and Worth (1990), Fasoyinu et al. (1998), Popescu et al. (1998) and Sadayappan et al. (1999) using, mainly, permanent mould test bars. These castings are not badly fed, so that the disappointing results by Cibula cannot be entirely ascribed to poor feeding. It seems likely that bifilms are at work once again.

There may be additional fundamental reasons why the copper-based alloys show poor ductility after grain refinement. Couture and Edwards noted that the lead- and tin-rich phases in coarse-grained alloys are distributed within the dendrites that constitute the grains. In grain-refined material the lead- and tin-rich phases occur exclusively in the grain boundaries. Thus it is to be expected that the grain boundaries are weak, reducing the strength of the alloy by: (i) offering little resistance to the spread of slip from grain to grain, and so effectively lowering the yield point; and (ii) allowing deformation in their own right, as grain boundary shear, like freshly applied mortar between bricks. However, this seems unlikely to be the whole story since some of the poor results are found in copper-based alloys that contain no lead or tin.

Many of the above studies of copper-based alloys have used Zr for grain refinement. Thus it seems possible that they may have been seriously affected by the sporadic presence of zirconium oxide bifilms at the grain boundaries. Thus the loss of strength and ductility and the variability in the results would be expected as a result of the overriding damage caused by surface turbulence during the casting of the alloys.

9.2 Dendrite arm spacing

Dendrite arm spacing (DAS) usually refers to the spacing between the secondary arms of dendrites. However, if tertiary arms were present at a smaller spacing, then it would refer to this. Alternatively, if no secondary arms were present, which occurs only rarely, the spacing would be that of the primary dendrite stems.

If the DAS is reduced, then the mechanical properties of the cast alloy are invariably improved. A typical result by Miguelucci (1985) is shown in Figure 9.2. Near the chill the strength of the alloy is high and the toughness is good. As the cooling rate is decreased (and DAS grows), the ultimate strength falls somewhat. Although the decrease in itself would not perhaps be disastrous, the fall continues until it reaches the yield stress (taken as the proof stress in this case). Thus failure is now sudden, without prior yield. This is disastrous. The alloy is now brittle, as is confirmed by elongation results close to zero.

Because of the effect of DAS, the effect of section size on mechanical properties is seen to be important even in alloys of aluminium that do not undergo any phase change during cooling. For ferrous materials, and especially cast irons, the effect of section size can be even more dramatic, because of the appearance of hard and possibly brittle non-equilibrium phases such as martensite and cementite in sections that cool quickly.

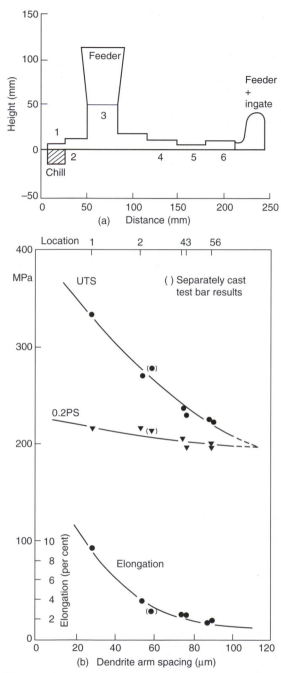

Figure 9.2 *(a) Al–7Si–0.4Mg alloy casting and (b) its mechanical properties, showing good strength and toughness near the chill, and expected brittle behaviour in the slowly solidified material. Data from Miguelucci (1985).*

The improvement of strength and toughness by a reduction in DAS is such a similar response to that given by grain refinement that it is easy to see how they have often been confused. However, the

effects cannot be the result of the same mechanisms. This is because no grain boundary exists between the arms of a single dendrite to stop the slide of a slip plane. A dislocation will be able to run more or less without hindrance across arm after arm, since all will be part of the same crystal lattice. Thus, in general, it seems that the Hall–Petch equation should not apply.

Why then does a reduction in DAS increase strength and toughness?

In the past, this question appears never to have been properly answered.

Classical physical metallurgy has been unable to explain the effect of DAS on mechanical properties. Curiously, this important failure of metallurgical science to explain an issue of central importance in the metallurgy of cast materials has been consistently and studiously overlooked for years.

In the first edition of *Castings* the author suggested that the answer seemed to be complicated and to be the result of the sum of a number of separate effects, all of which seem to operate beneficially. These beneficial processes are listed and discussed below. However, after these effects have been reviewed and assessed, it will become clear that the benefit from a refinement of DAS remains largely unexplained.

In this work, the action of bifilms will be presented as the dominant effect, capable of explaining for the first time the widely appreciated benefits of small DAS in castings, as we shall see.

9.2.1 Residual Hall–Petch hardening

Slight faults during growth will cause the dendrite arms within a grain to become slightly misoriented. This will result in a low-angle grain boundary between the arms. The higher the degree of misorientation, the greater the resistance will be to the passage of a slip plane.

When studying the structure of a cast alloy under the microscope, most dendrite arms are seen to be, so far as one can tell by unaided observation, fairly true to their proper growth direction. Thus any boundary between the arms will have an almost vanishingly small misorientation, presenting a minimal impediment to slip across the boundary. However, it is also usually possible to see a proportion of arms at slight deviations of several degrees, perhaps as a result of mechanical damage. If mechanical disturbance during freezing is increased, for instance by stirring or vibration, then the number of misaligned arms, and their degree of misalignment, would be expected to increase.

Thus one might expect some small resistance to slip even from rather well-aligned dendrites because of the lack of perfection; the result of the existence of subgrains within the grains. Some contribution from Hall–Petch hardening might therefore be expected to be present at all times.

Nevertheless, although the Hall–Petch mechanism is likely to be a contributor to increased strength, in most castings it will be negligibly small. In face-centred-cubic materials such as aluminium alloys the effect even for high-angle grain boundaries is usually only modest for the best achievable grain refinement as has been discussed in section 9.1.2. Thus for low-angle boundaries the effect can be dismissed as probably undetectable.

The final fact that eliminates the Hall–Petch effect as a contributor to the DAS effect is the fundamental fact that Hall–Petch strengthening affects only the yield strength. Figure 9.2 and many similar results in the literature indicate that yield strength is hardly affected by DAS. The main effect of changes in DAS is seen in the ductility and ultimate strength values.

We can therefore confidently and finally lay to rest any thought that the Hall–Petch effect makes any detectable contribution to the increased properties from finer DAS.

9.2.2 Restricted nucleation of interdendritic phases

As the DAS becomes smaller, the residual liquid is split up into progressively smaller regions. Although in fact these interdendritic spaces remain for the most part interconnected, the narrowness of the connecting channels does make them behave in many ways as though they are isolated.

Thus as solutes build up in these regions the presence of foreign nuclei to aid the appearance of a new phase becomes increasingly less probable as the number of regions is increased. As DAS decreases, the multiplication of sites exceeds the number of available nuclei, so that an increasing proportion of sites will not contain a second phase. Thus, unless the concentration of segregated solute reaches a value at which homogeneous nucleation can occur, the new phase will not appear.

Where the second phase is a gas pore, Poirier *et al.* (1987) have drawn attention to the fact that the pressure due to surface tension becomes increasingly

high as the curvature of the bubble surface is caused to be squeezed into progressively smaller interdendritic spaces. The result is that it becomes impossible to nucleate a gas pore when the surface tension pressure exceeds the available gas pressure. Thus as DAS decreases there becomes a cut-off point at which gas pores cannot appear. Effectively, there is simply insufficient room for the bubble! The model by Poirier suggests that this is at least part of the reason for the extra soundness of chill castings compared to sand castings. Later work by Poirier *et al.* (2001) and the theoretical model by Huang and Conley assuming no difficulty for the nucleation of pores confirms the improvement of soundness with increasing fineness of the structure.

In summary, therefore, we can see that as DAS is reduced, the interdendritic structure becomes, on average, cleaner and sounder. These qualities are probably significant contributors to improved properties.

9.2.3 Restricted growth of interdendritic phases

Meyers (1986) found that for alloys of the Al–7Si system the strength and elongation were controlled by the average size of the silicon particles, although where the particles were uniformly rounded as in structures modified with sodium, the strength and elongation were controlled by the number of silicon particles per unit volume. These conclusions were verified by Saigal and Berry (1984), using a computer model. This important conclusion may have general validity for other systems containing hard, brittle, plate-like particles in a ductile matrix.

The highly deleterious effect of iron impurities in these alloys is attributed to the extensive plate-like morphology of the iron-rich phases. Vorren *et al.* (1984) have measured the length of the iron-rich plates as a function of DAS. As expected, the two are closely related; as DAS reduces so the plates become smaller (Figure 9.4). From the work of Meyers, Saigal and Berry we can therefore conclude that the strength and toughness should be correspondingly increased, as was in fact confirmed by Vorren.

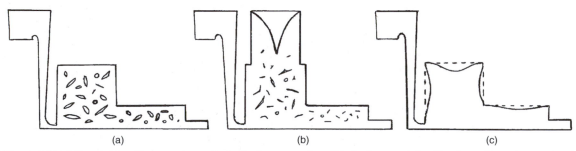

Figure 9.3 *(a) Loss of ductility seen in a poorly fed heavy section; (b) the improved ductility from the maintenance of pressure by a feeder; and (c) the excellent ductility, irrespective of pressure or feeding, expected in the absence of bifilms.*

It seems likely that although there may be an element of cause and effect in the restriction of the growth of second phases by the dendrite arms, the major reason for the close relation between the size of secondary phases and DAS is that both are dependent on the same key factor, the time available for growth. Thus local solidification time controls the size of both dendrite arms and interdendritic phases.

Later, in section 9.2.5 the reason underlying the importance of the large iron-rich plates in Al alloys will be explained further.

9.2.4 Improved response to heat treatment

For a single-phase alloy, Flemings (1974) describes a simple and elegant model of the microsegregation, or coring, present in the dendrites, and how it can be reduced by a high-temperature heat treatment termed homogenization. He defines a useful parameter that he calls the index of residual microsegregation, Δ, as

$$\delta = \frac{C_M - C_m}{C_M^0 - C_M^0} \qquad (9.2)$$

where C_M = maximum solute concentration of element (in interdendritic spaces) at time t, C_m = minimum solute concentration of element (in centre of dendrite arms) at time t, C_M^0 = maximum initial concentration of element, and C_M^0 = minimum initial concentration of element.

The parameter δ is precisely unity prior to any homogenization treatment. If homogenization could be carried out to perfection, then δ would become precisely zero. After any real homogenization treatment, δ would have some intermediate value that depends on the dimensionless group of variables Dt/l^2. Here D is the coefficient of diffusion of the homogenizing element, t is the time spent at the homogenizing temperature and l is the diffusion distance, of the order of the DAS. Assuming a sinusoidal distribution of the concentration of the element across the dendrite, Flemings finds the solution, approximately, as:

$$\delta = \exp(-\pi^2 Dt/l_0^2) \qquad (9.3)$$

where $l_0 = (DAS)/2$. Equation 9.3 is useful for the approximate prediction of times and temperatures required to homogenize a given cast structure. Flemings shows that for a low-alloy steel, carbon is always homogenized by the time that the steel is heated to about 900°C for all normal values of DAS because of its high value for D (see Figure 1.8). However, for the substitutional elements manganese and nickel, little homogenization occurs below 1100°C, and for homogenization to be about 95 per cent complete ($\delta = 0.05$) requires one hour at 1350°C and $DAS = 50$ μm. The fine DAS value is obtained by ensuring that such critical parts of

the casting are within 6 mm of a chill. If no chill is used and the $DAS = 200$ μm or more, then practically no homogenization is achieved ($\delta = 0.95$) even at temperatures of 1350°C and times of 1 hour.

Flemings emphasizes that the normal so-called homogenization treatments for steels based on temperatures of 1100°C achieve only the homogenization of carbon. The more recent use of vacuum heat-treatment furnaces capable of 1350°C and above has produced very large improvements in the mechanical properties of cast steels.

The term *homogenization treatment* is reserved for treatments designed to smooth out concentration gradients within a single-phase alloy.

The term *solution treatment* applies to those treatments designed to dissolve one or more second phases. These are also discussed by Flemings (1974). His presentation is summarized below.

Flemings considers a binary alloy containing a non-equilibrium eutectic. The dendrites are again assumed to be cored, having a sinusoidal distribution of solute as before, but containing interdendritic plates of divorced eutectic; for instance, in the case of the Al–4.5Cu alloy, a single plate of CuAl$_2$ phase separates the cored aluminium-rich dendrites. Dissolution of the interdendritic second phase is assumed to be limited by diffusion in the α-phase. 1f f and f_0 are the volume fractions of eutectic at times t and t_0 respectively, then the answer is similar to that seen in Equation 9.3, approximately:

$$f/f_0 = \exp(-2.5Dt/l_0^2) \qquad (9.4)$$

Flemings points out that a sand casting of moderate size in Al–4.5Cu alloy with $DAS = 200$ μm and given a solution treatment of 10 hours at 515°C will not have much of its eutectic phase taken into solution. More like 40 hours would have been required to eliminate the second phase. Conversely, if substantial chilling is applied to reduce the DAS below 100 μm, and if impurity levels in the alloys are kept low, so that solution temperatures within 10 or 20°C of the melting point can be employed without danger of the incipient melting of the alloy, then a 10 hour solution treatment is now more than ample to dissolve all the second phase.

Experimental tests of the theory show good agreement, particularly at short times (Figure 9.5). At long times the dissolution of the last traces of segregate require more time than the simple theory predicts. This is because more segregate exists between primary arms than between secondary arms, and so the last remnant of solute must diffuse over larger distances than simply the secondary DAS.

To summarize the effect of DAS on heat-treatment response: as DAS is reduced, so the speed of homogenization is increased, allowing more complete homogenization, giving more solute in solution and so greater strength from the subsequent

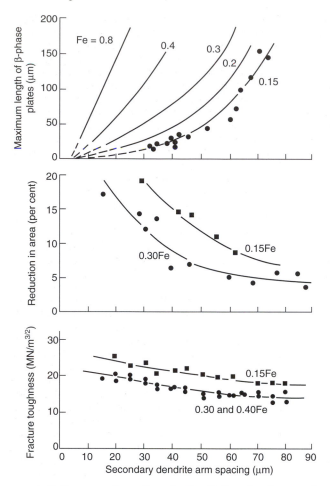

Figure 9.4 *Properties of Al–7Si–0.4Mg alloy as a function of dendrite arm spacing and iron content. Smoothed data from Vorren et al. (1984).*

precipitation reaction. Speed of solution is also increased, allowing a greater proportion of the non-equilibrium second phase to be dissolved. The smaller numbers and sizes of remaining particles, if any, and the extra solute usefully in solution, will bring additional benefit to strength and toughness.

Even when the second-phase particles are equilibrium phases, the high-temperature homogenization and solution treatments have a beneficial effect even though the total volume of such phases is probably not altered. This is because the inclusions tend to spherodize; their reduced aspect ratio, favouring improved toughness as discussed above. Any remaining pores will also tend to spherodize, with similar benefit.

Even so, it is salutary to note that improvements to heat treatment can improve yield strength, but Figure 9.2 displays no such benefit. Clearly, we still have some way to go to explain this important result.

There is no doubt therefore, the improved response to heat treatment is a valuable benefit from refinement to DAS. Thus although the following section will emphasize the role of bifilms,

the improvements due to heat treatment (if any) are quite a separate and additional factor.

9.2.5 Effect of bifilms

Examination of the above factors shows that only the restricted nucleation of interdendritic phases appears to be an independent aid to the improvement of mechanical properties as DAS is refined. The following examination of the effect of bifilms in the liquid alloy will explain how their presence is expected to be dominant.

The importance of the presence of bifilms only becomes serious as solidification time is extended. This is due to the fact that the bifilms arrive in the casting in a compact form, tumbled into compactness by the bulk turbulence during the filling of the mould cavity. In this form they are relatively harmless. They are small and rounded. Their effectiveness as defects grows as they slowly unfurl, gradually enlarging to take on the form of planar cracks up to approximately ten times larger diameter than the diameter of the original compacted shape. As we have seen in section 2.3, there are a number of driving forces for the straightening of the double films. These include the precipitation of gas inside, or second phases (particularly iron-rich phases in Al alloys) outside, or the action of shrinkage to aid inflation, or the action of growing grains to push and thereby organize the films into interdendritic or intergranular planes.

The bifilms are seen therefore to evolve into serious defects, given time. Time is a crucial factor. Thus if there is little time, the defects will be frozen into the casting in a compact form of diameter in the range of 0.1 to 1 mm in smaller castings. However, given time, the bifilms will open to their full size. This might be as large as 10 to 15 mm across in an Al alloy casting of 10 kg weight. In Al-bronze castings and some stainless steel castings weighing several tonnes bifilms up to 50 to 100 mm across are not uncommon. These constitute massive cracks.

The fall of ductility with increasing DAS becomes clear therefore. It is not the DAS itself that is important. *The DAS is merely the indicator of the time available* for the opening of bifilms. It is the opening of the bifilms into extensive planar cracks that is important.

The time required for the complete opening of bifilms depends, of course, on the rate at which they can open. This in turn depends on the various driving forces available. Thus the defects will open faster given a higher concentration of gas in solution, poorer conditions for feeding, and, in aluminium alloys, higher iron contents. A fuller discussion of this point is given in section 6.1.3.3.

A practical example is given in Figure 9.3. The sketch (a) shows the opening of bifilms in a poorly fed region whose solidification is delayed by its heavy section. In other regions of the casting the bifilms have no time to open, and are therefore

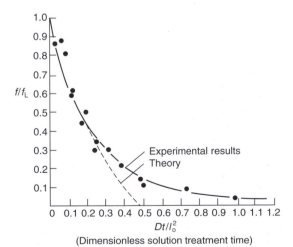

Figure 9.5 *Rate of solution of eutectic in Al–4.5Cu alloy as a function of (dimensionless) time. Data from Singh et al. (1970).*

frozen-in as compact and relatively harmless defects. The mechanical properties are good in the rapidly solidified regions, and poor in the poorly fed regions. Sketch (b) illustrates the benefit of keeping some pressure on the solidifying liquid, in this case by the planting of a feeder. The opening of the bifilms is thereby resisted to some extent, and improved properties are retained. Sketch (c) illustrates the case where the metal is superbly clean. The properties are good everywhere regardless of cooling conditions. (The reader will notice some solid feeding, however.)

We can therefore make an interesting prediction for results such as those presented in Figure 9.2. If there were no bifilms in the liquid metal, the curves of UTS and elongation to failure versus DAS would both be nearly horizontal lines (a slight downward slope might be expected as a result of other factors, such as interdendritic regions becoming less clean, as discussed earlier in section 9.2.2). Effectively, properties would become insensitive to DAS.

Some evidence for this utopia, of properties being independent of DAS, comes once again from early experience with the Cosworth process. When Cosworth Engineering made its racing car engines from castings poured in conventional foundries, at least 50 per cent of all the cylinder heads (four-valve-per-cylinder designs) failed by thermal fatigue between the exhaust valve seats. In 1980, when cylinder heads became available for the first time from the new process, characterized by quiescent transfer and counter-gravity filling, the failures fell to zero immediately. The DAS was hardly changed, and was in any case considerably larger than that available from permanent mould casting. Thus DAS was not controlling in this case. It is difficult to avoid the conclusion that the action of bifilms was crucial in this experience.

The action of some casting customers specifying DAS to avoid thermal fatigue in critical locations like the exhaust valve bridge in cylinder heads is seen therefore to miss a valuable potential benefit. Certainly control of DAS will be important for casting systems that provide poor filling control and thus contain a high density of bifilms. However, it is almost certainly an unnecessary complication and expense in a process designed to provide good quality metal. In general, for casting processes as a whole, it would be more constructive to specify the reduction in bifilms by good processing techniques. This would achieve benefits throughout the whole casting, not merely in those designated regions where a limited DAS has been specified.

9.3 Compact defects

In general, any non-metallic discontinuity in the structure, whether a hard, soft, brittle, gas or vapour phase, will probably impair the properties to some degree.

However, of course, the more compact the defect the less damage will be suffered. At the point at which a defect is smaller than the microstructural features of the alloy (i.e. smaller than the DAS) it probably ceases to be important. It is important therefore to make a distinction between the less harmful defects considered in this section and the more extensive planar defects considered in the next section. (Very large planar defects can measure 10 to 10 000 times the linear dimensions of the DAS.)

However, the number of defects present in a given volume, i.e. the density of defects, is also important, and in some cases will override the effects of the compactness of the defect. We need to keep this in mind, since we shall look into the twin aspects of defect size and defect density in the section on properties.

9.3.1 Inclusion types and diagnosis

In aluminium alloys the standard inclusions are isolated fragments of aluminium oxide (we shall consider films of oxide later). The fragments are too thick and chunky to be newly formed oxide, but are almost certainly particles carried over from the melting furnace. To be included in the alloy they will have necessarily undergone an entrainment event, so that they will themselves be entrained in a new oxide envelope. If this entrainment was some minutes or hours prior to casting there is a good chance that the defect can be regarded as a completely 'old' oxide, with its new envelop more or less welded nicely in place all over its surface as a result of continuing oxidation by the entrained layer of air, and the possible contribution of sintering. Thus it will remain permanently compact, being unable to reopen in the mould cavity.

Such alumina inclusions are extremely hard. During machining they are often pulled out of the surface, forming long tears on the machined face. At the same time they cause the machinist additional trouble by chipping the edge of the cutting tool.

In magnesium alloys the oxide and flux inclusions similarly have considerable nuisance value. However, in addition, because of its high reactivity, magnesium has a mould/metal reaction product that is unique, and consists of reacted sand inclusions. Lagowski (1979) carried out a detailed study of these defects and their effects. It seems that the sand grains are detached from the mould wall and swept into the interior of the melt. The majority sink and sit on the drag surface of the casting. Here the 0.1 to 0.2 mm diameter grains react with the magnesium, swelling and dissolving, until finally only a nebulous halo of the order of 1 mm diameter

remains. On a radiograph the residue can look like a snowstorm. However, despite the alarming appearance, the mechanical properties appear to be hardly affected, provided the radiographic rating does not deteriorate beyond 4 on the ASTM radiographic standards. (The ASTM standard radiographs represent increasing severity of defects ranging from the least severe, rated 1, equivalent to a single inclusion of approximately 1 mm diameter in the representative 50 mm × 50 mm area, to the most severe, rated 8, depicting a blizzard.)

The rather limited types of inclusions in the light alloys contrast with the huge variety of inclusions found in the higher-melting-point materials. We shall take a brief look at carbon steels as representing an example of a material showing most of the inclusion types to be found in ferrous materials, despite the fact that compared to the light alloys, they are usually impressively clean.

The sources of inclusions in steels are many and complicated. The liquid metal in the melting furnace is probably fairly clean at a late stage of melting because of the density difference between the non-metallic phases and the liquid steel, encouraging rapid flotation. However, prior to casting, the steel must be deoxidized by the addition of elements such as Si, Mn and Al, etc. known as deoxidizers. A proportion of the inclusions generated from this action separate by gravity. However, the transfer of the melt from the furnace into the transfer ladle and into the mould are pouring actions that create new surface, revealing the remnants of the newly added deoxidizing elements. In this way large fresh areas of oxide are created, which are then entrained by the surface turbulence. Huge numbers of inclusions, generally known in the trade as reoxidation products, result from these processes. They have been discussed in section 5.6.

However, in addition to these sources, steels are also noted for the number of very fine additional inclusions that form later, during solidification. The remaining unreacted deoxidizing elements are concentrated in the interdendritic liquid, where more inclusions pop into being by a nucleation and growth process, as also described in section 5.6. The interdendritic regions are small, limiting the size to which such inclusions can grow. Svoboda et al. (1987) observe that these inclusions are often also associated with small amounts of MnS. This is to be expected, since both manganese and sulphur will also be concentrated in the interdendritic spaces. It is possible that, because of their small size, some of the deoxidation products may be pushed ahead of the growing dendrites, and so become the nuclei for precipitates that arrive later.

As solidification proceeds, other inclusions may form by this concentration of segregated solutes in the interdendritic spaces. These may include nitrides such as TiN, carbides such as TiC, sulphides such as MnS and oxy-sulphides, etc. The formation of these products is discussed in section 5.6. In general they will be most concentrated and largest in size in the regions between grains, and in regions of the cast structure where segregation is highest, such as in channel segregates and the tops of feeders. They may also enjoy excursions into the casting as this remaining enriched metal from the feeder is sucked into the casting during the very last moments of feeding. The region under the feeder is known for its segregation problems.

Later still, further precipitation of inclusions will occur in the solid state. In general, these are even finer. The driving force for their appearance is the decreasing solubility of the elements as temperature falls. Many hardening reactions are driven in this way, for instance the formation of aluminium and vanadium carbides and nitrides in steels and the precipitation of $CuAl_2$ phase in Al–4.5Cu alloy. The hardening is the consequence of the very fine size and spacing of the precipitate, making the inclusions effective as impediments to the movement of dislocations.

9.3.2 Gas porosity: types and diagnosis

There are several different types of gas porosity, all of which are common in castings. They have to be diagnosed correctly, otherwise a wrong and consequently ineffectual treatment may be undertaken. The main categories are listed below.

9.3.2.1 Gas precipitated from solution

I hesitate to call this 'true gas porosity', but that would convey the correct sense. It is porosity that arises as a result of gas, in solution in the liquid metal, which precipitates out during solidification. The subject is covered in detail in Chapter 6.

Also, it is necessary to re-emphasize the message given in section 6.1. Pores cannot appear simply because the gas in solution has been rejected ahead of the advancing solidification front. If there were no nucleation sites present, the gas might exceed its solubility limit in the metal, but would still be unable to form a pore. The gas would simply remain in supersaturated solution. For aluminium alloys, the presence of supersaturated hydrogen in the solid does not appear to be harmful. For hydrogen in ferritic steels, however, the hydrogen may lead to the serious problem of hydrogen embrittlement.

In this section we shall assume that there are plenty of nuclei on which gases can precipitate during freezing. The most prolific nuclei will be expected to be bifilms, since they offer such an easy precipitation site and minimal resistance to unfurling.

In metals that solidify dendritically, the precipitated gas appears in the interdendritic spaces,

and so is a well-dispersed, fine precipitate. This is its usual form. For castings of a few kg in weight the pores are usually in the range of 0.01–0.5 mm in diameter, and uniformly distributed throughout the whole casting. They can be up to a millimetre or so in diameter for larger castings. In general, however, these pores are so small that they are often invisible on a machined surface of an Al alloy because a cutting tool with a carbide tip will smear the metal over, concealing such microscopic defects. On the other hand, a diamond-tipped tool will cut cleanly, and thus render the defect visible, perhaps with the aid of a magnifying glass.

On the introduction of polycrystalline diamond-tipped tools in the 1970s the automotive piston foundries in the UK were brought to an alarming stop. They were unable to solve the sudden increase in the apparent porosity in their castings. No customer was prepared to buy pistons that were clearly full of holes. The industry was forced to make a traumatic leap in the quality of its liquid metal, and lived to fight another day.

As a result of the requirement for the gas in solution to build up and concentrate before a precipitate can nucleate, there may be a sound skin over the outside of the casting. Thus the first pores appear usually 1 or 2 mm under the surface. This appearance of 'subsurface porosity' is simply a variant to be expected when conditions are right for the freezing of a sound skin, as discussed in section 6.2. It is a condition part-way between the completely uniform distribution of gas pores as discussed above and the highly directional form that is to be discussed below.

In metals that solidify on a planar front the gas pores may grow, keeping pace with the growth of the front, causing long, tunnel-like defects that are sometimes called wormholes. They are often observed in ice cubes because tap water freezes on a fairly planar front. The pores may be up to 1 mm or so in diameter and of various lengths of up to 100 mm or more. An example of pores nearly 20 mm long are seen in Figure 6.18. This radiograph of an aluminium-bronze plate cast in a sand mould using non-degassed metal contrasts with a similar radiograph of a plate (not shown) cast from properly degassed melt, in which no defects at all could be seen.

All of these forms of gas porosity can be identified easily and with certainty.

It has to be kept in mind that the source of gas in solution can be that in the metal from the melting conditions. For a sand casting, additional gas in solution will have been acquired during its journey through the runner, and from continuing reaction with the mould during and after mould filling. All of these sources cooperate, and may raise the gas content of the metal over the threshold at which pores will nucleate and grow.

9.3.2.2 Entrained bubbles

When the liquid metal tumbles into the mould in a ragged fashion, it necessarily entrains air or other mould gases. These will attempt to rise and separate from the metal if they have time and if they are not hindered in some way. However, the oxide trails attached to the bubbles, and the oxide film that they encounter on their journey, or when they reach the surface of the casting, will all conspire to retain the bubbles in the casting. Such bubbles are common in film-forming systems, such as aluminium alloy castings, because the smaller bubbles do not have sufficient buoyancy to burst through (i) the oxide film at the surface of the casting, and (ii) its own oxide film (notice it is a double film every time that has to be ruptured) for the bubble to escape to atmosphere. The smaller bubbles therefore come to rest at a depth of only a double oxide skin thickness under the surface. Their sizes range typically from 1 to 5 mm in diameter. The larger bubbles have sufficient buoyancy to break out and escape. Such defects are much less common in grey iron castings, where the surface film is a liquid, and so allows the easy escape of buoyant defects.

Any prior solidification of the surface will further add to the difficulty of escape, but such extra hindrance is less common because the rate of rise of bubbles is fast, taking place usually before freezing. Exceptions are, of course, in thin-walled sections of castings that freeze during filling. Pressure die castings are the extreme example that succeed to trap and retain the majority of the air in the die.

In gravity-filled castings, the bubbles will therefore be generally concentrated against the undersides of cores, or in the highest parts of the casting, to which bubbles will rise after being introduced via the ingates. The distribution is far from uniform; their concentration over ingates is highly characteristic, and the remainder of the casting will often be quite free from defects. Sometimes the bubbles are concentrated over the first ingate on the runner, and sometimes over the last, depending on whether the runner is sufficiently deep to encourage a reverse flow to roll over the underlying incoming flow.

These concentrated accumulations of severe defects are revealed after shot blasting, or, more inconveniently, after machining. They are typical consequences of a faulty running system.

Although often called 'gas holes', or 'blowholes', it is more helpful to call them 'air bubbles' to distinguish them from the finer and more well-distributed pores that precipitate from gas in solution in the liquid metal and from true core 'blow' defects. Naturally, although the majority of gas in many of such defects will be air, there will be an increasing content of binder outgassing vapours as sand moulds or cores heat up. Eventually, such bubbles may contain only mould vapours. Even so, the designation 'air bubbles' is sufficiently accurate for most purposes, and effectively describes their origin as entrainment defects.

9.3.2.3 Core blows

Blowhole defects have been dealt with in section 6.4. A brief review plus some additional data are included in this section.

Blowhole defects are caused by the outgassing of a core, or occasionally of a mould. They are characterized by huge size. The minimum diameter of a bubble that can form on a core is perhaps 10 to 20 mm (Figure 6.22). However, if the core contains gas at sufficient pressure to overcome the hydrostatic pressure and cause a blow defect, then it will usually have enough volume to continue to blow for some time. In this way strings of bubbles from the core can accumulate to create a massive pore near the top of the casting. The diameter of this cavity can easily reach 100 mm. The author has seen the whole top half of a casting made hollow by this process. (It suggests an interesting manufacturing technique for hollow ware!)

Because the core will usually take time to absorb heat, and build up its pressure prior to blowing, the top of the casting may have had sufficient time to become partly frozen. Thus because of its late arrival, a core blow is often some distance beneath, with its upper surface parallel to, the surface of the casting. (Note that this behaviour contrasts sharply with turbulence-entrained gases that arrive early, together with entry of the liquid metal into the mould, and which are prevented from escape only by the thickness of the surface oxide film.)

The site of a core blow is usually above the highest point of the core, especially if this is a sharply upward-pointing shape, like a steeple of a church. The bubbles will collect and detach preferentially upward from an upward-pointing shape. (It is the upside-down equivalent of the drop of water falling preferentially from the tip of a downward-pointing stalactite.) Such shapes are therefore best avoided, or the mould assembly turned upside-down to avoid the problem. Core bubbles will, of course, detach from flat surfaces but with greater difficulty. All cores benefit from being internally vented to reduce the build-up of pressure that might lead to blows as is dealt with quantitatively in section 4.5.1.

In film-forming alloys the passage of a bubble through the melt naturally creates a bubble trail. The passage of a large number of bubbles can therefore be highly damaging, leading to the development of a thickly oxidized trail that can form a highly efficient leakage path through the casting. Exfoliation-type defects can also be created where the bubbles break the surface of the casting (section 2.2.8).

In sand moulds it is common for blows to occur from chills or any other surfaces that are impermeable. For this reason the application of block chills to critical areas of mould sometimes gives disappointing results. The casting is often degraded by the boiling action of volatiles that have condensed on the chills.

Condensation on chills can occur prior to filling, especially if a sand mould is closed and stored overnight. The mould slowly cools during the night. The next morning the more rapid reheating of the mould as the foundry environment warms up drives moisture and other volatiles inwards, where the chill, still cold, surrounded by the insulating mould, now forms a convenient site for condensation.

Condensation on chills can also occur during filling. The progress of the hot metal through the mould produces clouds of volatiles that rush ahead, condensing on any cold surface.

The provision of surface vents on chills by (i) coating the chill with a permeable refractory, and (ii) creating V-grooves in a hatched pattern are both useful measures to reduce the problem. The grooves also seem to enhance the chilling action, possibly by their additional area, but probably mainly by the enhanced contact during the contraction of the casting against the grooves, the effect aided by the expansion of the chill.

There is a further danger of blows off some core repair pastes that are based on clays. The clays are composed of such fine particles that they are effectively completely impermeable. Although the repairs are dried, possibly by leaving in the air, or by gentle warming, the water of crystallization in the clay is not released until the clay reaches a temperature in the range 500 to 700°C. (The baking of the core at such high temperatures cannot, of course, usually be carried out because of the premature breakdown of the organic binder in the sand.) Thus the arrival of the melt causes this bonded moisture to boil out of the clay, and the released vapour, being unable to escape through the dried clay, is forced to escape through the melt. The creation of blow defects, and damage to mechanical properties and to leak tightness automatically results.

A summary of gas pore diagnostic information is given in Table 9.1.

9.4 Planar defects

All the planar defects whose diagnosis is discussed below are capable of constituting extensive faults in the casting, and so are of great concern with regard to its strength. Their effect on the various aspects of strength is discussed in section 9.5.

9.4.1 Layer porosity

Layer porosity is that type of shrinkage porosity that forms in long-freezing-range alloys under conditions intermediate between those for concentrated macroporosity and dispersed microporosity. This important intermediate condition is clear from Figure 7.13. Whereas macroshrinkage in the form of a shrinkage pipe, or large central cavity, can be viewed as a compact defect, and dispersed micropores can be viewed as individual compact defects, the intermediate condition is more severe, in the sense that a relatively small volume of porosity has formed itself into an extended defect by concentrating into a sheet or layer.

Recall the way in which layer porosity is formed. The pore spreads across a surface of constant tension that exists in the residual liquid in the dendrite mesh. This formation mechanism results in the unique structure of the defect. The layer is not exactly equivalent to a crack, since the dendrites are hardly disturbed during its growth; it is only the residual liquid that moves, opening up the planar defect. In fact the dendrites continue to bridge the pore as a dense array, effectively stitching the layer together with closely spaced threads. For this reason its effect on tensile properties seems to be

Table 9.1 Gas porosity diagnosis

Pore type	Size range	Spatial distribution	Special features
1. Pores precipitated from solution	0.05–0.5 mm	Uniform.	May develop 1mm pore-free rim under casting surface. Alternatively, may develop as 'worm hole'
2. Air bubbles from poor filling	0.5–5 mm	An oxide skin depth (<0.1 mm) under cope. Often aligned above ingates.	Always associated with oxide films (bubble damage).
3. Core blows from outgassing of binder	10–100 mm	Under solidified thickness (1–10 mm deep) of cope.	Always attached to originating core by thickened bubble trail. Large core blows faithfully parallel the cope profile of the casting, and have a horizontal base.

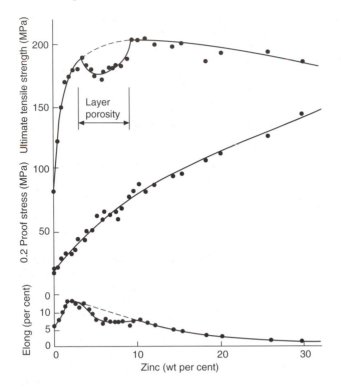

Figure 9.6 *Effect of layer porosity on the mechanical properties of Mg–Zn alloys. Data from Lagowski and Meier (1964).*

intermediate between that of individual pores and a fully open hot tear or crack. This intermediate position is seen in its effects on mechanical properties and will be discussed in section 9.5.

The presence of layers is usually not easily discerned optically on a polished section because there are so many bridging dendrites that the overall pattern is almost hidden, causing the porosity to appear as normal dispersed microporosity. However, it is clearly revealed on radiographs if the view direction is oriented along the plane of the porosity.

9.4.2 Tears and cracks

Hot tears and cold cracks form as a result of tensile stresses in the casting during cooling. They typically occur, therefore, across sections and in radii, where they will have an especially damaging effect on the serviceability of the casting. Their diagnostic features have already been discussed in Chapter 8.

Such defects can extend to the whole casting cross-section, in which case, of course, the casting is already broken, and therefore probably identifiable as defective before being put into service!

However, a surface crack that may have formed and opened at high temperature may close again as the casting cools because of the normal development of compressive stress in the casting surface during the final stages of cooling, as discussed in section 8.5. Such a crack, closed tightly under pressure,

would be extremely difficult to detect by normal methods such as dye-penetrant testing and radiography. It is possible, therefore, that this is a rarely detected but common defect. It represents a serious concern in safety-critical components. Its detection might require prestressing to open the crack while being viewed by real-time radiography, or while being investigated by a penetrant dye. The writer is not aware that such testing techniques are ever used. However, for the most important of applications they would be well worth careful consideration.

It is worth bearing in mind once again that the extensive crack-like defects are probably not so much initiated by bifilms, but actually *are* the bifilm. These entrainment defects can easily extend over the whole casting cross-section. The visible presence of one of these defects will almost certainly warn of the presence of many more, less easily seen neighbours, and is the reason why it is probably futile to attempt to repair such defects by welding, for instance.

The lesson remains clear. Intensive inspection and repair is no substitute for good metal, quiescently filled.

9.4.3 Bifilms

The double film defect, the bifilm, is formed by the folding-in of a solid surface film during surface turbulence. Its composition is usually an oxide, nitride or carbon. The film defect can be small and negligible, or extensive, occupying the whole of a cross-section. Similarly its placing and orientation are random, so that either its chance placing, or its orientation, or its size, or all factors together, in a highly stressed region, may be damaging. It is the unpredictable nature of the double film defect that is such a concern.

I am often questioned why a sporadic defect such as a crack always appears at the same place in a particular casting. Clearly the pattern of stress concentration in the component is fixed. The distribution of bifilms means that sometimes a bifilm will happen to be sited in the stressed region, and thus be opened by the strain. If there is a high concentration of bifilms, there will then always be at least one in this region, with the result that all castings will exhibit a similar crack.

Figure 9.19 shows two simple plates 10 mm thick cast in 99.5 per cent Al in the author's laboratory (Runyoro 1992). They were cast in a vertically parted mould via a bottom gate into the centre of the long side at ingate speeds of above and below 0.5 ms^{-1} respectively and were afterwards subjected to a three-point bend test (later work progressed to use the much improved four-point bend test). The turbulently filled casting exhibited a number of cracks, none of which were along the line of the maximum stress. In fact the cracks were aligned in random orientations. The randomness is a strong clue to their origin as the products of surface turbulence; since turbulence means randomness, chaos and essentially unpredictability. The material between the cracks is clearly highly ductile, as would be expected for rather pure cast aluminium. Also,

interestingly, under the microscope the tips of the cracks are found to be blunt. They have been unable to propagate in the ductile aluminium, which constitutes a crack-blunting matrix. It follows that the cracks have not propagated but must have pre-existed, and have merely opened. All these features are to be expected from entrained bifilms. The test plate (Figure 9.19b) cast at a speed at which entrainment is not possible is clearly free from bifilm defects.

If the bifilm is sufficiently large, and if it breaks the surface, then it *may* be detected by normal dye-penetrant methods. However, in film-forming alloys the presence of the oxide film on the casting surface will often prevent any extensive surface connection, so that the dye penetrant approach is not reliable, indicating perhaps only a minute pinpoint contact. The 'stars at night' appearance of defects on vacuum-cast Ni base alloys when the fluorescent penetrant dye is viewed under ultraviolet light are mostly defects of this kind. This is easily demonstrated by putting on a well-designed filling system, after which the 'stars' all disappear! Bifilm defects are also difficult to detect by radiography unless the incidence of radiation happens to be in or near the plane of the defect, or unless the central region is sufficiently open to constitute porosity, or the enclosing films sufficiently thick to appear as inclusions.

In contrast to the difficulty of observing bifilms non-destructively, they can sometimes be clearly seen on fracture surfaces. Very thin films require careful study of the fracture surface to reveal their minute characteristic signatures such as the tracery of fine folds. LaVelle (1962) studied the fracture surfaces of aluminium alloy pressure die castings and found layer upon layer of oxide films throughout the structure, giving the fracture a fibrous appearance. In some cases the surface contact of these extensive defects was minimal, indicating how unreliable their detection by surface penetrant dye techniques could be.

Because film defects are potentially so damaging, and because they are practically impossible to detect by non-destructive methods with any certainty, the only suitable method to deal with this problem is not to attempt to inspect for them, but to set in place manufacturing techniques that guarantee their avoidance. At the present time no specification calls for specific manufacturing techniques that would eliminate film defects. This unsatisfactory situation has to change.

Only control of the quality of the melt and control of the filling of the casting will prevent film defects entering the mould. Once good quality metal is in the mould, measures will be required to prevent the reintroduction of defects by such processes as the outgassing of cores, or other forms of surface turbulence after the filter (if any) in the runner.

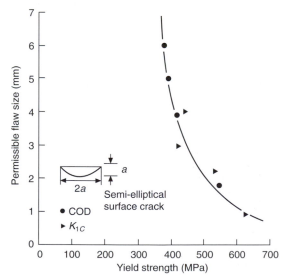

Figure 9.7 *Permissible edge defect sizes in ductile irons of increasing strength. Data from Seetharamu and Srinivasan (1985).*

Only by carefully planned and monitored processing will film problems be overcome with confidence.

Again, it is worthwhile to repeat, the answer lies not in inspection, but in process control.

9.5 Effects of defects on the properties of castings

It is salutary to reflect that in the past, castings that have contained large undetected defects have gone on to serve useful lives in critical applications. Others, however, have been less lucky. In the UK the Tay Bridge disaster is a lesson not to be forgotten. (The bridge over the River Tay, Scotland, built of wrought iron and cast iron, with 84 spans each 70 metres long, was considered one of the engineering wonders of the world. On the night of 29 December 1879, during a great storm, the bridge collapsed while a train was crossing, with the loss of 78 lives. The subsequent inquiry severely criticized the design that had made no allowance for side loads of wind and flood, and, more particularly, the poor quality of the iron castings, whose defects had been concealed with lead.)

The mixed performance of some castings arises because of the elementary point that the size of the defect is often of much less importance than its form and position. For instance, a large pore in a low-stressed area of the casting may be far less detrimental than a small bifilm in a sharp corner subject to a high tensile stress. To have blanket specifications requiring the elimination of all types of defect from every area of the casting is therefore not appropriate, and has resulted in the scrapping of many serviceable castings. The most logical and effective control over casting performance is achieved by specifying separate *designated regions* of the casting, each designated region being required to contain no defects above the critical size appropriate for that location.

Regarding the effect of pore size in particular on mechanical properties, there has been much claimed for the use of maximum pore length as a measure of defect size, as opposed to its area or its volume (Herrera and Kondic 1977; Surappa *et al.* 1986a; Jaquet 1988). However, later work by the same authors (Surappa *et al.* 1986b) has shown that there is an excellent linear relation between the maximum size of pore and the effective pore diameter. Thus these measures seem interchangeable; neither can claim to be more fundamental than the other.

The list of properties that a casting may be required to possess can be long. It might include, for instance, high-temperature properties such as creep resistance, high-temperature fatigue resistance, or oxidation resistance. Alternatively, room temperature properties such as resistance to corrosion in specific environments may be required. The different defects have quite different effects on each of these properties, as we shall examine in turn.

9.5.1 Yield strength

Only the ferrous materials have a well-defined yield stress. For other materials we shall assume that the 0.2 per cent proof stress (0.2PS) is sufficiently equivalent for our purposes. (The 0.2PS is that stress at which reversible elastic deformation has reached just exceeded its limit, and a small amount, 0.2 per cent, of permanent plastic deformation has occurred.)

Because no substantial deformation has taken place, it is logical to assume that the yield stress will be unaffected by most defects. The only effect will be that due to the reduction in area, and since most defects occupy at most only a few per cent of the area of the casting, this effect is usually hardly detectable.

For instance, the insensitivity of yield to layer porosity is seen in Figure 9.6. The 1 per cent or so of reduction in the proof stress because of the 1 per cent or so loss of area cannot be detected in the scatter of the results. Strength and elongation are, however, noticeably affected, as will be discussed below.

In those cases where the yield stress is found to be significantly reduced, it is a warning that the material had lost a significant fraction of its load-bearing area. In the past it has not been easy to accept that the reduction of 0.2PS by 90 per cent is the result of a crack occupying 90 per cent of the area, especially when the fault is not easily seen, even on the fracture surface! Figure 2.42 shows a fracture surface on a T-6 Al–4.5Cu alloy from a tensile specimen that was expected to show 10 per cent elongation to failure. However, by X-ray video techniques, the specimen was observed to suffer some surface turbulence during the filling of the mould, and the figure of 0.3 per cent (effectively near zero) elongation was found instead. Close inspection under the SEM confirmed that the whole surface was covered with an oxide film, one half of the original bifilm. No part of the surface was found to exhibit ductile dimple fracture. Thus the specimen was effectively 100 per cent cracked prior to testing. The only strength and elongation observed arose almost certainly as a result of a minor amount of plastic deformation as a result of the tensile failure having to propagate by plastic bridges between different bifilms.

9.5.2 Fracture toughness

Fracture toughness is a second material property that is generally independent of the presence of gross defects. This is because it is assessed by the force required to extend a crack that has been artificially introduced in the material usually by extending a machined notch by fatigue. Thus fracture toughness measures the properties of the matrix at the point at which the notch is placed, i.e. it is a material property. It is not a property like tensile strength or ductility in which the crack finds its own start location, ensuring failure from the largest defect. When preparing the toughness assessment test piece the machining of a notch into the specimen at a pre-fixed location would be unlikely to encounter a major defect by chance.

Fracture toughness is the material property that allows the prediction of the shapes and sizes of defects that might lead to failure. It is a basic tenet of fracture mechanics that fracture may begin when the stress-intensity factor K exceeds a critical value, the fracture toughness K_{1C}. A detailed presentation of the concept of fracture toughness is beyond the scope of this book. The interested reader is recommended to an introductory text like that by Knott and Elliott (1979). Here we shall simply assume some basic equations and the experimental results.

Stress intensity K as well as fracture toughness have the dimensions of stress times the square root of length, and is most appropriately measured in units of $MNm^{-3/2} = MPa.m^{1/2}$. (Care is needed with units of this property. Fracture toughness is sometimes measured less conveniently in $N.mm^{-3/2} = MPa\ mm^{1/2}$ that differ by a factor of $(1000)^{1/2} = 31.6$.) For a penny-shaped crack of diameter d in the interior of a large casting, the critical defect size at which failure will occur is approximately:

$$d = 2K_1C^2/\pi\sigma^2 \qquad (9.6)$$

from which, for an aluminium alloy of fracture toughness 32 Mpa m$^{1/2}$ at its yield point of 240 MPa, the critical defect size d is 11 mm. For an edge crack, Equation 9.6 is modified by a factor of 1.25, giving a corresponding critical defect size of 9 mm, indicating that edge cracks are somewhat more serious than centre cracks, but in any event defects of about a centimetre across would be required, even at a stress at which the casting is on the point of plastic failure. These are comparatively large defects, which is reassuring in the sense that such large cracks may have a chance to be found by non-destructive tests prior to the casting going into service. However, they underline the conclusion that most current radiography standards which state 'no linear defects' of any size, or which reject aluminium alloy castings for flaws of only approximately 1 mm in size, may not be logical.

The important message to be learned from the concept of fracture toughness is that the maximum defect size that can be tolerated can be precisely calculated if the fracture toughness and the applied stress are both known.

Equation 9.6 can be used to indicate that at the limit, when the applied stress σ equals the yield stress σ_y, the greatest resistance to crack extension that can be offered by a material is controlled by the value of K_{1C}/σ_y (units most conveniently in m$^{1/2}$). This parameter is a valuable measure of the defect tolerance of a material.

Figure 9.7 shows how the permissible defect size in ductile irons is large for low-strength irons, but diminishes with increasing strength because of the fall of fracture toughness.

Thus for the stronger irons the permissible defect size is below 1 mm. This is particularly difficult to detect, and sets a limit to the stress at which a strong ductile iron casting may be used with confidence.

Steels can have high fracture toughness, with correspondingly good tolerance to large defects at low applied stress. However, because they are used in highly stressed applications, the permissible defect size is reduced, as Equation 9.6 indicates. Jackson and Wright (1977) discuss the serious problem of detecting flaws when the permissible size is small. At the time of writing most non-destructive testing methods have been found wanting in some respect. However, the technology of non-destructive techniques has improved significantly and is expected to continue to improve, allowing the detection of smaller defects with greater certainty, and so extending the range over which castings can be stressed while reducing the risk of failure.

Nevertheless, Jackson and Wright's conclusion is good advice: choose an alloy with a large critical flaw size, which can be detected and measured accurately. Also, if the working load is fluctuating, it will be necessary to calculate from crack propagation rate curves the time for a flaw to grow to the critical size. This can then be compared to the design life of the component.

To be safe, they recommend the following assumptions:

1. A string of cracks is equivalent to one long crack.
2. A group of flaws is one large flaw of size comparable with the envelope that circumscribes them.
3. Unless clearly seen to be otherwise, all flaws have a sharp aspect ratio.
4. All flaws reside on the surface.

A casting designed on this basis is likely to be somewhat overdesigned, but perhaps not so much as if there had been no use of the principles of fracture mechanics.

So far all the discussion has related to linear elastic fracture mechanics (LEFM), in which the fracture resistance of a material is defined in terms of the elastic stress field intensity near the tip of the crack. In fact, the fracture toughness parameter K_{1C} is valid only when the region of plastic yielding around the tip is small.

For lower-strength materials, plastic deformation at the tip of the crack becomes dominant, requiring the application of yielding fracture mechanics (YFM) and the concept of either (i) crack opening displacement (COD or δ), or (ii) the J integral, as a measure of toughness.

The critical COD is the actual distance by which the opposite faces of the crack separate before unstable fracture occurs. This critical opening is known as δ_c. The critical flaw size is given by:

$$\delta_c = c(\delta_c/\varepsilon_y) \quad (9.7)$$

where c is a material constant and ε_y is the strain at yield. For materials that are on the borderline for treatment by LEFM or YFM, Jackson and Wright (1977) indicate that a unified test technique can be used to determine K_{1C} or δ_c from a single test piece, and that the following useful relation may be employed:

$$\delta_c/e_y = (K_{1C}/\sigma_y)^2 \quad (9.8)$$

Hence the ratio δ_c/e_y in YFM is comparable with $(K_{1C}/\sigma_y)^2$ in LEFM, i.e. it is a measure of the defect tolerance of a material. It deserves to be much more widely used in the design and specification of castings.

The transition between K_{1C} and COD, when properly applied, is seen to give consistent predictions of the critical defect size. The results

in Figure 9.7 for ductile irons show a region of smooth overlap between the two approaches.

Although the COD has a clear physical basis, in both its formulation and its application, there are empirical assumptions that are difficult to justify by the rigorous application of mechanics at this time. For this reason, more attention is being given to the use of the J integral in structural-integrity assessment, especially in the USA. J is a crack-tip driving force that, under certain conditions, can be applied to elastic/plastic situations to describe the onset (or small amounts) of stable crack growth. For a given geometry of loading and size of test piece (including size of defect), J is related directly to the area under the load–displacement curve obtained for the cracked body, where displacements are measured at the loading points.

A general overview of the fracture toughness and strength of cast alloys is given in Figure 9.8. This indicates the rather poor properties of flake irons, and the excellent properties of some steels and titanium alloys. Ductile irons occupy an interesting middle ground. In general it seems that most groups of alloys can exhibit either high strength and low toughness, or high toughness and low strength. The result is the hyperbolic shape of the curve as seen for ductile irons. As more results for other alloys are determined, Figure 9.9 will be expected to develop into a series of overlapping hyperbolic regimes containing the various alloy systems.

Figure 9.9 shows how the fracture toughness values can be translated into permissible flaw

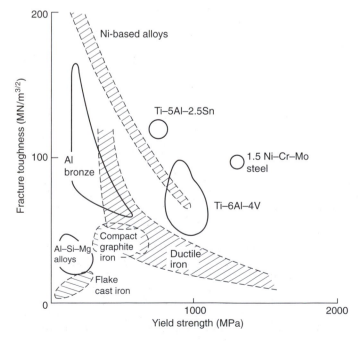

Figure 9.8 *Map of fracture toughness versus yield stress for various cast alloys. Data from Speidel (1982) and Jackson and Wright (1977).*

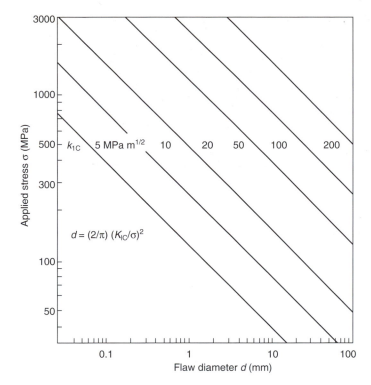

Figure 9.9 *Relation between fracture toughness, applied stress and the critical defect size for failure in a material containing a central circular flaw.*

diameters, assuming a central penny-shaped crack and a given level of applied tensile stress.

Figure 9.10 presents interesting results for the mechanical properties of a high-strength aluminium alloy that is not normally cast. In this work it was cast and given a range of heat treatments covering the peak response condition. In the optimally hard state it had an elongation close to zero, and would not normally be used in this condition because a designer or user would be nervous about brittle failure. It is in the nature of brittle failure (i.e. failure without general plastic deformation) that there is no benefit from either the redistribution of stress, nor any prior warning of impending failure by the general plastic yielding of the casting. Nevertheless, the fracture toughness is clearly little influenced by ageing, and retains a respectably high value regardless of the low ductility. This means that any crack or other defect will always have to exceed a certain critical size before failure occurs, even if such failure is eventually of a brittle character.

This general behaviour is confirmed for the more common casting alloy Al–7Si–0.4 Mg: Figure 9.4 shows that the fracture toughness falls with increasing iron content in the alloy, but significantly less than the fall in ductility, assessed by the reduction in area.

In those cases where general deformation of the casting cannot be tolerated, therefore, the use of fracture toughness is a more appropriate measure of the reliability of the casting than ductility. It

seems that for many years we have been using the wrong parameter to assess the reliability of many of our castings.

9.5.3 Fatigue

A definitive study of fatigue in a cast alloy Al–7Si–0.5Mg was carried out by Pitcher and Forsyth (1982). These workers were able to show that in general the fatigue performance of cast alloys was poor because the initiation of the fatigue crack during stage 1 of the fatigue process was short. This observation appears to have been confirmed for cast Al alloys many times, for instance recently by Wang *et al.* (2001a,b) who found no time was required for crack initiation, and that the fatigue life was the time merely for propagation of the crack. With the wisdom of hindsight, we can conclude that this was almost certainly a consequence of the presence of bifilms. Thus the cast specimens were effectively pre-cracked. This conclusion is strongly indicated by the rather poor casting technique used for the preparation of test specimens, and additionally confirmed by the pores that appeared to be associated with films in some of their micrographs. Thus the size of initiating defects in their castings effectively eliminated stage 1.

However, Pitcher and Forsyth found that stage 2 of the growth of the crack, which is its stepwise propagation across the majority of the section, was remarkably slow compared to high-strength wrought

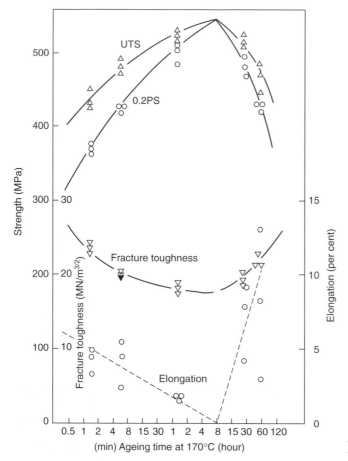

Figure 9.10 *Mechanical properties of a cast high-strength alloy, Al–6Zn–2.7Mg–1.7Cu, as a function of effective ageing time at 170°C. Data from Chen et al. (1984).*

alloys. The slow rate of crack propagation seemed to be the result of the irregular branching nature of the crack, which appeared to have to take a tortuous path through the as-cast structure. This contrasts with wrought alloys, where the uniformly even microstructure allows the crack to spread unchecked along a straight path.

The conclusion to be drawn from this work is that if crack initiation can be slowed, then fatigue lives of cast metals might be extended considerably, perhaps in excess of those of wrought alloys.

An indication of the general validity of this conclusion is confirmed by Pitcher and Forsyth. They shot-blasted their specimens to generate residual compressive stresses in the surface. This did help to slow initiation, and total fatigue lives were improved. For sand castings fatigue life was increased to that expected of chill cast material as shown in Figure 9.11. (It is a pity that the important high-stress/low-cycle part of the fatigue curve is often not investigated. It is interpolated here between the high-cycle results and the UTS values; i.e. the single cycle to failure result.) Chill-cast material is

subject to a similar substantial improvement when shot blasted. For steel castings Naro and Wallace (1967) also found that shot peening considerably improved the fatigue resistance.

Although the improvement of fatigue resistance following shot peening is perhaps to be expected, the effect of grit blasting is less easy to predict, because the fine notch effect from the indentations of the grit particles would impair, whereas the induced compressive stresses would enhance, fatigue life. From tests on aluminium alloys, Myllymaki (1987) finds that these opposing effects are in fact tolerably balanced, so that grit blasting has little net effect on fatigue behaviour.

Improvement is expected if cast defects are eliminated, so that the fatigue crack now needs to be initiated, thus introducing a lengthy stage 1. Figure 9.11 illustrates the case of a pearlitic ductile iron where consistently improved performance was obtained from the use of material with a reduced density of defects produced by filtering in the mould. The elimination of thermal fatigue cracking in the early Cosworth process Al–7Si–0.4Mg alloy

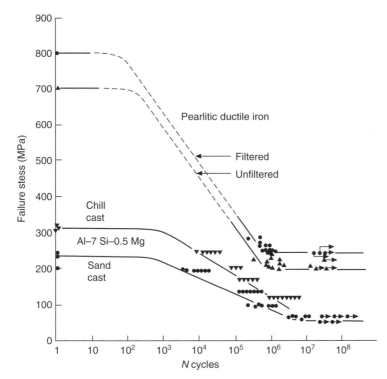

Figure 9.11 *Fatigue curves for filtered and unfiltered pearlitic ductile irons (data from Simmons and Trinkl 1987) and for an aluminium alloy (data from Pitcher and Forsyth 1982).*

cylinder heads is another instance of the importance of the elimination of defects, so as to introduce a long stage 1 to the fatigue crack propagation.

This industrial experience has now been confirmed by a careful laboratory study by Nyahumwa et al. (1998 and 2000) on Al–7Si–0.4Mg alloy castings. To vary the number density and size of oxide film defects in the castings, test bars were cast using bottom-gated filling systems with and without filtration. Test pieces were machined from the castings and were fatigue tested in pull–pull sinusoidal loading at maximum stresses of 150 MPa and 240 MPa under stress ratio $R = + 0.1$.

The use of the pull–pull mode of testing (positive R ratio) is important for two reasons. (i) The fracture surface is undamaged (in contrast to the hammering that normal laboratory fatigue specimens suffer) and so can be studied under the scanning electron microscope (SEM) in great detail. This is most important when searching for elusive features such as oxide films, and probably has been a significant factor explaining how such extensive defects have been overlooked until recently. (ii) In service conditions in many castings, the existence of residual tensile stress is a common feature because of inappropriate quenching practice after solution heat treatment. Some residual stress is to be expected even in carefully quenched castings. Thus fatigue failure is most likely to occur in regions already experiencing significant tensile loads. Thus pull–pull testing conditions will represent such regions

more accurately. It follows that fatigue testing in laboratories using reversed loading (push–pull, with negative R ratios) should be reconsidered in favour of pull–pull.

Test bars of an Al–7Si–Mg alloy (2L99) were cast in chemically bonded silica sand moulds. Two batches of test bars were cast using (i) a bottom filling unfiltered system through which a liquid metal entered the ingate at a velocity greater than $0.5\ \mathrm{ms}^{-1}$, and (ii) a bottom filling filtered system through a ceramic foam filter with 20 pores per inch (corresponding to an average pore diameter of approximately 1 mm); in this system a liquid metal entered the ingate at a velocity less than $0.5\ \mathrm{ms}^{-1}$. An SEM examination was carried out on the fracture surfaces of all the failed specimens to ascertain the crack initiator in every case. This was a huge exercise, probably never attempted previously.

Fatigue life data obtained from the filtered and unfiltered castings are plotted in Figure 9.12 for those tested at 150 MPa and Figure 9.13 for those tested at 240 MPa. The probability of failure for each of the specimens in a sample was defined as the rank position divided by the sample size (i.e. the test results were ranked worst to best in ascending order; so, for instance, the 20th sample from the bottom in a total of 50 samples exhibited a 40 per cent probability of failure).

Figure 9.14 shows the three main types of surfaces observed: (a) oxide film defects which

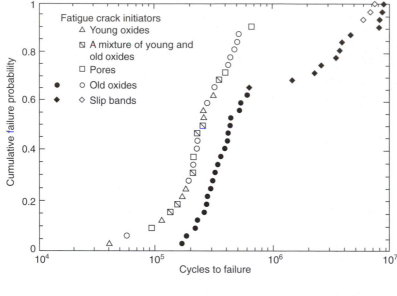

Figure 9.12 *Relation between the initiators of fatigue cracks and the fatigue life for unfiltered (open symbols) and filtered (solid symbols) Al–7Si–0.4Mg alloy castings tested at 150 MPa and R = +0.1 (Nyahumwa et al. 1998, 2001).*

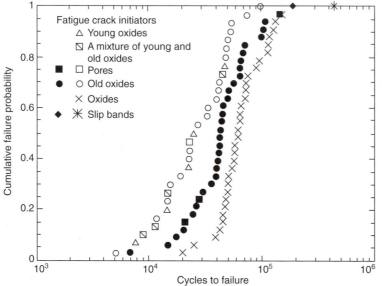

Figure 9.13 *Relation between the initiators of fatigue cracks and the fatigue life for unfiltered (open symbols), filtered (solid symbols) and unfiltered and HIPped (cross and star symbols) Al–7Si–0.4Mg alloy castings tested at 240 MPa and R = +0.1 (Nyahumwa et al. 1998).*

were categorized either as young, thin film or older, thicker film, based on the thickness of folds; (b) slip mechanisms indicated by a typical faceted transgranular appearance; and (c) fatigue striations, often called beach marks, denoting the step-by-step advance of the crack.

There was much to learn from this work. Figure 9.12 shows that for the unfiltered castings, most failures initiated from defects, sometimes pores, but usually oxides. The oxides were a mix of young and old. Only three specimens were fortunate to contain no defects. These exhibited ten times longer lives, and finally failed from cracks that had initiated by the action of slip bands. Thus these specimens

had lives limited by the metallurgical properties of the alloy.

The filtered castings (bold symbols) were expected to be largely free from defects, but even these were discovered to have failed mainly as a result of defects, the defects consisting solely of old oxides that must have passed through the ceramic foam filter. There were no young oxides and no pores. A check of the equivalent initial flaw size (determined from the square root of the projected area of fatigue defect initiators) showed that 90 per cent were in the range 0.1 to 1.0 mm, with a few as large as 1.6 mm. Thus most would have been able to pass through the filter without difficulty.

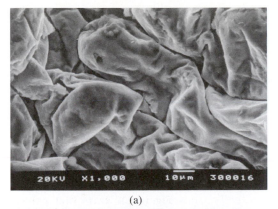

(a)

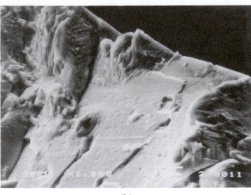

(b)

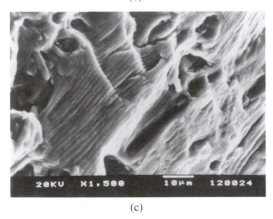

(c)

Figure 9.14 *Typical SEM images of fatigue fracture surfaces showing: (a) a new oxide film crack initiator; (b) a slip plane crack initiator; and (c) the beach mark striations made by the fatigue crack (Nyahumwa et al. 1998, 2001).*

Again, those few samples of cast material free from oxides displayed an order of magnitude improved life, the best of which agreed closely with the best of the results from the unfiltered tests. This agreement confirms the defect-free status of these few results.

The detrimental effect of mixed oxide films in the unfiltered castings is reflected by the lower fatigue performance compared to that of the filtered castings containing old oxide films. This indicates that the old oxide films, which were observed to act as fatigue crack initiators in the filtered castings, were less damaging than mixed oxide films. Clearly, we can conclude that the young films are more damaging, almost certainly as a result of their lack of bonding. This contrasts with the old oxide films that have benefited from a closing and partial re-bonding of the interface. This observation is consistent with the expectation that pores would be associated with new films, whose bifilm halves could separate to form pores, but not with old films, whose bifilm halves were (at least partly) welded closed.

These findings are confirmed in general by the results at the higher stress 240 MPa (Figure 9.13) The fatigue failures of unfiltered castings are initiated from a mixture of pores, and old and new oxides; the 33 filtered castings initiated from 29 old oxides, three pores and one slip plane; and the 33 unfiltered and HIPped tests initiated from 32 old oxides and one slip plane. The various types of fracture surface are shown in Figure 9.14.

An amazing coincidence occurred just as these results had emerged. Workers on the other side of the globe in New Zealand, Wakefield and Sharp (1992), published their findings on the fatigue of Al–10Mg alloy, poured interestingly badly (they do not claim to be foundrymen), and then tested in the as-cast condition and in the HIPped condition. Their results duplicated those seen in Figure 9.12 closely. The similarity was so compelling between these two very different alloy systems (a single-phase solid solution renowned for its ductility, versus a two-phase alloy in which one of the phases is not ductile) it suggested some underlying fundamental significance. This seems likely to be simply the overriding effect of bifilms on fatigue performance.

Other workers (for instance, Wang *et al.* 2001) using Al–7Si–0.4Mg alloy have in general confirmed that for a defect of a given area, pores are the most serious defects, followed by films of various types. This is to be understood in terms of the effect shown in Figure 2.39, where bifilms not only may be partly welded, but in any case always have some degree of geometrical interlocking as a result of their convoluted form.

However, the pre-eminence of pores in fatigue failure is not to be taken for granted. Nyahumwa found films to be most important in his work on this alloy. He had the advantage of having the latest techniques to search for and identify the full extent of films, whereas it is not known how thorough has been work elsewhere, and films are easy to overlook even with the best SEM equipment. In contrast, Byczynski (2002) working with the same techniques in the same laboratory as Nyahumwa found that pores were the most damaging defects in the more brittle A319 alloy used for automotive cylinder blocks. This difference may be explicable by the differences in ductility of the two alloys that were studied. Nyahumwa's alloy was ductile, and so required the stress concentration of bifilms acting as cracks. Byczynski's alloy was brittle, so that cracks could more easily occur from pores.

Thus we may tentatively summarize the hierarchy of defects that initiate fatigue in order of importance: these are pores and/or young bifilms, followed by old bifilms. It would be expected that in the absence of larger defects, progressively finer features of the microstructure would take their place in the hierarchy of initiators. However, it seems that such an attractively simple conceptual framework remains without strong evidence, even doubtful, at this time as is described below.

The considerable work now available on Al alloys illustrates the present uncertainties. In Al alloys it has been thought that silicon particles may become active in the absence of other initiators. However, the action of bifilms is almost certainly involved in the occasional observations of the nucleation of cracks from these sources. For instance, the observed decohesion of silicon particles from the matrix (Wang *et al.* 2001, part I) is difficult to accept unless a bifilm is present, as seems likely. The initiation of cracks from iron-rich phases occurs often if not exclusively from bifilms hidden in these intermetallics (Cao and Campbell 2001). The initiation of fatigue from eutectic areas, and reported many times (for instance, Yamamoto and Kawagoishi 2000 and Wang *et al.* 2001, part II) is understandable if bifilms are pushed by growing dendrites into these regions. The fascinating fact that Yamamoto and Kawagoishi observe silicon particles sometimes initiating fatigue cracks and sometimes acting as barriers to crack propagation strongly suggests bifilms are present sometimes and not at other times, as would be expected. It is not easy to think of other explanations for this curious observation.

Figure 9.15 is intended to illustrate the panorama of performance that can be seen in castings. The poor results can sometimes be seen in pressure die castings, in which the density and size of defects can cause failure to occur on the first cycle. However, it is to be noted that the unpredictability of this process sometimes will yield excellent results if the defects are, by chance or by manipulation, in an insensitive part of the casting, or where perhaps the bifilms are aligned parallel to the stress axis.

In terms of this panorama, the results of Nyahumwa are presented as 'fair' and 'good' respectively, showing a few tests that exhibit outstandingly good lives, reaching 10^7 cycles. Clearly, a series of ideal castings, free from defects, would display identical lives all at 10^7 cycles. The important lesson to draw from Figure 9.15 is that most engineering designs have to be based on the minimum performance. It is clear, therefore, that even the castings designated 'good' in this figure have a potential to increase their lowest results by 2 orders of magnitude, i.e. 100 times. It means that for most of the aluminium alloy castings in use today we are probably only using about 1 per cent of their potential fatigue life. To gain this hundredfold leap in performance we merely need to eliminate defects.

9.5.4 Thermal fatigue

Thermal fatigue is a dramatically severe form of fatigue. Whereas normal fatigue occurs at stresses comfortably in the elastic range (i.e. usually well below the yield point) *thermal fatigue* is driven by thermal *strains* that force deformation well into the plastic flow regime. The maximum stresses, as a consequence, are therefore well above the yield point.

Thermal fatigue is common in castings in which

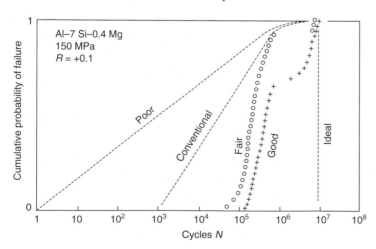

Figure 9.15 *Schematic overview of the extremes of fatigue performance that can be found in different kinds of castings.*

part of the casting experiences a fluctuating high temperature while other parts of the casting remain at a lower temperature. The phenomenon is seen in grey iron disc brakes, and aluminium alloy cylinder heads and pistons for internal combustion engines, particularly diesel engines, and air-cooled internal combustion engines. It is also common in the casting industry with the crazing and sometimes catastrophic failure of high-pressure die casting dies made from steels, and gravity dies made from grey cast iron.

The valve bridge between the exhaust valves in a four-valve diesel engine is an excellent example of the problem, and has been examined in detail by Wu and Campbell (1998). In brief, the majority of the casting remains fairly cool, its temperature controlled by water cooling. However, the small section of casting that forms a bridge, separating the exhaust valves, can become extremely hot, reaching a temperature in excess of 300°C. The bridge therefore attempts to expand by $\alpha \Delta T$ where α is the coefficient of thermal expansion and ΔT is the increase in temperature. For a value of α about $20 \times 10^{-6} \, K^{-1}$ for an Al alloy we can predict an expansion of $300 \times 20 \times 10^{-6} = 0.6$ per cent. This is a large value when it is considered that the strain to cause yielding is only about 0.1 per cent. Furthermore, since the casting as a whole is cool, strong and rigid, the bridge region is prevented from expanding. It therefore suffers a plastic compression of about 0.6 per cent. If it remains at this temperature for sufficient time (an hour or so) stress relief will occur, so that the stress will fall from above the yield point to somewhere near zero.

However, when the engine is switched off, the valve bridge cools to the temperature of the rest of the casting, and so now suffers the same problem in reverse, undergoing a tensile test, plastically extending by up to 0.6 per cent.

The starting and stopping of the engine therefore causes the imposition of an extreme high strain and consequent stress on the exhaust valve bridge. For those materials, such as poorly cast Al alloys, that have perhaps 0.5 to 1 per cent elongation to failure available, it is not surprising that failure can occur in the first cycle. What perhaps is more surprising is that any metallic materials survive this punishing treatment at all. It is clear that modern cylinder heads can undergo thousands of such cycles into the plastic range without failure.

The experience from the early days of setting up the Cosworth process provided an illustration of the problem as described earlier. In brief, before the new process became available, the Cosworth cylinder heads intended for racing were cast conventionally, via running systems that were probably well designed by the standards of the day. However, approximately 50 per cent of all the heads failed by thermal fatigue of the valve bridge when run on the test bed. These engines were, of course,

highly stressed, and experienced few cycles before failure. From the day of the arrival of the castings made by the new process (otherwise substantially identical in every way) no cylinder head failed again. The presence of defects is seen therefore to be critical to performance, particularly when the metal is subjected to such extreme strains as are imposed by thermal fatigue conditions.

Thermal fatigue tests can be carried out on nicely machined test pieces in the laboratory. One of the interesting observations is that for some ductile Al alloys the repeated plastic cycling for those specimens that survive causes them to deform into shapeless masses. This gross deformation appears to be resisted more successfully for higher strength alloys (Grundlach 1995).

9.5.5 Ductility

Figure 9.16 is a famous result showing the ductility (in terms of reduction of area) of a basically highly ductile material, pure copper, being reduced by the addition of various kinds of second-phase particles, including pores. It is clear that there is a large deleterious effect of the second phases, more or less irrespective of their nature. The lack of sensitivity to the nature of the particles or holes is almost certainly the result of the relatively easy decohesion of the particles from the matrix when deformation starts. Thus all particles act as holes.

This result is predictable if the particles are introduced into the melt by some kind of stirring-in process. As the particles penetrate the surface they necessarily take on the mantle of oxide that covers the liquid metal. Thus all immersed particles will be expected to be coated with a layer of the surface oxide, with the dry side of the oxide adjacent to the particle. The absence of any bonding across this interface will ensure the easy decohesion that is observed. In practice, the situation is usually rather worse than this, with the submerged particles appearing to remain in clumps despite intense and prolonged stirring. This seems to be most probably the consequence of the particles entering the liquid in groups, and being enclosed inside a packet of oxide. With time, the enclosure will gather strength as it thickens by additional oxidation, using up the enclosed air, and so gradually improve its resistance to being broken open.

In castings the volume of pores rarely exceeds 1 per cent. (Only occasionally is 2 or 3 per cent found.) Figure 9.16 indicates that the ductility will have fallen from the theoretical maximum (which will be 100 per cent reduction in area for a perfectly ductile material) to approximately 10 per cent, an order-of-magnitude reduction!

Why should an assembly of holes in the matrix affect the ductility so profoundly?

Figure 9.17 shows a simple model of ductile

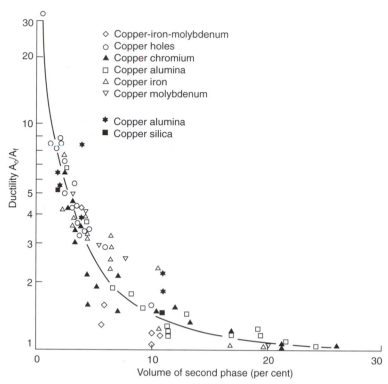

Figure 9.16 *Ductility of copper containing a dispersion of second phases. Data from Edelson and Baldwin (1962).*

failure. For the sound material the extension to failure is of the order of the width *l* of the specimen, because of the deformation of the specimen along 45-degree planes of maximum shear stress. For the test piece with the single pore of size *d*, the elongation to failure is now approximately $(l - d)/2$. In the general case for a spacing *s* in an array of micropores we have

Elongation = $s - d$

$$= 1/n^{1/2} - (f/n)^{1/2}$$

$$= (1 - f^{1/2})/n^{1/2} \qquad \textbf{(9.9)}$$

where *n* is the number of pores per unit area, equal to $1/s^2$ and *f* is the area fraction of pores on the fracture surface, equal to nd^2.

Equation 9.9 is necessarily very approximate because of the rough model on which it is based. (For a more rigorous treatment the reader is recommended to the pioneering work by Thomason 1968.) Nevertheless, our order-of-magnitude relation indicates the relative importance of the variables involved. It is useful, for instance, in interpreting the work of Hedjazi *et al.* (1976), who measured the effect of different types of inclusions on the strength and ductility of a continuously cast and rolled Al–4.5Cu–1.5Mg alloy. From measurements of the areas of inclusions on the fracture surface, Hedjazi reached the surprising conclusion that the film defects were less important than an equal area

fraction of small but numerous inclusions. His results are seen in Figure 9.18. One can see that for a given elongation, the microinclusions are about ten times more effective in lowering ductility. However, he reports that there were between 100 and 1000 times the number of microinclusions compared to film-type defects in a given area of fracture surface. From Equation 9.9, an increase in number of inclusions per unit area by a factor of 100 would reduce the elongation by a factor of 10, approximately in line with the observations.

The other observation to be made from Equation 9.9 is that ductility falls to zero when $f = 1$, for instance in the case of films which occupy the whole of the cross-section of the test piece. This self-evident result can easily happen for certain regions of castings where the turbulence during filling has been high and large films have been entrained. This is precisely the case for the example for the ductile alloy Al–4.5Cu that failed with nearly zero ductility seen in Figure 2.42a. This part of the casting was observed to suffer a large entrainment effect that had clearly created extensive bifilms. Elsewhere, other parts of the same casting had filled quietly, and therefore contained no new bifilms, but only its background scatter of old bifilms. In this condition the ductility of the cast material was 3.5 per cent (Figure 2.42b).

Pure aluminium is so soft and ductile that it is possible almost to tie a length of bar into knots.

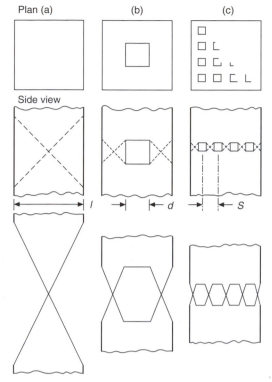

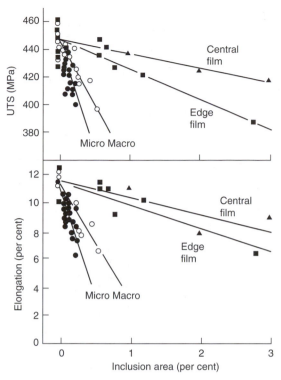

Figure 9.17 *Simple ductile failure model, representing a sound specimen in (a) which necks down to 100 per cent reduction in area; a single macropore in (b) which leads to a cup and cone fracture; and an array of micropores in (c) which effectively 'tear along the dotted line'. It is clear that extension to failure is directly related to sound length (d–a) in each case.*

Figure 9.18 *Strength and ductility of an Al–4.5Cu–1.5Mg alloy as a function of total area of different types of inclusions in the fracture surface. Data from Hedjazi et al. (1976).*

However, Figure 9.19 illustrates how the presence of bifilms has caused even this ductile material to crack when subjected to a three-point bend test. Notice the material close to the tips of the cracks is highly ductile, so the cracks could not have propagated as normal stress cracks, since the crack tips would have blunted, as they are seen to be under the microscope. Thus the only way for such cracks to appear in a ductile material like pure aluminium is for the cracks to have been introduced by a non-stress mechanism. The random accidents of the folding-in of the surface due to surface turbulence is the only conceivable mechanism, corroborated by the random directions of the cracks, not necessarily aligned along the direction of maximum strain.

Regardless of the inclusion content of a melt, one of the standard ways to increase the ductility is to freeze it rapidly. This is usually a powerful effect. Figure 9.2 illustrates an approximate tenfold improvement. As described in section 9.2.5, the effect follows directly from the freezing-in of bifilms in their compact form, reducing the time available

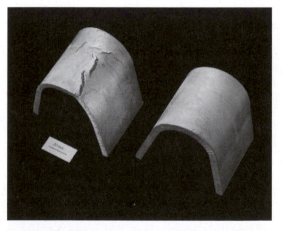

Figure 9.19 *Plates 10 mm thick cast in 99.5Al subjected to three-point bend (a) filled at an ingate speed greater than 0.5 ms⁻¹ and (b) less than 0.5 ms⁻¹ (courtesy Runyoro 1995).*

for the operation of the various unfurling mechanisms. (There may also be some contribution from the dendrites pushing the bifilms away from surface regions, effectively sweeping the surface regions clear, and concentrating the bifilms in the centre of the casting where they will be somewhat less damaging to properties. This effect has not been investigated, and, if real, may depend on whether the bifilms are not quite cleared from the surface regions, but are organized into planar sheets, as in Figure 2.42a, and whether therefore the benefits are now dependent on the direction of stress.)

The converse aspect of this benefit is that if the ductility of a casting from a particular melt quality is improved by chilling, this can be probably taken as proof that oxides are still present in the melt. A simple quality control test can be envisaged.

9.5.6 Ultimate tensile strength

Ultimate tensile strength (TS) is a composite property composed of the total of (i) the yield stress plus (ii) additional strengthening from work hardening during the plastic yielding of the material prior to failure. These two components make its behaviour more complicated to understand than the behaviour of yield stress or ductility alone.

TS equals the yield, or proof, stress when (i) there is no ductility, as is seen in Figure 9.2 and Figure 9.10, and (ii) when the work hardening is zero. The zero work hardening condition is less commonly met, but occurs often at high temperatures when the rate of recovery exceeds the rate of hardening.

The problem of determining the TS of a cast material is that the results are often scattered. The problems of dealing with this scatter are important, and are dealt with at length in section 9.6. Section 9.6 is strongly recommended reading.

Generally, for a given alloy, proof strength is fixed. Thus as ductility is increased (by, for instance, the use of cleaner metal, or faster solidification) so TS will usually increase, because with the additional plastic extension, work hardening now has the chance to accumulate and so raise strength. The effect is again clear in Figure 9.2. For a cast aluminium alloy, Hedjazi *et al.* (1975) show that TS is increased by a reduction in defects, as shown in Figure 9.20. However, it seems probable that the response of the TS is mainly due to the increase in ductility, as is clear from the strong shift of the property region to the right rather than simply upwards.

The rather larger effect that layer porosity is expected to have on ductility will supplement the smaller effect due to loss of area on the overall response of TS. Figure 9.6 shows the reduction in TS and elongation in a Mg–Zn alloy system where the reduction in properties seems modest. In Figure 9.21 the TS of an Al–11.5Mg alloy shows more serious reductions, especially when the porosity is in the form of layers perpendicular to the applied stress. Even so, the reductions are not as serious as

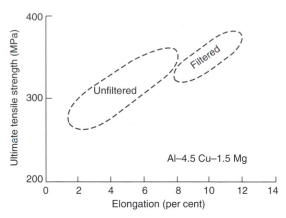

Figure 9.20 *Mechanical property regimes for an Al–4.5Cu–1.5Mg alloy in filtered and unfiltered conditions (Hedjazi et al. (1975).*

would be expected if the layers had been cracks, a result emphasizing their nature as 'stitched' or 'tack welded' cracks, as discussed in section 9.4.1.

When the layers are oriented parallel to the direction of the applied stress, then, as might be expected, Pollard (1965) has shown that layer porosity totalling even as high as 3 per cent by volume is not deleterious.

Finally, as for ductility, it is clear that cracks or films occupying the majority of the cross-section of the casting will be highly injurious. Clyne and Davies (1975) quantify the self-evident general understanding that the TS falls to zero as the crack occupies progressively more of the area under test (Figure 9.22).

9.5.7 Leak tightness

Leak tightness has usually been dismissed as a property hardly worthy of consideration, being merely the result of 'porosity'.

However, of all the list of properties specified that a casting must possess, such as strength, ductility, fatigue resistance, chemical conformity, etc., leak tightness is probably the most common and the most important. This might seem a trivial requirement to an expert trained in the metallurgy and mechanical strengths of materials. However, it is a requirement not to be underestimated.

A cylinder head for an internal combustion engine is one of the most demanding examples, requiring to be free from leaks across narrow walls separating pressurized water above its normal boiling point, very hot gas, hot oil at high pressure, and all kept separate from the outside environment. A failure at a single point is likely to spell failure for the whole engine. In this instance, as is common, leakage usually means 'through leaks', in which containment is lost because of a leak path completely through the containing wall.

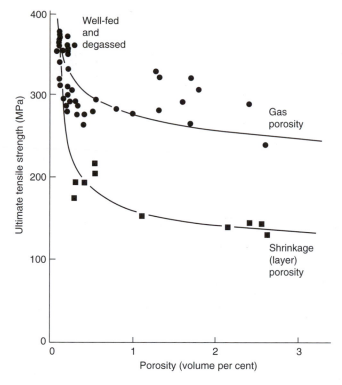

Figure 9.21 *Reduction in UTS of an Al–11.5Mg alloy by dispersed porosity and by layer porosity. Data from Jay and Cibula (1956).*

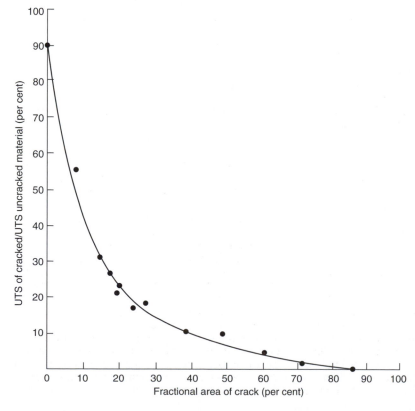

Figure 9.22 *UTS of a casting as a function of the area of the crack. From Clyne and Davies (1975).*

However, leakage sometimes refers to surface pores that connect to an enclosed internal cavity inside a wall or boss. Such closed pores give problems in applications such as vacuum equipment, where outgassing from surfaces limits the attainment of a hard vacuum. Problems also arise in instances of castings used for the containment of liquids, where capillary action will assist the liquid to penetrate the pore. If the pore is deep or voluminous the penetrated liquid may be impossible to extract. This is a particular problem for the food processing industry where bacterial contamination residing in surface-connected porosity is a concern. Similarly, in the decontamination of products used in the nuclear industry, aggressive mechanical and chemical processes fail to achieve 100 per cent decontamination almost certainly as a result of the surface contact with bifilms and possibly with shrinkage cavities. Such industries require castings made from clean metal, transferred into moulds with zero surface entraining conditions. Only then would performance be satisfactory.

It is true that leaks are sometimes the result of shrinkage porosity, especially if the alloy has a long freezing range, so that the porosity adopts a sponge or layer morphology. Clearly, any form of porous metal resulting from poorly fed shrinkage will produce a leak, especially after machining into such a region.

Leaks are seldom caused by gas porosity i.e. bubbles of gas precipitated from solution in the liquid metal. The following logic provides an explanation.

Gurland (1966) studied the connections between random mixtures of conducting and non-conducting phases by measuring the electrical resistance of the mixture. He used silver particles in Bakelite, gradually adding more silver to the mix. He found the transition from insulating to conducting to be quite abrupt, in agreement with stochastic (i.e. random) models. The results are summarized as:

per cent Ag	per cent $Conducting$
1	0
1.73	50
2.5	100

In the case of about 1–2 per cent gas porosity in cast metals the metal must surely therefore be permeable to gas. Why is this untrue? It is untrue because the distribution of gas pores is not random as in Gurland's mixtures. Gas pores are distributed at specific distances, dictated by the diffusion distance for gas. In addition, the pores are kept apart by the presence of the dendrite arms. Thus leakage due to connections between gas pores cannot occur until there are impossibly high porosity contents in the region of 20 to 30 per cent by volume (see Figures 6.14 and 6.16).

The only possible exception to this rule is the relatively rare occurrence of wormhole-type bubbles, formed by the simultaneous growth of gas bubbles and a planar solidification front. Such long tunnels through the cast structure naturally constitute highly effective leak paths (see Figures 6.17 and 6.18). Fortunately they are rare and easily identified, so that corrective action can be taken.

In the author's experience, most leaks in light-alloy and aluminium bronze castings are the result of oxide inclusions. These fall into two main categories:

1. Some are the result of fragments of old, thick oxide films or plates which are introduced from the melting furnace or ladle, in suspension in the melt, and which become jammed, bridging between the walls of the mould as the metal rises. The leak path occurs because the old oxide itself suffered an entrainment event; as it passed through the surface it would take in with it some new surface oxide as a thin, non-wetting film covering. The leak path is the path between the rigid old oxide fragment and its new thin wrapping.

2. The majority of leaks are the consequence of new bifilms introduced into the metal by the turbulent filling of the mould. These tangled layers of poorly wetted surface films, folded over dry side to dry side, constitute major leak paths through the walls of castings. The leaks are mainly concentrated in regions of surface turbulence. Such regions are easily identified in Al alloys as areas of frosted or grey striations down the walls of top-gated gravity castings, outlining the path of the falling metal. The remaining areas of walls, away from the spilling stream, are usually clear of any visible oxide striations, and are free from leaks. The reader should confirm, and take pride in, the identification of a de-gated top-poured aluminium alloy castings from a distance of at least 100 m! Unfortunately, this is not a difficult exercise, and plenty of opportunity exists to keep oneself in training in most light-alloy foundries! It is to be hoped that this regrettable situation will improve.

An example of a sump (oil pan) casting, top poured into a gravity die (permanent mould), is shown in Figure 9.23. The leakage defects in this casting are concentrated in the areas that have suffered the direct fall of the melt. The surface oxide markings are seen on both the outside and inside surfaces of these regions of the casting (Figures a and b). Other distant areas where the melt has filled the mould in a substantially uphill mode are seen to be clear of oxide markings and free from leaks. The precise points of leakage are found by the operator who

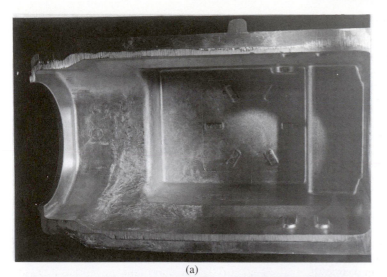

(a)

(b)

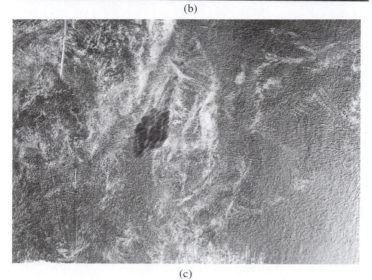

(c)

Figure 9.23 *Views of (a) the inside and (b) the outside of a top-poured oil pan (sump) casting showing the light traces of entrained oxides and the corresponding leak defects repaired by peening, as seen in close up at (c).*

inverts the casting, pressurizes it with air and immerses it under water. He is guided by the stream of air bubbles emerging from leaks and deals a rapid series of blows from a peening gun. This hammering action deforms the surface locally to close the leak. The peening marks are seen in close-up in Figure 9.23c.

The linkage between oxide films and leakage problems was noted by Burchell (1969), when he attempted to raise the hydrogen gas content of aluminium alloys by stirring with wood poles, dipped in water between use. The porosity of the castings increased, as was intended to counter feeding problems, but so did the number of leaks. Burchell identified the presence of oxide films on the fracture surfaces of tensile test bars that were cast at the same time.

It is unfortunate therefore that the folded form of entrained surface films creates ideal opportunities for leak paths through the casting. This source of leakage is probably more common than leaks resulting from other kinds of porosity such as shrinkage porosity (although, of course, leaks from entrained films as a consequence of a poor filling system are usually misdiagnosed as leaks from shrinkage because of poor feeding. This is a natural consequence of its appearance, because some interdendritic porosity usually grows from the entrainment defect.)

As an instance of the seriousness of leaks in castings that are required to be leak-tight, many foundries have been reluctant to cast aluminium manifolds and cylinder heads with sections less than 5 mm. This is because of the increased incidence of leaks that require the casting to be repaired or scrapped. The lack of pressure tightness relates directly to the presence of oxides whose size exceeds 5 mm (an interesting confirmation of the non-trivial size and widespread nature of these defects), and that can therefore bridge wall to wall across the mould cavity, connecting the surfaces by a leak path.

Leaks are often associated with bubble-damaged regions in castings. This is because all bubbles will have been originally connected to the surface as a necessary feature of their entrainment process. Some bubbles will have retained their bubble trail links to the outside world, whereas others will have broken away during the turmoil of filling. Bubble trails are particularly troublesome with respect to leak tightness, since they necessarily start at one casting surface and connect to the surface above, and as part of their structure have a continuous pipe-like hollow centre. The inflated bubble trails characteristic of high-pressure die castings (Figure 2.34) make excellent leak paths. A core blow also leaves a serious defect in the form of a collapsed bubble trail (Figure 6.20). Despite its collapsed form, the thickness and residual rigidity of its oxide will ensure

that the trail does not completely close, so that a leak path is almost guaranteed (Fig. 2.32).

In general, the identity of a leakage defect in a casting can be made with certainty by sawing the casting to within a short distance of the defect, and then breaking it open and studying the fracture surface under the microscope. A new oxide film (probably from surface turbulence during casting, or from a bubble trail) is easily identified from its folded and wrinkled appearance; an old oxide fragment (perhaps from the melting furnace or crucible) from its craggy form, like a piece of rock; and shrinkage porosity by its arrays of exposed dendrites. Fracture studies are a quick and valuable test and are recommended as one of the most powerful of diagnostic techniques. The reader is recommended to practise this often – despite its unpopularity with the production manager; the destruction of one casting will often save many.

9.5.7.1 Leak detection

Turning now from the nature of leakage defects to methods of detection. Bubble testing has already been mentioned in which the inside of the casting is simply pressurized with air, the casting immersed in water, and any stream of bubbles observed. Even this time-honoured and apparently simple technique is not to be underestimated, since providing effective and rapid sealing of all the openings from the cored internal cavity may not be easy in itself. It will almost certainly require careful planning to ensure that the correct amount of dressing has been carried out to eliminate troublesome flash and gating systems, etc. so that the sealing surfaces can be nicely accessed and sealed. Also, of course, the technique is slow, not quantifiable, and demands the constant attention of skilled personnel. After testing, the part often requires to be dried.

Hoffmann (2001) describes three basic methods of dry air leak testing suitable for production line applications: they measure (i) the rate of decay of gauge pressure, (ii) the rate of decay of differential pressure, and (iii) leakage rate directly in terms of mass flow. For highly specialized applications, helium mass spectrometry offers testing capability beyond the limits attainable by dry air methods.

The first technique is the simplest and lowest cost, and is generally suitable where pressures do not exceed 2 bar and volumes do not exceed 100 ml. The differential pressure method pressurizes a non-leaking reference volume along with the test part. A transducer reads any difference in pressure that occurs over time. The differential technique reduces errors due to temperature changes, and is more accurate and faster than the direct pressure decay method. The technique is also well suited to applications specifying higher test pressures, exceeding 10 bar, and where relatively small cavities must be tested to a very low leak rate.

The mass flow method pressurizes the test cavity, then allows any leakage to be compensated by actively flowing air into the cavity. The in-flowing air is measured directly by a mass flow meter in terms of volume per second. The method involves a single measurement, usually less than 1 second. (It avoids the taking of two measurements over a time interval – during which temperature may change for instance – that is required for the first two techniques, thus halving errors and increasing speed of response.) The method can tackle a wider range of volumes, and is accurate down to 0.001 mls^{-1}.

For castings that are required to be leak-tight to even greater standards, helium mass spectrometry can measure down to rates that are 10 000 times lower. This is principally because the helium atom is much smaller than molecules of nitrogen and oxygen, and so can penetrate much smaller pores.

Krypton gas has been used for the detection of very fine leaks, because of its content of 5 per cent radioactive krypton 85 (Glatz 1996). As before, the part to be tested is placed in an evacuated chamber to suck air out of the cavities. Krypton gas is then introduced to the chamber and allowed time to penetrate the surface pores. The krypton is then pumped out, ready for reuse, and air is admitted. The rate at which Kr is slowly released can be monitored to assess the volume of surface-connected internal pores. In addition, the spraying of the surface with a liquid emulsion of silver halide particles makes the surface sensitive to the low energy beta particles given off by the radioactive decay of Kr85. After the emulsion is developed by conventional photographic techniques the part reveals the site and shapes of surface pores and cracks. The beta particles can penetrate approximately 1 mm of metal, revealing subsurface cracks (if connected to the surface elsewhere of course) and magnifying the width of pores and cracks that otherwise would be too small to see. The technique is more sensitive than dye penetrant testing because the viscosity of gases is typically only 1/100 of that of liquids, making the test extremely searching.

Finally, it is worth emphasizing that a good melt quality combined with a good filling system will usually eliminate most of the leaks found in castings (providing core blows can be avoided by careful design or venting of cores). This conclusion is confirmed by foundries using intrinsically quiescent melt handling processes such as the Cosworth process. These operations are so confident of the quality of their products that they do not even bother to test for leak tightness; none are ever found to leak.

9.5.8 Residual stress

Unseen and often unsuspected, residual stress can be the most damaging defect of all. This is because the stress can be so large, outweighing the effect of all other defects. It is usually never specified to be low. This is a grave indictment of the quality of component specifications and of standards in general. It is also practically impossible to measure in a non-destructive way in the interior of a complicated casting. However, it can be controlled by correct processing – another vindication of intelligent manufacture compared to costly, difficult and unreliable inspection.

Also, of course, as we have noted already, in rare instances residual stress can be manipulated to advantage. However, in the general case it should be assumed for the sake of safety that somewhere in the finished part the retained stress will be in opposition to the strength of the casting. It will therefore add to the applied stress, and so put the casting near to its point of failure even at relatively small applied loads. Unfortunately, a conservative assessment of residual stress would have to assume that it reached the yield stress.

The remedy is, of course, either (i) the avoidance of stress-raising treatments such as quenching castings into water following solution treatment (and so accepting the somewhat reduced strengths available from safer quenchants such as forced air) or (ii) the application of stress relief as already discussed (remembering, of course, that stress relief will effectively negate much or all of the strength gained from heat treatment).

As an example of a part that suffered from internal stress, a compressor housing for a roadside compressor in Al–5Si–3Cu alloy (the UK specification LM4), was thought to require maximum strength and was therefore subjected to a full solution treatment, water quench and age (TF condition). Two housings blew up in service with catastrophic explosions in which, fortunately, no one was hurt. The manufacturer was persuaded to carry out a heat treatment that would reduce the internal stress, but also, of course, reduce the strength. This was very reluctantly agreed. However, on testing the parts to destruction, the implementation of a TB7 treatment followed by air cooling gave a part with roughly half the strength but twice the burst pressure resistance.

9.5.9 Elastic (Young's) modulus

In an engineering structure, the *elastic modulus* is the key parameter that determines the *rigidity* of the design. Thus steel with an elastic modulus of 210 GPa is very much preferred to the light alloys aluminium (70 GPa) and magnesium (45 GPa). However, of course, the specific modulus (the rigidity modulus divided by density) gives a somewhat more favourable comparison, since they are all closely similar at about $26\,MPam^3kg^{-1}$.

Since the elastic properties of metals arise directly as a result of the interatomic forces between the atoms, there is little that can normally be done to increase this performance by, for instance, alloying.

The one exception worth noting to this general rule is the Al–Li system of alloys, in which the very large volume fraction of Al–Li intermetallics does raise the elastic modulus by about 10 per cent, which, together with the density reduction of about 10 per cent, results in an overall increase in specific elastic modulus of 20 per cent. Even this triumph of metallurgical R&D is seen to be relatively modest when compared to the 200 per cent increase provided by moving to the much cheaper alternative, steel.

It was all the greater surprise to the author, therefore, when he arranged for a computer simulation to be carried out on the largest bell in the UK. This was Great Paul, the 17 000 kg bell in St Paul's Cathedral, London. The investigators, Hall and Shippen (1994), found that agreement with the fundamental note and the harmonics could only be obtained by assuming that the elastic modulus was only 87 GPa, not 130 GPa expected for alloys of copper. This alarming 33 per cent reduction was almost certainly the consequence of Great Paul being full of defects, mainly porosity and bifilms. This is true for most bells because, unfortunately, they are top poured via the crown of the bell. (The Liberty Bell in the USA is a famous example of a crack opened from the massive bifilm constituted by the oxide flow tube that can be seen to have formed around the sinuous flow of the falling metal stream, starting high on the shoulder of the bell and continuing down the side to its lip.) The one bell foundry in the world known to the author (Nauen, Tonsberg, Norway) that uses bottom filling for its bells, placed mouth upwards, claims that it has never suffered a cracked bell, and its bells sound for twice as long as conventional bells. The longevity of the sound is understandable in view of the reduced quantity of bifilms to be expected, so reducing the loss of energy by internal friction between the rubbing surfaces of the bifilms as the bell vibrates.

It is clear therefore, that many castings will suffer from a lower elastic modulus than expected because of their high content of bifilms. Furthermore, since elastic modulus is very rarely checked, this fact is not widely known. For the future, the full stiffness of the material will be gained only if the metal quality is good, and if the casting is poured with a minimum of surface turbulence.

The effect of a high density of bifilms in castings reducing the elastic modulus of the cast material is an interesting hypothesis that has a widely known and accepted metallurgical analogy that lends convincing comparison. The example is that of flake cast irons compared to steels. The graphite flakes in cast irons act as cracks since they are unable to withstand significant tensile stress. The iron, behaving like a steel matrix laced with a multitude of cracks, finds its modulus reduced from 210 GPa to only 152 GPa, a reduction of 28 per cent, a value comparable to the reduction expected for many defective cast materials. In addition, of course, flake graphite irons are renowned for their excellent damping capacity, their absorbtion of vibrations and deadening of noise making them an ideal choice for the beds of machine tools.

Further examples cannot be resisted.

Zildjian cymbals, much prized by percussionists and drummers, owe much of their unique sound and mechanical resilience (lesser cymbals can turn inside out when struck) to the fact that their original cast preform is tested by striking to see if the material has a ring in its cast state. If it does not, it is remelted. Clearly, this example confirms our expectations that the effects of the presence of bifilm defects survive in some alloys even after the extensive working, applied in this case by forging and spinning into a thin disc.

The writer reveals his age when he admits to recalling the wheel tapper, moving along the side of the rail track, tapping the wheels of the train with his long handled hammer, checking that the 'ring' was correct. A cracked wheel would give a different note. Less romantically, but perhaps more reliably, the wheel tapper has now been superseded by the use of automated checking by ultrasonics.

9.5.10 Creep

The gradual deformation of metals at high temperatures while under load takes place in typically in three stages.

1. Primary creep is the rapid early phase that gradually reduces in rate, eventually leading into stage 2.
2. Secondary creep is the steady-state regime, in which creep rate is constant.
3. Tertiary creep is the final stage in which the rate of strain increases because of the growth of microscopic internal pores and tears that gradually link to cause the fracture of the whole component.

Components are said to have failed by creep if, like a turbine blade, the length of the blade becomes too long under the centrifugal stress so that it scrapes the outer casing of the engine (although nowadays casings are designed to accept such problems). Alternatively, creep failure under tension might mean the fracture of the component at the end of tertiary creep.

Bifilms at grain boundaries are prime sites for the initiation of such catastrophic failures. Failure of the Ni-base alloy 718 described by Jo and colleagues (2001) appears to be precisely such a process initiated by oxide or nitride bifilms, judging from the micrographs published by these authors, although confirmatory evidence from the appearance and chemistry of the fracture surfaces was not carried out. Section 10.2 contains a description of a spectacular creep failure of HIPed turbine blades in which there is evidence for bifilm involvement.

In a related phenomenon of the superplastic deformation of Al–Zn alloys, it is widely known that the deformation of such materials is limited by the opening of grain boundary voids, that progressively link to cause complete failure. The presence of bifilms in such Al alloys is certain, so that it would be surprising if they were not involved in these failures too.

The implication of these creep and superplastic failures by the gradual accretion of grain boundary voids is that they would not occur if bifilms were not present. If true, the prospects for improvement in these materials is awe inspiring.

9.6 The statistics of failure

Unfortunately, the properties of materials are not accurately reproducible. Scatter of tensile and fatigue properties has been a particular problem with castings.

This is recognized by many of the world's standards for the testing of castings. If a tensile test bar fails below the strength value required by the specification, the standard normally allows a retest. The failed result is discarded. The discarding of low values is a practice that has crept into both foundries and laboratories. Somehow the low results are viewed as a mistake. We turn a blind eye to them.

The common approach to deal with scatter is simply to take an average. Occasionally, a standard deviation might also be calculated.

The attempt to assess scatter by the use of a standard deviation, usually denoted sigma, σ, is of course much better than nothing, but does make the implicit assumption that the distribution of results is Gaussian, or Normal. The familiar bell-shaped distribution is shown in Figure 9.24. The value $\pm\sigma$ encloses approximately 68 per cent of the scattered results. The values $\pm2\sigma$ includes 95 per cent and $\pm2.5\sigma$ includes 99 per cent. Higher multiples of sigma, for instance 3σ (99.73 per cent) and 4σ (99.994 per cent) are less commonly used.

If instead we retain all results, good and bad, and plot them to reveal their distribution, we can obtain a curve such as that in Figure 9.25. The shape of the distribution curve is usually not the symmetrical Gaussian form but is skewed, as are many strength and fatigue life distributions. The skew form arises naturally because strength values cannot be higher than when the alloy is perfect. This defines a cut-off to strengths at an upper limit. However, there is often no lower limit, so the low results exhibit a long 'tail'. A close approximation to the shape of the curve was derived theoretically by Weibull (1951). The curve (Figure 9.25) can be plotted as the familiar cumulative distribution (Figure 9.26) and by further mathematical manipulation the curved ends of this plot can be straightened in what is known as a Weibull plot (Figure 9.27). In other words, a Weibull plot is simply another way of presenting a cumulative distribution of data.

In general, the failure properties of metals seem more accurately described by the Weibull distribution. There are good fundamental reasons

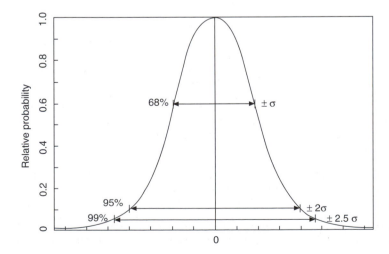

Figure 9.24 *Gaussian distribution of strengths, showing the assessment of scatter using different multiples of the standard mean distribution* σ.

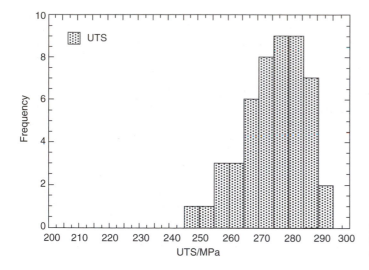

Figure 9.25 *Typical skewed distribution of tensile strengths as might be obtained from an Al–Si alloy (synthetic data) (Green 1995).*

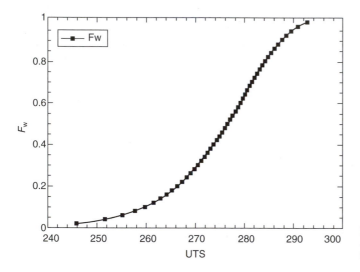

Figure 9.26 *Distribution identical to that in Figure 9.25 but replotted as a cumulative distribution (Green 1995).*

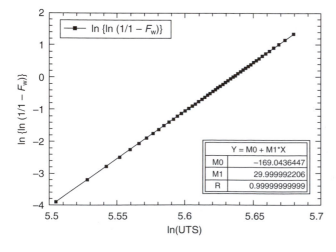

Y = M0 + M1*X	
M0	−169.0436447
M1	29.999992206
R	0.99999999999

Figure 9.27 *Weibull plot of the data in Figure 9.26, showing the simple straight line form of the cumulative distribution (Weibull modulus m = 30 and position parameter σ = 280 MPa) (Green 1995).*

for this. Although the original formulation by Weibull in 1951 was a purely mathematical description of scatter, later improved by Khalili and Kromp (1991), it was Jayatilaka and Trustrum in 1977 who explained that the formulation arose from the properties of a flaw size distribution of randomly inclined cracks. The strength distribution was accounted for by the probability of brittle fracture from the most severe defect, taking account of its size and orientation. It was a sophisticated development of the 'weakest link concept'.

In materials research, Weibull analysis was originally used almost exclusively for ceramics and glasses. The work by Green and Campbell (1993 and 1994) illustrated the usefulness of this approach to castings.

The slopes of the Weibull (pronounced 'Vybl' where 'y' is as in 'why') plots are a measure of the reproducibility of the data, and are known as the Weibull modulus, *m*. This parameter becomes more important than the position parameter (or average strength) when the reliability of the least reliable results are required. For instance, the value of strength of the weakest 1 in 100, or perhaps even the weakest 1 in 10 000 000. For pressure die castings *m* is often between 1 and 10, whereas for many gravity-filled castings it is between 10 and 30. For good quality aerospace castings a value between 50 and 100 is more usual. Values of 150 to 250 are probably somewhere near a maximum limit defined by the limits of accuracy of strength measurements (i.e. even if all strength results were perfectly identical, the expected modulus of infinity could not be demonstrated because the tensile testing machine would record slightly different strengths because of errors in the machine).

Figure 9.28 shows data for a single alloy, but cast in a variety of ways, showing the variation in reproducibility between different filling processes. This

kind of data is one of the simplest demonstrations of the importance of filling techniques on the reliability of cast products. It is also, of course, a clear demonstration of the action of entrainment defects.

Figure 9.29 shows the case of a bimodal distribution i.e. two separate distributions. Such multiple distributions are relatively common, and an indication of such an effect is to be seen in the poorest data shown in Figure 9.28. The various slopes are often found to correspond to different distributions of defects. Several slopes can sometimes be distinguished, from the lowest to highest results, corresponding to pores, new bifilms, old bifilms and sound material. The identification of the regimes of strength of each of these defect types can give a valuable insight into the causes of poor performance.

As a final caution, it is probably worth noting that all of the above analysis assumes reasonably brittle failure behaviour. This seems to be a good assumption for most castings, where the failure of one element will lead to the failure of the whole component. This is not the case for ductile materials where crack blunting will occur. Ductile failure may be better described by a Gamma distribution (Green 1994). The statistics of failure is an interesting and important area that holds many descriptions of distributions that would repay more widespread study.

9.6.1 Technical background to the use of Weibull analysis

This section follows an approach described by Green (1994).

The distribution of strengths of brittle materials was first formulated by Weibull (1951) who defined a probability of *survival* F_S. The Weibull distribution is given by the expression:

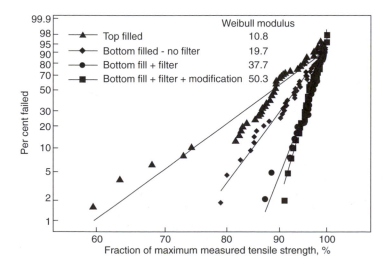

Figure 9.28 *Weibull plot of strength results for Al–7Si–0.4Mg alloy cast in various ways (Green and Campbell 1994).*

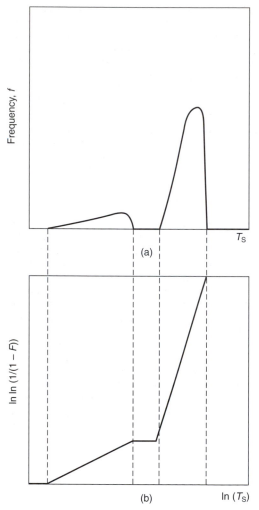

Figure 9.29 *(a) Bimodal distribution and (b) its corresponding Weibull plot.*

$$F_w = 1 - \exp\left\{\left(\frac{x - \mu}{\sigma}\right)^m\right\} \qquad (9.10)$$

where F_w is the cumulative fraction of *failures* up to a given value of x, say strength or fatigue cycles, σ is a position parameter where $1 - 1/e$ of the samples survive (~37 per cent), μ is a lower strength boundary below which no specimen fails and m is a width parameter, referred to as the Weibull modulus. The greater the value of m the narrower the range of strengths. The above expression is frequently used with the lower strength set to zero so that:

$$F_w = 1 - \exp\left\{-\left(\frac{x}{\sigma}\right)^m\right\} \qquad (9.11)$$

An example of a cumulative distribution of Weibullian failures described by Equation 9.11 is shown in Figure 9.26. Unlike the normal distribution there is no reflective symmetry about the mean. To obtain the Weibull modulus and position parameter from such a plot requires mathematical curve fitting. Alternatively, the distribution can be reduced to a straight line plot. First eliminate the minus sign in the exponential term:

$$\frac{1}{1 - F_w} = \exp\left(\frac{x}{\sigma}\right)^m \qquad (9.12)$$

and taking natural logarithms twice gives:

$$\ln\left\{\ln\left(\frac{1}{1 - F_w}\right)\right\} = m\ln(x) - m\ln(\sigma) \qquad (9.13)$$

This can now be presented as a straight line by plotting $\ln\left\{\ln\left(\dfrac{1}{1 - F_w}\right)\right\}$ as a function of $\ln(x)$ giving the straight line with slope m and intercept $-m \ln\sigma$. The data of Figure 9.26 is replotted in Figure 9.27 with a straight line fit regressed through it.

9.6.2 The volume effect

The probability of a specimen of unit volume (V_0) *surviving* to a given stress is $(1 - F_w)$, which can also be written as $F_S(V_0)$. If two samples of the same material were stuck perfectly together the probability of survival to a given stress is the product of their individual probabilities of survival, such that $F_S(2V_0) = F_S(V_0)^2$. Therefore, for any volume of specimen V, the probability of failure F_w is given by:

$$F_w = 1 - \exp\left\{-\frac{V}{V_0}\left(\frac{x}{\sigma}\right)^m\right\} \qquad (9.14)$$

9.6.3 Design using the Weibull distribution

It is implicit in the Weibull distribution that the failure strength of a material is determined by the distribution of defects resulting from its manufacture. The Weibull distribution can be of great use because it is possible to extrapolate back to the very low probability of failure (high reliability) region, and to select a design stress to give a desired failure rate. The following procedure can be used when the volume effect is small.

1. Select an acceptable failure rate.
2. Rank the data in ascending order of strength (fatigue life, etc.) and assign corresponding ascending failure rank j.
3. Make the extrapolation from a sample of

specimens to the population they are drawn from. This is required because the first specimen to fail in the sample tested has a probability of survival of 100 per cent at a stress below its failure stress. However, there is a finite probability that if more specimens were tested one of them would fail at a lower stress. The most probable value of the survival for the jth specimen in the ranked data is given by Khalili and Kromp (1991) as:

$$F_j = \frac{j - 0.5}{N}$$

where N is the total number of specimens tested.

4. Calculate $\ln\left\{\ln\left(\frac{1}{F_s}\right)\right\}$ for each F_j.
5. Calculate $\ln(x)$ for each x.
6. Plot $\ln\left\{\ln\left(\frac{1}{F_s}\right)\right\}$ as a function of $\ln(x)$ and regress through this a best-fit straight line. The Weibull modulus is the slope m of the line.
7. Calculate position parameter σ then back substitute into the Weibull equation (9.11) with F_s, σ and m to calculate the design stress x.

9.6.4 Extreme value distributions

The use of about 30 samples is usually sufficient to obtain a valid Weibull plot. The plot is often sufficiently accurate that an extrapolation to a failure rate of one in a hundred will be quite accurate enough for most purposes. However, further extrapolation to a rate of one in a thousand might involve unacceptable uncertainty. The uncertainty can often be assessed, at least to a first order approximation, by checking the standard deviation of the slope of the Weibull plot.

However, there is often much interest in the extrapolation of results of such a limited number of tests to assess the failure rate of a casting at a probability, for instance of one in ten thousand or one in ten million. The designs of safety-critical components on automobiles and aircraft fall into this category.

Beretta and Murakami (2001) developed a technique for the quantification of the probability of finding inclusion defects in very clean steels. This awesome task can be likened to the search for the occasional needle in the haystack, followed by the necessity of predicting precisely how many needles may be in other random haystacks. Their simple solution is the counting only of the maximum defects in any studied area of the sample, and analysing this data with the largest-extreme-value distribution (also known as the Gumbel distribution after its inventor). They successfully use the technique for the study of pores and cracks.

Although to the author's knowledge, this technique has not so far been used to assess the reliability of the strengths of castings, this elegant and powerful related approach studying inclusions in steels holds out hope that such techniques can be developed.

Chapter 10

Processing

10.1 Impregnation

The impregnation process is a sealing technique, designed to eliminate leakage problems in castings.

Clearly, good casting processes do not need impregnation.

However, many do.

For instance, high-pressure die castings are highly prone to leakage problems because of their high content of bifilms and inflated bubble trails. Similarly, low-pressure castings that have not had the benefit of the rollover action following the filling of the mould are usually prone to an interconnected type of shrinkage porosity as a result of convection problems.

Although some casting operations use impregnation only to seal those castings that are shown to leak, others do not carry out an initial sorting test, but simply apply impregnation to all castings. Although, like heat treatment, impregnation is often carried out by specialist operators at a price usually based on the weight of the casting, it is sometimes installed as an integral part of the foundry operation.

The impregnation process involves the placing of the casting in a vessel from which the air is evacuated. The casting is then lowered into the sealing liquid. When fully immersed, air is reintroduced into the vessel to restore atmospheric pressure. In this way the liquid is forced into the evacuated pores in the casting. The casting is then raised out of the liquid and is left to drain. Excess sealant is then washed off and the sealant is cured.

There are two main curing processes based on two different sealing systems. These are (i) sodium silicate cured with the addition of a solid catalyst to the liquid; and (ii) a thermosetting resin hardened by a subsequent low-temperature heat treatment.

A recent development is the streamlining of the process by the use of special sealants, allowing the process to be carried out in a single vessel with only a short treatment time (Young 2002).

In a small percentage of cases leakage of the casting is not cured by the first impregnation. Some casters and their buyers allow two or even three such attempts. Usually, if the casting still leaks after repeated attempts to seal it, it is finally scrapped.

10.2 Hot isostatic pressing

Coble and Flemings (1971) confirmed that pores, if sufficiently fine, and in a fine-grained matrix, would gradually disappear given a few tens of hours and a temperature high enough for a 'sintering reaction' to occur. They found that the application of modest pressure, about 20 atmosphere, greatly assisted the process. The development of hot isostatic pressing (HIPping) was the outcome. Much higher pressures were employed, usually nearer 1000 atmospheres, and temperatures as near the melting point as was practical. The conditions for HIP have been defined in an elegant study by Arzt *et al.* (1983) in which they characterize tool steel, superalloys, alumina and ice.

The significant improvement in mechanical properties, particularly average fatigue life, reported for certain alloys after HIPping is probably at least in part due to the contribution towards the deactivation of entrained double oxide film defects (bifilms) as fatigue crack initiators. We shall present the evidence here for HIPping as a solid-state process for the closing of pores and bifilms.

When a cast Al–7Si–0.3Mg alloy with oxide film defects, like that shown in Figure 10.1a, is subjected to HIPping treatment, the applied pressure at temperature close to its melting temperature induces a substantial plastic deformation in the casting causing the defects to collapse and their

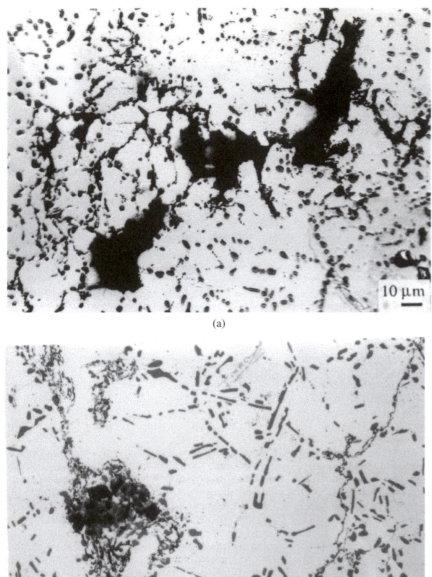

(a)

(b)

Figure 10.1 *Optical micrographs of an Al–7Si–0.4Mg alloy from a bottom-filled unfiltered casting showing (a) a network of oxide films and attached pores in the non-HIPped condition and (b) the network of oxide films and collapsed pore in the HIPped condition (Nyahumwa et al. 2000).*

surfaces to be forced into contact (Figure 10.1b). This plastic collapse phase probably requires only a few seconds or a few minutes at most. At this stage the pores and cracks are closed but, of course, not necessarily bonded. Thus only a modest, if any, benefit from HIPping may be expected at this early stage.

For many aluminium alloys of intermediate Mg content in the range of perhaps 0.05 to 0.5 per cent, as for instance typified by Al–7Si–0.4Mg alloy, an alumina (Al_2O_3) film is first formed during the rapid entrainment process. However, after an incubation period that is a function of time and temperature, the entrained film transforms to magnesium aluminate spinel, $MgAl_2O_4$. This involves volume change and atomic rearrangement of the crystal structure that would be expected to encourage diffusion bonding across the oxide/oxide interface. Analysis of bifilms that have acted as fatigue crack initiation sites have confirmed their conversion to a spinel and confirms that fatigue properties are improved by HIPping (Nyahumwa 1998 and 2000). A further positive finding from this work was that compared to filtered castings, the unfiltered but HIPped castings exhibited higher fatigue performance, despite larger maximum defect sizes, implying some degree of bonding across the crack. The application of HIP to castings shown in Figure 9.12 resulted in fatigue test samples that did not fail. The runouts at the stress 150 MPa reached nearly 10^8 cycles (not shown). At the higher stress of 240 MPa improved fatigue lives were still recorded (Figure 9.13) although it is of interest that in all except the most resistant specimen, the fatigue failures still occurred from oxides, probably healed or partly healed bifilms.

The liquid-state healing mechanism as described in section 2.4, and the HIP solid-state healing mechanism, are suggested to be analogous. However, there are interesting differences. In the liquid state the healing mechanism operates at high temperature (i.e. 700°C for an aluminium alloy) and with only the moderate external applied pressure due to depth (<0.1 MPa). The solid-state oxide film healing process operates at very high-applied pressure but lower temperature (i.e. 100 MPa at 500°C for Al–7Si–Mg alloy). In both the liquid and solid conditions the pressure drop will be expected to occur internally within the bifilm as the oxidation reaction proceeds. When all the oxygen is consumed, it is expected that the nitrogen may subsequently be consumed to form nitrides. Pores in this alloy before and after HIPping are shown in Figure 10.1.

Whether bifilms are actually 'healed' (i.e. effectively welded) may relate to the chemistry of the alloy and its films. Some bifilms appear to heal as we have seen above. However, others appear to be resistant. These important factors are not well researched at this time. Thus HIPping has limitations that do not appear to be widely known or understood. However, evidence for possible mechanisms is discussed below.

In contrast to this beneficial action of HIPping found by Nyahumwa and others as described above, Wakefield and Sharp (1992) observed HIPping to have no beneficial effect on the fatigue properties of Al–10Mg alloy castings, despite the closure of pores and cracks. The inference is that the bifilms in this alloy proved impossible to deactivate, resisting effective bonding. This is attributed to the magnesia (MgO) film formed during oxidation of Al–10Mg and entrained during casting. The magnesia film is (i) thicker, and (ii) has a stable structure that does not transform during HIPping (Nyahumwa 1998 and 2000). This lack of any

substantial atomic movements explains the inert nature of the highly stable MgO compound.

Similarly, it would be expected that Al alloys with very low Mg contents, which would be expected to contain entrained alumina films, would be similarly resistant to HIPping because of the great stability of alumina in the absence of sufficient Mg to convert it to a spinel structure.

During the early development of the Pegasus engine for the Harrier Jump Jet 25 polycrystalline Ni-based alloy turbine blades that had previously been scrapped because of their content of porosity were subjected to HIPping, and were fitted to a test engine alongside sound blades to evaluate whether HIPping might be a satisfactory reclamation technique for blades that otherwise would be scrapped. The HIPped blades failed within a few hours, damaging the engine and forcing a rapid shutdown of the test. The failures had occurred by creep cavitation at the grain boundaries of recrystallized regions in the centre of the castings. Almost certainly the original porosity would be caused by aluminium oxide or aluminium nitride films entrained by the severe turbulence that is usual during the vacuum casting process. (The vacuum is known to contain plenty of residual air to ensure the growth of surface films.) The great stability of the films, formed at the high casting temperature would ensure that they were resistant to any re-bonding action. The recrystallization would have happened because of the large plastic strains that were a necessary feature of the collapse of the porosity. However, the subsequent grain growth would expand grains up to local barriers such as bifilms. Thus the bifilms, effectively unbonded, and so acting as efficient cracks, were automatically located at the grain boundaries from where the failures were seen to occur.

The closure of internal cavities usually causes negligible changes to the overall dimensions of the casting if the pores are small and/or deep seated. For large or near-surface pores, however, the collapse of the surface of the casting in the form of a localized sink may scrap the casting if the depression exceeds the machining allowance. In a severe case, the surface may puncture, opening up the internal cavity to the surface (Zeitler and Scharfenberger 1984).

Naturally, HIPping cannot work if the pores are already connected to the outer surface of the casting. Such pores will never heal. Unfortunately this is all too common. The existence of the various forms of surface-connected porosity is well known to those who HIP castings. In particular, pressure die castings are woefully resistant to the benefits of HIPping because of their many surface connected bifilms (even traversing the so-called 'dense' outer layers of the casting). However, many gravity-filled or even counter-gravity-filled castings exhibit surface-connected pores as a result of bifilms intersecting the surface.

However, although an excellent start, even good filling of the castings will not guarantee a satisfactory response to HIP treatment. As we have seen in section 7.5.1 poor feeding in a long-freezing-range alloy can easily create surface-connected pores.

For a reliable HIP response, the surface of the casting must be sound.

Finally, therefore, although the mechanical properties of castings usually exhibit an improvement, in the sense that their average properties are raised, the Weibull modulus most often falls. This is a direct result of the healing of many defects, but leaving a few unaffected, for the various reasons we have seen. Thus the scatter of properties is increased. Regrettably, this is one of the greatest disadvantages of HIPping, often overlooked.

10.3 Working (forging, rolling, extrusion)

It seems curiously perverse that most metallurgical texts continue to foster the erroneous assumption that working eliminates casting defects. Usually, the strains involved in most working operations are too low to effect any significant welding of faults. Bifilms in general are merely pushed around, if anything, growing worse before (if ever) getting better. The growth of casting defects by plastic working is well known to those who work in the forging industry. There are good reasons for this behaviour.

During the working of cast material, for instance by rolling, it is to be expected that the defects will be elongated in the direction of working. The elongation of the defect necessarily rotates it, to align it along the rolling direction.

In addition, of course, its elongation increases the area of the defect. The newly extended surfaces in the bifilm would necessarily be oxidized by the remnant of air entrained in the defect. The entrapped air would be contained partly amid the microscopic pores between the crystals of the oxide, and partly in macroscopic reservoirs formed by folds and bubbles. During rolling, the continued oxidation of the expanding surfaces would hinder the welding of the interfaces that were being newly created, and by this means assisting the defect to grow as a crack, possibly to several times its initial size. The great effectiveness of the creation of this newly oxidized extension to the crack would be a consequence of the small oxygen requirement; the oxide would grow to a thickness of only nanometres at the hot working temperature (which is, naturally, much lower than the temperature at which the film

was first formed on the melt). In other words, the entrained residual oxygen is highly efficient at these temperatures to continue the oxidation of new area as it is formed.

Miyagi et al. (1985) used ultrasonics to observe this increased area of cavities in 5083 alloy (Al–Mg type) during the early stages of hot rolling. Microcavities near the surface of the rolled plate seemed to close early, but those nearer the centre were long-lived. Micro-examination revealed that they were smooth sided, and appeared to have opened and expanded along grain boundaries. (The association with grain boundaries is a feature to be expected of bifilms.) They were seen to reduce in size only after reductions of over 50 per cent.

Later, if the extension were sufficiently large to consume the remaining air, further extension would result in the welding up of new extensions to the crack. It seems likely that normal forging and rolling do not reach this stage for most alloys, and so do little to heal defects, simply extending and realigning them as described above. Processes such as extrusion and multi-pass rolling may sometimes account for a more complete elimination of the casting defects because of the much greater strains that can be involved. Even so, it seems that many defects remain, as indicated by the evidence from corrosion (section 11.2) among others.

Work by Harper (1966) on the hot rolling of wire-bar copper showed that rounded bubbles of gas are collapsed asymmetrically, forming convex bubbles, from which are pinched off strings of microscopic bubbles only 30 nm diameter. These minute bubbles appear to be completely stable in the solid. They can only be removed by remelting. It seems they do reappear in welds.

Conversely, for continuously cast steel slabs, Leduc et al. (1980) found the wide scatter in porosity was eliminated by working, the steels becoming fully dense at about 75 per cent reduction. Examination of the microstructure indicated that all the pores were eliminated. This is probably understandable in view of the low residual levels of Al in the steels (0.02–0.04 per cent) and absence of other strong oxide-forming elements. Any surface oxide would therefore have been expected to be a liquid iron–manganese silicate, and so any bifilms, if present, would have been easily welded shut.

10.4 Machining

The effect of oxides in castings is well known to machinists. The alumina inclusions in light alloys and mixed oxides in iron and steel castings are all known to cause defects on the machined surface by dragging out, leaving unsightly grooves.

Worse still, the cutting edge of the tool is often chipped or blunted by encounters with such hard particles, which are in general vastly harder than the cast alloy, and even harder, in many cases, than the tool tip.

In the author's experience, those working in a machine shop drilling ductile iron crankshafts could tell the difference between those castings made with good and bad filling system designs simply by listening as they walked among the machines. Those that were cast with significant surface turbulence contained inclusions rich in magnesium oxide, creating the characteristic high-pitched singing noise that signalled a poor finish and short tool life.

Grey iron castings are not free from the problems of poor machinability caused by turbulent filling. The presence of various kinds of bubble trails with their consequential residual bubbles admixed with oxides can cause reductions in tool life by a substantial margin in an alloy that is specifically chosen for its normally excellent machinability.

For sand castings of all types, the presence of residual sand on the surface of the casting is often a problem. For iron castings it is widely appreciated that the first cut should be sufficiently deep to get below the glassy surface inclusions. In light alloys cast in sand the complete removal of all sand is practically impossible short of dissolving away part of the surface in an acid or alkaline bath. Normally, some residual surface grains are broken and hammered into the surface by the so-called cleaning processes such as shot blasting. Grit blasting can be even worse, since particles of grit (often hard, abrasive grains of alumina grit) remain embedded in the surface. A freshly grit blasted casting, simply picked up and lightly dropped on to a clean iron surface plate will be found to have left a dusty outline of its shape on the plate.

Because of this residual surface aggregate problem, the slightly impaired machinability of sand castings compared to die (permanent mould) castings is one of the very few potential disadvantages of sand castings that requires to be acknowledged and accepted by all the parties in the casting supply chain.

10.5 Painting

For those painting processes where the paint is cured by heat, such as powder coating, the curing cycle is usually integrated as the final heat-treatment stage of the product. Aluminium alloy wheels for cars are usually heat treated and painted in this combined process.

However, the incidence of porosity linked to the surface can cause a defect in the smooth surface of the paint known as a paint crater.

During the heating of the casting for the baking cycle, air expands from surface-connected cavities in the casting. Usually at this early stage of the process the paint powder is not fully melted and the air can escape harmlessly. Only rarely will a hemispherical blister be blown and survive to spoil the paint surface.

During the cooling of the casting after the baking cycle, the paint coat is sucked into the internal pore as the gas in the pore contracts. If the surface pore is linked to extensive shrinkage porosity, the crater can have a central hole, and the hole will connect to the internal shrinkage cavity. More often, however, the surface defect is merely the point at which a bifilm intersects the surface. In this case the depth of the crater will depend on the volume enclosed by the bifilm. Such craters are often a serious source of scrap for aluminium alloy road wheels for cars, since the cosmetic requirement for such castings is, unfortunately, near perfection.

Chapter 11

Environmental interactions

During the cooling of the casting in the mould, and to some extent after it has been extracted from the mould, and especially during high-temperature heat treatment, air can gain access to and can attack the interior of the casting. Similarly, if the casting experiences a corrosive environment, internal access of the corrodant can lead to a variety of problems. The role of the bifilm is central to the understanding of these important problems.

11.1 Internal oxidation

When a metal such as an aluminium alloy is heat treated, the external surface naturally develops a thick oxide skin. However, in addition, some researchers have noted the development of internal oxides.

How is this possible? Oxygen is insoluble in aluminium and its alloys, and so, in principle, is not able to penetrate the alloy.

Clearly, such internal oxidation can only occur if the oxidizing environment has access to the interior through some kind of hole. The natural access will be those bifilms that connect to the outside surface. In the case of machined castings, particularly test bars, the machined surface will certainly cut into many bifilms, opening up probably multitudes of access points for gaseous penetration and attack deep into the interior of the casting.

Working with an Al–4Mg alloy subjected to a solution treatment at 520°C for several hours Samuel and co-workers (2002) expected to find the usual benefits including not only the taking of solutes into solution, but fragmentation and spherodizing of intermetallics, resulting in an improvement in mechanical properties. However, they reported the development of spinel (the $Al_2O_3.MgO$ mixed oxide) in the interior of the casting, and the reduction in tensile properties.

Fenn and Harding (2002) in the author's laboratory treated some cast iron test bars at 950°C for 1 hour to eliminate carbides to allow them to be machined. The tensile properties turned out to be unexpectedly low, especially the ductility. Elongations in the region of 20 per cent or more for this grade were to be expected, since ductile iron, as its name implies, is expected to be reliably ductile. It was a shock therefore to discover that elongations were highly scattered in the range of only 1 to 10 per cent. Many specimens appeared to fail brittlely. SEM (scanning electron microscope) studies revealed thick uniform carpets of oxide on the fracture surfaces (Figure 11.1). These oxides clearly did not have the appearance of entrained films, and did not contain the content of Mg that would be expected if the oxides were bifilms entrained in the liquid metal during casting. It seems that the oxygen had penetrated the core of the bifilm, and added to the original magnesium silicate from which the original bifilm was constituted. The additional oxide appeared to be nearly pure iron oxide, as would be expected from a solid-state reaction. At the relatively low temperature of 950°C (low compared to the casting temperature of about 1450°C) the magnesium in solution in the solid alloy would be fixed, being unable to diffuse to the surface of the bifilm to enhance the magnesium concentration in the oxide. A further confirmation of the solid-state thickening of these oxides is the fact that the graphite nodules appear through holes in the carpet. Clearly the iron oxide cannot thicken above the nodules, giving the curiously dimpled or quilted appearance of the surface of the oxide seen in Figure 11.1a.

These internal oxidation problems are almost certainly common in cast alloys that have been subjected to heat treatment.

The same effect will occur, of course, automatically during the cooling of the casting in

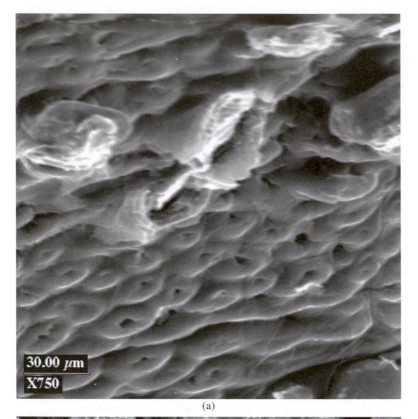

(a)

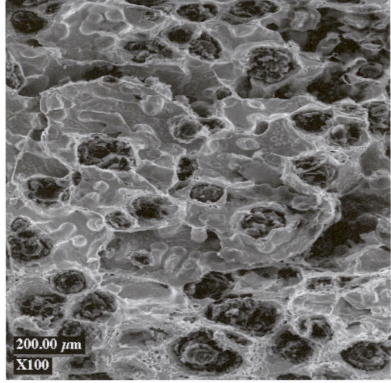

(b)

Figure 11.1 *SEM image of a fracture surface of a ductile iron casting damaged by growth during annealing at 950°C for 1 hour on a surface-connected bifilm of (a) carpet of solid-state grown oxide and (b) possible additional oxidation of graphite nodules (Fenn and Harding 2002).*

the mould. Additional oxidation of the originating bifilm will occur for surface-connected bifilms, thickening the film and possibly masking its original form. The original composition of the oxide may also be diluted and/or hidden by overgrowth of new oxide resulting from the solid state reaction.

The solution to the avoidance of such internal oxidation is the avoidance of bifilms. Although this would be a complete solution, it may not be always practical. A next-best solution might be one of the many techniques to keep the bifilms closed, since, clearly, any action to open the bifilm, for instance by shrinkage, will enhance the access routes to the interior. It follows that a well-fed casting (i.e. pressurized by the atmosphere) or a casting artificially pressurized during solidification (as provided by some casting processes) will be less susceptible to the ingress of air during high temperature treatment. It will therefore retain its mechanical properties relatively unchanged (whether originally good or bad of course).

11.2 Corrosion

Corrosion of metals, particularly aluminium alloy castings, and wrought products such as alloy plate and sheet, is a troublesome feature that has attracted much research in an effort to understand and control the phenomenon. Naturally, no comprehensive review of such a vast discipline can be undertaken here. The reader is referred to some recent reviews (Leth-Olsen and Nisancioglu 1998). The purpose of this section is to present the evidence that most corrosion problems, not only in shaped castings, but also in wrought alloys, arise from casting defects. The defects are the surface pits that are the sites where bifilms happen to meet the surface. In the absence of bifilms it is proposed that there would probably be no corrosion of metals from surface pits. Corrosion might still be expected, but would probably be vastly reduced, and might be forced to occur by quite different mechanisms. It could be envisaged to occur from other inclusions, or grain boundaries, or, finally, from dislocations that intersect the surface.

Many of the current theories of the corrosion of metals have been principally concerned with environmental attack on an essentially continuous unbroken planar substrate, regarding the surface of the metal as a uniform reactive layer (Leth-Olsen and Nisancioglu 1998). The result has been that theories of filiform and intergranular corrosion of aluminium alloys are at a loss to explain many of the observed features of these phenomena, since these corrosion processes clearly do not exhibit uniformity of attack; the attack is extremely localized and specific in form.

The presence of bifilms generated in the pouring process has not, of course, been considered up to now as a factor contributing to the severity of corrosion. It will become clear in this section how bifilms help to explain many of the observed features of metallic corrosion. The link occurs since bifilms are, of course, often connected to the surface, allowing them to be detected by dye-penetrant techniques. Similarly, in a corrosive environment such bifilms will allow the local ingress of corrosive fluids between their unbonded inner surfaces. In aluminium alloys, the presence of intermetallic phases precipitated on the outer surfaces of bifilms will be expected to act as a further enhancement of the corrosion process, explaining the major differences observed between Al alloys of different iron, manganese and copper contents.

11.2.1 Pitting corrosion

Although there are many instances in which the corrosion of metals occurs uniformly across the whole surface, the special case of concentrated corrosion at highly localized sites, generating deep pits, is sometimes a serious concern. Most of the studies of pitting corrosion have been carried out on steels. However, we cannot in this short work survey this vast subject. We shall take Al and its alloys as an example, following the review by Szklarska-Smialowska (1999), and see how pitting corrosion relates to the cast structure.

The main message of this section is that, in general, the familiar corrosion pit is not, originally, the product of corrosion. It pre-exists, being the product of poor casting technology. This pre-existence appears to have been generally overlooked until now. Naturally, the corrosion process develops the pit, which is originally usually practically, if not actually, invisible, into a highly visible and deleterious feature.

The corrosion proceeds as illustrated in Figure 11.2 (Bailey and Davenport 2002). The intermetallic particle acts as a cathode, the electrical current passing through the electrolyte to anodic areas of the surface. It has been generally thought that the intermetallic particles provide the conductive path through the insulating alumina film. However, it is probable that the bifilm itself is sufficiently thin to be conductive, and so will aid this effect. The cathodic pit is the bifilm pit containing the intermetallic, whereas the anodic pits may be part of the same bifilm pits but distant from the intermetallic, or may be quite separate surface-intersecting bifilms that do not happen to contain intermetallics.

Oxygen is reduced at the cathode, demanding electrons, and so forming hydroxyl ions according to:

$$O_2 + 2H_2O + 4e^- = 4OH^-$$

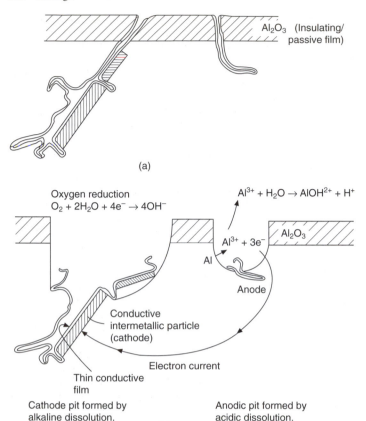

Oxygen reduction
$O_2 + 2H_2O + 4e^- \rightarrow 4OH^-$

$Al^{3+} + H_2O \rightarrow AlOH^{2+} + H^+$

$Al^{3+} + 3e^-$

Al_2O_3 (Insulating/passive film)

Al_2O_3

Al

Anode

(a)

Conductive
intermetallic particle
(cathode)

Electron current

Thin conductive
film

Cathode pit formed by
alkaline dissolution.

Anodic pit formed by
acidic dissolution.

(b)

Figure 11.2 *Mechanism of pitting corrosion (a) prior to corrosion and (b) during corrosion (adapted from Bailey and Davenport 2002).*

The alkaline conditions created by the hydroxyl ions assist to dissolve the material around the intermetallic, enlarging the pit. Conversely, at the anodic pit, conditions are acidic because of the generation of hydrogen ions as follows:

$$Al = Al^{3+} + 3e^-$$

$$Al^{3+} + H_2O = AlOH^{2+} + H^+$$

Thus this pit also enlarges as matrix material is dissolved. The electrical circuit is, of course, completed by electrons travelling though the aluminium matrix from the anode pit to the cathode pit.

The random nature of the creation of such defects, being linked to the action of surface turbulence at several stages of manufacture of the sheet, explains why the corrosion behaviour is so variable, changing in severity from one supplier of metal to another, and from one production batch of alloy to the next. Also, of course, every pit will be different because of the random nature of the oxide tangles. The tangled geometry is indicated in Figure 11.2. This randomness has been a major problem to investigators.

The bifilms are expected to survive, and even grow, during plastic deformation, as discussed in section 10.3. Thus surface-linked cracks, possibly plated with intermetallics, will be not only characteristic of castings but also of wrought products.

11.2.2 Filiform corrosion

In a standard test, filiform corrosion takes the form of a high surface density of superficial corrosion paths, called filaments, which propagate rapidly and extensively from a scribe mark on a test plate. The corrosion proceeds away from the scratch along filamentary lines aligned with the original rolling direction. They travel under any protective layer such as paint, occasionally tunnelling beneath the metal surface, only to break out at the metal surface once again after a few millimetres or so. The lengthwise growth and subsequent sideways spreading of the filaments eventually causes any protective coating, such as a paint layer, to exfoliate. The length of filaments has been found to be generally in the range 1 to 10 mm. However, reviewers confirm (Leth-Olsen and Nisancioglu 1998) that quantification of the phenomenon suffers from significant scatter that has hampered these studies.

The concentration of corrosion at strictly

localized sites (the filaments) is clear. However, it is important to observe that the great majority of the metal surface remains completely free from attack (despite the long and deep breach of the protective coating by the scratch). Also clear is the different behaviour of different casting batches of nominally identical material, on different occasions giving filaments shallow or deep, or short (1 mm) or long (10 mm).

Growth of filaments stops when the length reaches some value between 1 and 10 mm. This has been suggested to be the result of chloride depletion in the head of the filament (Leth-Olsen and Nisancioglu 1998) but is clearly more likely to be that the bifilms that provide the easy path for corrosion are simply only that long, as is seen in direct observations of the melt (section 2.7). In other words, the corrosion stops when it reaches the end of the bifilm.

In his review of the subject, Nordlien (1999) describes how the filaments of corrosion can grow at up to 5 mm per day. They occur on all families of aluminium alloys (1000, 2000, 3000, 5000, 6000, 7000 and 8000 series) and on all product forms (sheet, foil, extrusions).

Interestingly, a surface of rolled aluminium alloy sheet can be sensitized to the formation of filiform corrosion (in corrosion jargon it is 'activated') by annealing at 400°C. This effect can be understood as the growth of oxidation products on the internal surfaces of cracks, which will assist to open the cracks (see section 11.1). The deactivation by etching probably corresponds to the preferential attack and removal of surface cracks and laminations. Reactivation by subsequent annealing seems likely to be the result of the opening of slightly deeper defects by oxidation. The removal of defects by etching removes only a few μm of depth of the surface. Considering the defects are commonly 1 mm to 10 mm in size, there will be no shortage of new defects to open on a subsequent reactivation cycle.

In severe cases of surface corrosion, the frequent observations of delamination (Leth-Olsen and Nisancioglu 1998) can be understood as the lifting of irregular fragments of bifilm that lie just under the metal surface. Other related observations of blistering can also be understood as the inflation of just-subsurface bifilms by hydrogen evolved from the chemical reaction between the corrodent and the intermetallic compounds associated with the bifilm.

Direct and clear observations of oxide film tangles associated with corrosion sites has been made by Nordlien et al. (2000) and Afseth and co-workers (2000).

11.2.3 Intergranular corrosion

Intergranular corrosion in its various forms is also proposed to be associated in some cases with the newly identified bifilm defects, as a result of the natural siting of bifilms at grain boundaries in the cast structure.

Metcalfe (1945) records studies of the inter-crystalline corrosion of the heads of rivets in an Al–Mg alloy from an aircraft that has been flown near marine environments. He concludes that the effect is one of stress corrosion cracking. Undoubtedly there would be both applied and residual stress, and both may have played a part in the failures that are described. More especially so since the cracks were observed to follow grain boundaries sensitized by prolonged in-service ageing, and the convoluted form of the crevices was due to the fact that the flattened grains themselves were distorted in this fashion by the complex flow pattern involved. Even so, a look at a section of one of the decapitated rivets in his work reveals a convoluted crack that can hardly have been propagated by stress. The stress would have been reduced to near zero after the spread of the first crack across the neck of the rivet. In fact, there is the trace of a crack which has repeatedly turned, spreading back and forth across the neck of the rivet at least five or six times. This type of crack is typical of a folded oxide defect. Its presence would ensure the stability of the convoluted form of the grain boundaries, which it would pin. Furthermore, in this vintage of alloy a high density of entrainment defects would be the norm. The defect has provided an easy path for the attack of corrodent.

Forsyth (1995 and 1999) describes seawater corrosion leading to intergranular cracking in 7010 alloy. Corroded surfaces that have been polished back through the worst of the surface layer are presented in Figure 11.3. The intergranular and subgrain boundary cracks were, once again, typical of the localized tangled arrays of films that are normal in aluminium alloys produced via the melting and casting route. The cracks exhibit the typical irregular branching and changes of direction on a number of different size scales, often unrelated to the general size of the grain size of the matrix. Alloy material between such damaged regions was recorded to be completely free from attack. It is suggested that these observations are difficult to explain without the existence of random entrainment defects from the original casting.

When etching to reveal the dendrite structure, the cracks were seen (Forsyth 1999) to be confined to the interdendritic regions (Figure 11.4). This corroborates with work on solidifying aluminium alloys described at several points in this book (for instance, section 2.3); during growth, the dendrites are found to push the double films ahead. The defects are therefore concentrated in the residual liquid in the interdendritic regions and in grain boundaries.

Forsyth (1999) also investigated the corrosion

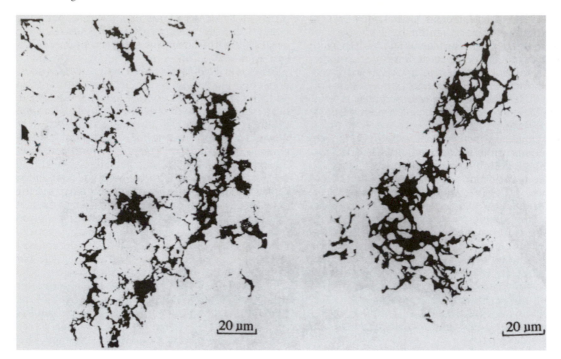

Figure 11.3 *Two typical views of forged 7010-T736 alloy subjected to seawater corrosion (courtesy P.J.E. Forsyth 2000).*

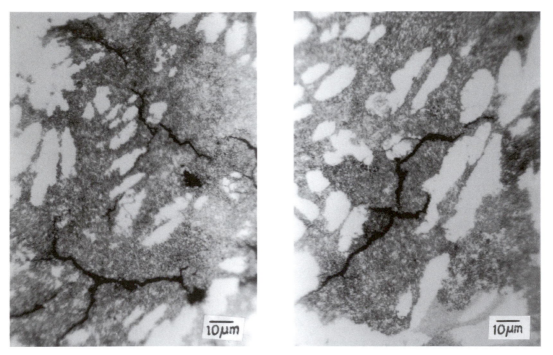

Figure 11.4 *Two typical views of polished and etched 7010 alloy in the solutionized condition subjected to seawater corrosion, illustrating the interdendritic nature of the cracks (courtesy P.J.E. Forsyth 2000).*

of 7010 alloy in seawater as a result of machining or bruising of the surface. In the case of bruising, the deformation of the surface would be expected to open any entrained defects at or near the surface, creating highly localized and deeply penetrating intergranular pathways for attack.

Forsyth also draws attention to the especially damaging nature of the attack, in that despite rather little dissolution of material, complete blocks of material could be removed simply by the penetration of the attack along narrow planes in different directions. This observation corroborates his earlier report (Forsyth 1995) in which subsequent anodizing of the surface led to the incorporation of unanodized grains of metal in the corrosion debris remaining from such localized attack. The metal grains remained unanodized because they were found to be electrically isolated from their surroundings. This would not be surprising if double oxide films, separated by their interlayer of air, surrounded the grains.

In conclusion, it seems there is considerable evidence that in the absence of bifilms, some types of intergranular corrosion might be reduced or eliminated. In addition, the localized pitting corrosion of metals will probably be reduced, and in many cases, eliminated.

The elimination of bifilms would revolutionize metals and improve the quality of our lives in many ways.

References

Adefuye Segun (1997). 'The fluidity of Al–Si alloys', PhD Thesis, University of Birmingham UK.

Afseth A., Nordlien J. H., Scamans G. M., Nisancioglu K. (2000). ASST 2000 (2nd Internat. Symp. Al Surface Science Technology), UMIST Manchester UK, Alcan + Dera, pp. 53–58.

Agema K. S., Fray D. J. (1988). Agema PhD Thesis, Dept Materials, Cambridge, UK.

Allen D. I., Hunt J. D. (1979). Solidification and Casting of Metals, Sheffield Conference, 1977, *Metals. Soc.,* 39–43.

Allen N. P. (1932). *J. Inst. Metals*, **49**, 317–346.

Allen N. P. (1933). *J. Inst. Metals*, **51**, 233–308.

Allen N. P. (1934). *J. Inst. Metals*, **52**, 192–220.

American Foundrymen's Society (1987). *Green Sand Additives*, AFS, Detroit, USA.

Anderson J.V., Karsay S. I. (1985) *Brit Foundryman*, 492–498.

Anderson S. H., Foss J. W., Nagan R. M., Jhala B. S. (1989). *TAFS*, **97**, 709–722.

Andrew J. H., Percival R. T., Bottomley G. T. C. (1936). *Iron and Steel Special Report*, **15**, 43–64.

Angus H. T. (1976). *Cast Iron*, Butterworths, London.

Anson J. P., Gruzleski J. E. (1999). *Trans American Foundry Soc.,* **107**, 135–142.

Arnold F. L., Jorstad J. L. and Stein G. E. (1963). *Current Engineering Practice*, **6**, 10–15.

Asbjornsonn (2001). PhD Thesis, Department of Materials, University of Nottingham.

Ashton M. C. (1990). In SCRATA 34th Conference, Sutton Coldfield, paper 2, Steel Castings Research and Trade Assoc., Sheffield, UK.

Ashton M. C., Buhr R. C. (1974). Phys. Met. Div. Internal Report PM-1-74-22, Canada Dept Energy Mines and Resources.

Askeland D., Holt M. L. (1975). *TAFS*, **83**, 99–106.

Atwood R. C., Lee P. D. (2000). Modeling of Casting Welding and Advanced Solidification Processing Conf., Aachen, eds P. Sahm, P. Hansen.

Avey M. A., Jensen K. H., Weiss D. J. (1989). *TAFS*, **97**, 207–212.

Bachelot F. (1997) MPhil Thesis, University of Birmingham, UK.

Bachmann P. K., Messier R. (1984). *Chemical and Engng News*, **67**(20), 24–39.

Bailey A., Davenport A. J. (April 2002) Final year project, University of Birmingham, Dept Metall., UK.

Baker W. A. (1945). *J. Inst. Metals*, **71**, 165–204.

Baker W. F. (1986). *TAFS*, **94**, 215–218.

Baliktay S., Nickel E. G. (1988). Seventh World Conf. Invest. Casting, Munich, paper 10.

Barbe L., Bultink I., Duprez L., Cooman B. C. De. (2002). *Materials Science & Technology*, **18**, 664–672.

Barkhudarov M. R. and Hirt C. W. (1999). *Die Casting Engineer,* Jan–Feb, 44–47.

Barlow G. (1970). PhD Thesis, University of Leeds.

Bastien P., Armbruster J. C., Azou P. (1962). In 29 Internat. Found. Congress Detroit, pp. 400–409.

Bates C. E., Monroe R. W. (1981). *TAFS*, **89**, 671–686.

Bates C. E., Scott W. D. (1977). *TAFS*, **85**, 209–226.

Bates C., Wallace J. F. (1966). *TAFS*, **74**, 174–185.

Batty O. (1935). *TAFS*, **43**, 75–106.

Bean X., Marsh L. E. (1969). *Metal Progress*, **95**, 131–134.

Beck A., Schmidt W., Schreiber O. (1928). US Patent 1, 788, 185.

Beech J. (1974). *The Metallurgist and Materials Technologist*, April, 129–232.

Beeley P. R. (1972). *Foundry Technology*, London: Butterworths, pp. 104, 125, 128, 501.

Benaily N. (1998). 'Inoculation of flake graphite iron', MPhil Thesis, Univ Birmingham, UK.

Benson L. E. (1938a). *Foundry Trade J.,* June 527–528.

Benson L. E. (1938b). *Foundry Trade J.*, June 543–544.

Benson L. E. (1946). *J. Inst. Metals*, **72**, 501–510.

Beretta S., Murakami Y. (2001) *Met. and Mat. Trans B*, **32B**, 517–523.

Berger R. (1932). *Fonderie Belge*, **17**.

Berry J. T., Kondic V., Martin G. (1959). *TAFS*, **67**, 449–476.

Berry J. T. and Taylor R P. (1999). *TAFS*, **107**, 203–206.

Berry J. T. and Watmough T. (1961). *TAFS*, **69**, 11–22.

Berthelot M. (1850). *Ann. Chim.*, **30**, 232.

Bertolino M. F., Wallace J. F. (1968). *TAFS*, **76**, 589–628.

Betts B. P., Kondic V. (1961). *Brit. Found.*, **54**, 1–4.

Bindernagel I., Kolorz A., Orths K. (1975). *TAFS*, **83**, 557–560.

Bishop H. F., Ackerlind C. G., Pellini W. S. (1952). *TAFS*, **60**, 818–833.

Biswas P. K., Rohatgi P. K., Dwarakadasa E. S. (1985). *Brit. Found.*, **78**, 511–516.

Boenisch D. (1967). *TAFS*, **75**, 33–37.

Boenisch D., Patterson W. (1966). *TAFS*, **74**, 470–484.

Bonsak W. (1962). *TAFS*, **70**, 374–382.

Boom R., Dankert O., Veen van A., Kamperman A. A. (2000). *Metall. and Materials Trans B*, **31B**, 913–919.

Bounds S. M., Moran G. J., Pericleous K. A., Cross M. (1998). Modelling of Casting, Welding and Solidification Processing Conf VIII, San Diego, eds B. G. Thomas and C. Beckerman, *TMS*, pp. 857–864.

Boutorabi S. M. A., Din T., Campbell J. (1990). University of Birmingham, unpublished work.

Bower T. F., Brody H. D., Flemings M. C. (1966). *Trans AIME*, **236**, 624.

Bramfitt B. L. (1970). *Met. Trans*, **1**, 1987–1995.

Brandes E. A. and Brook G. B. (eds) (1992). *Smithells Metals Reference Book*, 7th edition, Butterworths.

Bridge M. R., Stephenson M. P., Beech J. (1982). *Metals Technol.*, **9**, 429–433.

Bridges D. (1999). 'Wheels and axles', I. Mech. E. Seminar, London.

Briggs L. J. (1950). *J. Appl Phys.*, **21**, 721–722.

Briggs C. W., Gezelius R. A. (1934). *TAFA*, **42**, 449–476.

Brimacombe J. K., Sorimachi K. (1977). *Met. Trans*, **8B**, 489–505.

Brondyke K. J. and Hess P. D. (1964). *Trans Met. Soc. AIME*, **230**(7), 1542–1546.

Bryant M. D., Moore A. (1971). *Brit. Found.*, **64**, 215–229, 306–307.

Burchell V. H. (1969). *Brit Found.*, **62**, 138–146.

Burton B., Greenwood G. W. (1970). *Met. Sci. J.*, **4**, 215–218.

Butakov D. K., Mel'nikov L. M., Rudakov I. P., Maslova Yu N. (1968). *Lit. Proizv.*, **4**, 33–35.

Butler C. J. (1980). UK Patent GB 2020714.

Byczynski G. E., Cusinato D. A. (2001). First International Conf Filling and Feeding of Castings 4–5, April, University of Birmingham, UK; and *International J. Cast Metals Research* (2002). **14**(5), 315–324.

Caine J. B., Toepke R. E. (1966). *TAFS*, **74**, 19–22.

Caine J. B., Toepke R. F. (1967). *TAFS*, **75**, 10–16.

Campbell J. (1967). *Trans Met. Soc. AIME*, **239**, 138–142.

Campbell J. (1968a). The Solidification of Metals, *ISI Publication*, **110**, 19–26.

Campbell J. (1968b). *Trans Met. Soc. AIME*, **242**, 264–268.

Campbell J. (1968c). *Trans Met. Soc. AIME*, **242**, 268–271.

Campbell J. (1968d). *Trans Met. Soc. AIME*, **242**, 1464–1465.

Campbell J. (1969a). *Cast Metals Res. J.*, **5** (1), 1–8.

Campbell J. (1969b). *Trans AIME*, **245**, 2325–2334.

Campbell J. (1971). *Metallography*, **4**, 269–278.

Campbell J. (1980) 'Solidification Technology in the Foundry and Casthouse', Warwick Conference, Metals Soc. Publication (1981) **273**, 61–64.

Campbell J. (1981). *International Metals Reviews*, **26**(2), 71–108.

Campbell J. (1988). *Materials Science and Technology*, **4**, 194–204

Campbell J. (1991a). *Cast Metals*, **4**(2), 101–102.

Campbell J. (1991b). *Castings*, Butterworth–Heinemann.

Campbell J. (1994). *Cast Metals*, **7**(4), 227–237.

Campbell J. (2000). *Ingenia*, **1**(4), 35–39.

Campbell J., Bannister J. W. (1975). *Metals Technol.*, **2**(9), 409–415.

Campbell J., Caton P. D. (1977). Institute of Metals Conference on Solidification, Sheffield UK, pp. 208–217.

Campbell J., Clyne T. W. (1991). *Cast Metals*, **3**(4), 224–226.

Campbell J., Olliff I. D. (1971). *AFS Cast Metals Research J.*, June, 55–61.

Campbell H. L. (1950). *Foundry*, 78, 86, 87, 210, 212, 213.

Cao X. and Campbell J. (2000). *Amer. Found. Soc. Trans*, **108**, 391–400.

Cao X., Campbell J. (2000). *Internat. J. Cast Metals Research* **13**, 175–184.

Cao X., Campbell J. (2001). *Trans. AFS.* **109**, 269–283.

Capello G. P., Carosso M. (1989). AGARD Report, May, No. 762.

Cappy M., Draper A., Scholl G. W. (1974). *TAFS*, **82**, 355–360.

Carlson G. A. (1975). *J. Appl. Phys.*, **46**(9), 4069–4070.

Carte A. E. (1960). *Proc. Phys. Soc.*, **77**, 757–769.

Carter S. F., Evans W. J., Harkness I. C., Wallace J. F. (1979). *TAFS*, **87**, 245–268.

Celik M. C., Bennett G. H. J. (1979). *Metals Technology*, April, 138–144.

Chadwick G. A. (1991). Southampton: Hi-Tech Metals R&D Ltd. Personal communication.

Chadwick H. (1991). *Cast Metals*, **4**(1), 43–49.

Chalmers (1953). Quoted by Flemings M.C. (1974).

Chamberlain B., Zabek V. J. (1973). *TAFS*, **81**, 322–327.

Charles J. A., Uchiyama I. (1969). *J. Iron and Steel Inst.*, **207**, July, 979–983.

Chechulin V. A. (1965). In *Gases in Cast Metals*, ed. B. B. Gulyaev, Consultants Bureau Translation, pp. 214–218.

Chen C. O., Ramberg F., Evensen J. D. (1984). *Metal Sci.*, **18**, 1–5.

Chen X. G., Engler S. (1994). *Trans Amer. Found. Soc.*, **102**, 673–682.

Chung Y. and Cramb A. W. (2000). *Met. and Materials Trans B*, **31B**, 957–971.

Church N., Wieser P., Wallace J. F. (1966). *TAFS*, **74**, 113–128. Also in *Brit. Found.*, 1966, **59**, 349–363.

Chvorinov N. (1940). *Giesserei*, **10**, 177–186, 201, 222 and **27**, 31 May, 201–208.

Cibula A. (1955). *Proc. IBF*, **45**, A73–A90. Also in *Foundry Trade J.* (1955), **98**, 713–726.

Claxton K. T. (1967). *The Influence of Radiation on the Inception of Boiling in Liquid Sodium*. UK Atomic Energy Authority Research Group Report AERE-R5308. Also in Proc. Internat. Conf on the Safety of Fast Reactors (CAE). Aix-en-Provence, Sept. 1967, p. II-B-8-1.

Claxton K. T. (1969). UKAEA, Harwell, UK. Private communication.

Clyne T. W. (1977). PhD Thesis, University of Cambridge.

Clyne T. W., Davies G. J. (1975). *Brit. Found.*, **68**, 238–244.

Clyne T. W., Davies G. J. (1979). In *Solidification and Casting of Metals*, Metals Soc. Conference, Sheffield, 1977, pp. 275–278.

Clyne T. W., Davies G. J. (1981). *Brit Found.*, **74**, 65–73.

Clyne T. W., Kurz W. (1981). *Metall. Trans*, **12A**, 965–971.

Clyne T. W., Wolf M., Kurz W. (1982). *Met. Trans*, **13B**, 259–266.

Coble R. L., Flemings M. C. (1971). *Metall. Trans*, **2**, Feb., 409–415.

Cochran C. N., Belitskus D. L., Kinosz D. L. (1977). *Metall. Trans B*, **8B**, 323–332.

Cottrell A. H. (1964). *The Mechanical Properties of Matter*. Wiley, p. 82.

Couture A., Edwards J. O. (1966). *TAFS*, **74**, 709–721, 792–793.

Couture A., Edwards J. O. (1973). *TAFS*, **81**, 453–461.

Cox M., Wickins M., Kuang J. P., Harding R. A., Campbell J. (2000). *Materials Science and Technology*, **16**, 1445–1452.

Creese R. C., Sarfaraz A. (1987). *TAFS*, **95**, 689–692.

Creese R. C., Sarfaraz A. (1988). *TAFS*, **96**, 705–714.

Cunliffe E. L (1996). '*The minimal gating of aluminium alloy castings*', PhD Thesis, Metals and Materials Department, University of Birmingham, UK.

Cunningham M. (1988). Stahl Speciality Co., Kingsville, MO, USA. Private communication.

Cupini N. L., Galiza J. A. de, Robert M. H., Pontes P. S. (1980). *Solidification Technology in the Foundry and Cast House*. Metals Soc. Conf., pp. 65–69.

Cupini N. L., Prates de Campos Filho M. (1977). Sheffield Conf. '*Solidification and Casting of Metals*', Metals Soc., 1979, pp. 193–197.

Dasgupta S., Parmenter L., Apelian D. (1998). *AFS 5th Internat. Conf. Molten Aluminum Processing*. pp. 285–300.

Davies G. J., Shin Y. K. (1980). In *Solidification Technology in the Foundry and Casthouse*, Metals Soc. Conf., Warwick, paper 78.

Davies I. G., Dennis J. M., Hellawell A. (1970). *Met. Trans*, **1**, 275–280.

Davies K. G. (1977). *AFS Internat. Cast Metals J.*, March, 23–27.

Davies K. G., Magny J.-G. (1977). *TAFS*, **85**, 227–236.

Davies V. de L. (1963). *J. Inst. Metals*, **92**, 127.

Davies V. de L. (1964–65). *J. Inst. Metals*, **93**, 10.

Davies V. de L. (1970). *Brit Found.*, **63**, 93–101.

Dawson J. V. (1962). *BCIRA J.*, **10**(4), 433–437.

Dawson J. V., Kilshaw J. A., Morgan A. D. (1965). *TAFS*, **73**, 224–240.

Daybell E. (1953). *Proc. Inst. Brit. Found.*, **46**, B46–B54.

Delamore G. W., Smith R. W., Mackay W. B. F. (1971). *TAFS*, **79**, 560–564.

Denisov V. A., Manakin A. M. (1965). *Russian Castings Production*, 217–219.

Devaux H. (1987) Measurement and Control in Liquid Metals Processing, ed. R. J. Moreau, Nijhoff, The Netherlands, pp. 107–115.

Dietert H. W., Doelman R. L., Bennett R. W. (1970). *TAFS*, **78**, 145–156.

Dimayuga F. C., Handiak N., Gruzleski I. E. (1988). *TAFS*, **96**, 83–88.

Dimmick T. (2001). *Modern Castings*, **91**(3), 31–33.

Dion J. L., Fasoyinu F. A., Cousineau D., Bibby C., Sahoo M. (1995). *TAFS*, **103**, 367–377.

DiSylvestro G., Faist C. A. (1977). *TAFS*, **85**, 627–642.

Divandari M. (1998). University of Birmingham UK. Unpublished work.

Divandari M., Campbell J. (1999). *AFS 1st Internat Conf. on Gating Filling and Feeding of Aluminum Castings*, 11–13 Oct. 1999, pp. 49–63.

Divandari M., Campbell J. (2000). *Aluminum Trans*, **2**(2), 233–238.

Dixon B. (1988). *Solidification Processing Conference* 1987, Inst. Metals, pp. 381–383.

Dodd R. A. (1950). PhD Thesis, Dept Industrial Metallurgy, University of Birmingham, UK.

Dodd R. A. (1955a). *Hot Tearing of Casting: A Review of the Literature*, Dept Mines, Ontario, Canada, Research Report PM184. Also in *Foundry Trade J.* (1956), **101**, 321–331.

Dodd R. A. (1955b). *Hot Tearing of Binary Mg–Al and Mg–Zn Alloys*, Dept Mines, Ontario, Canada, Research Report PM191.

Dodd R. A., Pollard W. A., Meier J. W. (1957). *TAFS*, **65**, 100–117.

Dong S., Niyama E., Anzai K. (1995). *ISIJ Internat.*, **35**, 730–736.

Doremus G. B., Loper C. R. (1970). *TAFS*, **78**, 338–342.

Double D. D. and Hellawell A. (1974). *Acta Metall.*, **22**, 481–487.

Draper A. B. (1976). *TAFS*, **84**, 749–764.

Drezet J. M., Commet B., Fjaer H. G., Magnin B. (2000). *Modeling of Casting, Welding and Advanced Solidification Processes IX*, eds P. R. Sahm, P. N. Hansen and J. G. Conley, pp. 33–40.

Drouzy M. and Mascre C. (1969). *Metall. Reviews*, **14**, 25–46.

Durrans I. (1981). Thesis, University of Oxford.

Edelson B. J., Baldwin W. M. (1962). *Trans ASM*, **55**, 230.

El-Mahallawi S., Beeley P. R. (1965). *Brit. Found.*, **58**, 241–248.

Ellison W., Wechselblatt P. M. (1966). *TAFS*, **74**, 350–356.

Emamy G. M., Campbell J. (1995a). *Internat. J. Cast Metals Research*, **8**, 13–20.

Emamy G. M., Campbell J. (1995b). *Internat. J. Cast Metals Research*, **8**, 115–122.

Emamy G. M., Campbell J. (1997). *Trans AFS*, **105**, 655–663.

Enright P. (2001). Private communication.

Enright P., Hughes I. R. (1996). *The Foundryman*, November, 390–395.

Evans E. P. (1951). *BCIRA J.*, **4**(319), 86–139.

Evans J., Runyoro J., Campbell J. (1997). *Solidification Processing* (eds J. Beech and H. Jones), Published by Dep. Engineering Materials, University of Sheffield, pp. 74–78.

Fang O. T., Granger D. A. (1989). *TAFS*, **97**, 989–1000.

Fasoyinu F. A., Sadayappan M., Cousineau D., Zavadil R., Sahoo M. (1998). *TAFS*, **106**, 327–337.

Feliu S. (1962). *TAFS*, **70**, 838–844.

Feliu S. (1964). *TAFS*, **72**, 129–137.

Feliu S., Flemings M. C., Taylor H. F. (1960). *Brit. Found.*, **53**, 413–425.

Feliu S., Luis L., Siguin D., Alvarez J. (1962). *TAFS*, **70**, 838–844; (1964), **71**, 145–157; (1964), **72**, 129–137.

Fenn D. and Harding R. A. (2002). University of Birmingham, UK. Unpublished work.

Feurer U. (1976). *Giesserei*, **28**, 75–80 (English translation by Alusuisse).

Fischer R. B. (1988). *TAFS*, **96**, 925–944.

Fisher J. C. (1948). *J. App. Phys.*, **19**, 1062–1067.

Flemings M. C. (1974). *Solidification Processing*, McGraw-Hill, USA.

Flemings M..C. (1963). *30th Internat. Found. Congr.* Prague, pp. 61–81 and *Brit. Found.* (1964), **57**, 312–325.

Flemings M. C., Conrad H. F., Taylor H. F. (1959). *TAFS*, **67**, 496–507.

Flemings M. C., Kattamis T. Z., Bardes B. P. (1991). *TAFS*, **99**, 501–506.

Flemings M. C., Mollard F. R., Niyama E. F. (1987). *TAFS*, **95**, 647–652.

Flemings M. C., Niiyama E. and Taylor H. F. (1961). *TAFS*, **69**, 625–635.

Flemings M. C., Poirier D. R., Barone R. V., Brody H. D. (1970). *J. Iron Steel Inst.*, April, 371–381.

Fletcher A. J. (1989). *Thermal Stress and Strain Generation in Heat Treatment*, Elsevier.

Fomin V. V., Stekol'nikov G. A., Omel'chenko V. S. (1965). *Russian Castings Production,* 229–231.

Forest B., Berovici S. (1980). In *Solidification Technology in the Foundry and Casthouse*, Metals Soc. Conf., Warwick, paper 93, pp. 1–12.

Forgac J. M., Schur T. P., Angus J. C. (1979). *J. Appl. Mech.*, **46**, 83–89.

Forsyth P. J. E. (1995). *Materials Science and Technology*, **11**(3), 1025–1033.

Forsyth P. J. E. (1999). *Materials Science and Technology*, **15**(3), 301–308.

Forward G., Elliott J. F. (1967). *J. Metals*, **19**, 54–59.

Fox S., Campbell J. (2000) *Scripta Mat.* **43** (10) 881–886.

Fox S., Campbell J. (2002). *Internat. J. Cast Metals Research*, **14**, (6) 335–340.

Franklin A. G., Rule G., Widdowson R. (1969). *1. Iron Steel Inst.*, September, 1208–1218.

Frawley I. J., Moore W. F., Kiesler A. I. (1974). *TAFS*, **82**, 561–570.

Freti S., Bornand J. D., Buxmann K. (1982). *Light Metal Age*, June, 12–16.

Fruehling J. W., Hanawalt J. D. (1969). *TAFS*, **77**, 159–590.

Fredriksson H. (1984). *Mat. Sci. Eng.*, **65**, 137–144.

Fredriksson H. (1996). *Solidification Science and Processing*, eds I. Ohnaka and D. M. Stefanescu, The Minerals, Metals and Materials Society.

Fredriksson H., Lehtinen B. (1977). In *Solidification and Casting of Metals*, Metals Soc. Conf., Sheffield. Metals Soc., (1979) pp. 260–267.

Fuji N., Fuji M., Morimoto S., Okada S. (1984). *J. Japan Inst. Light Met.*, **34**(8), 446–453. Also *Met. Abstr.*, 1651–1657.

Fuji M., Fuji N., Morimoto S., Okada S. (1986) *J. Japan Inst. Light Met.*, **36**(6), 353–360 (transl. NF 197).

Fuoco R., Correa E. R., Andrade Bastos M. de (1998). *TAFS*, **106**, 401–409.

Gagne M., Goller R. (1983). *TAFS*, **91**, 37–46.

Gallois B., Behi M., Panchal J. M. (1987). *TAFS*, **95**, 579–590.

Garbellini O., Palacio H., Biloni H. (1990). *Cast Metals*, **3**(2), 82–90.

Gelperin N. B. (1946). *TAFS*, **54**, 724–726.

Genders R., Bailey G. L. (1934). *The Casting of Brass Ingots*, The British Non-Ferrous Metals Research Association.

Gernez M. (1867). *Phil. Mag.*, **33**(4), 479.

Geskin F. S., Ling E., Weinstein M. I. (1986). *TAFS*, **94**, 155–158.

Ghomashchi M. R. (1995). *Journal of Materials Processing Technology*, **52**, 193–206.

Ghomashchi M. R., Chadwick G. A. (1986). *Metals and Materials,* 477–482.

Girshovich N. G., Lebedev K. P., Nakhendzi Yu A. (1963). *Russian Castings Production*, April, 174–178.

Glatz J. (1996). *Materials Evaluation*, December, 1352–1362 and *INCAST* (1997), June, 1–7.

Godding R. G. (1962). *BCIRA J.*, **10**(3), 292–297.

Goria C. A., Serramoglia G., Caironi G., Tosi G. (1986). *TAFS*, **94**, 589–600 (these authors quote Kobzar A. I., Ivanyuk E. G. (1975). *Russian Castings Production*, **7**, 302–330).

Godlewski L. A., Zindel J. W. (2001). *TAFS*, **109**, 315–325.

Goklu S. M., Lange K. W. (1968). *Proc. Conf. Process Technology*, vol. 6, Washington, USA, April, Iron and Steel Society; pp. 1135–1146.

Graham A. L., Mizzi B. A., Pedicini L. J. (1987). *TAFS*, **95**, 343–350.

Green N. R. (1994). University of Birmingham, UK. Personal communication.

Green N. R. (1995). University of Birmingham, UK. Unpublished work.

Green N. R., Campbell J. (1993). *Materials Science and Engineering*, **A173**, 261–266.

Green N. R., Campbell J. (1994). *Trans AFS*, **102**, 341–347.

Greer A. L., Bunn A. M., Tronche A., Evans P. V. and Bristow D. J. (2000). *Acta Materialia*, **48**, 2823.

Greaves R. H. (1936). *ISI Special Report*, **15**, 26–42.

Groteke D. E. (1985). *TAFS*, **93**, 953–960.

Gruznykh I. V., Nekhendzi Yu A. (1961). *Russian Castings Production*, **6**, 243–245.

Gurland J. (1966). *Trans AIME*, **236**, 642–646.

Guthrie R. I. L. (1989). *Engineering in Process Metallurgy*, Clarendon Press, Oxford, UK.

Guven Y. F., Hunt J. D. (1988). *Cast Metals*, **1**(2), 104–111.

Hall F. R., Shippen J. (1994). *Engineering Failure Analysis*, **1**(3), 215–229.

Hallam C. P., Griffiths W. D., Butler N. D. (2000). *Modeling of Casting, Welding and Advanced Solidification Processes IX*, eds P. R. Sahm, P. N. Hansen and J. G. Conley.

Halvaee A., Campbell J. (1997). *TAFS*, **105**, 35–46.

Hammiti F. G. (1974). In *Finite-Amplitude Wave Effects in Liquids*, ed. L. Bjorno, Proc. 1973 Symp. Copenhagen, IPC Science and Technology Press, paper 3.10, pp. 258–262.

Hansen P. N. (1975). PhD Thesis Part 2, Technical University of Denmark, Copenhagen (lists 283 publications on hot tearing).

Hansen P. N., Sahm P. R. (1988). In *Modelling of Casting and Welding Processes IV*, eds A. F. Giamei, G. J. Abbaschian, The Mineral, Metals and Materials Society.

Harding R. A., Campbell J., Saunders N. J. (1997). *Solidification Processing 97 Sheffield Conference*, 7–10 July 1997, paper 'Inoculation of ductile iron'.

Harinath U, Narayana K. L., Roshan H. M. (1979). *TAFS*, **87**, 231–236.

Harper S. (1966). *J. Inst. Metals*, **94**, 70–72

Harsem O., Hartvig T., Wintermark H. (1968). *35th Internat. Found. Cong.*, Kyoto, paper 15.

Hart R. G., Berke N. S., Townsend H. E. (1984). *Met. Trans*, **15B**, 393–395.

Hayes J. S., Keyte R., Prangnell P. B. (2000). *Materials Sci. & Technology*, **16**, 1259–1263.

Hebel T. E. (1989). *Heat Treating*, Fairchild Business Publication, USA.

Hedjazi D., Bennett G. H. J., Kondic V. (1975). *Brit. Found.*, **68**, 305–309.

Hedjazi D., Bennett G. H. J., Kondic V. (1976). *Metals Technol.*, December, 537–541.

Heine R. W. (1951). *TAFS*, **59**, 121–138.

Heine R. W. (1968). *TAFS*, **76**, 463–469.

Heine R. W., Loper C. R. (1966a). *TAFS*, **74**, 274–280.

Heine R. W., Loper C. R. (1966b). *TAFS*, **74**, 421–428 and *Brit Found.* (1967), **60**, 347–353.

Heine R. W., Uicker I. I., Gantenbein D. (1984). *TAFS*, **92**, 135–150.

Henschel C., Heine R. W., Schumacher J. S. (1966). *TAFS*, **74**, 357–364.

Hernandes-Reyes B. (1989). *TAFS*, **97**, 529–538.

Herrera A., Kondic V. (1979). In Solidification and Casting of Metals. Conf., Sheffield, 1977. Metals Soc. London, pp. 460–465.

Hill H. N., Barker R. S., Willey L. A. (1960). *Trans Am. Soc. Metals*, **52**, 657–671.

Hiratsuka S., Niyama E., Funakubo T. and Anzai K. (1994). *Trans Japan Foundryman's Soc.*, **13**(11), 18–24.

Hiratsuka S., Niyama E., Anzai K., Hori H., Kowata T. (1966). 4th Asian Foundry Congress, pp. 525–531.

Ho K., Pehlke R. D. (1984). *TAFS*, **92**, 587–598.

Ho P. S., Kwok T., Nguyen T., Nitta C., Yip S. (1985). *Scripta Metall.*, **19**(8), 993–998.

Hoar T. P., Atterton D. V. (1950). *J. Iron & Steel Inst.*, **166**, 1–7.

Hoar T. P., Atterton D. V., Houseman D. H. (1953). *J. Iron & Steel Inst.*, **175**, 19–29.

Hoar T. P., Atterton D. V., Houseman D. H. (1956). *Metallurgia*, **53**, 21–25.

Hochgraf F. G. (1976). *Metallography*, **9**, 167–176.

Hodaj F., Durand F. (1997). *Acta Materialia*, **45**, 2121.

Hodjat Y., Mobley C. E. (1984). *TAFS*, **92**, 319–321.

Hofmann F. (1962). *TAFS*, **70**, 1–12.

Hofmann F. (1966) *AFS Cast Metals Research J.*, **2**(4), 153–165.

Hoffmann J. (2001). *Foundry Trade Journal*, **175**(3578), 32–34.

Holtzer M. (1990). *The Foundryman*, March, 135–144.

Hu D., Loretto M. H. (2000). University of Birmingham, UK. Personal communication.

Hua C. H, Parlee N. A. D. (1982). *Met. Trans*, **13B**, 357–367.

Huang J., Conley J. G. (1998). *TAFS*, **106,** 265–270.

Huang L. W., Shu W. J., Shih T. S. (2000). *TAFS*, **108**, 547–560.

Hultgren A., Phragmen G. (1939). *Trans AIME*, **135**, 133–244.

Hull D. R. (1950). *Casting of Brass and Bronze*, ASM.

Hummer R. (1988). *Cast Metals*, **1**(2), 62–68.

Hunsicker H. Y. (1980). *Met. Trans*, **11A**, 759–773.

Hunt J. D. (1980) University of Oxford, UK. Personal communication.

Hutchinson H. P. Sutherland D. S. (1965). *Nature*, **206**, 1036–1037.

IBF Technical Subcommittee TS 17 (1948). Symp. Internal Stresses in Metals and Alloys, London 1947. Published by the Inst. of Metals 1948, pp. 179–188.

IBF Technical Subcommittee TS 18 (1949). *Proc. IBF*, **42**, A61–A77.

IBF Technical Subcommittee TS 32 (1952). *Brit. Found.*, **45**, A48–A56; *Foundry Trade J.*, **93**, 471–477.

IBF Technical Subcommittee TS 32 (1956). *Foundry Trade J.*, **101**, 19–27.

IBF Technical Subcommittee TS 32 (1960a). *Brit. Found.*, **53**, 10–13 (but original work reported in *Brit. Found.* (1952), **45**, A48–A56).

IBF Technical Subcommittee TS 35 (1960b). *Brit. Found.*, **53**, 15–20.

IBF Technical Subcommittee TS 61 (1964). *Brit. Found.*, **57**, 75–89, 504–508.

Iida T., Guthrie R. I. L. (1988). *The Physical Properties of Liquid Metals*, Oxford: Clarendon Press, p. 14.

Impey S., Stephenson D. J., Nicholls J. R. (1993). Microscopy of Oxidation 2. Cambridge Conference, pp. 323–337.

Ionescu V. (2002). *Modern Castings*, **92**(5), 21–23.

Isaac J., Reddy G. P., Sharman G. K. (1985). *TAFS*, **93**, 29–34.

Isobe T., Kubota M., Kitaoka S. (1978). *J. Japan Found. Soc.*, **50**(11), 671–676.

Iyengar R. K., Philbrook W. O. (1972). *Met. Trans*, **3**, 1823–1830.

Jackson K. A., Hunt J. D., Uhlmann D. R., Seward T. P. (1966). *Trans AIME*, **236**, 149–158.

Jackson R. S. (1956). *Foundry Trade J.*, **100**, 487–493.

Jackson W. J. (1972). *Iron and Steel*, April, 163–172.

Jackson W. J., Wright J. C. (1977). *Metals Technol.*, September, 425–433.

Jacobi H. (1976). *Arch. Eisenhuttenwesen*, **47**, 441–446.

Jacobs M. H., Law T. J., Melford D. A., Stowell M. J. (1974). *Metals Technol.*, **1**(11), 490–500.

Jaquet J. C. (1988). In 8th Colloque Européen de la Fonderie des Metaux Non Ferreux du CAEF.

Jay R., Cibula A. (1956). *Proc. Inst. Brit. Found.*, **49**, A126–A140.

Jayatilaka A. de S., Trustrum K. (1977). *J. Mat. Sci.*, **12**, 1426.

Jeancolas M., Devaux H. (1969). *Fonderie*, **285**, 487–499.

Jeancolas M., Devaux H., Graham G. (1971). *Brit. Found.*, **64**, 141–154.

Jelm C. R., Herres S. A. (1946). *Trans Amer. Found. Assoc.*, **54**, 241–251.

Jirsa J. (1982). *Foundry Trade J.*, Oct. 7, 520–527.

Jo C.-Y., Joo D.-W., Kim I.-B. (2001). *Materials Science and Technology*, **17**, 1191–1196.

Johnson A. S., Nohr C. (1970). *TAFS*, **78**, 194–207.

Johnson R. A., Orlov A. N. (1986). *Physics of Radiation Effects in Crystals*, Elsevier, North Holland.

Johnson T. V., Kind H. C., Wallace J. F., Nieh C. V., Kim H. J. (1989). *TAFS*, **97**, 879–886.

Johnson W. H., Baker W. O. (1948). *TAFS*, **56**, 389–397.

Jones D. R., Grim R. E. (1959). *TAFS*, **67**, 397–400.

Jorstad J. L. (1971). *TAFS*, **79**, 85–90.

Kaiser G. (1966). *Berichte der Bunsengesellschaft fur physikalische Chemie*, **70**(6), 635–639.

Karsay S. I. (1980). *Ductile Iron: the state of the art 1980*, QIT-Fer et Titane Inc., Canada.

Karsay S. I. (1985). *Ductile Iron Production Practices*, published by AFS.

Karsay S. I. (1992). *Ductile Iron Production Practices; the state of the art 1992*, QIT-Fer et Titane Inc., Canada.

Kaspersma J. H., Shay R. H. (1982). *Met. Trans*, **13B**, 267–273.

Katgerman L. (1982). *J. Metals*, **34**(2), 46–49.

Kato E. (1999). *Metall. and Materials Trans A*, **30A**, September, 2449–2453.

Kay J. M., Nedderman R. M. (1974). *An Introduction to Fluid Mechanics and Heat Transfer*, 3rd edn, CUP, pp. 115–119.

Kearney A. L., Raffin J. (1987). *Hot Tear Control Handbook for Aluminum Foundrymen and Casting Designers*, American Foundrymen's Soc., Des Plaines Illinois, USA.

Khalili A., Kromp K. (1991). *J. Mat. Sci.*, **26**, 6741–6752.

Khan P. R., Su W. M., Kim H. S., Kang J. W., Wallace J. F. (1987). *TAFS*, **95**, 105–116.

Kiessling R. (1987). *Non-metallic Inclusions in Steel*, The Metals Society.

Kim M. H., Loper C. R., Kang C. S. (1985). *TAFS*, **93**, 463–474.

Kilshaw J. A. (1963). *BCIRA J.*, **11**, 767.

Kilshaw J. A. (1964). *BCIRA J.*, **12**, 14.

Kirner J. F., Anewalt M. R., Karwacki E. J., Cabrera A. L. (1988). *Met. Trans*, **19A**, 3045–3055.

Klemp T. (1989). *TAFS*, **97**, 1009–1024.

Knott J. F., Elliott D. (1979). *Worked Examples in Fracture Mechanics*, Inst. Metals Monograph 4, Inst. Metals, London.

Kolorz A., Lohborg K. (1963). In *30th Internat. Found. Congress*, pp. 225–246.

Kondic V. (1959). *Foundry*, **87**, 79–83.

Koster W., Goebring K. (1941). *Giesserei*, **28**(26), 521.

Kotschi R. M., Loper C. R. (1974). *TAFS*, **82**, 535–542.

Kotschi R. M., Loper C. R. (1975). *TAFS*, **83**, 173–184.

Kotschi T. P., Kleist O. F. (1979). *AFS Internat. Cast Metals J.*, **4**(3), 29–38.

Kotsyubinskii O. Yu (1961–62). *Russian Castings Production*, 269–272.

Kotsyubinskii O. Yu (1963). *30th Internat. Found. Cong.*, pp. 475–487.

Kotsyubinskii O. Yu, Gerchikov A. M., Uteshev R. A., Novikov M. I. (1961). *Russian Castings Production*, **8**, 365–368.

Kotsyubinskii O. Yu, Oberman Ya I., Gerchikov A. M. (1962). *Russian Castings Production*, **4**, 190–191.

Kotsyubinskii O. Yu, Oberman Ya I., Gini E. Gh. (1968). *Russian Castings Production*, **4**, 171–172.

Krinitsky A. I. (1953). *TAFS*, **61**, 399–410.

Kubo K., Pehlke R. D. (1985). *Met. Trans*, **16B**, 359–366.

Kubo K., Pehlke R. D. (1986). *Met. Trans*, **17B**, 903–911.

Kujanpaa V. P., Moisio T. J. I. (1980). *Solidification Technology in the Foundry and Cast House*, Warwick Conf., Metals Soc. 1983, pp. 372–375.

Lagowski B. (1979). *TAFS*, **87**, 387–390.

Lagowski B., Meier J. W. (1964). *TAFS*, **72**, 561–574.

Lane A. M., Stefanescu D. M., Piwonka T. S., Pattabhi R. (1969). *Modern Castings*, October, 54–55.

Lang G. (1972). *Aluminium*, **48**(10), 664–672.

LaVelle D. L. (1962). *TAFS*, **70**, 641–647.

Ledebur A. (1882). *Stahl und Eisen*, **2**, 591.

Leduc L., Nadarajah T., Sellars C. M. (1980). *Metals Technology*, July, 269–273.

Lee R. S. (1987). *TAFS*, **95**, 875–882.

Lees D. C. G. (1946). *J. Inst. Metals*, **72**, 343–364.

Leth-Olsen H., Nisancioglu K. (1998). *Corrosion Science*, **40**, 1179–1194 and 1194–1214.

Levelink H. O. (1972). *TAFS*, **80**, 359–368.

Levelink H. G., Berg H. van den (1962). *TAFS*, **70**, 152–163.

Levelink H. G., Berg H. van den (1968). *TAFS*, **76**, 241–251.

Levelink H. O., Berg H. van den (1971). *TAFS*, **79**, 421–432.

Levelink H. O., Julien F. P. M. A. (1973). *AFS Cast Metals Research J.*, June, 56–63.

Lewis G. M. (1961). *Proc. Phys. Soc. (London)*, **71**, 133.

Li Y., Jolly M. R., Campbell J. (1998). *Modeling of Casting, Welding and Advanced Solidification Processes VIII*, eds B. G. Thomas and C. Beckermann, The Minerals, Metals and Materials Soc., pp. 1241–1253.

Liddiard E. A. G., Baker W. A. (1945). *TAFS*, **53**, 54–65.

Lin H. J., Hwang W.-S. (1988). *TAFS*, **96**, 447–458.

Lin J., Sharif M. A. R., Hill J. L. (1999). *Aluminum Trans*, **1**(1) 72–78.

Ling Y., Mampaey F., Degrieck J., Wettinck E. (2000). *Modeling of Casting, Welding and Advanced Solidification Processes IX*, eds P. R. Sahm, P. N. Hansen and J. G. Conley, pp. 357–364.

Liu L., Samuel A. M., Samuel F. H., Dowty H. W., Valtierra S. (2002). *Trans AFS* (paper 02–139, 14 pp.).

Livingston H. K., Swingley C. S. (1971). *Surface Sc.*, **24**, 625–634.

Locke C., Ashbrook R. L. (1950). *TAFS*, **58**, 584–594.

Locke C., Ashbrook R. L. (1972). *TAFS*, **80**, 91–104.

Longden E. (1931–32). *Proc. IBF*, **25**, 95–145.

Longden E. (1939–40). *Proc. IBF*, **33**, 77–107.

Longden E. (1947–48). *Proc. IBF*, **41**, A152–A165.

Longden E. (1948). *TAFS*, **56**, 36–56.

Loper C. R. (1981). *TAFS*, **89**, 405–408.

Loper C. R., Kotschi R. M. (1974). *TAFS*, **82**, 279–284.

Loper C. R., LeMahieu D. L. (1971). *TAFS*, **79**, 483–492.

Loper C. R., Newby M. R. (1994). *TAFS*, **102**, 897–901.

Loper C. R., Saig A. G. (1976). *TAFS*, **84**, 765–768.

Low J. R. (1969). *Trans AIME*, **245**, 2481–2494.

Maidment L. J., Walter S., Raul G. (1984). *8th Heat Treating Conf.*, Detroit.

Majumdar I., Raychaudhuri B. C. (1981). *Internat. J. Heat Mass Transfer*, **24**(7). 1089–1095.

Mansfield T. L. (1984). Proc. Conf. 'Light Metals', *Met. Soc. AIME*, 1305–1327.

Marek C. T., Keskar A. R. (1968). *TAFS*, **76**, 29–43.

Matsubara Y., Suga S., Trojan P. F., Flinn R. A. (1972). *TAFS*, **80**, 37–44.

Mazed S., Campbell J. (1992). University of Birmingham, UK. Unpublished work.

McCartney D. G. (1989). *Internat. Mat. Rev.*, **34**(5), 247–260.

McClain S. T., McClain A.S., Berry J.T. (2001). *TAFS*, **109**, 75–86.

McDonald R. J., Hunt J. D. (1969). *Trans AIME*, **245**, 1993–1997.

McDonald R. I., Hunt J. D. (1970). *Trans AIME*, **1**, 1787–1788.

McGrath C., Fischer R. B. (1973). *TAFS*, **81**, 603–620.

McParland A. J. (1987). Sheffield Conf *'Solidification Processing'*, Institute of Metals, London 1988, pp. 323–326.

Medvedev Ya I., Kuzukov V. K. (1966). *Russian Castings Production*, 263–266.

Metcalf G. J. (1945). *J. Inst. Metals*, (1029), 487–500.

Mertz J. M., Heine R. W. (1973). *TAFS*, **81**, 493–495.

Merz R., Marincek B. (1954) *21st International Foundry Congress*, paper 44, pp. 1–7.

Meyers C. W. (1986). *TAFS*, **94**, 511–518.

Mi J., Harding R. A., Campbell J. (2002). *Internat. J. Cast Metals Research*, **14**, in press (10 pp.).

Micks F. W., Zabek V. J. (1973). *TAFS*, **81**, 38–42.

Middleton J. M. (1953). *TAFS*, **61**, 167–183.

Middleton J. M. (1970). *Brit. Found.*, **64**, 207–223.

Middleton J. M., Canwood B. (1967). *Brit. Found.*, **60**, 494–503.

Miguelucci E. W. (1985). *TAFS*, **93**, 913–916.

Miles G. W. (1956). *Proc. Inst. Brit. Found.*, **49**, A201–A210.

Miller G. F. (1967). *Foundry*, **95**, 104–107.

Minakawa S., Samarasekera I. V., Weinberg F. (1985). *Met. Trans*, **16B**, 823–829.

Mintz B., Yue S., Jonas J. J. (1991). *International Materials Reviews*, **36**(5) 187–217.

Miyagi Y., Hino M., Tsuda O. (1985). *'R&D' Kobe Steel Engineering Reports 1983*, **33**, July (3), also published in *Kobelco Technical Bulletin* (1985), **1076**.

Mizuno K, Nylund A, Olefjord I. (1996). *Materials Science and Technol.*, **12**, 306–314.

Mohanty P. S., Gruzleski J. E. (1995). *Acta Metall. Mater.*, **43**, 2001–2012.

Mohla P. P., Beech J. (1968). *Brit. Found.*, **61**, 453–460.

Molaroni A., Pozzesi V. (1963). In *30th Internat. Found. Congress*, pp. 145–161.

Mollard F. R., Davidson N. (1978). *TAFS*, **78**, 479–486.

Mondloch P. A., Baker D. W., Euvrard L. (1987). *TAFS*, **95**, 385–392.

Morgan A. D. (1966). *Brit. Found.*, **59**, 186–204.

Morgan P. C. (1989). *Metals and Materials*, **5**(9), 518–520.

Mountford N. D. G., Calvert R. (1959–60). *J. Inst. Metals*, **88**, 121–127.

Mountford N. D. G., Sommerville I. D., From L. E., Lee S., Sun C. (1992–93). *A Measuring Device for Quality Control in Liquid Metals*, Scaninject Conf., Lulea, Sweden.

Mulazimoglu M. H., Handiak N., Gruzleski J. E. (1989). *TAFS*, **97**, 225–232.

Muller F. C. G. (1887). *Zeit. ver dent. Ingenieure*, **23**, 493.

Mutharasan R., Apelian D., Romanowski C. (1981). *J. Metals*, **83**(12), 12–18.

Mutharasan R., Apelian D., Ali S. (1985). *Met. Trans*, **16B**, 725–742.

Myllymaki R. (1987). In *'Residual Stress in Design, Process and Materials Selection'*, ASM Conf., USA, ed. W. B. Young, pp. 137–141.

Nai S., M. L., Gupta M. (2002). *Materials Science & Technology*, **18**, 633–641.

Nakagawa Y., Momose A. (1967). *Tetsu-to-Hagane*, **53**, 1477–1508.

Naro R. L. (1974). *TAFS*, **82**, 257–266.

Naro R. L., Pelfrey R. L. (1983). *TAFS*, **91**, 365–376.

Naro R. L., Tanaglio R. O. (1977). *TAFS*, **85**, 65–74.

Naro R. L., Wallace J. F. (1967). *TAFS*, **75**, 741–758.

Neiswaag H., Deen H. J. J. (1990). *57th World Foundry Congress*, Osaka, Japan.

Nicholas K. L., Roberts R. W. (1961). *BCIRA J.*, **9**(4), 519.

Nishida Y., Droste W., Engler S. (1986). *Met. Trans*, **17B**, 833–844.

Niyama E., Ishikawa M. (1966). *4th Asian Foundry Congress*, pp. 513–523.

Niyama E., Uchida T., Morikawa M., Saito S. (1982). In *Internat. Found. Congress 49*, Chicago, paper 10.

Noesen S. J., Williams H. A. (1966). In *4th National Die Casting Congress*, Cleveland, Ohio, paper 801.

Nordland W. A. (1967). *Trans AIME*, **239**, 2002–2004.

Nordlien J. H. (1999). *Aluminium Extrusion*, **4**(4), 39–41.

Nordlien J. H., Davenport A. J., Scamans G. M. (2000). *ASST 2000 (2nd Internat. Symp. Al Surface Science Technology)*, UMIST, Manchester, UK, Alcan + Dera, pp. 107–112.

Northcott L. (1941). *J. Iron Steel Inst.*, **143**, 49–91.

Novikov I. I. (1962). *Russian Castings Production*, April, 167–172.

Novikov I. I., Novik F. S., Indenbaum G. V. (1966). *Izv. Akad. Nauk. Metal*, **5**, 107–110 (English translation in *Russ. Metall. Mining* (1966), **5**, 55–59).

Novikov I. I., Portnoi V. K. (1966). *Russian Castings Production*, **4**, 163–166.

Nyahumwa C., Green N. R., Campbell J. (1998). *TAFS*, **106**, 215–223.

Nyahumwa C., Green N. R., Campbell J. (2000). *Metall. Mat. Trans A*, **31A**, 1–10.

Oelsen W. (1936). *Stahl u Eisen*, **56**, 182.

Ohno A. (1987). *Solidification: The Separation Theory and its Practical Applications*, Springer-Verlag.

Ohsasa K.-I., Takahashi T., Kobori K. (1988a). *J. Japan Inst. Met.*, **52**(10), 1006–1011 (Materials Abstr. 51-0785).

Ohsasa K.-I., Ohmi T., Takahashi T. (1988b). *Bull. Fac. Eng. Hokkaido Univ.*, **143** (in Japanese).

Ostrom T. R., Biel J., Wager W., Flinn R. A., Trojan P. K. (1982). *TAFS*, **90**, 701–709.

Ostrom T. R., Trojan P. K., Flinn R. A. (1975). *TAFS*, **83**, 485–492.

Ostrom T. R., Trojan P. K., Flinn R. A. (1981). *TAFS*, **89**, 731–736.

Outlaw R. A., Peterson D. T., Schmidt F. A. (1981). *Met. Trans*, **12A**, 1809–1816.

Owen M. (1966). *Brit. Found.*, **59**, 415–421.

Owusu Y. A., Draper A. B. (1978). *TAFS*, **86**, 589–598.

Pakes M., Wall A. (1982). Zinc Development Assoc., UK.

Pan E. N., Hsieh M. W., Jang S. S., Loper C. R. (1989). *TAFS*, **97**, 379–414.

Pan E. N., Hu J. F. (1996). *4th Asian Foundry Congress*, pp. 396–405.

Panchanathan V., Seshadri M. R., Ramachandran A. (1965). *Brit. Found.*, **58**, 380–384.

Papworth A., Fox P. (1998). *Materials Letters*, **35**, 202–206.

Parkes T. W., Loper C. R. (1969). *TAFS*, **77**, 90–96.

Parkes W. B. (1952). *TAFS*, **60**, 23–37.

Paray F., Kulunk B., Gruzleski J. E. (2000). *Internat. J. Cast Metals Research*, **13**, 147–159.

Patterson W., Koppe W. (1962). *Giesserei*, **4**, 225–249.

Patterson W., Engler S., Kupfer R. (1967). *Giesserei-Forschung*, **19**(3), 151–160.

Pelleg J., Heine R. W. (1966). *TAFS*, **74**, 541–544.

Pellerier M., Carpentier M. (1988). *Hommes et Fonderie*. **184**, 7–14.

Pellini W. S. (1952). *Foundry*, **80**, 124–133, 194, 196, 199.

Pellini W. S. (1953). *TAFS*, **61**, 61–80 and 302–308.

Pell-Walpole W. T. (1946). *J. Inst. Metals*, **72**, 19–30.

Petro A., Flinn R. A. (1978). *TAFS*, **86**, 357–364.

Petrzela L. (1968) *Foundry Trade J.*, Oct. 31, 693–696.

Pillai R. M., Mallya V. D., Panchanathan V. (1976). *TAFS*, **84**, 615–620.

Pitcher P. D., Forsyth P. J. E. (1982). *The Influence of Microstructure on the Fatigue Properties of an Al Casting Alloy*, Royal Aircraft Establishment Technical Report 82107, Nov. 1982.

Piwonka T. S., Flemings M. C. (1966). *Trans Met. Soc. AIME*, **236**, 1157–1165.

Poirier D. R. (1987). *Met. Trans*, **18B**, 245–255.

Poirier D. R., Sung P. K., Felicelli S. D. (2001). *TAFS*, **105**, 139–155.

Poirier D. R., Yeum K. (1987). *Solidification Processing Conf.*, Sheffield, Institute of Metals, London.

Poirier D. R., Yeum K. (1988). In *Light Metals Conf.*. USA, pp. 469–476.

Poirier D. R., Yeum K., Maples A. L. (1987). *Met. Trans*, **18**, 1979–1987.

Pollard W. A. (1965). *TAFS*, **73**, 371–379.

Pollard W. A. (1984). *TAFS*, **72**, 587–599.

Polodurov N. N. (1965). *Russian Casting Production*, **5**, 209–210.

Pope J. A. (1965). *Brit. Found.*, **58**, 207–224.

Popel G. I., Esin O. A. (1956). *Zh. Fiz. Khim.*, **30**, 1193.

Portevin A., Bastien P. (1934). *J. Inst. Metals*, **54**, 45–58.

Prates M., Biloni H. (1972). *Met. Trans*, **3A**, 1501–1510.

Prible J., Havlicek F. (1963). *30th Internat. Foundry Congress*, pp. 394–410.

Pumphrey W. I. (1955). *Researches into the Welding of Aluminium and its Alloys*, Research Report 27, Aluminium Development Association, UK.

Pumphrey W. I., Lyons J. V. (1948). *J. Inst. Metals*, **74**, 439–455.

Pumphrey W. I., Moore D. C. (1949). *J. Inst. Metals*, **75**, 727–736.

Rabinovich A. (1969). *AFS Cast Metals Research J.*, March, 19–24.

Ragone D. V., Adams C. M., Taylor H. F. (1956). *TAFS*, **64**, 653–657.

Ramseyer J. C., Gabathuler J. P., Feurer U. (1982). *Aluminium*, **58**(10), E192–E194 and 581–585.

Ransley C. E., Neufeld H. (1948). *J. Inst. Metals*, **74**, 599–620.

Rao G. V. K., Panchanathan V. (1973). *Cast Metals Res. J.*, **19**(3), 135–138.

Rao G. V. K, Srinivasan M. N., Seshadri M. R. (1975). *TAFS*, **83**, 525–530.

Rappaz M. (1989). *Internat. Mat. Rev.*, **34**(3), 93–123.

Rauch A. H., Peck J. P., Thomas G. F. (1959). *TAFS*, **67**, 263–266.

Rege R. A., Szekeres E. S., Foreng W. D. (1970). *Met. Trans*, **1**, 2652–2653.

Richmond O., Tien R. H. (1971). *J. Mech. Phys. Solids*, **19**, 273–284.

Richmond O., Hector L. G., Fridy J. M. (1990). *Trans ASME J. Appl. Mechanics*, **57**, 529–536.

Rickards P. J. (1975). *Brit Found.*, **68**, 53–60.

Rickards P. J. (1982). *Brit. Found.*, **75**, 213–223.

Rivas R.A.A., Biloni H. (1980). *Zeit. Metallk.*, **71**(4), 264–268.

Rivera G., Boeri R., Sikora J. (2002). *Materials Science and Technology*, **18**, 691–697.

Roberts T. E., Kovarik D. P., Maier R. D. (1979). *TAFS*, **87**, 279–298.

Rogberg B. (1980). *Solidification Technology in the Foundry and Cast House*, Warwick Conf., pp. 365–371 (Metals Soc. 1983).

Rosenberg R. A., Flemings M. C., Taylor H. F. (1960). *TAFS*, **68**, 518–528.

Roth M. C., Weatherly G. C., Miller W. A. (1980). *Acta Met.*, **28**, 841–853.

Rouse J. (1987). *54th Internat Foundry Congress*, New Delhi, paper 30, p. 16.

Ruddle R. W. (1956). The Running and Gating of Sand Castings, *Monograph & Report Series No. 19*, Institute of Metals, London.

Ruddle R. W. (1960). *TAFS*, **68**, 685–690.

Ruddle R. W., Cibula A. (1957). *Inst. Metals Monograph*, Report, 22, 5–32.

Ruddle R. W., Mincher A. L. (1949–50). *J. Inst. Metals*, **76**, 43–90.

Runyoro J. (1992). PhD Thesis, University of Birmingham, UK.

Runyoro J., Boutorabi S. M. A., Campbell J. (1992). *TAFS*, **100**, 225–234.

Ruxanda R., Sanchez L. B., Massone J., Stefanescu D. M. (2001). *Trans Amer. Found. Soc.*, **109** (cast iron division), 37–48.

Sadayappan M., Fasoyinu F. A., Thomson J., Sahoo M. (1999). *TAFS*, **107**, 337–342.

Sadayappan M., Sahoo M., Liao G., Yang B. J., Li D., Smith R. W. (2001). *TAFS*, **109**, 341–352.

Saeger C. M., Ash E. J. (1930). *TAFS*, **38**, 107–145.

Sahoo M. and Whiting L V. (1984). *TAFS*, **92**, 861–870.

Sahoo M. and Worth M. (1990). *TAFS*, **98**, 25–33.

Saigal A., Berry J. T. (1984). *TAFS*, **92**, 703–708.

Sakamoto M., Akiyama S., Ogi K. (1996). *4th Asian Found. Cong. Proc.*, Australia, pp. 467–476.

Samarasekera I. V., Anderson D. L., Brimacombe J. K. (1982). *Met. Trans*, **13B**, 91–104.

Sambasivan S. V., Roshan H. Md. (1977). *TAFS*, **85**, 265–270.

Samuel A. M., Samuel F. H. (1993). *Metall. Trans A*, **24A**, 1857–1868.

Samuel F. H., Samuel A. M., Doty H. W., Valtierra S. (2002).

Submitted to *Metall. and Materials Transactions A*, 15 November 2001, paper 01-616-A.

Sandford P. (1988). *The Foundryman*, March, 110–118.

Santos R. G., Garcia A. (1998). *Internat. J. Cast Metals Research,* **11**, 187–195.

Saucedo I. G., Beech J., Davies G. J. (1980). *Conf. 'Solidification Technology in Foundry and Cast House'*, Warwick University, Metals Society Publication 1983, pp. 461–468

Schumacher P., Greer A. L. (1993). *Key Eng. Mater.*, 81–83, 631

Schumacher P., Greer A. L. (1994). *Materials Science and Engineering*, **A181/A182**, 1335–1339.

Schurmann E. (1965). *Arch. Eisenh.*, **36**, 619–631 (BISI translation 4579).

Sciama G. (1974). *TAFS*, **82**, 39–44.

Scott A. F. *et al.* (1948). *J. Chem. Phys.*, **16**, 495–502.

Scott D., Smith T. J. (1985). Personal communication.

Scott W. D., Bates C. E. (1975). *TAFS*, **83**, 519–524.

Scott W. D., Goodman P. A., Monroe R. W. (1978). *TAFS*, **86**, 599–610.

SCRATA (1981). *Hot Tearing – Causes and Cures, Tech. Bull. No. 1*, Steel Castings Research and Trade Assoc., Sheffield, UK.

Seetharamu S., Srinivasan M. W. (1985). *TAFS*, **93**, 347–350.

Sherby O. D. (1962). *Metals Engng Quarterly (ASM)*, May, 3–13.

Ship Department Publication 18 (1975). Design & Manufacture of Nickel–Aluminium–Bronze Sand Castings, Ministry of Defence (Procurement Executive), Foxhill, Bath, UK.

Sicha W. E., Boehm R. C. (1948). *TAFS*, **56**, 398–409.

Sigworth G. (1984). *Met. Trans*, **15A**, 227–282.

Sigworth G. K., Engh T. A. (1982). *Met. Trans*, **13B**, 447–460.

Simard A., Proulx J., Paquin D., Samuel F. H., Habibi N. (2001). *Amer. Found. Soc. Molten Al Processing Conf.*, Orlando, Florida, November.

Simensen C. J. (1993). *Zeit Metallkunde*, **84**(10), 730–733.

Simmons W., Trinkl G. (1987). *BCIRA Conference*, British Cast Iron Research Assoc., UK.

Singh S. N., Bardes B. P., Flemings M. C. (1970). *Met. Trans*, **1**, 1383.

Sinha N. P., Kondic V. (1974). *Brit. Found.*, **67**, 155–165.

Sinha N. P. (1973). PhD Thesis, University of Birmingham, UK.

Smith C. S. (1948). *Trans AIME*, **175**, 15–51.

Smith C. S. (1949). *Trans AIME*, **185**, 762–768.

Smith C. S. (1952). *Metal Interfaces*, ASM: Cleveland, Ohio, pp. 65–113.

Smith D. D., Aubrey L. S., Miller W. C. (1988) *'Light Metals' Conf.*, ed. B. Welch, The Minerals, Metals and Materials Soc., pp. 893–915.

Smith D. M. (1981). UK Patent Application GB 2 085 920 A.

Smith R. A. (1986). UK Patent GB 2 187 984 A; priority date 21 Feb. 1986.

Smith T. J., Lewis R. W., Scott D. M. (1990). *The Foundryman*, **83**, 499–507.

Solberg J. K., Onsoien M. I. (2001). *Materials Science and Technology*, **17**, 1238–1242.

Sosman R. B. (1927). *The Properties of Silica*. The Chemical Catalogue Co. USA, *Am. Chem. Soc. Monograph Series*, pp. iv–45.

Southin R T. (1967). The Solidification of Metals, Brighton Conf., *ISI Publication 110*, pp. 305–308.

Southin R T., Romeyn A. (1980). Warwick Conf., *'Solidification Technology in the Foundry and Cast House'*, Metals Soc. 1983, pp. 355–358.

Speidel M. O. (1982). *Sixth European Non-Ferrous Metals Industry Colloquium* CAEF, pp. 65–78.

Spitaler P. (1957). *Giesserei,* **44**, 757–766.

Spittle J. A., Brown S. G. R. (1989a). *J. Mat. Sci.*, **23**, 1777–1781.

Spittle J. A., Brown S. G. R. (1989b). *J. Mat. Sci.*, **5**, 362–368.

Spittle J. A., Cushway A. A. (1983). *Metals Technology*, **10**, 6–13.

Spraragen W., Claussen G. E. (1937). *J. Am. Welding Soc.* (Supplement: Welding Research Committee), **16**(11), 2–62.

Stahl G. W. (1961). *TAFS*, **69**, 476–478.

Stahl G. W. (1963). *TAFS*, **71**, 216–220.

Stahl G. W. (1986). *TAFS*, **94**, 793–796.

Staley J. T. (1981). *Metals Handbook*, 9th edn, Vol. 4, *Heat Treating*, American Society for Metals, USA, pp. 675–718.

Staley J. T. (1986). *Aluminium Technology 86*, Inst. Metals UK Conference, pp. 396–407

Steel Founders Soc. of America (1970). *Steel Castings Handbook*, 4th edn.

Stein H., Iske F., Karcher D. (1958). *Giesserei-Technisch-Wissenschaftliche Beihefte*, **21**, 115–1124.

Steiger R. von. (1913). *Stahl und Eisen*, **33**, 1442.

Stolarczyk J. E. (1960). *Brit. Found.*, **53**, 531–548.

Stoltze P., Norskov I. K., Landman U. (1988). *Phys. Rev. Lett.*, **61**(4), 440–443.

Street A. C. (1977). *The Diecasting Book*, Portcullis Press, Redhill, UK.

Sullivan E. J., Adams C. M., Taylor H. F. (1957). *TAFS*, **65**, 394–401.

Sumiyoshi, Ito N., Noda T. (1968). *J. Crystal Growth*, **3 and 4**, 327 only.

Surappa M. K., Blank E., Jaquet J. C. (1986a) *Conference 'Aluminium Technology'*, ed. T. Sheppard, Inst. Metals, pp. 498–504.

Surappa M. K., Blank E., Jaquet J. C. (1986b) *Scripta Met.*, **20**, 1281–1286.

Suutala N. (1983) *Met. Trans*, **14A**, 191–197.

Suzuki S. (1989) *Modern Castings*, October, 38–40.

Svoboda J. M., Monroe R. W., Bates C. E., Griffin J. (1987). *TAFS*, **95**, 187–202.

Svoboda J. M., Geiger G. H. (1969). *TAFS*, **77**, 281–288.

Sugden A. A. B., Bhadeshia H. K. D. H. (1987). University of Cambridge, UK.

Sy, A. de (1967). *TAFS*, **75**, 161–172.

Szklarska-Smialowska Z. (1999) *Corrosion Science*, **41**, 1743–1767.

Tadayon M. R., Lewis R. W. (1988). *Cast Metals*, **1**, 24–28.

Tafazzoli-Yadzi M., Kondic V. (1977). *AFS Internat. Cast Metals J.*, **2**(4), 41–47.

Talbot D. E. G., Granger D. A. (1962). *J. Inst. Metals*, **91**, 319–320.

Taylor L. S. (1960). *Foundry Trade J.*, **109**(2287), 419–427.

Thiele von W. (1962) *Aluminium* (Germany), **38**, 707–715 and 780–786 (English translation by H. Nercessian 29 April 1963, 5229, Alcan, Banbury, UK, and Electicity Council Research Centre, UK).

Thomas B. G., Parkman J. T. (1997). Conf. '*Solidification*', Indianapolis, Indiana, The Minerals, Metals and Materials Society, 1998, pp. 509–520.

Thornton D. R. (1956). *J. Iron and Steel Inst.*, **183**(3), 300–315.

Tiberg L. (1960). *Jerkont. Ann.*, **144**(10), 771–793.

Tien R. H., Richmond O. (1982). *Trans ASME. J. Appl. Mechanics*, **49**, 481–486.

Timmons W. W., Spiegelberg W. D., Wallace J. F. (1969). *TAFS*, **77**, 57–61.

Tiryakioglu M. (2001). Personal communication.

Tiryakioglu M.., Askeland D. R., Ramsay C. W. (1993). *TAFS*, **101**, 685–691.

Tiwara S. N., Gupta A. K., Maihotra S. L. (1985). *Brit. Found.*, **78**(1), 24–27.

Tokarev A. I. (1966). *Izv VUZ Chern. Met.,* **3**, 193–200. BCIRA translation T1190, January 1966.

Thomason P. F. (1968). *J. Inst. Metals*, **96**, 360–365.

Tomono H., Ackermann P., Kurz W., Heinemann W. (1980). *Solidification Technology in Foundry and Cast House*, Warwick University Conf. Metals Society Publication 1983, pp. 524–531.

Townsend D. W. (1984). *Foundry Trade J.*, **24**, May, 409–414.

Trbizan Katerina (2001). *Casting Simulation*, World Foundry Organization (paper 4), pp. 83–97.

Travena D. H. (1987). *Cavitation and Tension in Liquids*, Bristol: Adam Huger, Inst. Phys.

Trojan P. K., Guichelaar P. J., Flinn R. A. (1966). *TAFS*, **74**, 462–469.

Tsai H. L., Chiang K. C., Chen T. S. (1988). In *Modeling of Casting and Welding Processes IV*, ed. A. F. Giamei and O. J. Abbaschian, The Minerals, Metals and Materials Society, USA, 1989.

Tschapp M. A., Ramsay C. W., Askeland D. R. (2000). *TAFS*, **108**, 609.

Tucker S. P., Hochgraph F. G. (1973). *Metallography*, **6**, 457–464.

Turkdogan E. T. (1988). *Foundry Processes, Their Chemistry and Physics* (International Symposium (1986), Warren, Mich., USA, eds S. Katz and C. F. Landefeld, Plenum Press, pp. 53–100.

Turner A., Owen F. (1964). *Brit, Found.*, **57**, 55–61 and 355–356.

Turner G. L. (1965). *Brit. Found.* **58**, 504–505.

Turpin M. L., Elliott I. F. (1966). *J. Iron and Steel Inst.*, **204**, 217–225.

Twitty M. D. (1960). BCIRA I, Report 575, pp. 844–856, ed. C. F. Walton (1971), *Gray and Ductile Iron Casting Handbook*, Cleveland, Ohio. Gray and Ductile Iron Founders' Soc. Inc.

Tyndall J. (1872). *The Forms of Water*, New York, D. Appleton and Co.

Uto Y., Yamasaki D. (1967). UK Patent Specification 1,198,700.

Vandenbos S. A. (1985). *TAFS*, **93**, 871–878.

Venturelli G., Sant'Unione G. (1981). *Alluminio*, Feb., 100–106.

Vigh L., Bennett G. H. J. (1989). *Cast Metals*, **2**(3), 144–150.

Vincent R. S., Simmons G. H. (1943). *Proc. Phys. Soc. (London)*, 376–382.

Vogel A., Doherty R D., Cantor B. (1977). Univ Sheffield Conf. '*The Solidification and Casting of Metals*', published Metals Soc., 1979, pp. 518–525.

Vorren O., Evensen J. E., Pedersen T. B. (1984). *TAFS*, **92**, 459–466.

Wakefield G. R., Sharp R. M. (1992). *J. Materials Sci. and Technol.*, **8**, 1125–1129.

Wakefield G. R., Sharp R. M. (1996). *J. Materials Sci. and Technol.*, **12**, 518–522.

Wallace J. F. (1988). *TAFS*, **96**, 261–270.

Wang Q. G., Apelian D., Lados D. A. (2001 part I). *J. Light Metals*, **1**(1), 73–84.

Wang Q. G., Apelian D., Lados D. A. (2001 part II). *J. Light Metals*, **1**(1), 85–97.

Ward C. W., Jacobs I. C. (1962) *TAFS*, **70**, 332–337.

Wardle G., Billington J. C. (1983). *Metals Technology*, **10**, October, 393–400.

Warrington D., McCartney D. G. (1989). *Cast Metals*, **2**(3), 134–143.

Watmough T. (1980). *TAFS*, **88**, 481–488.

Way L. D. (2001). *Materials Science and Technology*, **17**(10), 1175–1190.

Webster P. D. (1964). *Brit. Found.*, **57**, 520–523.

Webster P. D. (1966). *Brit. Found.*, **59**, 387–393.

Webster P. D. (1967). *Brit. Found.*, **60**, 314–319.

Wen S. W., Jolly M. R., Campbell J. (1997). *Solidification Processing Conference SP97*, Sheffield, 7–10 July 1997.

Weibull W. (1951). *J. Applied Mechanics*, **18**, 293–297.

Weiner J. H., Boley B. A. (1963). *J. Mech. Phys. Solids*, **11**, 145–154.

Weins M. J., Bottom J. L. S. de, Flinn R. A. (1964). *TAFS*, **72**, 832–839.

Whittenberger E. J., Rhines F. N. (1952). *J. Metals*, **4**(4), 409–420 and *Trans AIME*, **194**, 409–420.

Wieser P. F. (1983). *TAFS*, **91**, 647–656.

Wieser P. F., Dutta I. (1986). *TAFS*, **94**, 85–92.

Wieser P. F., Wallace J. F. (1969). *TAFS*, **77**, 22–26.

Wightman G., Fray D. J. (1983). *Met. Trans*, **14B**, 625–631.

Wildermuth J. W., Lutz R. H., Loper C. R. (1968). *TAFS*, **76**, 258–263.

Williams D. C. (1970). *TAFS*, **78**, 374–381, 466–467.

Williams J. A., Singer A. R. E. (1968). *J. Inst. Metals*, **96**, 5–12.

Winter B. P., Ostrom T. R., Sleder T. A., Trojan P. K., Pehlke R. D. (1987). *TAFS*, **95**, 259–266.

Wlodawer R. (1966). *Directional Solidification of Steel Castings*, English translation by L. D. Hewit and R. V. Riley, Pergamon Press.

Wojcik W. M., Raybeck R. M., Paliwoda E. J. (1967). *J. Metals,* **19**, Dec., 36–41.

Woodbury K. A., Ke Q., Piwonka T. S. (2000a). Tr*ans AFS*, **108**, 259–265.

Woodbury K. A., Piwonka T. S., Ke Q. (2000b). *Modeling of Casting, Welding and Advanced Solidification Processes IX*, eds P. R. Sahm, P. N. Hansen and J. G. Conley, pp. 270–277.

Worman R. A., Nieman J. R. (1973). *TAFS*, **81**, 170–179.

Wray P. J. (1976). *Acta Met.*, **76**, 125–135.

Wray P. J. (1976). *Met. Trans*, **7B**, 639–646.

Wright T. C., Campbell J. (1997). *Trans AFS*, **105**, 639–644.

Yamaguchi K., Healy G. (1974). *Met. Trans*, **5**, 2591–2596.

Yamamoto N., Kawagoishi N. (2000). *Trans AFS*, **108**, 113–118.

Yan Y., Yang G., Mao Z. (1989). *J. Aeronaut. Mater. (China)*, **9**(3), 29–36.

Yarborough W. A., Messier R. (1990). *Science*, **247**, 688–696.

Yonekura K.. *et al.* (1986). *TAFS*, **94**, 277–284.

Youdelis W. V., Yang C. S. (1982). *Metal Sci.*, **16**, 275–281.

Young P. (2002). Quoted anon. *Foundry Trade Journal*, April, 27–28.

Zeitler H., Scharfenberger W. (1984). *Aluminium (Germany)*, **60**(12), E803–E808.

Zhao L., Baoyin, Wang N., Sahajwalla V., Pehlke R. D. (2000). *Internat. J. Cast Metals Research*, **13**, 167–174.

Zildjian Avedis Company cited in *Modern Castings* (2002), May, p. 68.

Ziman J. (2001). *Non-instrumental Roles of Science*, Physics Department, University of Bristol.

Zuithoff A. I. (1964). *31st Internat. Foundry Congress Amsterdam*, 1964, paper 29; and in *Geisserei* (1965), **52**(9), 820–827.

Index